BALSAM FIR

ABIES BALSAMEA (LINNAEUS) MILLER

A Monographic Review

E. V. BAKUZIS AND H. L. HANSEN

School of Forestry, University of Minnesota

with contributions by

F. H. KAUFERT, D. B. LAWRENCE,

D. P. DUNCAN, S. S. PAULEY, R. M. BROWN,

R. E. SCHOENIKE, AND L. W. KREFTING

THE UNIVERSITY OF MINNESOTA PRESS · MINNEAPOLIS

Printed in the United States of America at the
North Central Publishing Company, St. Paul

Library of Congress Catalog Card Number: 65-17539

PUBLISHED IN GREAT BRITAIN, INDIA, AND PAKISTAN BY THE
OXFORD UNIVERSITY PRESS, LONDON, BOMBAY, AND KARACHI, AND IN CANADA BY THE
COPP CLARK PUBLISHING CO. LIMITED, TORONTO

FOREWORD

BALSAM FIR is one of the most important constituents of the forests in the Quetico-Superior wilderness and adjoining areas. Uncertainty as to the successional role of this species was the basis for the recommendation to the University of Minnesota School of Forestry by the Quetico-Superior Wilderness Research Center Advisory Committee in 1953 that a complete literature survey be made of all research conducted on this species. The preparation and publication of this book were supported by the Wilderness Research Foundation. The sponsor of this Foundation, a firm advocate of the important role of basic biological research in solving wilderness area problems, maintained continuous interest and provided encouragement throughout the course of the prolonged study. Without his dedicated interest and support, completion of this project would not have been possible.

Additional support has been received from the Agricultural Experiment Station and the Graduate School of the University of Minnesota under various grants which aided in certain aspects of the work.

One of the most valued experiences shared by the senior authors was a visit to educational institutions, research centers, and forest industries in the northeastern United States and Canada, where balsam fir problems are of particular concern. While it is impossible to enumerate all persons visited, special appreciation is expressed to the many foresters and others who contributed valuable information and assistance during the course of this trip.

The first draft of the chapter on insect problems was reviewed by J. H. MacAloney and that of the chapter on utilization by W. B. Wallin. J. W. Hall provided material on paleobotany, and Elizabeth G. Lawrence made contributions to the two chapters on ecology. The final checklist on fungi was reviewed by N. A. Anderson, and L. K. Paulsell edited one of the later revisions of the complete manuscript.

The authors and co-authors are indebted to their many university associates, to graduate students, to the staff of the Lake States Forest Experiment Station of the United States Forest Service, and to field foresters and biologists in the United States and Canada who have been helpful in the various activities of literature search and evaluation, data summarizing, counseling, and editing.

FRANK H. KAUFERT, Project Leader

v

PREFACE

THE objective in the preparation of this book has been to collect scattered information relating to balsam fir and to integrate this diversified knowledge into a unified presentation from which both broad theoretical generalizations and practical applications can be made. The book is intended to be a source of information for specialists, in areas outside their own particular fields of interest. It should also serve as a helpful reference for students of various disciplines, for research workers, and for forest practitioners by providing background information and suggesting new approaches to particular problems.

The first draft of the manuscript was prepared by E. V. Bakuzis in cooperation with H. L. Hansen. They are largely responsible for the several revisions and, together with the co-authors of individual chapters, for the organization of the completely rewritten final draft. F. H. Kaufert served as the project leader. R. E. Schoenike assisted in the editing of the final draft.

The major functions of the co-authors were to give organization and balance to the various topics in their fields of competence, to provide additional references, and to rewrite wholly or in part various sections of the chapters. These contributions were as follows: Chapter 1 (Botanical Foundations), S. S. Pauley and R. E. Schoenike; Chapter 2 (Geography and Synecology), D. B. Lawrence and R. E. Schoenike; Chapter 3 (Ecological Factors), D. B. Lawrence, R. E. Schoenike, and L. W. Krefting; Chapter 4 (Microbiology), F. H. Kaufert; Chapter 5 (Entomology), R. E. Schoenike; Chapter 6 (Reproduction), D. P. Duncan; Chapter 7 (Stand Development), D. P. Duncan; Chapter 8 (Growth and Yield), R. M. Brown; Chapter 9 (Utilization), F. H. Kaufert. D. B. Lawrence is a member of the faculty of the Department of Botany, University of Minnesota; L. W. Krefting is a wildlife research biologist, United States Bureau of Sport Fisheries and Wildlife; R. E. Schoenike, formerly of the University of Minnesota School of Forestry, is at present a member of the faculty at Clemson University, South Carolina. The other contributors are faculty members of the School of Forestry, University of Minnesota.

The early drafts were prepared primarily as a review of the literature. In the present text an attempt was made to evaluate each reference within the framework of the whole book. New material was incorporated into the text to the cut-off

BALSAM FIR

TABLE A. LITERATURE CITED AND PUBLICATIONS EXAMINED BUT NOT CITED °

Period of Publication	U.S.A.		Canada		Great Britain		Germany		Others		Total	
	Cited	Not Cited	Cited	Not Cited	Cited	Not Cited	Cited	Not Cited	Cited	Not Cited	Cited	Not Cited
1961–63	52	14	66	10	2	2	4	3	3	1	127	30
1951–60	319	183	184	109	9	11	15	30	14	41	541	374
1941–50	183	103	79	58	3	2	3	17	4	47	272	227
1931–40	136	79	29	20		4	8	30	3	37	176	170
1921–30	83	35	10	9	1		10	21	4	10	108	75
1911–20	51	18	4	6	1		4		1	2	61	26
1901–10	21	12	3				6	3		1	30	16
19th century	35	10	1		7	3	5	9	10	1	58	23
18th century					3		1		1		5	
17th century					2				1		3	
Total	880	454	376	212	28	22	56	113	41	140	1381	941

°The table does not include 12 cited annual reports and encyclopedic publications.

date — July 1, 1963. The final draft includes 1393 references. An additional 941 references were consulted. Distribution of the references by period of publication and by contributing country is shown in Table A.

The period covered by the literature extends over 300 years. It includes the observations of early travelers and investigations made during the period of initial settlement of the North American continent, followed by the beginning of forestry and its development to the present time. The great acceleration in the rate of publication is demonstrated in the table. The table also indicates that European references played a relatively greater role in the early years, while in the last two decades the number originating in Canada has increased considerably. Canadian authors have contributed to books and journals published in the United States, and to some extent in other countries, perhaps more than foreign authors have contributed to Canadian publications. Investigators of different countries frequently pass over national boundaries, however, and quite a few publications have resulted from international cooperation such as that provided by the Quetico-Superior Wilderness Research Center and international exchanges organized by various agencies.

The types of publication are diversified. In this study special attention was directed to the collection of valuable information of the kind which appears in temporary form as mimeographed notes, leaflets, and similar items of limited distribution. Even printed bulletins become difficult to obtain after a decade or two. The number of cited references by type of publication and by country is shown in Table B.

The relatively large number of Canadian publications in forestry journals is partially explained by inclusion of the bimonthly progress reports of the Forest Entomology and Pathology Branch (formerly Forest Biology Division, Department of Agriculture) of the Canada Department of Forestry. Most Canadian forestry publications contain relatively greater amounts of information about

TABLE B. NUMBER OF REFERENCES BY TYPE OF PUBLICATION AND BY COUNTRY*

Type of Publication	U.S.A.		Canada		Great Britain		Germany		Others		Total	
	Cited	Not Cited	Cited	Not Cited	Cited	Not Cited	Cited	Not Cited	Cited	Not Cited	Cited	Not Cited
Books	113	79	3	3	14	9	16	31	12	9	158	131
Periodicals												
Forestry	113	69	145	96	6	4	30	57	5	48	299	274
Botany	130	59	6		2		2	3	1	6	141	68
Others	115	76	75	45	1	8	6	16	8	41	205	186
Printed bulletins	207	50	84	16	5		2	6	15	33	313	105
Mimeo and leaflets	202	121	63	52		1				3	265	177
Total	880	454	376	212	28	22	56	113	41	140	1381	941

*The table does not include 12 cited annual reports and encyclopedic publications.

balsam fir than do similar publications from the United States. This is a reflection of the greater importance of the species in Canada than in the United States. Although the majority of printed bulletins are pamphlets, several are extended monographs which approach the size and scope of books. About a third of the sources reviewed have been published in temporary form.

While the search for information led in many directions, the greatest possible use was made of *The Bradley Bibliography, Forestry Abstracts, Biological Abstracts, A Catalogue of Books Represented by Library of Congress Printed Cards,* the *Dictionary Catalogue of the Yale Forestry Library,* and indexes of the various journals. The contributions by Sargent (1884, 1898), Graves (1899), Zon (1914), and Sudworth (1916) can be considered as the classics on the species. Reviews of the literature by Roe (1948) and Hart (1959) were helpful in checking omissions. A University of Michigan master's thesis by H. J. Heikkenen (1957), "Balsam Fir, an Analysis of Literature," was also consulted.

Organization and evaluation of the material has presented problems. For this monograph, a major one was gathering and organizing numerous small sources of information, produced over an extensive period, written from many points of view, and analyzed by different methods. Although large organizations are now involved in research concerned with the problems of balsam fir, the direction and goals of such research are often complex, and much of the information is derived as a by-product of some other activity. The review of published information revealed gaps in some areas as well as extensive concentrated activity on certain specific problems. An attempt has been made to find a plan of organization which fitted the material available rather than to fill a predetermined system borrowed from the general classification patterns of scientific and technical information. An attempt has also been made to reconcile the points of view of the past in the light of more recent developments by looking for common denominators. This frequently required regrouping original data followed by the preparation of new tables and figures which sometimes resulted in interpretations quite different from those used in the original. However, in the absence of contradictory information from other sources, an effort has been made to preserve the author's viewpoint as far as possible. On occasion it has been necessary to discuss problems and give details of techniques not specific to balsam fir to provide a framework for integrating the accumulated knowledge of the species. Such excursions were undertaken to clarify problems connected with life cycles, ecological models, wood decay, dynamics of spruce budworm epidemics, wildlife considerations, and others.

Tables, figures, and specific listings or catalogs are important means for the condensation and storage of information. Originally over 250 tables and five appendixes were prepared; of these 149 tables and two appendixes were finally included. Of the 13 figures and 16 plates included, seven have been obtained through the courtesy of the original authors or the copyright owners. The remainder have been redrawn and recombined to give more specific interpretations to problems of balsam fir.

Each of the nine chapters begins with an introductory statement relating the chapter to the rest of the book, defining its objective and indicating the specific characteristics of the information available and the most important considerations of organization. The introduction is followed by an analysis of the subject matter and organization of the material with an attempt to unify the findings in generalizations. The chapters conclude with an evaluation and some recommendations which involve partial summarization, conclusions, criticisms, hypotheses, and suggestions for future research. Depending on the material available the formulations are made in either empirical or more abstract form. The need for answers to urgent practical problems has frequently necessitated the acceptance of temporary solutions which really deserve more fundamental scientific treatment.

It would be presumptuous to say that generalizations of a new order have been attained. Rather, it is hoped that a more coherent picture of balsam fir and its place in nature and in forestry practice has emerged.

INTRODUCTION

THE inherent characteristics of balsam fir (*Abies balsamea* (Linnaeus) Miller), its environmental relationships and management in forest stands, and the use of products derived from it constitute the scope of this study. A series of biological disciplines including their application in forestry is involved. Balsam fir has not served as an object of research in which new laws of nature have been developed. Rather, research relating to this species has served to strengthen scientific and technological developments over a wide field.

Historically, the increase in knowledge about the species has been slow. As might be expected, information on its distribution and dendrological characteristics appears in the earliest literature. The use of Canada balsam in optical work brought the species name to the attention of the entire scientific world. When a series of catastrophic spruce budworm epidemics destroyed large forest areas in eastern Canada and the United States, and led to the development of aerial control measures, balsam fir again received international attention. Establishment of a modern forest biology laboratory at Sault Ste. Marie, Ontario, and the organization and modernization of other forest laboratories in Canada and the United States introduced a new era in forestry research.

The first five chapters of the book are largely restricted to the basic fields, while the remaining four may be considered as dealing with the application of knowledge and techniques. Ninety per cent of the references are specific to only one chapter. The remainder, which are used in two or more chapters, point to some inherent relationships among the chapter topics. Table C indicates that the chapters on Microbiology and Entomology are more self-contained than the others.

Botanical information discussed in the first chapter is of a basic kind, knowledge *per se*, and is more broadly considered in the treatment of geographical and synecological problems, silvicultural techniques, and utilization in subsequent chapters. Information on taxonomy relates closely to genetics, but is necessarily limited because the study of genetics is only beginning with respect to balsam fir. Further study is needed of the anatomical and morphological characteristics which largely determine the properties of its wood, as most of the anatomical work reported on balsam fir was done before modern instruments became available. The

TABLE C. REFERENCES CITED IN EACH CHAPTER AND THEIR CITATION IN OTHER CHAPTERS

Chapter Title	Number of References		Number of References Cited in Other Chapters								
	Total Used	Used in Only One Chapter	Botanical Foundations	Geography and Synecology	Ecological Factors	Microbiology	Entomology	Reproduction	Stand Development	Growth and Yield	Utilization
Botanical Foundations	190	156		17	2	7	5	5	3	3	9
Geography and Synecology	264	192	17		28	3	2	21	21	14	7
Ecological Factors	197	150	2	28		4	6	16	13	7	5
Microbiology	203	179	7	3	4		9	1	3	4	6
Entomology	270	240	5	2	6	9		3	7	1	2
Reproduction	128	82	5	21	16	1	3		26	8	3
Stand Development	121	74	3	21	13	3	7	26		12	1
Growth and Yield	75	52	3	14	7	4	1	8	12		5
Utilization	149	126	9	7	5	6	2	3	1	5	

physiological aspects of drought and flooding, abundance and deficiency of nutrients, frost and excessive heat, fire and wind, attacks from insects and diseases, effect of competition, seed production, and the vegetative growth of different parts of the tree provide a whole research program for the future.

Since this is a monographic study of a species, autecological information might be expected to prevail. However, most of the observations on the life history of the species have been made with little consideration of the environment, and this type of information is therefore treated in the chapter on Botanical Foundations. On the other hand, since geographical and synecological information on balsam fir is abundant, more autecological information has been derived from geographical and synecological studies than from experimental ecology.

The chapter on Geography and Synecology deals first with the broad geographical aspects of the past and present distribution. Forest geographical classification is more advanced in Canada than in the United States. The pioneering work done by Halliday has been recently updated by Rowe and intensive regional studies are in progress in Ontario, New Brunswick, and elsewhere. In the United States the investigations by Braun on the Deciduous Forest Formation of eastern North America, and the beginnings of some detailed studies by Westveld in the northeastern United States are the major regional additions to the large-scale contributions of such pioneers as Sargent, Merriam, Mayr, Zon, Clements, Harshberger, and others.

A comprehensive geographic-synecological classification of forests provides the basis for their systematic investigation and the integration and exchange of both scientific knowledge and practical experience. Many pioneers in the dynamic school of American ecology have worked within the range of balsam fir. Important early work in forest synecology was carried out in the New York Adirondacks and in the Lake Edward Experimental Forest in Quebec. Recent progress has been made by the Wisconsin school in the United States and by the Ontario and Quebec schools in Canada. The concept of ecosystem is at the center of current ecological thought. It establishes a link with physiology in its emphasis on matter-energy flow through a community.

The literature on forest communities including balsam fir is rich. In geographical and synecological classification systems balsam fir appears as a distinctly boreal species. It retains this character in the southern parts of its range by occupying edaphic positions of a boreal nature or by existing as an understory species below pine and hardwood overstories. This greatly modifies the watershed effect of the overstory cover, the wildlife and other conditions, and has great influence on silviculture, reproduction, and the production processes as well.

In investigations of the relationships between balsam fir and environmental factors, use is made of two types of factor classification. The common type which recognizes climatic, edaphic, physiographic, and biotic factors has been most widely used, and the observations made respecting them are rather simple. Generalization from such an approach, however, is difficult. The physiological type of factor classification which recognizes such factor complexes as moisture and nutrients, heat, light, and other types of radiation, and mechanical force, is more difficult to apply although its potential for making more widely applicable generalizations is much greater. This latter approach requires, for its full use, construction of multidimensional ecosystem models with positive and negative feedback systems. Only a few elementary approaches involving balsam fir have been made thus far. With the advances made recently in construction of models of epidemic and endemic populations of the spruce budworm, and with a tendency to rejuvenate the models of stand structure and growth in the new growth and yield studies, it can be expected that in the coming years the approach to many problems discussed in this monograph will take on a different and more fundamental viewpoint that will result in more effective applications.

The effects of fire and wind, unfavorable weather conditions, and to some extent wildlife, together with the material discussed in the chapters on Microbiology and Entomology, constitute the field of protection for balsam fir. Since the technology of balsam fir protection does not differ much from that of general forest protection, greater attention is given to the ecological aspects than to the techniques. General ecological and microenvironmental studies are of particular importance in the investigation of insects and fungi. Considerable recent progress in this direction has been made in the studies of the spruce budworm. The more detailed site classification used by some authors indicates that success can also be achieved in delineating conditions favorable for the detection and prevention of rots.

The chapter on Reproduction introduces the technological part of this study. Together with the chapter on Stand Development it contains most of the information on the silviculture of balsam fir. The great abundance of balsam fir reproduction frequently has stimulated unwarranted optimism in some areas about the recovery of the forest after destructive treatments in the past. In other areas it has caused great concern as an aggressive substitute for economically more desirable species.

Some fundamental aspects of balsam fir reproduction relating to site and stand conditions have been investigated by Place and other workers. An outstanding reproduction survey carried out by Candy and several associates effectively demonstrated the geographical gradient of balsam fir reproduction over its entire range in Canada. Studies of natural reproduction predominate. After the failure of early plantations, interest in artificial regeneration of balsam fir lagged and only recently has shown signs of reviving.

Although chemical silviculture has made marked progress during the last decade, the treatments have been primarily limited to the release effects on established reproduction. Ground preparation for new reproduction has been largely neglected. Both the physiological and synecological aspects of release need more serious consideration in the application of chemical techniques. In work with chemical silvicultural measures, balsam fir has been the object of only incidental attention.

Even though much of the early interest in balsam fir silviculture has been directed toward the development of partial cutting systems, in many instances instead of response to release, reaction to exposure has resulted with a consequent loss of growing stock. The results of partial cutting apparently have not been too successful. In addition, it has been difficult to apply mechanized logging to partial cuts. This has resulted in increasing attention being given to regulated clearcutting or shelterwood systems for the management of spruce-fir forests.

The classical disputes over site determination and measurement have also involved balsam fir. Future developments in ecosystem research will shed more light on a number of controversial problems. The great abundance of the available tables on balsam fir volume and yield in the literature is not an indication that further research is unnecessary. Most such tables were originally prepared with techniques considered out of date by present standards.

Attempts have been made to give some of the many different mensurational indexes presented here a biological interpretation, although their biological significance has been a matter of frequent conjecture. The recently developed method for rapid measurement of basal area of forest stands by Bitterlich has eliminated the need for some of these indexes; nonetheless, the thought processes which were used to develop these indexes are still rather stimulating, in having not only clarified certain problems in greater depth but in having also produced a number of residual values. Balsam fir has served as the study object in several cases.

The transition from exploitation of the virgin forests to management of second-growth stands has greatly changed the roles of individual species. Pines, particularly white pine and red pine, have declined, giving place to aspen and birch. Balsam fir, in spite of its short life span and even with its insect and disease problems, has improved its relative position because of its ability to reproduce under adverse conditions. Together with jack pine and aspen, balsam fir will probably be of considerable importance over large areas in northern forests in the future. Intensive research is needed, however, to avoid the creation of an offsite balsam fir problem similar to that of aspen.

Resource information on balsam fir and its products is concentrated primarily in two chapters: Growth and Yield, and Utilization. In Canada, balsam fir is a major species on a national scale; in the United States, in parts of the Northeast and the Lake States, it is also a major species. Although it has largely disappeared from the lumber picture, at least in the United States, its use for fiber continues to increase. In some instances pulp and paper mills formerly geared to spruce as the predominant raw material for quality products have reoriented their processing technology and their over-all operations to greatly increase the use of balsam fir.

TABLE OF CONTENTS

1. BOTANICAL FOUNDATIONS (with the aid of S. S. Pauley and R. E. Schoenike) 1

Taxonomy and Cytology (Historical Records and Nomenclature, Taxonomic Units and Identification, Cytology), 1. Life Cycle, Morphology, and Anatomy (Life Cycle, General Morphological Description, Leaf Anatomy, Wood Anatomy, Bark Anatomy), 9. Physiology and Genetics (Photosynthesis and Respiration, Vegetative Growth, Flowering and Seed Production, Phenology, Growth Correlations, Genetic Variability and Tree Breeding, Hybridization, Vegetative Propagation), 20. Conclusions, 29.

2. GEOGRAPHY AND SYNECOLOGY (with the aid of D. B. Lawrence and R. E. Schoenike) 31

Distribution (Past Distribution, Contemporary Natural Distribution, Distribution outside the Natural Range), 31. Geographical and Synecological Classification and Models (Regional Classifications and Mapping, Schools of Thought and the Development of Models), 41. Geographical and Synecological Descriptions of Regions (The Boreal Forest Region, The Great Lakes–St. Lawrence Forest Region, The Acadian-Appalachian Forest Region), 50. Conclusions, 81.

3. ECOLOGICAL FACTORS (with the aid of D. B. Lawrence, R. E. Schoenike, and L. W. Krefting) 84

Ecological Models (Wisconsin Models, Saskatchewan-Manitoba Models, New Brunswick Models, Minnesota Models), 84. Climatic Factors (Heat, Light, Precipitation, Wind, Fire), 91. Edaphic Factors (Physical Properties of Soil, Chemical Properties of Soil), 101. Wildlife: A Biotic Factor (General Ecological and Taxonomic Problems, Moose and Woodland Caribou, White-Tailed Deer, Hares, Mice, Red Squirrels, Beaver, Porcupine, Bear, Grouse, Other Birds), 110. Conclusions, 122.

4. MICROBIOLOGY (with the aid of F. H. Kaufert) 124

Microorganisms Involved (Miscellaneous Microorganisms, Fungi Associated with Balsam Fir, Fungi Involved in Symbiotic and Other Relationships), 124. Seedling and Sapling Diseases (Young Seedling Diseases, Needle Casts and Blights, Rusts, Cankers), 128. Decay (Decay of Living Trees, Decay of Injured, Dead, and Down Trees, and Wood Products), 132. Descriptions of the Principal Heart Rot Fungi, 144. Conclusions, 147.

5. ENTOMOLOGY (with the aid of R. E. Schoenike) 149

Taxonomic and Ecological Problems (Taxonomic Survey, General Ecological Relationships), 149. Spruce Budworm (Historical Records, Bionomics, Predators and Parasites, Conditions Associated with Outbreaks, Effects of Outbreaks, Population Dynamics and Surveys, Control), 152. Balsam Woolly Aphid (Life History, Control), 169. Other Associated Insects (Primary Insects, Secondary Insects, Control), 173. Conclusions, 177.

6. REPRODUCTION (with the aid of D. P. Duncan) 178

Seed and Seedbeds (Seed Characteristics and Procurement, Seedbeds and Seedling Survival), 179. Reproduction Methods (Advance and Subsequent Reproduction, Reproduction Cuts, Effect of Logging on Reproduction, Promotion of Natural Reproduction, Seeding and Planting), 189. Regional Problems (Boreal Forest Region, Great Lakes–St. Lawrence Forest Region, Acadian-Appalachian Forest Region), 199. Conclusions, 208.

7. STAND DEVELOPMENT (with the aid of D. P. Duncan) 210

Virgin Conditions and Natural Disturbances (Undisturbed Virgin Conditions, Natural Disturbances), 211. Cleaning (Competition in Seedling and Sapling Stages, Mechanical Control of Competing Vegetation, Chemical Control of Competing Vegetation), 218. Thinning (Stand Structure, Thinning in One-Storied Stands, Thinning in Two-Storied Stands, Girdling, Chemical Thinning), 224. Clearcutting and Partial Cutting (Clearcutting, Release and Exposure, Cutting Cycle, Selective Cutting of Conifers, Cut of Conifers and Hardwoods), 232. Conclusions, 244.

8. GROWTH AND YIELD (with the aid of R. M. Brown) 246

Tree Form, Tree Volume, and Stand Volume (Stem Form, Stem Volume, Use of Aerial Photographs, Unused and Defective Material), 246. Yield Tables (Measurement of Site, Normal Yield Tables, Special-Use Yield Tables), 261. Regional Growth and Yield Data (Area Distribution, Growing Stock, Yield and Growth per Acre), 283. Conclusions, 291.

9. UTILIZATION (with the aid of F. H. Kaufert) 293

Wood Properties (Physical Properties, Mechanical Properties, Chemical Composition), 293. Pulping Properties (Groundwood or Mechanical Pulping, Chemical Pulping, Semichemical Pulping), 302. Lumber and Pulpwood Products (Harvest and Procurement, Lumber, Pulpwood), 307. Minor Products (Bark Properties and Uses, Canada Balsam, Needle Oils), 318. Miscellaneous Uses and Values (Christmas Trees, Ornamental and Recreational Use), 322. Conclusions, 327.

APPENDIX I. A TENTATIVE LIST OF MYXOMYCETES AND FUNGI ASSOCIATED WITH BALSAM FIR 349

APPENDIX II. A TENTATIVE LIST OF INSECTS ASSOCIATED WITH BALSAM FIR 362

LITERATURE CITED 368

INDEX 416

ILLUSTRATIONS

LIST OF FIGURES

Balsam Fir Woodland Scene, by Francis Lee Jaques .... *Frontispiece*

1. *Abies balsamea*: pollen grain at time of shedding 10

2. *Abies balsamea*: foliage, strobili, and seedling characteristics 16

3. Cross section and tangential section of balsam fir stem 17

4. Photosynthetic capacities of *Abies balsamea* foliage 22

5. The place of balsam fir in forest classification systems based on primary succession 43

6. The place of balsam fir in forest classification systems based on edaphic relationships 45

7. Analysis of vegetation of the principal forest types in the Lake Edward Experimental Forest, Quebec 64

8. Mean light intensity in foot-candles (shown by main forest types) in the Adirondacks 96

9. Soil profiles of major spruce-fir types in the Adirondacks 103

10. Heavy mortality following severe spruce budworm infestation 154

11. Height, basal area, and volume growth of balsam fir and spruces in the northeastern United States, composite growth of spruce-fir in Michigan, and height growth in the Lake States 266

12. Production of balsam fir lumber from 1905 to 1943 in the United States 312

13. Production of balsam fir pulpwood from 1905 to 1940 in the United States 316

LIST OF PLATES (pages 329–346)

1. Distribution of *Abies balsamea* in relation to other *Abies* species

2. Distribution of balsam fir in the forest regions and sections

3. Relation of the potential forest cover to different soil groups in Michigan, Wisconsin, and Minnesota

4. Site regions, soil development, and changes in species composition in the humid eastern part of Ontario

5. The position of balsam fir in regional edaphic fields as outlined by a modified forest cover type system and local forest classification systems for the Labrador Peninsula, Acadian Experimental Forest, and Wisconsin

6. Climographs characterizing the range of balsam fir

7. Major soil groups in the natural range of balsam fir

8. Distribution of balsam fir in Minnesota

9. Distribution of balsam fir and of community types in moisture-heat, moisture-light, and phytosociological coordinate axes in Wisconsin

10. Importance values of balsam fir in phytosociological and environmental coordinate systems in northwestern New Brunswick

11. Frequency distribution of balsam fir in bivariate combinations of moisture, nutrient, heat, and light synecological coordinates in Minnesota

12. Distribution of balsam fir in the edaphic field of the Central Minnesota Pine Section

13. Distribution of balsam fir, and soil nitrogen percentages, in upland forest communities of the Central Minnesota Pine Section

14. Degrees of frost injury sustained by balsam fir during late spring freeze in Minnesota

15. Percentage of stocked milacre quadrats of balsam fir advance and subsequent reproduction following logging in different forest regions and sections of Canada

16. Dynamics of advance reproduction and the position of balsam fir in comparison with competing species in the edaphic field of the Central Minnesota Pine Section

BALSAM FIR
A Monographic Review

1

BOTANICAL FOUNDATIONS

THE botanical characteristics and variations of a species determine its relationships with other species and its reaction to the environment. These relationships
are basic to the development of sound silvicultural and forest management practices. In spite of modern technological advances, botanical characteristics, especially the morphological, determine the technical properties of primary forest
products and affect the processing of secondary products. Much of the information
on the taxonomy and morphology of balsam fir is rather old. On the other hand,
specific work on the physiology and genetics of the species has been started only
recently.

It is the purpose of this chapter to summarize the known botanical knowledge
on balsam fir, to point out some of the shortcomings of past approaches, and to
help clarify botanical problems that need to be resolved in adapting the species
to man's growing needs. To do this it will be profitable not only to consider balsam
fir, but to examine some of the available information for other species in the genus
Abies as well.

Taxonomy and Cytology

In this section the nomenclature, historical record, and classification of balsam
fir are reviewed. The historical record is particularly revealing in giving an insight
into the methods developed and used by plant systematists. The review begins with
folklore and early observations, followed by the pioneer gropings for a natural
system of classification. More detailed descriptions over the years brought out
the need for a unified system of nomenclature. Misidentifications, confusion of
names, and other "growing pains" are recorded. Gradually taxonomists worked
out their difficulties by developing international codes and rules of nomenclature.
Modern classification is based on detailed and thorough study involving related
disciplines such as morphology, anatomy, genetics, and ecology. The history of
balsam fir records all these different approaches. In addition to taxonomy a short
section on cytology is added here because of its relevance to classification.

The botanical names for American plant taxa in this book follow the Latin
nomenclature as given in the eighth edition of Gray's *Manual of Botany* (Fernald,

1

1950) unless specifically indicated. The exotic plant species are listed as given by Krüssmann (1960). The common names for trees of North America follow the United States Forest Service checklist by Little (1953) or *Standardized Plant Names* by the American Committee on Horticultural Nomenclature (Kelsey and Dayton, ed. 1942).

HISTORICAL RECORDS AND NOMENCLATURE

Theophrastus (ca. 371–286 B.C.) in his *Enquiry into Plants* recorded a considerable amount of information on the fir trees of his region. Hort's (1916) translation shows that this early scientist had a surprising knowledge of the morphology, ecological characteristics, phenology, distribution, and wood properties of the firs (genus *Abies*). Theophrastus recognized "male" and "female" firs, the distinguishing feature being that the wood of male firs had more knots than the wood of female trees.

The concept of sex in plants originated in ancient Assyria. Very often this concept was applied to different species. For example, Evelyn (1664) in his *Sylva* wrote: "There are of the fir two principal species: the picea or male, which is the bigger tree, very beautiful and aspiring, and of an harder wood and hirsute leaf; and the silver-fir or female." A lingering trace of this idea still finds expression in the southern Appalachians where the mountaineers refer to Fraser fir as the "she-balsam" and red spruce as the "he-balsam" (Sargent, 1898).

The early explorers and settlers in eastern Canada and New England were the first to become acquainted with balsam fir. In his diary for August 19, 1535, Jacques Cartier made references to "pruches," which according to Pickering (1879) meant three species: balsam fir, white spruce, and black spruce, seen by the explorer during his voyage on the St. Lawrence River. There are early notes from New England in the period 1607–1622 in which balsam fir is referred to as "firre." In 1669 Governor John Winthrop of Connecticut sent specimens of "fir balsam" to the Royal Society of London, reporting that they were taken from trees in Nova Scotia, but that the same species was reputed to grow in the more eastern parts of New England (Pickering, 1879).

Published reports on balsam fir appeared somewhat later. Apparently the earliest report was prepared by Pierre Boucher (1664; cited in Sudworth, 1916), who observed the plant in eastern Canada. Sudworth also indicated that John Josselyn (1674) had noted the species in New England.

The earliest botanical description of balsam fir was made by John Ray (1704), but the first scientific name, *Pinus balsamea* Linnaeus, was assigned by Linné (1753) in his *Species Plantarum*.

The following chronological sequence of scientific names for balsam fir is taken from Sargent (1898).

> *Pinus balsamea* Linnaeus (1753)
> *Abies balsamea* Miller (1768)
> *Pinus abies balsamea* Münchhausen (1770)
> *Pinus taxifolia* Salisbury (1796)
> *Abies balsamifera* Michaux (1803)
> *Pinus balsamea* var. *longifolia* Lawson & Son (1836)
> *Picea balsamea* Loudon (1838)

Picea balsamea var. *longifolia* Loudon (1838)
Picea balsamifera Emerson (1846)
Picea fraseri Emerson (1846)
Abies americana Provancher (1862)

To this list Lamb (1914) added *Abies excelsa fraseri* Hort. ex Carr., *Abies fraseri hudsoni* Carr., and *Abies hudsonia* Bosc. ex Carr. Dallimore and Jackson (1931) further recorded *Abies aromatica* Rafinesque and *Abies minor* Dunham (Duhamel du Monceau?). Rehder (1949) cited *Peuce balsamea* Richard (1810) and *Abies balsamea* var. *brachylepis* Willkomm (1863; see also Willkomm, 1869). In a rather extensive treatment of the genus *Abies*, Viguié and Gaussen (1928–29) noted 17 Latin names for balsam fir and 28 names for its many supposed varieties. These are not treated in detail here.

The above names refer to the present *Abies balsamea* (Linnaeus) Miller. However, very similar names have been given to Fraser fir (*Abies fraseri* (Pursh) Poiret). Lamb (1914) gives the following for this closely related species: *Abies balsamea* Bigel, *A. balsamea* var. *fraseri* Nutt., and *A. balsamea* var. *fraseri* Spach.

The generic name *Abies* was early accepted as the botanical name for firs in continental Europe, although *Picea* was used in England until the late nineteenth century (Gordon, 1875; Engelmann, 1878). The binomial *Abies balsamea* (L.) Mill. was originally published in the eighth edition (1768) of Miller's *Gardener's Dictionary*. *Abies balsamea* (L.) Mill. is in agreement with the International Code of Botanical Nomenclature and can be considered stable. It has been listed in its present form in all checklists of forest trees of the United States (Sudworth, 1898, 1927; Little, 1953).

Although A. Michaux (1803) considered balsam fir and Fraser fir as one species, his son, F. A. Michaux (1810–13, 1818–19) recognized the differences between them and treated them as two separate species. This distinction has been maintained by later authors.

For completeness, brief mention should be made of other workers who supplied identifying characteristics for balsam fir. Botanical information was supplied by Browne (1832), Hooker and Spratt (1832), Gosse (1840), Pritzel (1866), Hereman (1868), Trelease and Gray (1887), and others. The horticultural viewpoint was emphasized by a number of writers, international in scope, including Pepin (1860), Regel (1884), Codman (1889), Wesmael (1890), Masters (1895), Sargent (as C.S.S., 1889, 1897), Huntington (1904), McAdams (1909), Rothrock (1910). Later, the forestry viewpoint was added. References to the latter appear in succeeding chapters.

In older literature and in local use a wealth of common names are known for balsam fir. According to Edlin (1956) the old Norse name *fura* was introduced by Norsemen into Scotland during the period 800–1263 A.D. In England "fir" was the common name given to "pine" (*Pinus sylvestris* L.) long before species of *Abies* were introduced. In the German language *Föhre* is synonymous with *Kiefer*, both referring to pine. When Norway spruce (*Picea abies* (L.) Karst.) was introduced into the British Isles about 400 years ago, it was called "pruce fir," meaning Prussian fir. The name became modified as "spruce fir" and was later shortened to spruce (Jacombe, 1920). It appears that historically, the name "fir" has gradually lost its composite meaning and become more and more limited to the genus *Abies*. Pines were the first to establish their specific identity in the English language probably under the influence of the French word *pin*. Spruce firs (*Picea*) and hemlock firs (*Tsuga*) were the next to lose their identity as firs. At present only "Douglas fir" (*Pseudotsuga*) retains the name of fir among those that are not true firs (*Abies*). In the German language the name *Douglasie* is already well established, and in time this term or an equivalent may spread to other languages.

The following names have been or are presently in use for *Abies balsamea*: balsam fir, balm of Gilead, American silver fir, balsam, blister pine, fir pine, fir tree, silver pine, white fir, white spruce, Canadian fir, Canada balsam, balm of fir, single spruce, and eastern fir (Loudon, 1838; Dallimore and Jackson, 1931; Canada Forestry Branch, 1956; Sudworth, 1897, 1927). Indians in New York State called balsam fir *cho-koh-tung*, meaning "blisters." Standardized plant names (Kelsey and Dayton, 1942) and the current U.S.D.A. checklist (Little, 1953) recognize balsam fir as the preferred common name.

Common names for firs in other languages are: *Tanne* (German); *sapin* (French), (*sapin baumier* = balsam fir); *ädelgran* (Swedish); *abéte* (Italian); and *pikhta* (Russian).

TAXONOMIC UNITS AND IDENTIFICATION

The genus *Abies* was first given thorough monographic treatment by Endlicher (1847) in his *Synopsis Coniferarum*. Parlatore (1868) and McNab (1877) also treated the genus at some length. A synopsis of American firs was prepared by Engelmann (1878). Mattfeld (1925) dealt with the species of *Abies* in the Mediterranean area. Modern treatments begin with Pilger (1926), who recognized 40 species. A more liberal approach was taken by Viguié and Gaussen (1928–29), who listed 52 species and 12 varieties. Dallimore and Jackson in the third edition of their handbook (1948) listed 42 species. The most recent monograph (Franco, 1950) treats 43 species, of which the 20 cultivated in Portugal are described in great detail. Recent summaries of *Abies* cultivated in America have been given by Bailey (1933) and Rehder (1940), and in Germany by Krüssmann (1955, 1960).

The place of *Abies balsamea* (L.) Mill. in higher taxonomic units has varied with different classification systems. In the latest revision of Engler's *Syllabus der Pflanzenfamilien* (Melchior and Werdermann, 1954), the species is placed in the subgenus *Sapindus* and the genus *Abies*, in the subfamily Abietoidea of the family Pinaceae (Abietaceae). Engelmann (1878) placed the American firs in four sections of the genus, whereas Sargent (1898) classified them in three sections. Florin (1931) criticized the divisions of *Abies* into sections, and felt these were not well founded on either morphological or paleobotanical grounds. Nevertheless, for taxonomic purposes they may be useful. In Franco's recent monograph (1950), the genus *Abies* is subdivided into two subgenera: *Pseudotorreya* (Hickel) Franco, and *Sapindus* (Endl.) Franco. *Abies balsamea* (L.) Mill. belongs to the section *Balsameae* Engelm., one of seven in *Sapindus*, and to the series *Lasiocarpae* Franco, one of two in *Balsameae*. *Abies lasiocarpa* (Hook.) Endl., and *Abies fraseri* (Pursh) Poir. also belong in *Lasiocarpae*.

The three firs, *Abies balsamea*, *A. fraseri*, and *A. lasiocarpa* are commonly known as the balsam firs. The range of *A. balsamea* comes in close geographical proximity with the two other balsam firs (see Chapter 3), and one of the primary taxonomic problems is the distinction between the three species. In this work balsam fir refers to *Abies balsamea* unless indicated otherwise. The distinguishing characteristics of the three species are outlined in Table 1.

Several subspecific categories have been proposed for balsam fir. Kent (1900) described *Abies balsamea* var. *macrocarpa* Kent, which was found by Robert Douglas on the Wolf River in Wisconsin, and later propagated in Waukegan, Illinois, nurseries (Sargent (S.), 1892). The variety was described as having more

TABLE 1. DISTINGUISHING TAXONOMIC CHARACTERISTICS OF *ABIES BALSAMEA* IN COMPARISON WITH *A. FRASERI* AND *A. LASIOCARPA*

Characteristic	Abies balsamea	Abies lasiocarpa	Abies fraseri
Leaf color	Dark green and lustrous above, paler below	Blue-green and glaucous	Dark green and lustrous above, paler below
Leaf shape on sterile shoots	Rounded or obtusely pointed, and occasionally emarginate	Obtusely pointed and occasionally emarginate	Obtusely short-pointed and occasionally emarginate
Leaf shape on fertile shoots	Acute or acuminate	Thickened and acute	Acute or acuminate
Leaf arrangement (on lower branches)	Two-ranked and wide spreading	Imperfectly two-ranked, and directed forward and upward	Two-ranked and wide spreading
Leaf hypodermal cells	None to few above, few on edges or on keel	Interrupted above, abundant on edges and on keel	Continuous above, crowded on the edges
Stomata	Mostly on lower surface, two narrow bands, each with 4–8, usually 6 rows	On upper surface above the middle, two bands on lower surface	Mostly on lower surface, two broad bands, each with 8–12, usually 10 rows
Color of male strobili	Yellow-red, tinged with purple	Blue-violet	Blue-violet
Relative size of scales and bracts	Scales 1.5 cm wide, twice as long as the short-awned bracts	Scales 2–2.5 cm wide, three times as long as the long-awned bracts	Bracts longer than cone scales
Bract exposure (mature cones)	Bracts enclosed, occasionally the points protruding	Bracts entirely enclosed	Bracts largely exserted and reflexed, protruding
Bract shape	Oblong, emarginate, short-pointed at the broad serrulate apex	Oblong-obovate, laciniate, rounded, emarginate, long-pointed at the apex	Oblong, rounded, short-pointed at the broad denticulate apex
Cone shape	Slender, long	Slender, long	Somewhat oval, shorter
Bark	Thin, scaly on older trees	Thin or thick, fissured on older trees	Thin, scaly on older trees
Wood odor	Pleasantly aromatic	Disagreeably aromatic	Pleasantly aromatic

Sources: after Engelmann (1878), Sargent (1898), Zon (1914), Fulling (1934), Wyman (1943), Taubert (1926), Laing (1956), Moss (1959), Boivin (1959).

persistent branches, longer and more crowded leaves, and larger cones than the common balsam fir, thus making it of considerable value for landscape purposes. Although this is a horticultural variety, Rehder (1940) suggested that it might be a transitional form between *Abies balsamea* and *Abies lasiocarpa*. The variety was not included in the eighth edition of Gray's *Manual of Botany* (Fernald, 1950) or in the recent checklist of American trees (Little, 1953). The variety *macrocarpa* is probably best considered as a horticultural variant (cultivar) of balsam fir, maintained by vegetative propagation (see the listing by Krüssmann, 1960).

Evidence of introgression between balsam fir and subalpine fir was considered by Raup (1946) and Moss (1959). Raup (1946) maintained that the natural ranges of balsam and subalpine firs have expanded slowly and have not as yet met in Alberta. On the other hand, Moss (1959) considered the mixing of the two species in the region of Lesser Slave Lake and the Athabasca River in Central Alberta as a distinct possibility.

The variety *Abies balsamea* var. *phanerolepis* Fern. was described by Fernald (1909) from collections in Quebec, Newfoundland, and Maine. Because of its exposed bracts, the variety has been called the "bracted" balsam fir. In contrast to balsam fir, the female strobili are subcylindric, and the laminas of the ovuliferous scales are suborbicular or reniform. In many ways the variety resembles Fraser fir, but the cones of the latter are more ovoid, and the bracts are very much longer and have strongly recurved, broad tips. *A. balsamea* var. *phanerolepis* is the only subspecific taxon of balsam fir recognized in the checklist of American trees (Little, 1953). Krüssmann (1960) lists this taxon as *A. balsamea* f. *phanerolepis* (Fern.) Rehd.

In 1936 Fulling described *Abies intermedia* as a new species of fir from the Blue Ridge mountains in the southern Appalachians. He considered this species to be intermediate with respect to cone characters between *A. balsamea* and *A. fraseri* and possibly represents hybridization between them. Though somewhat similar in morphological characters to that of *A. balsamea* var. *phanerolepis*, Fulling did not consider the two taxa to be equivalent. At present the status of *A. intermedia* as a separate species is best considered questionable.

A recent study by Myers and Bormann (1963) throws doubt on the validity of the taxon, *phanerolepis*, as a separate entity. These workers studied cone scale-bract ratios and leaf length of a large number of trees along an altitudinal gradient on Mt. Washington and a latitudinal gradient from the east coast to the Lake States. Forms resembling *phanerolepis* were found at high altitudes and forms resembling *balsamea* at low altitudes on Mt. Washington, yet there was a regular gradient between them over the whole slope. Though less distinct than the latter, the low altitude gradient has forms more nearly like *phanerolepis* at the east coast and like *balsamea* in the interior. This suggested to the authors that "var. *phanerolepis* and var. *balsamea* are ends of a cline that represents a continuous population." *Abies fraseri* which resembles var. *phanerolepis* is considered to be part of the same population, but due to the restriction of its range and its isolation since the Pleistocene, it has undergone reduction in its gene pool, and now appears to be quite distinct.

Fernald and Weatherby (1932) described *Abies balsamea* f. *hudsonica*

(Jacques) Fern. & Weath., as a new combination from New England. This form has been variously listed as *A.b.* var. *hudsonia* Sarg. & Engelm. (Schenck, 1939), and *A.b.* var. *hudsonica* (Jacques) Sarg. (Rehder, 1940). The variant, f. *hudsonica* is a dwarf form found on mountain peaks in New England. Its short and numerous branches spread broadly, and it is not known to produce cones. Krüssmann (1960) listed this form as *A.b.* f. *hudsonia* (Bosc.) Fern. & Weath.

Boivin (1959) described *Abies balsamea* var. *phanerolepis* f. *aurayana* B. Boivin as a dwarf form of the bracted balsam fir occurring at high elevations in southeastern Canada.

In his taxonomic study of *Abies balsamea* and the closely related *A. lasiocarpa* Boivin (1959) recognizes one species, *A. balsamea* with two subspecies, *A.b.* *balsamea* and *A.b. lasiocarpa*. Subspecies *balsamea* has two varieties, i.e., var. *balsamea* (including f. *hudsonia*) and var. *phanerolepis* (including f. *aurayana*). Subspecies *lasiocarpa* includes two varieties, i.e., var. *lasiocarpa* (including f. *compacta*) and var. *arizonica*. Although this is an attempt to relate closely connected forms, it is evident that much additional study in natural variation is needed before the taxonomic problems can be solved.

European horticulturists have propagated a great number of ornamental forms of *Abies balsamea*. These have been described repeatedly by different authors under various names, but have not been treated systematically since Sudworth (1897), who recognized ten horticultural varieties and gave their synonyms. A summary of horticultural variants (cultivars) in the following list is taken from Gordon (1875), Dallimore and Jackson (1931), Rehder (1928, 1940), Sudworth (1897), Schwerin (1903), Beissner (1906), and Krüssmann (1960). No attempt has been made to check all references for synonyms.

In accordance with the *International Code of Botanical Nomenclature* (Lanjouw, 1956) and the *International Code of Nomenclature for Cultivated Plants* (Int. Union Biol. Sci., 1961), all cultivars are to be treated separately from naturally occurring variants. Cultivars may be recognized by the symbol "cv." preceding the cultivar name or by a set of single quotation marks enclosing the cultivar name and placed after the species name. Authorities for cultivar names are recognized only when their identity can be established with the cultivar in question. For the sake of uniformity, the quotation system for naming cultivars has been adopted here although the various taxa were not always listed this way in the original references. In the list the species name, *Abies balsamea*, is entered as *A.b.*

Cultivar	References
A.b. 'Angustata' (a handsome pyramidal form with short ascending leaves)	Rehder (1928)
A.b. 'Argentea' (needles with white tips)	Krüssmann (1960), Sudworth (1897)
A.b. 'Brachylepis' (similar to the natural form, *A.b.* var. *phanerolepis*)	Sudworth (1897)
A.b. 'Coerulea' (a dwarf shrub, conical, with short branches) . .	Krüssmann (1960), Sudworth (1897)
A.b. 'Columnaris' (with ascending branches)	Krüssmann (1960), Schwerin (1903)

A.b. 'Glauca' (more bluish than the type) .... Krüssmann (1960)
A.b. 'Globosa' (similar to 'Nana') .... Beissner (1903)
A.b. 'Longifolia' (with long needles) .... Sudworth (1897),
                                                            Gordon (1875)
A.b. 'Lutescens' (needles of a light yellow, strawlike color) ...... Schwerin (1903)
A.b. 'Macrocarpa' (described in text) .... Krüssmann (1960),
                                                            Rehder (1940)
A.b. 'Marginata' (with yellow margins on the leaves of new shoots) Beissner (1894)
A.b. 'Nana' (a dense globose form with short dark needles) Krüssmann (1960),
                                              Rehder (1940), Sudworth (1897)
A.b. 'Nudicaulis' .... Sudworth (1897)
A.b. 'Paucifolia' .... Sudworth (1897)
A.b. 'Prostrata' .... Sudworth (1897)
A.b. 'Pyramidalis' .... Beissner (1906)
A.b. 'Variegata' (with variegated leaves) .... Dallimore and Jackson (1931)
A.b. 'Versicolor' .... Sudworth (1897)

The term "double balsam," frequently encountered in the horticultural trade, has a variety of meanings. Emerson (1846) noted that double balsam fir has its leaves much more crowded than the normal form, and as a special variety may refer to Fraser fir or the bracted balsam fir (var. *phanerolepis*). Double balsam may also refer to abnormally dense branching due to a variety of conditions. Cash (1941) observed such balsam fir in shipments from Newfoundland, Nova Scotia, and New Hampshire, and suggested that a fungus may be involved since sclerotia were observed on some of the branches. The names double fir or double balsam are used in the Christmas tree trade to refer to the types of older trees which have somewhat twisted branches and thick, dense, brushlike needles. The secondary shoots are also located closer together than the shoots of young trees. Double fir is also known in Europe as *Doppeltanne*, and has even been given a Latin name, *Abies duplex*, a practice strongly criticized by Schwerin (1929).

CYTOLOGY

The basic number of chromosomes in the genus *Abies* is 12 (Sax and Sax, 1933). In *Abies balsamea* some discussion has arisen because of differing counts by Miyake (1903) and Hutchinson (1915). Miyake's early work showed that the haploid number of chromosomes in *Abies balsamea* was 12. This was disputed by Hutchinson (1915), who counted 16. Using more refined techniques, Sax and Sax (1933) reestablished the haploid number as 12. In recent summary treatments (Richens, 1945; Seitz, 1951), the basic number has been maintained. Of the 12 chromosomes, 5 are isobrachial (Sax and Sax, 1933).

Natural polyploids have not been reported. In an attempt to induce polyploidy, Mergen and Lester (1961b) treated nine species of *Abies*, including *A. balsamea*, with aqueous colchicine solutions in concentrations of 250 to 20,000 ppm. The most successful treatment consisted of burying freshly germinated seedlings in fine quartz sand that had been moistened in colchicine solution having a concentration of 4000 ppm. Although seedlings with polyploid cells were obtained, these cells were unable to compete with diploid tissue. The highest polyploidy obtained was 8n (96) chromosomes in cells located at the base of some needles. In some cases much higher numbers were observed but the exact amounts could not be determined.

Aside from chromosome counts and cellular aspects of embryology and wood anatomy, treated in the following section, little cytological work has been carried out with balsam fir.

Life Cycle, Morphology, and Anatomy

Form and structure have occupied the attention of many plant scientists during the past three centuries. Whether from the points of view of the systematist hoping to uncover evolutionary relationships, of the physiologist studying the influence of growth substances on meristematic tissues, or of the forester and wood technologist who hope to utilize and improve the wood product, form and structure play an important role in the botany of forest trees.

This section will review the life cycle of balsam fir. General morphological characteristics and specific information on the anatomy of leaves, wood, and bark are also considered.

LIFE CYCLE

Certain phases of the life cycle of balsam fir have been described in great detail. Some reported findings are in need of confirmation by additional research, and other phases specifically referring to balsam fir have not been covered at all. To present a somewhat coherent summary, it was necessary to consult available information at the level of the genus, *Abies,* and the family, Pinaceae (Abietaceae). For the latter, the standard work of Chamberlain (1935) was used as the principal source. Other studies consulted are cited in the discussion.

Buds differentiating into male and female strobili become recognizable microscopically in the spring, about one year before pollination. The male structures develop more rapidly at first, and can generally be distinguished before the female.

The male or microsporangiate strobilus consists of numerous microsporophylls spirally attached to the central axis. Each microsporophyll has a short stalk and bears two microsporangia. The development of a microsporangium and its contained microspores follows the usual ontogeny of an eusporangiate plant. At first the microsporangium consists of a single initial or row of initials (this has not been determined specifically for *Abies*) which may be either hypodermal or superficial in origin (for discussion of this point, see Foster and Gifford, 1959). A periclinal division of the first initials gives rise to two cells or series of cells, the outer by further division forming the parietal or wall layers including the tapetum, and the inner giving rise to the sporogenous tissue. The development of the microsporangium is very slow. Developing strobili become recognizable late in spring, the microsporangium appears in early summer, and the development of sporogenous tissue continues until late in the fall.

By repeated division, the sporogenous tissue becomes very massive, eventually composed of small rounded cells, poorly differentiated structurally and staining homogeneously (Mergen and Lester, 1961a). These latter develop into relatively large, angular, primitive archesporial cells containing prominent nucleoli. The archesporial stage lasts several weeks but differs among the various species of

Abies. Rounding of the archesporial cells and their filling with starch grains marks the formation of the microspore or pollen-mother cells.

Hofmeister (1848) found that *Abies balsamea* in Germany reached the pollen-mother-cell stage in November. Mergen and Lester (1961a) disputed his findings. Examining Hofmeister's descriptions and drawings, they found that, most likely, he observed the archesporial stage in which most details are masked by the presence of large numbers of starch grains.

In the spring the microspore mother-cells undergo meiosis, producing four microspores with the haploid number of chromosomes. The microspores thus become the first cells of the microgametophyte. The microspores are retained in the microsporangium where, after several internal cell divisions, they develop into pollen grains. At the time of pollen shedding, a normal pollen grain of *Abies balsamea* contains 2–4 prothallial cells, a stalk cell, a body cell, and a tube nucleus imbedded in a vacuolated protoplasmic mass, the latter containing numerous starch grains (Hutchinson, 1914). Each pollen grain has two layers, an outer or exine layer and an inner or intine layer, as shown in Figure 1. The wings or bladders of the pollen grains are formed from the exine layer. *Abies balsamea* pollen grains are very large, about twice the size of most *Pinus* species.

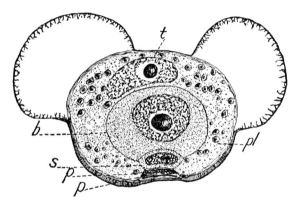

Figure 1. *Abies balsamea* pollen grain at time of shedding: (b) body cell; (s) stalk cell; (p) prothallial cell; (pl) starch; (t) tube nucleus. X535. (From Chamberlain, 1935. Courtesy of the University of Chicago Press.)

Wodehouse (1935) described the pollen grains of *Abies balsamea* as follows: "Grains various, some with three and some with four bladders, and in size ranging from 50 to 104μ in diameter, but averaging about 90μ. Exine of cap smooth, its boundary sharply defined, and marginal crest absent. Triradiate streak faint or absent. Bladders generally bulbous."

Details of pollen grain structure and development are given by Hutchinson (1914) and Mergen and Lester (1961a). The pollen tube nucleus is large and usually compressed. The stalk cell is much smaller than the body cell. The periclinal division into stalk and body cells takes place six or seven days before pollination. There are usually three prothallial cells, but this varies frequently. The maximum number is four. Hutchinson (1914) found that in 10 per cent of the material examined the nucleus of the body cell divided into two male nuclei before pollination. Mergen and Lester (1961a) were unable to confirm this finding. The male nuclei (sperms or gametes) are large, well developed, and imbedded in a common mass of protoplasm which is enclosed by the walls of the former body cell. Occasionally each male nucleus again divides, forming derivative cells. Thus there may be as many as four derivatives of the generative cell.

The female or megasporangiate strobilus is larger than its male counterpart and is composed of an axis and several spirally arranged appendages. These latter are of two kinds, bracts and megasporophylls (or ovuliferous scales). The megasporophylls are borne in the axils of the minute bracts. On the upper surface of each megasporophyll are two ovules or megasporangia. The megasporangium develops eusporangiately and becomes very massive. This structure eventually consists of a nucellus, surrounded by an integument. The opening in the latter extending to the nucellus is the micropyle. Within the nucellus a megaspore mother cell forms, although it does not become recognizable until very late in its development.

Meiosis in the megaspore mother cell has been difficult to observe. Reduction in number of chromosomes usually gives rise to four linearly arranged megaspores, the lower of which functions while the remaining three abort. The functional megaspore is the first cell of the megagametophyte. The megaspore nucleus then divides repeatedly without subsequent cell division, giving rise to a free nucleate stage. The megaspore membrane thickens during the free nucleus development, reaching a thickness of about 4.6μ in *Abies balsamea*. Eventually cell differentiation takes place and prothallial tissue is formed near the micropylar end of the ovule. The prothallial tissue bears several archegonia, each consisting of an ovum or egg, a ventral canal cell, and two neck canal cells. The remaining undifferentiated portion of the gametophyte becomes the endosperm or food-storage tissue. The above-described megasporogenesis has been generalized, as the process leading to maturation of the egg has not been specifically investigated in *Abies balsamea*.

When matured, the male strobilus, which has been enlarging throughout the spring, expands to expose the microsporangia. The latter then rupture, releasing the pollen grains, which are light and windborne. Some eventually reach the female strobilus, sift down between the ovuliferous scales and rest on the micropylar tip of the ovule. At the time of pollination, a pollination drop appears at the tip of the micropyle. Pollen grains falling on this drop are drawn onto the nucellus where the growth of the pollen tube begins. At first the growth of the tube is slow, about four weeks elapsing before fertilization takes place. Toward the end of this period the tube develops very rapidly, reaching the egg, and discharging the sperms in a period of two or three days. The sperms together with the stalk and tube nuclei are discharged into the cytoplasm of the egg. One sperm then fuses with the egg nucleus and effects fertilization. According to Hutchinson's (1915) observations at Lake Joseph, Ontario, fertilization occurred on June 25.

Hutchinson (1915) discusses in great detail the nature of chromosome pairing, duplication, and division. Following the fusion of the gametic nuclei, two groups of chromosomes are formed separately in the prophases of the first division of the zygote, each group containing the haploid number of chromosomes. This interpretation was strongly criticized by Sax (1918). Such type of division is fundamentally different from the generally accepted ideas of chromosome behavior and genetic theory (Beal, 1934). No such pairing occurs in *Pinus*, and it would seem unlikely that two conifers in the same family would differ so decidedly (Chamberlain, 1935). Hutchinson (1915) also claimed that double fertilization occurs occasionally in *Abies balsamea*. By this is meant the fertilization of the ventral canal cell nucleus with a sperm nucleus in addition to fertilization of the egg.

Embryogeny in *Abies* has been investigated by Miyake (1903), Hutchinson (1924), Buchholz (1920, 1942), Johansen (1950), and Wardlaw (1955). As with the discussion on earlier phases of the life cycle, some consideration of embryogenesis at levels higher than the species is presented in order to give a complete account.

In the earliest phase of embryogenesis, a proembryo develops from the fertilized egg or zygote. A free nuclear proembryo is characteristic of the Coniferales. After the second mitosis, the four nuclei migrate from the middle of the proembryo to the base of the archegonium, where a third division takes place which terminates with wall formation. The succeeding division forms four tiers of cells. The lowest tier will give rise to the embryo proper, the middle to the primary suspensor, and the third tier to the rosette. The upper tier soon degenerates.

Development of the proembryo in *Abies* is probably similar to that in *Pinus* and *Picea*. *Abies* differs from *Picea* in that the three tiers, rosette, suspensor, and embryonic, do not divide until the suspensor tier has elongated considerably to produce the primary suspensor. The tier of relict nuclei vanishes early. The rosette tier has been described as being absent in *Abies balsamea*, but Johansen (1950) disputed this. Rosette cells may form small rosette embryos after the primary suspensor has fully elongated.

The primary suspensor elongates to a considerable extent in *Abies venusta* and *A. pinsapo*, but less so in *A. balsamea*. By the time the first division occurs in the embryonic tier, each cell in the suspensor tier has elongated to approximately 60 times its original diameter. Before their collapse the suspensor cells attain lengths 150–200 times the original diameter. The nuclei are no longer recognizable when elongation ceases.

Meanwhile the embryo tier divides successively to produce three tiers of four cells each, the upper tiers known as embryonal tubes. Elongation of the embryonal tubes produces the secondary suspensors, which become very massive and before their collapse comprise much of the tissue enclosed in the megasporangium. More than one embryo develops. In *Abies*, simple polyembryony predominates. However, cleavage polyembryony has been observed in 10–13 per cent of the cases examined. As many as four cleavage embryos may attain the 8–16 cell stage before elimination commences.

The primary suspensor and secondary suspensor force the embryo through the base of the egg and into the gametophyte material (endosperm). Competition sets in among the various embryos derived from several fertilized eggs. In *Abies* usually only one embryo survives and develops to maturity. The suspensors collapse in order of their formation and their cell contents are eventually resorbed by the growing embryo. Differentiation now proceeds in the embryo and the cotyledons and hypocotyl develop, the former at the basal portion of the ovule. The hypocotyl terminates in a radicle formed at the micropylar end of the ovule. Surrounding the radicle is a calyptroperiblem, a type of root cap. A minute growing tip, the plumule, develops at the base of the cotyledons. The mature embryo consists of 4–5 cotyledons (Masters, 1891) and the partially differentiated hypocotyl that eventually develops into the stem and root of the sporophyte plant. The embryo, endosperm, a thin papery layer of nucellus, and a hard three-layer seed coat formed from the integument, make up the seed.

GENERAL MORPHOLOGICAL DESCRIPTION

Under favorable conditions the young sporophyte germinates, becomes a young seedling, and through normal growth processes goes through several stages before it becomes a mature tree. Details of the developmental and maturation process are presented in Chapters 6 and 7. General morphological descriptions of the mature sporophyte of balsam fir have been published by many authors, but original sources of information are relatively few. Unless otherwise indicated, the morphological descriptions presented in this section were compiled from the following standard dendrological sources: Sargent (1884, 1898, 1926), Otis (1913), Zon (1914), Sudworth (1916), Schenck (1939), Rehder (1940), Bailey and Bailey (1941), and Rosendahl (1955).

Balsam fir, whether grown in deep or shallow soils, produces a very superficial root system, penetrating to a depth of only 2–2.5 feet (Zon, 1914). Taproots, if present at all, soon die and rot away, especially in soils lacking an abundance of moisture. The strongly developed lateral roots extend horizontally in all directions for distances of 4–5 feet or more. The bark of the roots is bright red and separates in thin scales. Hopkins and Donahue (1939) confirmed the shallow rooting habit of balsam fir on podzol and brown podzolic soils in the Adirondacks of New York. They also observed that balsam fir roots are less branched than those of spruce. Bannan (1940, 1942) reported that balsam fir roots are uniformly distributed in the top 2–4.5 feet in sandy soils of northern Ontario.

At maturity balsam fir is 15–27 m (40–90 ft) tall, depending on the location and growing conditions. On the Green River watershed in New Brunswick where the species is in its growth optimum, individual trees grow to 90 feet in height with diameters of 30–75 cm (12–30 in.), reaching close to 200 years of age.

The straight and slender trunk has spreading branches. The crown is dense, dark green, and narrowly pyramidal, characteristically terminating in a slender, rigid, spire-like top. The bole is usually marked by small, dead, persistent branches. The major branches occur in whorls of four to six with occasional internodal branching. The young twigs are covered with very short hairs which may persist up to three years.

The bark is smooth and grayish with numerous blisters containing resin, the "Canada balsam" of commerce. The bark is comparatively thin, 2.5–11 mm (0.10–0.43 in.), according to Bonner (1941). On older trees the bark becomes brown and scaly.

The mature leaves are deep blue-green, shiny on the upper side, with conspicuous whitish rows of stomata on the under surface. The stomatal bands become less bright after the leaves are two to three years old. Leaves on the lower part of the crown often differ in their form and arrangement on the twigs from those on the middle crown and from the upper or cone-bearing branches. The thinly set leaves of the terminal shoot are strikingly different from all the other foliage.

The leaves of the lateral branches of balsam fir are more or less flattened, notched, or obtuse at the tip, grooved on the upper surface and with a prominent midrib or keel below (Anderson, 1897). They are twisted at the base, giving the needles a distinctive two-ranked arrangement peculiar to species of the genus *Abies*. The leaves on upper shoots are not transversely heliotropic nor twisted at the base, but grow in all directions from the shoot. They are shorter and thicker than those of the lower branches, more or less awl-shaped, and are sharply pointed at the tip. The leaves are resinous, pleasantly aromatic, flat, blunt, and 1–3 cm (0.4–1.2 in.) long. Leaves persist 8–13 years (Anderson, 1897).

Buds are 3–6 mm (⅛–¼ in.) long, subglobose, with orange-green or reddish scales covered with a varnish-like resin. According to Anderson (1897), no epidermal hairs are present on the bud scales. All the scales are fringed with thin-walled, hypha-like, marginal hairs through which resin diffuses to the exterior of the scales. Eventually the bud is covered with a layer of thick resin that effectively prevents transpiration. From two to six resin canals are usually present in the bud scales.

Balsam fir is monoecious, in that male and female strobili appear on the same tree, usually on the outermost parts of the branches. The ovuliferous strobili occupy the extreme top of the crown and are borne perpendicularly in the leaf axils on the upper sides of the previous year's branches. Polliniferous strobili are borne mostly on the lower branches. Wright (1953) reports that male strobili are distributed over the entire crown if light conditions permit.

At maturity the polliniferous strobili are oblong-cylindrical, 3 mm (⅛ in.) long, yellowish-red and tinged with purple; and, as stated previously, are each composed of an axis to which are attached spirally-arranged microsporophylls. The ovuliferous strobili are oblong-cylindrical, 25 mm (1 in.) long, purplish, and made up of two kinds of appendages (bracts and ovuliferous scales), spirally arranged upon a central axis. The strobili appear between late April and early June, depending on the locality and the season.

The mature female cones are oblong-cylindrical, erect, puberulous, 3–8.5 cm (1.2–3.3 in.) long, and 2–3 cm (0.8–1.2 in.) in diameter. Dark purple at first, in late summer the ripening cones become paler and turn somewhat grayish. In fall the scales dry, loosen, and fall away from the central axis. The bare, spikelike axis frequently persists for many years (Morris, 1951).

The bracts attached to the dorsal surface of the cone scales are usually about half the length of the scales. Occasionally the points of the bracts protrude slightly beyond the cone scales. In rare instances there may be cones with exserted bracts and cones with hidden bracts on a single tree. The bracts are obovate, serrulate, and abruptly and slenderly awned.

Two winged seeds are attached to the face of the cone scale, each about 3 mm

(⅛ in.) long and shorter than the light purplish-brown or brown wings. The seeds are pale yellowish-brown. The seed coat is soft and has three or more resin vesicles on the surface, giving the seed a pronounced resinous odor. The shape of the seed is roughly quadrangular (Roe, 1948a).

Several morphological features of balsam fir are illustrated in Figure 2.

LEAF ANATOMY

Early studies on leaf anatomy in *Abies* have been reviewed by Fulling (1934). Most of the basic information was provided by Anderson (1897) and additional data have been supplied by Dorner (1899), Zon (1914), and Taubert (1926). The description that follows is a synthesis of the aforementioned studies.

The leaf or needle consists of three major parts, an outer protective covering composed of epidermis and hypodermal strengthening cells, a central chlorophyll-bearing spongy tissue, and an inner conducting tissue.

Stomata are distributed equally on all sides of leaves on the terminal shoots. On the lateral shoots, stomata are found in greater abundance toward the tips and on the lower surfaces of the leaves.

Hypodermal cells are always present in the leaves, but are seldom found in cross sections made above the middle of needles taken from lateral branches. In the basal portions of the needles hypodermal cells are usually isolated on the upper leaf surface. They appear in continuous layers on the lower surface. The short, rigid, and rounded leaves of terminal shoots have a greater development of hypoderm. The number of hypodermal cells decreases from the base toward the tip of the leaf, but with the decrease in number of cells there occurs a corresponding increase in number of stomata.

The mesophyll or green tissue is composed of palisade and spongy parenchyma cells. In the center of the mesophyll and midway between the endoderm and the outer angles of the leaf lie the two circular resin canals. The resin canals are lined with thin-walled epithelial cells, which are themselves surrounded by one layer of thick-walled strengthening cells. These strengthening cells differ from the sub-epidermal (hypoderm) cells in that they are shorter and not fiber-like. Taubert (1926) states that resin canals are located differently in light and shade needles. In shade needles, resin canals are located nearer the epidermis, whereas in well-lighted needles resin canals are located deeper within the mesophyll. The change in position may also be related to fertile and sterile branches.

The central portion of the balsam fir needle consists of two fibrovascular bundles surrounded by an imperfect bundle sheath of transfusion tissue and an endodermis. In the needles of lateral shoots two small areas of transfusion tissue lie dorsal to the outer half of the two phloem areas. In the needles of terminal shoots, the two areas of transfusion tissue are united to form one large area on the dorsal side of the phloem beneath the endodermis (Anderson, 1897).

WOOD ANATOMY

Detailed descriptions of the anatomy of the wood of balsam fir have been presented by Penhallow (1907), Brown and Panshin (1934), Brown, Panshin, and

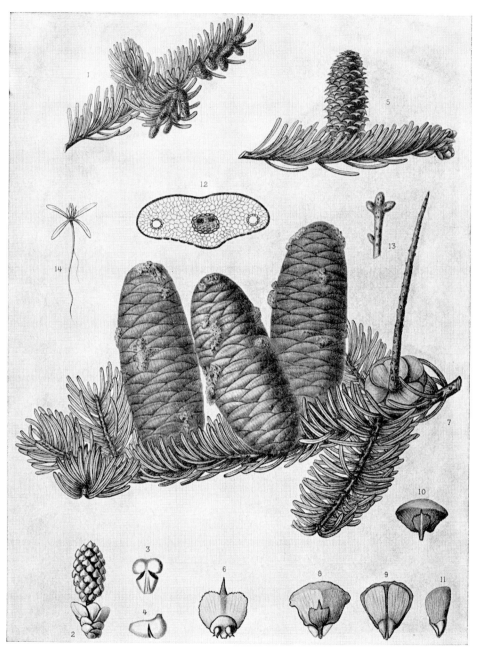

Figure 2. Abies balsamea: (1) branch with male strobili; (2) male strobilus; (3) microsporophyll, seen from below; (4) microsporophyll, side view; (5) branch with female strobilus; (6) ovuliferous scale, upper side with bract and ovules; (7) fruiting branch; (8) cone scale, lower side, with bract; (9) cone scale, upper side, with seeds; (10) cone scale of the long-coned Wisconsin form, upper side with bract; (11) seed; (12) cross section of leaf; (13) winter buds; (14) seedling plant. (From Sargent [1898], *The Silva of North America.* Courtesy of Houghton Mifflin Company.)

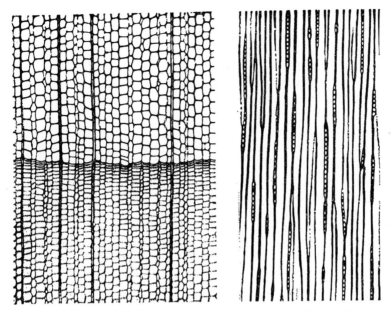

Figure 3. Cross section (left) and tangential section (right) of balsam fir stem. X75. (By permission from *Textbook of Wood Technology* by Brown, Panshin, and Forsaith. Copyright 1949 by the McGraw-Hill Book Company, Inc.)

Forsaith (1949), and others. The following wood description is taken from the *Textbook of Wood Technology* (Brown et al., 1949).

Growth rings distinct, delineated by the contrast between the somewhat denser summerwood and the springwood of the succeeding ring, medium wide to wide or narrow in the outer portion of mature trees. Springwood zone usually occupying two-thirds or more of the ring; transition from spring- to summerwood very gradual; summerwood zone distinct to the naked eye, somewhat darker than the springwood, generally narrow. Parenchyma not visible. Rays very fine, not distinct to the naked eye, forming a fine, close, inconspicuous fleck on the quarter surface. Normal resin canals wanting; longitudinal wound (traumatic) canals sometimes present, sporadic and often in widely separated rings, arranged in a tangential row which frequently extends for some distance along the ring, appearing as dark streaks along the grain. Tracheids up to 50 microns (average 30–40) in diameter; bordered pits in one row or very rarely paired on the radial walls; tangential pitting in the last few rows of summerwood tracheids; pits leading to ray parenchyma taxodioid, small, quite uniform in size, with distinct border, 1–3 (generally 2–3) per ray crossing. Longitudinal parenchyma wanting. Rays uniseriate, very variable in height (1–30 plus cells and up to 400 plus microns), consisting wholly of ray parenchyma or rarely with a row of ray tracheids on the upper and lower margins.

Cross sections and tangential sections are illustrated in Figure 3.

The presence or absence of ray tracheids in *Abies balsamea* has been disputed for a considerable time (Penhallow, 1907; Thompson, 1912; Barghoorn, 1940). Lately Kukachka (1960), in his description of balsam fir wood, showed that ray tracheids were absent.

Myers (1922) emphasized the importance of ray volume in determining the technical properties of wood. He found that the ray volume of balsam fir was 5.7 per cent of total volume. Comparable figures for conifers in general was 7.1 per cent, and for hardwoods, 17.0 per cent. Photomicrographs of tracheids of balsam fir and other species have been published by Carpenter and Leney (1952). Bailey (1954), in his highly productive research in plant anatomy, included the wood of balsam fir in his studies. Protoplasmic streaming in the fusiform cambial initials was observed.

Greguss (1955), in his use of xylotomy for the identification and classification of gymnosperms, presented illustrations, photomicrographs, and descriptions of balsam fir wood. The wood of *Abies balsamea* was described as follows:

Rays 6–8 (or 10) cells high, uniseriate, some of the higher ones with 1–2 paired cells in the middle of the body. Height of ray cells 10–21μ, in marginal and single-row rays up to 27μ; width 6–8μ. Bordered pits in tangential walls of tracheids 5–8μ, in radial walls 10–16μ. In the cross field generally 2 minute pits side by side, 5–6μ in diam. Pit apertures subcircular. Tangential walls of ray cells with sieve-like pitting in tangential view; 40–50 rays and 180–210 ray cells per sq. mm. Calcium oxalate crystals abundant.

Comparative data for closely related species are shown in Table 2.

TABLE 2. SOME CHARACTERISTICS OF WOOD STRUCTURE OF *ABIES BALSAMEA* COMPARED WITH *A. FRASERI*, *A. LASIOCARPA*, AND *A. LASIOCARPA* VAR. *ARIZONICA*

Characteristics	*Abies balsamea*	*Abies fraseri*	*Abies lasiocarpa*	*Abies lasio-carpa* var. *arizonica*
IN TRANSVERSE SURFACE				
Number of tracheids per sq mm..........	2000	6000	2000	8000
IN TANGENTIAL SURFACE				
Number of rays per sq mm...............	45–50	60–70	35–40	110–120
Number of ray cells per sq mm............	180–210	180–200	450–460	400–410
Number of superposed cell rows in a ray.....1–6 (10)	1–6 (10)	1–20 (49)	1–5 (10)
Number of juxtapositional cells in a ray.....	1 (2)	1	1	1
Height of ray cells in μ...................10–21 (27)	10–16	10–26	10–18 (25)
Width of ray cells in μ....................	6–8	5–8	8–10	4–8
Diameter of pits in tracheids in μ..........	5–8	6–9	5–8	4–5
IN RADIAL SURFACE				
Diameter of pits in tracheids in μ..........	10–16	10–13	12–16	10–13

Source: Greguss, 1955.

Previously, Wiesehuegel (1932) indicated that the pits on the radial walls of ray parenchyma of balsam fir are wider than those of Fraser fir. However, Table 2 shows that a single characteristic is inadequate for species identification.

Kukachka (1960) divided the woods of *Abies* species into two broad groups. One is formed by the two eastern American species and the subalpine fir; the other includes the remaining western firs. In the first group the ray cell contents are color-less or slightly pale yellowish and frequently form a reticulum within the cells so that detection of the end walls may be somewhat difficult. Contrary to the findings of Greguss (1955), Kukachka found no crystals in the ray parenchyma. The western firs have crystals, and the contents of the ray cells are distinctly brown.

Anatomical studies of balsam fir wood have been confined mainly to the stem. Bannan (1941) studied the anatomy of the secondary xylem of roots of various Abietineae but gave no specific information on *Abies*. Jeffrey (1917) presented illustrations of cross sections of young and old balsam fir roots; however, no discussion was provided. Fegel (1941) compared anatomy and varying physical properties of stem, branch, and root wood of different coniferous species including balsam fir. Specific figures for balsam fir were not presented, but for a group of conifers the average tangential diameters of tracheids was highest in roots and least in branches. Tracheid length was greatest in stems and least in branches.

Several miscellaneous investigations have been conducted on the wood of balsam fir. Anderson (1897) noted that resin canals, usually lacking in normal wood, were always present in the wood of tumors, mostly in the springwood portion of the annual ring. Hypertrophy of root lenticels caused by excess soil water was reported by Hahn et al. (1920). Changes in annual rings in connection with climatic conditions, defoliations, and seed crops have been investigated by several authors (McLeod, 1924; Morris, 1951), and Balch (1952) has studied "compression wood" caused by the balsam woolly aphid (see Chapter 5).

BARK ANATOMY

The term "bark" has various meanings but is applied most commonly to the tissues lying outside the vascular cambium or outside the xylem. Chang (1954a) was the first to give a systematic account of the bark anatomy of North American conifers.

Bannan (1936) studied albuminous cells in phloem rays. Bark anatomy was mentioned by Marriott and Greaves (1947) in connection with their discussion of Canada balsam. Information on the chemical composition of bark is relatively abundant and is reviewed in Chapter 9. General information on fir bark may be found scattered through the European literature, the work by Holdheide (1951) being representative.

The anatomy of balsam fir bark was given intensive study by Chang (1954a,b). In the earlier publication he presented keys based upon macroscopic and microscopic structure of coniferous bark that could be used to identify the genus *Abies*. He also noted that chemical methods for bark identification are in an early stage of development, and that such techniques may be better suited for bark than for wood identification. In his second publication, Chang (1954b) presented a thorough description of balsam fir bark, summarized in a somewhat abridged form as follows:

The bark has essentially three zones, the periderm, cortical region, and the secondary phloem. The phellem cells of the new periderm are mainly thin-walled suberized cork cells. Phelloderm merges into the cortical region. The cortical region is the broadest portion of branches or young stems. Some of the cortex cells become enlarged and contain "resinous" substances. As the bark grows older, some of the cortical cells become "lignified". In the middle of the cortex there are always one or two layers of resin canals that later become the "blisters". Secondary phloem of a 1-year-old branch shows regularly aligned sieve cells and phloem rays, and scattered parenchyma cells. In the second year, the initial stage of resin passages appears. The sclereids, however, are formed much later. In mature bark, the three zones are in different proportions, and the sclerotic and lignified cells increase. The blisters enlarge.

As compared with related species, balsam fir bark is characterized by (1) the continuous and well-developed phellem composed mostly of thin-walled cells; (2) sieve cells shorter than in most species of softwood studied, and often interrupted by tangential lines of parenchyma at an interval of every 5–9 sieve cells along the radial rows; (3) sclereids in groups and aligned more or less in tangential rows; (4) parenchyma with isodiametric crystals of calcium oxalate and tanniferous substance; (5) absence of fusiform rays, and resin passages originating from the erect phloem ray cells present in the secondary phloem.

Variation in balsam fir bark structure is probably influenced mainly by the age of the trees and their habitats. The smoothness of the young bark is due to the persistent periderm and the cortex. The roughness of old bark is caused by an increase in sclereids and resin passages and the formation of rhytidomes.

Physiology and Genetics

Over the past fifteen years there has been a continual increase in studies of tree physiology and genetics. These areas of research are currently contributing the greatest new knowledge to the understanding of life processes of forest trees and the mechanisms underlying such processes. Physiological studies in balsam fir have been concentrated in relatively few areas, notably on photosynthesis, respiration, and on flowering. Genetic studies are just beginning, and only a few scattered reports on hybrids and tree breeding are available. This section reviews the present knowledge pertaining to balsam fir in these related study areas.

PHOTOSYNTHESIS AND RESPIRATION

Clark (1956, 1961) investigated the different performances of light and shade leaves, the effect of temperature, the effect of needle age, the seasonal variation in dark-respiration rates, and the effect of soil moisture. He measured primarily the changes in CO_2-amount, expressing it in mg used or evolved per cc of foliage volume per hour.

Shade-foliage showed higher rates of apparent photosynthesis and lower rates of respiration than sun-foliage. At 200 ft-c, shade needles assimilated about 150 per cent more CO_2. However, at 4000 ft-c the differential in favor of shade needles was only 25 per cent. At intensities below 800 ft-c, balsam fir assimilated in absolute amounts more CO_2 than white spruce. The saturation light intensities of shade foliage at a constant temperature of 20C were near 3000 ft-c for fir and 4000 ft-c for spruce. The rates of apparent photosynthesis in sun foliage of both species were quite similar at all light intensities from 50 to 4000 ft-c.

The effect of temperature was investigated for 1-year-old foliage at a light intensity of 2400 ft-c. The rate of apparent photosynthesis for shade foliage reached its maximum at 22–23C. The rate of total photosynthesis culminated at 24C. Respiration rates continued to increase with increasing temperature. The performance of sun needles was in general very similar, although the rates were lower. The total photosynthesis rate of sun needles did not reach a distinct maximum but leveled off at 24C. At temperatures above 20C, net CO_2 assimilation of white spruce foliage was relatively more efficient than that of balsam fir, but at temperatures below 20C the opposite was true.

Balsam fir retains its needles for 8 years and even longer. However, the contribution to the amount of foliage of individual trees declines rapidly for increasing needle-age classes. In open-grown trees 25 feet tall and 30 years of age, the first-year foliage contributed 27 per cent, the second-year 20 per cent, with declining amounts for increasing age classes until in the eighth year only 2 per cent was contributed to the total foliage volume. In an experiment, twigs bearing needles of uniform age were illuminated at 960 ft-c and the apparent photosynthetic rates measured periodically during the whole season. The results of these experiments are shown in Figure 4.

The apparent photosynthetic rates were fairly uniform from May to October for all needle age classes except the current year's foliage. As the young shoots developed, the photosynthetic rates increased. In the early part of the growing season the new needles used more food for respiration than they were able to produce by photosynthesis. This may explain the lag in the effect of defoliation when new needles are first consumed by insects.

The dark-respiration rates of the current year's foliage were measured weekly from early May to late July. The rates were high at first but decreased steadily with maturation of the needles. Respiration was most active when 25 per cent of the current growth was attained. At this time the rate was twice as great as the rate when maturation was reached.

The maximum photosynthetic rate was reached at a soil moisture slightly below field capacity. When subjected to artificial drought, white spruce showed a greater initial drought resistance or endurance than did balsam fir. Balsam fir was also subjected to severe drought. The measurements showed that before permanent wilting took place there was a period of increased respiration. This occurred after photosynthesis had ceased.

The studies of Clark (1956, 1961) indicated that balsam fir is more adapted to photosynthetic production than white spruce under low temperatures and low light conditions. However, balsam fir requires more moisture. This is in agreement with other information presented in Chapter 3.

VEGETATIVE GROWTH

Information on vegetative growth is scattered in the literature, and the material available is of sketchy nature. A portion of this topic was treated in the previous section, and a discussion on phenology appears later. Those aspects of leaf, shoot, and root growth which are primarily of a physiological nature are treated here.

Needles continue to grow throughout the first and into the second season. Morris (1951), working in the Green River area of New Brunswick, noted an increase in

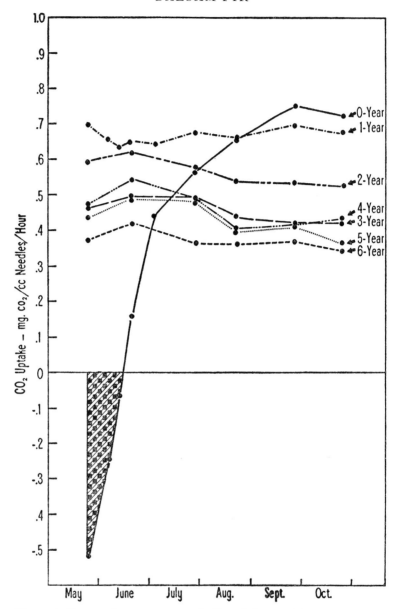

Figure 4. Photosynthetic capacities of *Abies balsamea* foliage of different
ages during the vegetation season. (From Clark, 1956.
Courtesy of the Department of Forestry, Canada.)

needle length of about 8 per cent during the second growing season. The weight
of needles increased 20 per cent during the second season.

Stone (1953) reported that meristems in the axils of needles of *Abies balsamea*
commonly persist as minute dormant buds that often escape casual notice. The
duration of dormancy varies. Some buds elongate and form lateral branches within
the first ten years; others remain quiescent for several years after needle fall, but

are capable of elongation upon adequate stimulation. Stone (1953) further stated that buds of adventitious origin had not been observed on balsam fir.

Clark and Bonga (1961, 1963) have recently begun a study of growth substances in balsam fir, investigating buds, needles, and stems of one-year-old twigs at periodic intervals. In the earlier study they reported the presence of an indole compound, apparently acting as an auxin precursor, in the needles. In the later study, an ether-extractable auxin was found in the inner bark. Characterization of the auxin by paper chromatography, *Avena* bioassay, and chromogenic tests indicated that it was indole-3-acetic acid. In addition to the auxin, a strong growth inhibitor was extracted.

Facey (1956) investigated leaf abscission in balsam fir. When the leaves are shed, the needles break off at the point of attachment. A layer of tissue composed of ligno-suberized cells protects the living parenchyma cells underneath. Balsam fir apparently lacks an anatomically differentiated abscission layer. No chemical changes in the cell walls of the abscission zone were found. Facey (1956) concluded that abscission is a mechanical process in balsam fir.

Cook (1945) reported on an aberrant 20-year-old balsam fir growing on shallow limestone soil on Valcour Island, Lake Champlain, New York. The tree apparently had never produced a lateral bud. Though not erect, the branchless stem had reached a length of 85 inches. The internodes were marked by distinct ridges in the bark and by a frill of needles when present. Needles were of normal length but were strongly keeled and had pointed tips.

Balsam fir seedlings develop strong, slightly branched lateral roots in the surface humus. Frequently the seedlings develop a heavy central root which initially appears to be a taproot, but then splits at the bottom of the humus into a number of laterals which remain in the organic layers. One rather small and comparatively insignificant root may continue downward into mineral soil (Moore, 1922).

Mature balsam fir in producing anchorage roots depends on the humus layer for nutrients as do other species. The absorbing roots are much less branched but also thicker than those of spruce. The total root length of balsam fir is less than that of spruce, and the latter species has greater absorbing root-tip surface as well as a better root distribution (Moore, 1922).

Humus conditions are believed to have a relatively greater influence on the development of roots of balsam fir seedlings than those of red spruce and white pine. Moore (1917) found that the average seasonal root elongation of young seedlings growing in mild humus was 106 mm for balsam fir, 76 mm for red spruce, and 90 mm for white pine. In mineral soil the average seasonal root elongation was 39 mm, 30 mm, and 79 mm, respectively. Raw humus soils showed intermediate effects on the same species. This research also demonstrated that humus soils reduced damping-off losses. In general, balsam fir roots penetrated more deeply than red spruce.

Anderson (1932) concluded that taproot development of balsam fir seedlings is determined by soil texture rather than by soil moisture. This conclusion also applies to the number and length of lateral roots and rootlets.

White (1951), in his studies of balsam fir heart rots, noted that infections may occur through root grafting. Although root grafting in balsam fir has been observed rather frequently, the first authentic report is of very recent origin. Kozlowski and Cooley (1960, 1961) excavated balsam fir roots to a depth of two feet. Stems 4–6 inches dbh and 1–2 feet apart were found to be bound together with numerous grafts among the larger roots. Since root grafting also occurred on upland soils, they concluded that inter-root compatibility and growth pressure apparently suffice to produce grafting. Abrasion of the bark of roots due to swaying of trees in the wind on lowland soils is not a critical prerequisite for root fusion as claimed by earlier authors.

Layering of balsam fir was first described by Chittenden (1905), who observed rooting of basal branches. In the White Mountains of New Hampshire prostrate balsam fir occurs above 5500 feet (1700 m), and reproduces almost entirely by such vegetative means. Cooper (1911) also observed balsam fir layering on Isle Royale in Lake Superior and described it as crooked, stunted, and short-lived.

Layering apparently occurs at any age. Sometimes the ramets are only a few years younger than the parent trees and often produce a second generation. In many cases the layers become independent by the decay of the connecting portions.

Layering is more common in northern and in mountainous regions. Moisture and the absence of light stimulate the branches to root, but layers are also found in xerophytic situations. Potzger (1937) stated that layering is apparently stimulated by rigorous habitat conditions such as low temperatures, sand dunes, and rocky and shallow soils. Roy (1940) observed balsam fir layering on open swamps and deep mossy areas in Quebec. Balsam fir layers are found under dense white pine and jack pine overstories with pine needles supplying the necessary covering.

The roots of layered balsam fir arise from dormant buds and are distributed irregularly among the layered branches. In black spruce, however, they are restricted to the neighborhood of the terminal bud scars (Bannan, 1942).

FLOWERING AND SEED PRODUCTION

A comprehensive study of flowering in balsam fir was completed by Morris (1951), who studied the species in the Green River area of New Brunswick.

The male strobili are borne in the axils of the leaves on the undersides of the shoots. As many as 22–70 strobili may appear on a single shoot. Male strobili usually do not occur on primary branch terminals. After the pollen is shed, the male strobili drop off, but at their point of attachment there remain woody cup-like bracts, which persist on the twigs 8–14 years, and occasionally as long as 25 years.

The female strobili are borne erect in groups of 1–3 on the upper surfaces of single shoots in the upper portion of the crown. The majority of female strobili develop into cones. Shortly after reaching maturity the cones disintegrate; however, their bare spikelike axes may persist for many years. In the Green River study, the oldest cone axes found were 44 years of age.

Flowering in balsam fir is influenced by exposure to light (Morris, 1951). In the Green River study area, female strobili were observed on 83 per cent of dominants, 59 per cent of codominants, 6 per cent of intermediates, and none at all on suppressed trees.

According to Zon (1914) balsam fir begins to bear seed at about age 20 or when trees have reached a height of 15 feet. Regular seed production occurs somewhat later, at about 25 years in mountainous areas and 30 to 35 years elsewhere. Morris (1951) reported that flowering in balsam fir begins after 20 to 30 years and increases with age. Roe (1948b) reported that a balsam fir tree on the Kawishiwi Experimental Forest in northeastern Minnesota bore cones at age 15 when the tree was only 6.5 feet tall. This early seed production record is probably not typical.

Periodicity of seed production was investigated by Zon (1914) and Morris (1951). Zon stated that plentiful yields occur at intervals of two to four years. Morris noted a regular two-year interval during the period 1920 to 1950 in his Green River study, which was based on the persistence of male and female strobili remnants.

PHENOLOGY

To obtain comparative phenological data it would be necessary to organize a long-range program over extensive geographical areas. The data available at present for balsam fir are mostly local in origin and somewhat contradictory when viewed on a large scale. Phenological data are mainly concerned with seasonal measurements of leader elongation and radial expansion, time of flowering, and cessation of growth.

The shape of the seasonal growth curve is apparently largely determined by inherent factors rather than by the environment. This was indicated by an experiment in which nursery-grown trees from Syracuse, New York were transplanted to Durham, North Carolina in 1937 (Kramer, 1943). The plants completed 37 per cent of their total seasonal height growth in April, 47 per cent in May, 3 per cent in June, and 12 per cent in July. Although balsam fir made less growth and showed higher mortality in North Carolina than in New England, the general pattern of growth in the two areas was very similar. Baldwin (1931) in New England also observed an increase of growth in July. Kozlowski (1961) pointed out that balsam fir shows much variability in the seasonal pattern of height growth. Late season growth involving lammas shoots is not known in this species.

Phenological observations on leader elongation and seasonal height growth have been reported by at least six investigators (Kozlowski and Ward, 1957; Cook, 1941; Kienholz, 1934; Baldwin, 1931; Moore, 1917; and Morris et al., 1956). Their data have been summarized in Table 3. Direct comparisons are not possible because of variation in material, locality, and season; however, the general pattern of growth is quite clear. Leader growth usually begins in early to mid-May, culminates in June and terminates in late July or early August. Most of the height growth is completed in about 65 days.

Belyea et al. (1951) measured the radial growth of balsam fir at Cedar Lake in northwestern Ontario. Radial growth began on May 26, was 50 per cent completed by July 10, and ceased on August 30. In the Green River, New Brunswick, area the radial growth of balsam fir lasted for 86 days.

Lamb (1915) attempted to summarize the occurrence of the main phenological processes of balsam fir in the northern and southern parts of its range. In the south leafing and flowering began about May 1; in the north, about June 1. Seed ripening extended from September 1 in the south to October 15 in the north.

TABLE 3. PHENOLOGICAL OBSERVATIONS ON THE INITIATION AND
CESSATION OF HEIGHT GROWTH IN BALSAM FIR

Observer	Material	Locality	Dates of Initiation	Dates of Cessation	Length of Growth Period	Remarks
Kozlowski and Ward (1957)	3-yr-old seedlings	Amherst, Mass.	4-15	8-18	18 weeks (126 days)	90% completed in 82 days; 4-30 to 7-21
Cook (1941)	7-ft saplings	Stephentown, N.Y. (1400 ft elev.)	5-28	8-1	65 days	
Kienholz (1934)	15-ft saplings	Keene, N.H.	5-20	7-23	65 days	Maximum elongation of leader on 6-23
Baldwin (1931)	0.2–5m saplings	Berlin, N.H. and Cupsutic Lake, Me.	5-17 to 6-7	8-9 to 8-29	11–14 weeks (77–98 days)	Variation over a 3-yr period
Moore (1917)	Young seedlings transplanted from forest	Mt. Desert Id., Me.	6-1	7-10	40 days	Shortest elongation period reported
Morris et al. (1956)	(Not specified)	Fredericton, N.B.	5-1 to 5-20	7-1 to 7-15	60–75 days	

Sources: as listed in first column.

Seed fall began in the south about September 15 and in the north about October 31. Leaf fall in the two areas ranged from September 15 in the south to October 15 in the north.

Data from the *Woody Plant Seed Manual* (U.S. Forest Service, 1948), which summarizes average conditions, show that flowering of balsam fir takes place in May; cones ripen from late August to early September; and seed dispersal takes place in October.

At Green River, New Brunswick, flower buds open before vegetative buds. Pollen is normally shed in early June. Shoot growth extends from the last week of May to the last week of July. Flower buds for the next year become apparent before the shoot growth is completed. This suggests that June and July are the critical months for flower conditions (Morris, 1951).

Bonner (1941) found balsam fir flowering during the last week of May at Kapuskasing in the Clay Belt of northern Ontario. The foliage buds opened about a week later. The cones ripened in August and began to disintegrate in early September. Much of the seed was released within a few weeks. Some seed was still falling in early winter.

Seed fall of balsam fir started in the Superior National Forest in Minnesota in late August, 1944. About 1 per cent was released before September 1, 10 per cent by September 16, 50 per cent by October 19, and 62 per cent by November 17. Some seed (0.1 per cent) had not fallen by May 29, 1945 (Roe, 1946).

The phenological observations of Ahlgren (1957) covered all of the main phases of growth and extended over a five-year period (1951–55) at the Quetico-Superior Wilderness Research Center in northeastern Minnesota. Balsam fir bud-swell began between April 4 and May 20; initial leaf activity was observed between May 17 and June 3; strobili appeared between April 30 and May 8; initial shedding of pollen began between May 25 and June 4; earliest seed fall was observed October 3. Cambial activity at breast height started between April 6 and May 20, and ceased between July 28 and September 10. Bud-swell and flowering of balsam fir occurred simultaneously after maximum temperature had risen above 70F and the minimum temperatures remained above freezing for one week or more. Pollen release and leafing were almost simultaneous. The vegetative and reproductive responses were closely associated with the abrupt increase in stem expansion that began in spring. Abrupt changes in cambial activity were frequent in periods of sudden temperature change. Rapid stem expansion of balsam fir occurred after periods of heavy precipitation, but only a part of this could be attributed to growth.

GROWTH CORRELATIONS

Balch (1946) noted that flowering shoots bear only half the foliage of vegetative shoots. Morris (1948, 1951) found that flower production was associated with reduction of the number and length of shoots and the number and size of needles. The effect was greatest in the top 15–20 feet of crown where the heaviest flowering occurs.

In his later study Morris (1951) found that the primary branch terminals, leaders, and annual rings of mature balsam fir generally showed reduced growth in years of heavy flowering, although annual rings were also smaller in drought years. In 1947 flowering trees produced only 27 per cent of the weight of foliage produced in 1946, as compared with 84 per cent for nonflowering trees. Reduction in needle length in flowering years is an indicator of reduced weight of the total foliage production.

Zon (1914) reported a reduction of balsam fir bark blister size in years of heavy seed production.

Bannan (1940, 1942) studied the developments of balsam fir roots along the northeast shore of Lake Superior. He concluded that there is no correlation between root depth and rate of stem growth.

GENETIC VARIABILITY AND TREE BREEDING

Since *Abies balsamea* is widely distributed, and grows under a variety of climatic, edaphic, and biotic conditions, a number of genotypic variations can be expected. Some general views on balsam fir were expressed by Heimburger and Holst (1955), who considered the *Abies balsamea* found in Virginia to be morphologically the same species that is found in eastern Canada, although possibly of a different ecotype.

Balch (1935) observed a balsam fir which was apparently resistant to the black-headed budworm (*Peronea variana* Fern.). The possibilities of a genetic approach to spruce budworm (*Choristoneura fumiferana* (Clem.)) control were discussed by Heimburger (1945). He suggested that resistant material suitable for Canada could probably be found in eastern Asia, where *Abies sibirica* Ledeb. (which is resistant to the spruce budworm) has contact with two other species of firs, *A. nephrolepis* Maxim. and *A. holophylla* Maxim.

In 1953 a search was begun in the Atlantic Provinces for specimens of balsam fir exhibiting resistance to the spruce budworm (Canada Forestry Branch, 1955). Four trees were selected for possible resistance.

Rudolf (1956) suggested that attention should be given to developing balsam fir trees desirable for Christmas trees and resistant to spruce budworm, rots, and frost damage.

In 1956 researchers at the Acadia Forest Experiment Station in New Brunswick initiated a balsam fir provenance study and a tree selection program for the purpose of improving wood production and resistance to the spruce budworm and balsam woolly aphid. Breeding programs with *Abies* species have also been started in the United States by the Pacific Northwest and the Northeastern Forest Experiment stations, the universities of Wisconsin and Kansas, Yale University, and probably other institutions.

HYBRIDIZATION

The first reported hybrids between *Abies balsamea* and *A. sibirica* were obtained by R. Schröder from open pollinated seed of both species growing together in the arboretum of the Agricultural Institute of Moscow. The hybrids were originally named *A. sibirica parvula* (Beissner, 1897, 1901). Schröder considered that there were six different hybrid forms between *A. sibirica* and *A. balsamea*. Four of these were upright growth forms, i.e., *A. pendula*, *A. hibrida* Sr. (= *A. sibirica* x *A. balsamea columnaris*), *A. conica*, and *A. pyramidalis*; and two were dwarf forms, i.e., *A. nana* and *A. parvula*. However, all of these may have been only forms of *A. sibirica* (Meyer, 1914), although the latter species is closely related to *A. balsamea* (Mattfeld, 1925).

Johnson (1939) reported that there were no known natural or artificial hybrids of balsam fir in North America. A preliminary report of the Canada Forest Research Branch (1961) noted an attempt to hybridize *A. balsamea* with *A. alba*. A few seed were obtained but further confirmation of the hybrid is necessary.

In 1936 Fulling described *A. intermedia* as an intermediate between *A. fraseri* and *A. balsamea*. As indicated earlier in this chapter, this taxon is of doubtful status and has also been reported as *A. balsamea* var. *phanerolepis*.

Klaehn and Winieski (1962) reported that two artificial crosses, *A. balsamea* x *A. lasiocarpa*, and *A. fraseri* x *A. balsamea*, were made in 1956 at the Petawawa Forest Experiment Station, Ontario. Attempts were made to cross *A. balsamea* with several species of *Abies* by workers at Antigonish, Nova Scotia. Three putative hybrids were grown, although six other apparently successful crosses were subsequently destroyed.

VEGETATIVE PROPAGATION

Bailey (1950) stated that *Abies balsamea* is often used as a stock for grafting *A. balsamea* 'Columnaris,' *A. balsamea* f. *hudsonica*, and *A. balsamea* 'Lutescens.' Scion material from seven species of *Abies* was successfully grafted to balsam fir root stock in greenhouse and field experiments at the State University College of Forestry at Syracuse University, New York (Pitcher, 1960). In a test of hetero-plastic grafts, *Abies* gave better results than *Picea*, *Acer*, or *Fraxinus* grown under similar conditions. *Abies* yielded a higher percentage of grafting "take," formed better unions, and became established earlier than any of the other genera. *Abies* was also the first to break dormancy in both the greenhouse and the field, and required the least care. The root stocks were cut back to the point of union on most of the greenhouse stock, three months after grafting. The average successful "take" was 85 per cent for greenhouse grafts and 80 per cent for field grafts. When studied, the grafts were still in the first growing season.

MacGillivray (1957) successfully rooted balsam fir cuttings under intermittent mist without the use of growth substances at the Acadia Forest Experiment Station in New Brunswick. Cuttings were selected from immature, open-grown trees. Seven-inch-long shoots of the previous year's growth were planted in pots con-taining coarse, moist vermiculite. Needles were stripped from the lower 3–4 inches. Of eight cuttings, four developed roots between February and mid-May. During the first part of June, cuttings were placed in a cold frame outdoors. Normal shoot growth started the following season.

Chouinard and Parrot (1958) attempted, unsuccessfully, to root balsam fir twigs enclosed in plastic covers (air layers). All 54 branches on which 3–4 year wood was girdled developed wound calluses. However, the calluses barely ex-tended beyond the surface of the bark and never downward more than half the width of the exposed wood of the girdled areas. Treated areas, including the wound calluses, were completely covered by thick crusts of resin. The aerial layers were applied in June. By early fall none of the branches had produced adventitious roots, and three of the treated branches had died. No growth substances were applied.

Conclusions

1. Study of the historical records of a species is interesting but difficult because of the inaccessibility of much of the data. A comprehensive review of early dendrological information on North American species would be desirable.

2. Although early authors prepared outstanding detailed descriptions of balsam fir and other tree species, there is urgent need for a modern taxonomic study covering the genetic and phenotypic variations of balsam fir throughout its wide range of distribution.

3. More complete morphological descriptions of balsam fir and other tree species would be highly desirable. Root morphology is most frequently neglected. Greater knowledge of root growth habits, including the role of natural root grafting, may suggest changes in present silvicultural techniques.

4. While considerable work has been done on wood anatomy, the development of new equipment opens new avenues of research. Research on the anatomy of bark, leaves (e.g., studies on juvenile and adult leaf anatomy, etc.), twigs, and roots may suggest uses of these materials, which now have little value. Research in wood anatomy may also provide a better understanding of physiological processes, with resulting possibilities of increasing wood production.

5. Systematic studies on assimilation, transpiration, and respiration in balsam fir under different carefully described environmental conditions, should produce information needed for the development of sound silvicultural management techniques for the species. Specifically, studies of this type might help determine what microclimatic conditions should be established to compensate for departures from the favorable macroclimatic conditions.

6. Systematically collected phenological data, classified according to the physical and biotic conditions of the environment over the wide range of the species might help to evaluate local balsam fir potentials, to adjust silvicultural techniques, and to aid in combating insect enemies.

7. In the future, continued work with growth substances may lead to techniques for control of branching and for initiating the development of new terminal buds after logging and other damage. Moreover, the prevention of flower development, for better control of seed production and consequent increase of wood production, could result from increased knowledge of growth substances and their manipulation artificially.

8. Comparative studies on the ecotypes to be discovered in the new taxonomic study suggested above could provide information on the most productive and suitable seed sources for the cultivation of balsam fir in pure stands, in mixed spruce-fir stands, or as a second story under a pine or hardwood overstory. A highly aggressive ecotype superior in combating shrubs and undesirable hardwoods might be found. Ecotypes more resistant to diseases and insects might also be located.

2

GEOGRAPHY AND SYNECOLOGY

THE range of balsam fir extends over wide areas that have been intensively investigated from the points of view of plant geography and synecology. Study has been largely centered in the Great Lakes–St. Lawrence Region, where balsam fir is for the most part a minor species. Balsam fir comprises a far greater proportion of the forest composition in the Boreal and Appalachian regions, but in these areas geographical and synecological investigations have been much less intensive.

There is a great lack of autecological information on balsam fir. Such information has been largely extracted from geographical and synecological studies and will be presented in Chapter 3. Certain geographic and synecological relationships with microorganisms and insects, reproduction characteristics, species competition, growth patterns, and productivity are discussed with forest management and economic problems in subsequent chapters. This chapter is concerned with the historical distribution of balsam fir, regional descriptions, community interrelationships, and forest succession.

The generalizations arrived at here depend primarily upon the classification and systematization of local descriptive information. Balsam fir has not yet been examined in the more advanced biological models that attempt to interpret ecosystems. It has, however, been considered in improved models describing genetic or developmental relationships between communities and in primary succession patterns that eventually will form the basis for ecosystem models. These new approaches will be outlined in this chapter and further expanded in Chapter 3.

Distribution

The ancestral history of the genus *Abies* as based on fossil remains may be separated into the preglacial on the one hand and glacial and postglacial on the other. Its modern history is bisected into two periods, the prehistoric and the historic, the latter beginning in the fifteenth century in North America with the arrival of European man. In its broad scope the study of present-day distribution includes not only the natural range of balsam fir but also its distribution as an exotic species.

The pre-Tertiary history of the genus *Abies* is not well documented. The needle impressions attributed to *Abietites linkii* Röm. first discovered in the Wealden coal beds of the Lower Cretaceous in Germany resemble both the present *Abies* and certain members of the Podocarpaceae (Mägdefrau, 1953). Several cones ascribed to *Abietites* have been described by Fontaine (1889) from the Potomac group (Lower Cretaceous) of Virginia. In 1898 Newberry described *Abietites cretacea* Newb. from branchlet and needle fossils found in New Mexico. Other reports refer to such form or organ genera as *Abietipites* and *Abietoxylon*, which may or may not be related to what is today recognized as *Abies*. A catalog of the Cretaceous and Tertiary plants of North America including a number of *Abietites* and similar species was prepared by Knowlton (1898).

It has been recognized that the present species, *Abies alba* or similar forms, existed in the Pliocene in Germany as indicated by the fossils found near Frankfort-on-Main (Mägdefrau, 1953) and in some localities in France. The species *Abies ramesi* of the Miocene can be considered as a possible ancestor of *A. alba* in France (Emberger, 1944). Engler (1879), in his historical review of the development of the North American flora since the Miocene, noted that remains of Tertiary *Abies* have been found in Montana, Wyoming, and Banksland (74°27′N lat.). Fossil remains attributable to *Abies* occur at least as early as the Eocene. Wodehouse (1933) has identified *Abies*-like pollen from the Green River formation (Eocene) of Colorado. The species, *Abies concolipites* Wodehouse, had pollen grains very similar to *Abies concolor*. He also described a new genus and species, *Abietipites antiquus* Wodehouse, a form intermediate between *Pinus* and *Tsuga*. LaMotte (1952) prepared a catalog of the Cenozoic plants of North America described through 1950, which contains a number of *Abies* species attributed to the Miocene and Pliocene by different authors.

Opinions have been expressed that the differentiation of the North American flora into eastern and western parts had its beginning in the Cretaceous period when the Cretaceous sea occupied the central part of the continent (Harshberger, 1911). The differentiation could then have continued during the Tertiary. During the Pleistocene, glaciation was more extensive in eastern North America thus increasing the floristic differences between east and west. Knowledge of the history of floristic changes is necessary to an understanding of the distribution of the present vegetation. A recent paper by Braun (1955) does much to clarify these points.

Fernald (1925) considered that the Laurentian ice sheet was so thin that it did not cover a number of highlands in the vicinity of the Gulf of St. Lawrence, the Gaspé Peninsula, and parts of northern Labrador. Hultén (1937) accepted this interpretation, considering that during glaciation these areas, in addition to the Driftless area which lies southwest of the Great Lakes, served as refugia for many plants. Important refugia were also said to be located in the vicinity of the Bering Sea ("Beringia"), the Yukon Valley, and the Pacific Coast. Antevs (1932) presented evidence that Mt. Washington in New Hampshire and Mt. Katahdin in Maine had been glaciated. Flint (1957) summarized the evidence for glaciation in the eastern refugia and concluded that they had been glaciated. Refugia that did exist in eastern North America must have been located on the Atlantic coastal shelf, in the far north, and south of the glacial border.

Halliday and Brown (1943) suggested that four main biotas in North America were available as sources for the establishment of plants on newly deglaciated surfaces following the Pleistocene ice recession:

1. The eastern biota on the ancient Appalachian land mass, mainly a mixed forest of *Acer* sp., *Betula* sp., *Quercus* sp., *Populus grandidentata*, *Pinus strobus*, *Pinus resinosa*, *Pinus banksiana*, *Tsuga canadensis*, *Abies balsamea*, and *Picea rubens*.

2. The western biota on the Cordilleran land mass, beyond the Cretaceous depression of the center of the continent, predominantly coniferous and including such species as *Pseudotsuga menziesii*, *Pinus monticola*, *Pinus ponderosa*, *Abies grandis*, *Picea engelmannii*, and *Larix occidentalis*.

3. The boreal biota contained *Picea glauca*, *Picea mariana*, and *Larix laricina*, mainly around the Bering Sea. *Betula papyrifera*, *Pinus contorta*, and *Abies lasiocarpa* may have found a refuge in the Yukon Valley.

4. The biota to the south of the Bering Sea contained *Picea sitchensis*, *Tsuga heterophylla*, *Thuja plicata*, and *Abies amabilis*.

Each of the four biotas had its own species of *Abies*. The location and composition of these biotas explain some otherwise obscure relationships among species. The work by Potzger (1953b) in New Jersey supports the hypothesis of a balsam fir refugium in eastern biota.

The oldest fossils from within the present natural range of balsam fir belong to several interglacial periods of the Pleistocene. Winchell (1884) discovered an interglacial peat deposit in Mower County, Minnesota, in 1882, and indicated the presence of similar deposits in Fillmore County and neighboring areas in Iowa. Nielsen (1935) considered the Mower County deposit to be pre-Kansan in age. He identified pollen of *Abies balsamea*, *Picea* sp., *Pinus* sp., *Acer rubrum*, *Juniperus* sp. or *Larix* sp., *Juglans* sp. or *Carya* sp., and pollen of several angiosperm species. Wilson and Kosanke (1940) identified *Abies* pollen grains in pre-Kansan peat deposits in Tama County, Iowa. Rosendahl (1948) noted *Abies balsamea* macrofossils from a soil dated as Aftonian interglacial at Faribault, and from a pre-Nebraskan deposit near Springfield, both localities in south-central Minnesota. One of the oldest findings of the remains of balsam fir wood was reported by Penhallow (1907) at Scarborough, near Toronto, Ontario. Flint (1957) described this site, known as the Toronto formation, as one of the best known units of the Sangamon interglacial. The Scarborough beds consist of stratified clay, silt, sand, and peat layers and contain the remnants of 14 tree species indicative of a climate cooler than that of the present.

Abies balsamea pollen has been reported from deposits of Yarmouth, Sangamon, and the early Wisconsin age in Illinois (Fuller, 1939). Cain (1944) referred to fossil soils of unknown age in the Piedmont of South Carolina in which pollen grains of *Abies*, unlike that of *A. balsamea* and *A. fraseri*, were found. Whitehead and Barghoorn (1962) investigated deposits similar to those described by Cain in the Piedmont of western North Carolina and South Carolina. The size measurements indicated that the fir pollen may still have been those of *Abies fraseri*.

Most reports of *Abies* pollen in peat bogs and lake sediments come from late-glacial and postglacial deposits. Species identification has rarely been made; but, by inference from the geographic locality where studies have been made, the species is usually considered to be *Abies balsamea*. A geographic summary of the most important published papers on pollen analysis in which *Abies* pollen is mentioned is presented in the accompanying tabulation.

Alberta	Hansen (1949a,b, 1952)
British Columbia	Heusser (1960)
Labrador	Wenner (1948)
New Brunswick	Auer (1927)
Nova Scotia	Auer (1927)
Ontario	Auer (1927), Potzger (1949, 1950, 1953a)
Quebec	Auer (1927), Potzger (1953b), Potzger and Courtemanche (1954)
Alaska	Heusser (1960)
California	Heusser (1960)
Florida	Davis (1946)
Illinois	Voss (1933, 1934), Fuller (1939), Sears (1942)
Indiana	Sears (1942), Keller (1943), Potzger (1944)
Maine	Deevey (1951)
Massachusetts	Davis (1958, 1960)
Michigan	Potzger (1942), Sears (1942)
Minnesota	Cooper (1932), Voss (1934), Artist (1939), Rosendahl (1948), Potzger (1949, 1950, 1953a), Jelgersma (1962), Fries (1962), Winter (1962), Wright et al. (1963), Cushing (1963)
North Carolina	Buell (1945), Whitehead and Barghoorn (1962)
Ohio	Sears (1941, 1942), Potter (1947), Sudia (1952)
Oregon	Heusser (1960)
South Carolina	Cain (1944), Whitehead and Barghoorn (1962)
Texas	Potzger and Tharp (1947), Graham and Heimsch (1960)
Washington	Heusser (1960)
Wisconsin	Voss (1934), Wilson (1938), Hansen (1939), Potzger (1942), Wilson and Webster (1942)

Potzger (1944) noted that errors have been introduced in a number of reports on pollen profiles because *Picea* and *Abies* were separated on basis of size as presented in a key by Sears (1930), which included only *Picea mariana* but not *Picea glauca* or other *Picea* species. Early pollen studies (e.g., Sears, 1938) indicated that *Abies* shared dominance with *Picea* in the basal sediments of three major forest regions, the Northern Lake, the Central Deciduous, and the Northeastern Oceanic forests. But recent studies in northern Minnesota (Fries, 1962) and central Minnesota (Cushing, 1963), both in the Northern Lake forest, have shown *Abies* pollen to be very rare or absent from the basal sediments where *Picea* was of special importance. In neither of these two studies has *Abies* pollen exceeded 5 per cent of the tree pollen, until very recently, and even then only at the more northeasterly station in the uppermost few centimeters of sediment. Davis (1958) investigated three pollen deposits from the time of retreat of pre-Valders ice to the time just previous to European colonization in central Massachusetts. She found that even in periods when spruce pollen amounted to 80 per cent of total tree pollen, fir pollen did not exceed 5 per cent. Graham and Heimsch (1960) reexamined the pollen samples of Potzger and Tharp (1947) from Texas bogs, found much less *Picea* pollen, and questioned the presence of *Abies* pollen.

Recent studies of pollen rain have improved the interpretation of pollen deposits. Davis and Goodlett (1960) compared the present vegetation with pollen-spectra in surface samples from Brownington Pond, Vermont, and concluded that oak, pine, birch, and alder pollen are highly overrepresented with respect to the tree basal area percentages. Hemlock, beech, ash, elm, hop-hornbeam, and spruce were fairly well proportionally represented while balsam fir, maples, white-cedar, poplars, basswood, and tamarack were underrepresented. King and Kapp (1963) investigated 15 sites from Toronto north to Lake Timagami, Ontario. Balsam fir pollen was found present on 11 sites ranging from 0.5 to 4.0 per cent of the total pollen deposit, increasing northward. This illustrates the difficulties in reconstructing the history of vegetation from pollen analyses only and points out the need for further studies.

Griggs (1946) considered the Arctic timberline as advancing in Alaska, retreating in eastern Canada, and remaining stationary in the interior. Marr (1948) pointed out that atmospheric conditions cannot always be credited for lack of forest vegetation. He noted that at Richmond Gulf on the east coast of Hudson Bay, forest vegetation is absent because of a scarcity of suitable soils, a consideration that must be taken into account in interpreting vegetation patterns in borderline areas and tension zones. In the humid conditions present in southeastern Canada and New England, balsam fir and spruce reproduction is very abundant and vigorous in taking over sites occupied previously by other species, especially the pines (Potzger, 1953b). In the tension zone between northern hardwood and conifer communities at Itasca Park, Minnesota, the fluctuating status of spruce and fir was confirmed by the studies of Buell (1956), Buell and Martin (1961), and others. In the Great Lakes–St. Lawrence Region balsam fir, though absent from the mesic sugar maple–basswood forest type, is present in its counterpart, the sugar maple–beech–yellow birch type, in the more humid Acadian-Appalachian Region.

Red spruce gradually replaces white spruce in the Appalachian extension of the eastern coniferous forest; although, according to some authors, the former species extends much further to the northwest than has been previously reported. Fraser fir substitutes for balsam fir in southern Virginia, North Carolina, and Tennessee. The new community of *Abies fraseri–Picea rubens–Betula alleghaniensis* is expanding at the expense of grass and rhododendron balds on Roan Mountain in North Carolina (Brown, 1941). Oosting and Billings (1951) compared in detail the northern and southern spruce-fir types. The lack of aggressiveness of balsam fir in West Virginia, near the southern extremity of its range, is demonstrated by the fact that its reproduction does not suppress red spruce. Just the opposite occurs in the north.

CONTEMPORARY NATURAL DISTRIBUTION

The genus *Abies* is distributed rather widely in the northern hemisphere. Most of the species are found in four major areas: (1) eastern Asia, 18 species; (2)

Mediterranean Basin, 8 species; (3) eastern North America, 2 species; and (4) western North America including Mexico and Guatemala, 15 species. This total of 43 species has been recognized by Franco (1950) in a recent monograph. Varying interpretations regarding some of the more widespread species have caused some authors to recognize a greater or lesser number, e.g., Viguié and Gaussen (1928–29) list 52 species. The work of Martinez (1953) on the Mexican coniferous flora has recently pointed to the importance of that region as an active center of speciation in the genus.

The latitudinal limits of the genus are reached by *Abies sibirica* (67°40′N) and *A. lasiocarpa* (64°30′N) in the north and by *A. guatemalensis* (14–15°N) and *A. kamakamii* (23–24°N) in the south (Hustich, 1952; Kapper, 1954; Alexander, 1958; Rehder, 1939; Franco, 1950). The most widely distributed species are *Abies sibirica, A. balsamea,* and *A. lasiocarpa* (see Plate 1).

The greatest localized concentrations of species are found in the mountain forests of southwest China (Wang, 1961), the eastern Mediterranean (Mattfeld, 1925), and the Pacific coastal region of North America from northern California to Washington.

The relative density of fir species in the United States and Canada can be gleaned from forest survey data. The contribution of firs to the growing stock of forests in California is about 20 per cent, in Washington 15 per cent, and in Oregon 6 per cent (U.S. Forest Service, 1958). In Newfoundland, by contrast, the one species, balsam fir, accounts for 60 per cent of the total growing stock.

The range of balsam fir is extensive. Often confused with *Abies lasiocarpa* and *A. fraseri* in the peripheral areas, balsam fir is lumped with these and other firs in forest surveys in those portions of the range where it is commercially unimportant.

Sargent (1884) prepared the first description of the natural range of balsam fir. His later description (1898) was more complete and more accurate as it excluded areas around Great Bear Lake and the Rocky Mountains where balsam fir previously was thought to be present:

From the interior of the Labrador peninsula, in about 56° north, *Abies balsamea,* ranging southeastward, reaches the Atlantic Coast near Cape Harrison, a degree farther south, and extends southwestward to . . . Hudson Bay, near the mouth of the Great Whale River; west of Hudson Bay it ranges from latitude 54° north to northern Manitoba, and crossing by the hills of western Manitoba, the basin of the Saskatchewan, near Cumberland House, to the valley of the Churchill, extends down the Churchill to the divide which separates the waters of that river from those of the Athabasca, down this stream to the shores of Lake Athabasca, and up the Athabasca to the neighborhood of Fort Assiniboine and Lesser Slave Lake, the most northern point where it has been observed being in latitude 62° north. Southward the balsam fir is spread over Newfoundland, the Maritime Provinces of Canada, Quebec and Ontario, over northern New England, and through north- ern New York, northern Michigan and Minnesota to northeastern Iowa; leaving the Atlantic Coast near Portland, in southern Maine, it ranges along the Appalachian Mountains through western Massachusetts, over the Catskills of New York and western Pennsylvania to the high mountains of southwestern Virginia.

The report originally made by Richardson in 1851 of balsam fir at 62° north latitude near Great Slave Lake has never been confirmed. Subsequently, Raup (1935) reported the furthest northward occurrence for this species at Reed Portage, Upper Embarrass River, Alberta (58°30'N, 111°30'W).

Hough (1907) published the first range map of balsam fir; this agrees in general with the description given by Sargent (1898). The erroneous inclusion of Yukon territory and northeastern British Columbia within the range of balsam fir as originally given by Sargent in 1884 was perpetuated by a large number of authors including Sudworth (1898, 1916, 1927), Zon (1914), Munns (1938), Preston (1948), Peattie (1950), and others. In the current U.S. Forest Service checklist Little (1953) gives separate range descriptions for *A. balsamea* var. *balsamea* (the "typical variety") and *A.b.* var. *phanerolepis*, the bracted balsam fir (see Chapter 1 for the distinguishing features). The typical variety ranges from Newfoundland and Labrador, west to northeastern Alberta, and south to Minnesota, Wisconsin, Michigan, southern Ontario, northern Pennsylvania, New York, and New England It is local in northeastern Iowa. The variety *phanerolepis* ranges from Newfound· land and Labrador to Ontario and Maine and the high mountains of New Hampshire, Vermont, and New York. It also occurs on the higher mountains of northern Virginia and West Virginia. The trees of Virginia and West Virginia may be intermediate between *Abies balsamea* and *A. fraseri* (Little, 1953).

Maps incorporating recent knowledge of the range of balsam fir have been prepared by Halliday and Brown (1943), Betts (1945), Little (1949), and Hart (1959). The following information on the distribution of balsam fir near the borders of its natural range has been taken from detailed presentations made by many authors.

In north-central Alberta balsam fir occurs near Lesser Slave Lake. Subalpine fir is also reported in the area. A few balsam fir have also been found in other parts of west-central and north-central Alberta as, e.g., in the Saddle Hills, near the Smoky River, and at Sturgeon Lake (Moss, 1953). The species was previously reported from the upper valleys of the Peace and Liard rivers in northwestern Alberta (Rydberg, 1912), but neither Moss (1953) nor Rowe (1959) list balsam fir from these areas. A confusion exists as to the identity of a species of fir located about 50 miles southwest of Edmonton, Alberta (Moss, 1953). Balsam fir is absent from the extensive aspen grove communities of south-central Alberta, Saskatchewan, and Manitoba.

Balsam fir is present on Charlton and George islands in James Bay, but is absent from the immediate coastline of both James Bay and Hudson Bay. Further north on the east coast of Hudson Bay the species occurs on the islands of Richmond Gulf (Hustich, 1950, 1954, 1955). It extends to the Koksoak River in the Ungava Bay area, from which its range line turns south to the Knob Lake area (lat. 54°50'N and long. 66°42'W), whence again north toward the Atlantic Ocean where it is found on a few small islands off the coast.

Balsam fir reaches its greatest proportion of total growing stock in southeastern Canada, especially in Newfoundland and the Maritime Provinces. In the United States the tree is most important in Maine and to a lesser degree in New Hampshire, Vermont, and New York. It is present in swamps and on moist soils in western Massachusetts, where it has been reported from Berkshire and Worcester counties (Churchill, chmn. 1933; Knowlton, 1900). The species is not known to occur in

Rhode Island, New Jersey, Delaware, or Maryland except for two questionable reports from the latter state. The American Forestry Association (1945) reported the largest balsam fir in the United States to be at Pocomoke City, Maryland. A later report by Dixon (1961) gave a similar "champion's rating" to a tree at Colesville, Maryland. That the largest balsam fir trees should occur outside its natural range is a rather curious phenomenon, particularly since the tree does not usually thrive in cultivation (see Chapter 9).

Balsam fir has been reported from western Connecticut near Griswold (Harper, chmn. 1922) and from the northwestern part of the state (Nichols, 1913). In New York it is generally absent south of the Catskills. In Pennsylvania, it is confined almost entirely to the swamps and lake shores of eleven north-central counties (Illick, 1914, 1928; Grimm, 1950).

Apparently absent from southern Pennsylvania, western Maryland, and northern West Virginia, balsam fir appears again in the higher mountains of southern West Virginia and Virginia, with four known locations in West Virginia: (1) River Dam in Grant County; (2) Canaan Valley in Tucker County; (3) Blister Swamp in Pocahontas County; and (4) Cheat Bridge in Randolph County (Core, 1940). The southernmost occurrence is on the higher mountains of Virginia, i.e., Crescent Rock on the Blue Ridge in Page and Madison counties, and the summit of Mt. Rogers in Grayson County (Coker and Totten, 1934). No authentic reports of balsam fir are known from North Carolina. Browne's (1832) description of *Abies balsamifera* in the mountains of North Carolina fits *Abies fraseri*.

The species is not found in Ohio, Indiana (Deam, 1921), or Illinois, but does occur in the northern Lake States of Michigan, Wisconsin, and Minnesota. In Michigan balsam fir is mostly absent from the southern half of the Lower Peninsula although communities of northern conifers including tamarack, black spruce, and white pine occur quite frequently. The Michigan forest survey (Essex et al., 1955) indicates the presence of balsam fir in 13 southeastern counties.

Roth (1898) listed 21 counties in Wisconsin where balsam fir was abundant. At that time he envisioned no future for the species in that state since it grew on soils well suited for agriculture. Fassett (1929) presented a detailed range map of the distribution in Wisconsin. In general, its range follows the limits of the conifer-hardwood forest as presented by Curtis (1959). In southwestern Wisconsin outliers occur in tamarack bogs in Trempealeau and LaCrosse counties (Hartley, 1960) and in the Kickapoo River valley in Vernon County (E. I. Roe, Lake States Forest Experiment Station, personal communication).

In Minnesota the distribution of balsam fir roughly follows the boundary line of the northern coniferous forest as drawn by Rosendahl (1955), which is in close agreement with the original map prepared by Upham (1884). Upham noted balsam fir occurring outside this boundary at Mantorville in Dodge County and northeast of Spring Valley in Fillmore County. This latter locality is probably identical with the balsam fir community described by Moyle (1946) from near Wykoff. Roe (personal communication) also recorded a locality from Fillmore County.

The balsam fir relicts in Iowa have caused much discussion as they represent outliers of boreal forest far removed from the present-day conifer forest zone (Conard, 1939). Sargent (1898) reported a community of balsam fir near Decorah in Winneshiek County. Other reports from Winneshiek and Allamakee counties were made by MacBride (1895), Shimek (1906), Pammel (1920), and others. Conard (1939) listed five localities along the Upper Iowa River in Winneshiek County and one locality in the Yellow River valley near Postville, in Allamakee County. Reports of balsam fir from near New Albin in Allamakee County (E. I. Roe, personal communication) and from near Elkader in Clayton County (A. L. McComb, personal communication) are in need of further investigation.

A provisional range map of balsam fir is shown in Plate 2. In addition to the range the map also shows the location of spruce-fir forest types in the United States, and the proportion of balsam fir by volume in the forests of Canada, according to the most recent survey data. The spruce-fir forest type was defined by the U.S. Forest Service (1958) as follows: "forests in which 50 per cent or more of the stand is spruce or true firs or in combination. Common associates include white-cedar, tamarack, maple, birch, and hemlock." The highest altitudinal limits of firs occur in the Himalayas and mountains of southwest China, where several species are found at altitudes of 4500–4600 m (14,800–15,100 ft) (Franco, 1950). *Abies lasiocarpa* extends up to 1050 m (3500 ft) at the northern end of its range and from 3200 to 3800 m (10,500–12,500 ft) at its southern end (Sudworth, 1916; Rydberg, 1906). Preston (1948) gave the highest elevation for *A. magnifica* as 3350 m (11,000 ft) in California. Rehder (1939) noted the occurrence of *A. religiosa* at 3500 m (11,500 ft) in Mexico, and Franco (1950) stated the altitudinal limit of *A. guatemalensis* was 3800 m (12,500 ft) in Guatemala. Altitudinal limits depend on geographic location, particularly latitude. Hustich (1951) observed a balsam fir tree two feet high, two inches in diameter, and about 60 years old, at sea level on the George Islands in Hudson Bay. The scrub forests of the subalpine zone extend above 600 m (2000 ft) in the Laurentian Highlands (Heimburger, 1934). On Cape Breton Island, balsam fir reaches its best development at altitudes of 700–1100 feet (Fuller, 1919). In New England balsam fir dominates the forest at 2500–3000 feet and even extends into the alpine zone in association with black spruce. The alpine tundra on Mount Washington in the White Mountains is found from 500 to 2000 feet below the summit (1917 m or 6288 ft), but prostrate balsam fir can be found in sheltered places only 50–75 feet below the summit (Antevs, 1932; Griggs, 1946). Further south balsam fir appears on the summit of Mt. Rogers (1734 m or 5720 ft) in Virginia (Sargent, 1898).

DISTRIBUTION OUTSIDE THE NATURAL RANGE

Except in scattered arboreta and botanical gardens, balsam fir is seldom planted outside its natural range in the United States or Canada. Aughanbaugh et al. (1958) compared the performance of balsam fir with 27 other species and varieties of *Abies* at the Secrest Arboretum, Ohio. At 29 years of age, balsam fir averaged 28.7 feet in height and 4.2 inches in diameter. The growth of balsam fir was better than that of Fraser fir and about equal to that of European silver fir (*A. alba*), but was less than that of several European and Asiatic species. Balsam fir also suffered from wind, snow, and ice.

In the Pacific Northwest balsam fir has been planted in the Wind River Arboretum at Carson, Washington, and in the Hoyt Arboretum, Portland, Oregon

(Munger and Kolbe, 1932, 1937; Munger, 1947; Silen and Woike, 1959). Generally speaking, eastern trees planted in the Cascade Range of Washington and Oregon do not thrive, mainly owing to the extended drought period in summer and the drying, cold, easterly winds in winter. Balsam fir obtained from a Massachusetts nursery was transplanted in 1929 to the Wind River Arboretum. The arboretum lies in a valley at 1150 feet (350 m) elevation. The soil is a deep, alluvial, coarse sand over loam, an average site (site index 130) for Douglas fir (*Pseudotsuga menziesii*). Of the original 28 species of *Abies* planted, 24 survived. The growth of balsam fir was poor in comparison with native true firs and Douglas fir. In 1956, 27 years after transplanting, the average height was only 24.4 feet. This compares with 44 feet for *Abies grandis* and 62 feet for *Pseudotsuga menziesii* during the same period. In recent years balsam fir was attacked by the balsam woolly aphid, and its growth was reduced. The Portland planting originally consisted of 86 trees set out in 1931. A 1960 report by Munger (personal communication) showed only ten trees surviving.

Balsam fir survival and growth is affected by drought in shelterbelt plantations on the prairies of Minnesota (Deters and Schmitz, 1936), Nebraska (Pool, 1939), and North Dakota (George, 1953).

In the east, restrictions on the planting of balsam fir were given by Tryon et al. (1951) for West Virginia. Planting balsam fir below 1700 feet elevation was not recommended. In the elevation zone, 1700–2500 feet, plantings should be made only on moist sites on north- and east-facing slopes.

Balsam fir has been cultivated in Europe since the latter part of the seventeenth century. Loudon (1838) and Wein (1931) indicate that interest in this species was widespread and considerable plantings were made in English parks during the period 1683–1720. Since the latter date, interest in this species has diminished, and it is seldom planted today. The seed sources of early plantings are rarely indicated; because of the concentration of English colonists in the Maritime Provinces and New England, most seed probably originated from these areas. In recent years experimental plantings of balsam fir on peat in Great Britain have been made (Zehetmayr, 1954). Success would seem to be limited, however, as the species is often damaged severely by late spring frosts (Great Britain Forestry Commission, 1955).

No records of balsam fir plantings in France or the Mediterranean countries were found. Undoubtedly the poor adaptability of the species to warm, dry climates makes it unsuitable for those areas.

In Norway balsam fir was cultivated as early as 1772 (Sudworth, 1916). Balsam fir has made good growth, said to be the best in Europe, at Uppsala, Sweden (Ilvessalo, 1926). In Finland, the Baltic States, and Russia, it was already being cultivated in 1810. In Finland balsam fir is considered less frost hardy than *Abies sibirica*, but is more resistant to insects. Cajander (1923) recommended that seed for Finland should be selected from natural stands in southeastern Canada (excluding Newfoundland) or from the mountains of New England above 500 m (1600 ft) elevation. Balsam fir has been found to be frost hardy in Latvia. Early height growth is rapid, but it ceases after reaching about 20 m (63 ft) at 60–70 years (Sivers, 1911).

Balsam fir has been fairly successful in Hungary (Roth, 1920). Its satisfactory growth in the Ukraine was noted by Lipa (1940). Gegelsky (1953) reported on two plantings in Russia, a 55-year-old plantation at Chernigovsk and an 80-year-old plantation at Trostjantc in the Ukraine. At the latter location balsam fir reached 23 m in height and 51 cm in diameter. Natural reproduction proceeded satisfactorily. Kapper (1954) considered that balsam fir could be grown with fair success west of a line connecting Leningrad, Moscow, Tambov, and Voronesh. Some damage to young trees by frost and sunscald on exposed sites was noted. Later, balsam fir becomes more resistant. At 80 years, the largest trees were 27 m tall and had diameters of 45 cm at breast height.

The species was introduced a century ago in Germany. The first plantation, apparently, was at Harburg in 1860 (Hoff, 1931). Of the many reports concerning balsam fir in Germany, most deal with plantations less than 40 years of age (Forster, 1905; Schelle, 1920; Grundner, 1921; Döring, 1927, 1930; Schwerin, 1929; Müller, 1938; Schenck, 1939). These reports agree that early growth is rapid, but begins to drop off at 50–60 years. The species is subject to insect damage, although frost damage has not been reported. The silvicultural qualities of the tree and the technical properties of the wood are considered inferior to such preferred exotics as Douglas fir and *Abies grandis* (Müller, 1938). Schenck (1939) and Nachtigall (1943) recommended trial plantings of balsam fir in the swamplands of East Prussia.

Scattered reports of balsam fir plantings in other countries have been made. Spaulding (1956) noted that it had been planted in Australia. Sylvain (1940) reported that it had been tried on the mountains near Kenscoff in Haiti. From unconfirmed reports, it appears that balsam fir was planted by the U.S. Army on the Aleutian Islands shortly after the end of the Second World War.

Geographical and Synecological Classification and Models

There is a great abundance of geographical and synecological information concerning the distribution, physiognomy, and performance of balsam fir under differing environmental conditions. Difficulties exist in the reduction and evaluation of this material because of the great variety of rather loose classification systems used, and because geographical and synecological models are at present only in a very early stage of development.

REGIONAL CLASSIFICATIONS AND MAPPING

Sargent (1884) and Fernow (1912) provided much of the early geographical information about the forests of North America. Later, Clements (1916) outlined a system for the regional classification of vegetation. In his scheme balsam fir occurs as a member of the eastern part of the Boreal Forest Formation where it is a codominant species of the climatic climax association. Balsam fir is also widely distributed in the Lake Forest where it is an important component of subclimax communities. The species rarely penetrates into the oak-beech, oak-hickory and oak-chestnut associations of the Deciduous Forest Formation.

A map of the natural vegetation of the United States was prepared by Shantz and Zon (1924). In 1949 the U.S. Forest Service prepared a map showing the areas characterized by major vegetation types in the United States. Braun (1950) prepared a monograph on the eastern deciduous forests accompanied by a map showing the major forest regions in the eastern United States. Plate 2 shows the distribution of spruce-fir types within the range of balsam fir, based on these maps and additional information.

The Society of American Foresters has worked out several revisions of a regional cover-type system for North America exclusive of Mexico. The types are numbered consecutively for the entire system. According to this classification, balsam fir occurs in the Boreal Forest Region, in the Northern Forest Region, and to a limited extent, mostly along the borders, in the Central Forest Region. In the western forests balsam fir occurs in the Northern Interior area, where it is found as a minor constituent of the white spruce type (No. 201), mostly in Alberta.

The classification system developed by Braun (1950) allows for a rather detailed geographical approach to the United States portion of the balsam fir range, as may be seen in Plate 2. Balsam fir is an important component within the Hemlock–White Pine–Northern Hardwoods Forest (Region 9). Some balsam fir also occurs in the Mixed Mesophytic Forest (Region 1: Section B), the Oak-Hickory Forest (Region 3: Section C), the Oak-Chestnut Forest (Region 4), the Beech-Maple Forest (Region 7) in central and southeastern Michigan, and in the Maple-Basswood Forest (Region 8).

SCHOOLS OF THOUGHT AND THE DEVELOPMENT OF MODELS

The literature in which balsam fir has been considered as an important object of investigation is fragmented both geographically and among the different schools of thought.

The Clementsian school emphasized a trend in vegetation development toward a single regional climax, controlled by climate, which was intermediate with respect to moisture and temperature conditions between those of initial bare rock and open water. Moisture was recognized as the most important factor complex in the development of primary succession sequences within a climatic region. The role of nutrients has been considered in relation to the mesification process during the passage of time. Most of the vegetation studies have used the present moisture-nutrient status to evaluate the position of a community in successional sequences. The time axis originally used by this school is in actuality a nutrient axis. A true evaluation of the changes in time would require the addition of a third axis to the de facto existing moisture-nutrient model. Such diagrams have not been made.

The three diagrams presented here (Fig. 5) illustrate primary succession in three forest communities. The first diagram shows the succession trends of forest communities in northern Minnesota and Wisconsin after Kittredge (1938). It applies to an area where a transition occurs from the Deciduous Forest Formation represented by the sugar maple–basswood climax association to the Boreal Forest Formation represented by the white spruce–balsam fir climax association. Kittredge has pointed out that the indicators of the spruce-fir climax indicate a lower nutrient level than the indicators of the maple-basswood climax. The paper by Kittredge contains abundant information demonstrating the increase of soil fertility with the advance of the climax. Other data by Kittredge from Star Island, Cass Lake, Minnesota, together with additional information obtained by other workers investigating successional phases, have repeatedly shown the increasing soil fertility with advancing status of the plant communities toward the climatic climax.

The second diagram shows the primary succession trends of forest communities at Douglas Lake, Cheboygan County, Michigan. This area is located east of the

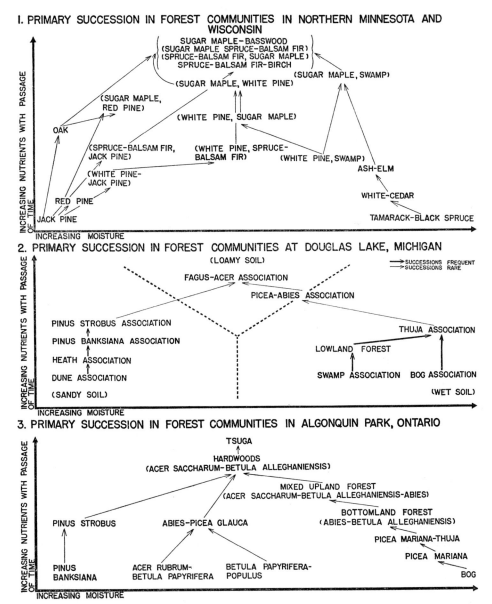

Figure 5. The place of balsam fir in forest classification systems based on primary succession trends. (1) Primary succession in forest communities in northern Minnesota and Wisconsin, after Kittredge (1938); (2) primary succession in forest communities at Douglas Lake, Cheboygan County, Michigan, after Gates (1926); (3) primary succession in forest communities in Algonquin Park, Ontario, after Martin (1959).

preceding and within the range of *Fagus grandifolia*. It is clearly indicated that the early stages of the succession are found on soils with less nutrients than the climax stage. The third diagram shows the primary succession in Algonquin Park, Ontario, as investigated by Martin (1959). In this area *Tsuga canadensis* plays an important role in the development of forest communities. The species listed in these diagrams are dominants and codominants. In less significant amounts they can occur over a much wider range of conditions.

The polyclimax school, championed by Nichols and others emphasized the importance of such local factors as soil, relief, aspect, and others that could bring about stability in vegetation. The original thoughts go back much further. Kerner, as early as 1863, was searching for "genetic" relationships between the differences in vegetation and differences in environment. This led to the investigations of genetic or ecological series of vegetation with surrounding environments. At first, attention was paid primarily to morphological features of the environment which appeared to be highly diversified. The search was later extended by looking for the physiological basis of vegetation development as has been pointed out by Schimper (1903), Livingston (1921), and others. Different sets of ecological series were combined to obtain a genetic scheme for the regional vegetation. Of particular importance in forest classification is the genetic scheme devised by Sukachev (1932), which emphasizes the changes in moisture and nutrient conditions. The studies made by Pogrebniak (1930, 1955) and others led to the consolidation of a relative moisture-nutrient coordinate system.

As far as the investigations of balsam fir are involved, the works by Graves (1899), Zon (1914), Heimburger (1934), Westveld (1931, 1953), Westveld, chmn. (1956), Wilde et al. (1949), Wilde (1958), Linteau (1955), and others are of particular importance in the development of ideas which come close to the polyclimax school. These works will be discussed together with other regional information later in this monograph.

The two diagrams in Figure 6 illustrate the place of balsam fir in edaphic classification systems. Diagram 1 shows the forest types of the Adirondack area, New York, in a genetic classification system by Heimburger (1934) patterned after Sukachev. Diagram 2 shows an arrangement of the forest types (site types) of the northeastern coniferous section of the Boreal Forest Region in Quebec by Linteau (1955). Both classification systems are presented in edaphic coordinate systems with moisture and nutrient axes; the basic data are provided in Tables 5 and 7. The modern continuum ideas, whether approached from the vegetational or physiographic sides, have led to substantially the same results.

The hierarchic systems for classification of vegetation as developed by the Zürich-Montpellier school under the leadership of Braun-Blanquet (1951) can also be organized into multidimensional coordinate systems as shown by Wagner (1954, 1958).

The different schools of thought have emphasized different aspects of a vegetation-environment complex that can lead to a deepened understanding of the ecosystem. The fundamental model emerging from a century of work in forest synecology is the regional edaphic field or the vegetation-environment complex in moisture and nutrient axes. At the same time it is also the outline of the primary succession process. The time element can be shown with the aid of vectors within the edaphic field. For more complete investigations it is necessary to introduce time as the third axis. Mesification has been the dominant trend in the geographic area in which the original studies of the monoclimax, Clementsian school were

I. SYSTEM OF FOREST TYPES IN ADIRONDACKS, NEW YORK

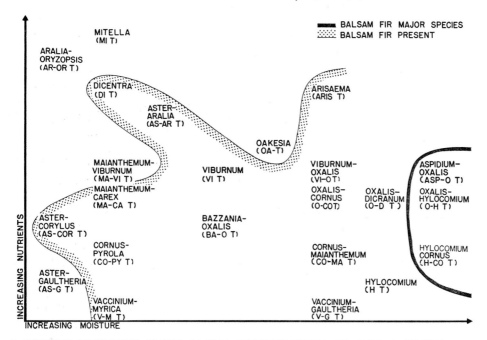

2. SYSTEM OF FOREST TYPES IN THE NORTHEASTERN CONIFEROUS SECTION IN QUEBEC

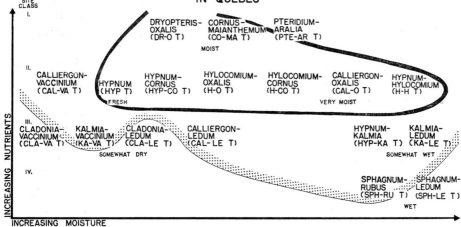

Figure 6. The place of balsam fir in forest classification systems based on edaphic relationships. (1) Genetic diagram of forest types in the New York Adirondacks, after Heimburger (1934); (2) edaphic system of forest types (site types) in the northeastern coniferous section of the Boreal Forest Region (Sections B-1a and B-1b of Rowe, 1959) in Quebec, after Linteau (1955).

made, and has been recognized by workers in the Soviet Union (Zonn, 1955), especially in areas near the steppe region. In more humid areas within the range of balsam fir, the podzolization process proceeds much faster and the general trend may be paludification or bog formation. Equilibrium can be reached, however, at any point of the edaphic field, i.e., the site complex in moisture-nutrient axes. Within the field, changes occurring with the passage of time may have a definite or a variable direction, and they can proceed with different speeds.

A closely related type of model for demonstrating a regional edaphic field was suggested by Roe (1935). He investigated the relationship between the potential forest and various soil groups in Michigan, Wisconsin, and Minnesota. Roe recognized a northern hardwood climax region, and a white spruce–balsam fir region. Within each region he distinguished six soil moisture classes based on the texture of the subsoil and relative depth of the water table. Plate 3 shows the model by Roe (1935). Moisture increases in two directions, from left to right and from top to bottom. Additional data indicate that nutrients in general increase from top to bottom and towards medium moisture conditions.

A model by Hills (1960a) provided information about the performance of balsam fir under changing climatic and edaphic conditions, and associated changes in species in undisturbed forest stands in Ontario. Plate 4 shows the site regions of Ontario (diagram 1), the soil development in the different areas (diagram 2), and illustrates the changing pattern of associated species. Diagram 3 indicates a general expansion of the forest site complex (synecological field) and the increasing diversity of vegetation within individual classes as the optimum conditions for forest growth are attained. Hills' model can be visualized as a multidimensional space having two major macroclimatic axes: heat and humidity. Within this framework is placed a second coordinate system which analyzes the landform with the aid of moisture, ecoclimate, and, as shown by Hills elsewhere (1960b), nutrients. In the first coordinate system, the heat axis is oriented primarily in a north-south direction, and the humidity axis in an east-west direction. In the second coordinate system moisture appears as an edaphic factor while ecoclimate represents primarily the local variations in heat due to physiography. The nutrients are evaluated from the base contents of the parent material.

Hills' model (1960a), based on the work of many years in Ontario forests, represents one of the most detailed approaches to a regional synecological classification of forests that has been presented. Similar work in other forest regions could add substantial knowledge to the understanding of synecological and geographical distribution of forest communities. Some insight into the ecological distribution of balsam fir can be obtained by evaluating already existing information within a simplified framework, similar to Hills' classification, as discussed below.

The Society of American Foresters (1954) has worked for nearly thirty years on the development and improvement of the forest cover-type system. The types are grouped by regions and numbered in a sequence according to moisture conditions (see Table 4). Information on other edaphic conditions and successional status is also provided. A few cover types composed mainly of ubiquitous species occur on very heterogeneous edaphic sites, e.g., the black spruce type (No. 12), the red spruce type (No. 32), and the white spruce–balsam fir–paper birch type (No. 36). Subsequently, these latter types have been subdivided by various authors. The regions as outlined by the Society of American Foresters (1954) are too broad for more specific purposes. The first need is to distinguish between the western and eastern parts of the Northern Region. With some aid from additional sources, the permanent forest cover types could be placed in regional edaphic (or more broadly synecological) coordinates as shown in Plate 5. Abbreviated descriptions of the forest cover types are presented in Table 4.

The three pairs of diagrams in Plate 5 show the arrangement of the permanent forest cover types of three regions and three local forest classification systems in edaphic (moisture-nutrient) coordinate systems. The local systems are prepared on a more detailed scale. The scales at present are indeterminate and highly relative since the status of forest communities is known only in a general way. The nutrient problem is the more complicated.

The position of balsam fir in the models (Plate 5) is based on the information provided in the descriptions of the various forest types. In the Boreal Region balsam fir is spread widely over the edaphic field and occupies the most mesic-rich sites. In the Great Lakes–St. Lawrence Region balsam fir cannot compete with tolerant hardwoods on these sites, and its edaphic field is more restricted. The Acadian Appalachian Region is intermediate between the other two regions in respect to the position of balsam fir.

TABLE 4. LIST OF RELATIVELY STABLE FOREST TYPES IN
EASTERN NORTH AMERICA

Type Name and Number *	Sites Occupied	Associated Species		Remarks
		Major	Minor	
BOREAL FOREST REGION				
Jack pine (1)	Driest outwash sands and rock outcrops	Jack pine, quaking aspen, and paper birch	Black spruce	No *balsam fir*, except as undergrowth
Black spruce–white spruce (2)	Tundra borderline under mesic conditions	Black spruce and white spruce	Quaking aspen	North of the range of *balsam fir*
White spruce–balsam fir (4)	Moist or fresh soils †	White spruce and *balsam fir*	Jack pine, quaking aspen, and black spruce	Regional climax, in the Clementsian sense
Upland black spruce (12a)	Slopes, and, as a successor of SAF type 6 (jack pine–black spruce), fairly coarse soils	Black spruce	*Balsam fir*, white spruce, and paper birch	SAF groups 12a and 12b into one type, 12, with two variants
Black spruce–balsam fir (7)	Sandy loams, silty loams, and sands; alluvial flats, and mountain slopes	Black spruce and *balsam fir*	Tamarack, quaking aspen, paper birch, white spruce	

* Numbers are those applied by the Society of American Foresters.
† Fresh = moderately dry, intermediate between dry and moist.
Source: after Society of American Foresters (1954).

TABLE 4. Continued

Type Name and Number[*]	Sites Occupied	Associated Species		Remarks
		Major	Minor	
Lowland black spruce (12b)	Low flats and shallow swamps	Black spruce	*Balsam fir*, jack pine, quaking aspen, paper birch, tamarack, and white-cedar	
Black spruce–tamarack (13)	Poor low wet sites	Black spruce and tamarack	*Balsam fir*, white-cedar and black willow	
	GREAT LAKES–ST. LAWRENCE FOREST REGION			
Jack pine (1)	Driest outwash sands and rock outcrops	Jack pine	Red pine, pin oak, quaking aspen, bigtooth aspen, paper birch, and black spruce	No *balsam fir* in the permanent jack pine type
Black spruce (12)	Poorly drained peat swamps	Black spruce	*Balsam fir*, paper birch, tamarack, white-cedar, black ash, and red maple	
Red pine (15)	Sandy and gravelly locations or dry sandy loams and on shallow-soiled rocky knolls	Red pine	White pine, jack pine, pin oak, white oak, red maple, paper birch, and quaking and bigtooth aspens	*Balsam fir* may appear as a secondary story (not indicated by SAF, 1954)
White pine (21)	Sandy loams and loams	White pine	*Balsam fir*, red pine, paper birch, yellow birch, sugar maple, aspen, oaks, basswood, and hemlock	
Sugar maple–basswood (26)	Rich loamy upland soils	Sugar maple and basswood	American elm, green ash, yellow birch, and red oak	*Balsam fir* and white spruce (not mentioned by SAF, 1954) have a spotty occurrence
White spruce–balsam fir–paper birch (36)	Moist to wet loamy upland soils	White spruce, *balsam fir*, and paper birch	Quaking and bigtooth aspen, white pine, balsam poplar, white-cedar, sugar maple, black ash or jack pine, and black spruce	This type should probably be split, as by Zon (1914) and Cheyney (1942) into spruce flat and spruce slope
Northern white-cedar (37)	Poorly drained mucks	White-cedar	*Balsam fir*, tamarack, yellow birch, paper birch, black ash, red maple, black spruce, white pine, and hemlock	
Tamarack (38)	Wet swamps and mucks	Tamarack	Black spruce, white-cedar, red maple, black ash, and quaking aspen	*Balsam fir* not mentioned by SAF (1954), occurs occasionally
Black ash–American elm–red maple (39)	River valleys, alluvial flats, muck and shallow peat soils	Black ash, American elm, and red maple	Balsam poplar, *balsam fir*, yellow birch, white pine, tamarack, white-cedar, and basswood	

TABLE 4. Continued

Type Name and Number*	Sites Occupied	Associated Species		Remarks
		Major	Minor	

ACADIAN–APPALACHIAN FOREST REGION

Type Name and Number*	Sites Occupied	Major	Minor	Remarks
Red pine (15)	Sandy and gravelly locations or dry sandy loams and on shallow-soiled rocky knolls	Red pine	White pine, pitch pine, jack pine, white oak, chestnut oak, red oak, and hemlock	*Balsam fir* not mentioned by SAF, occurs occasionally
White pine (21)	Sandy loams and loams	White pine	Red pine, pitch pine, gray birch, quaking and bigtooth aspens, red maple, pin cherry, white oak, paper birch, sweet and yellow birches, black cherry, white ash, red oak, sugar maple, basswood, hemlock, red spruce, and white-cedar	No *balsam fir* in the permanent type
Sugar maple– yellow birch– beech (25)	Rich, moist, loamy soils	Sugar maple, beech, and yellow birch	Basswood, red maple, hemlock, red spruce, *balsam fir*, white pine, American elm, and white ash	
Red spruce– sugar maple– beech (31)	Deep, well-drained, fertile soils	Sugar maple, beech, and red spruce	*Balsam fir*, hemlock, and red maple	
Red spruce (32a) (spruce slope)	Well-drained rocky slopes	Red spruce	*Balsam fir*, yellow and paper birches, hemlock, and black spruce	
Red spruce (32b) (spruce flat)	Moist level or rolling lands	Red spruce and *balsam fir*	White spruce, black spruce, white pine, hemlock, red maple, sugar maple, beech, yellow birch, and white ash	
Red spruce (32c) (spruce swamp)	Mucks and peats	Black spruce, red spruce, and *balsam fir*	White-cedar, tamarack, black ash, red maple, and yellow birch	
Northern white-cedar (37)	Slowly drained sites (not stagnant bogs)	White-cedar	*Balsam fir*, tamarack, yellow and paper birches, black ash, red maple, black spruce, white pine, and hemlock	
Tamarack (38)	Wet peats and mucks	Tamarack	Black spruce, white-cedar, red maple, black ash, and quaking aspen	*Balsam fir* not mentioned by SAF. Type rare
Black ash– American elm– red maple (39)	River valleys and alluvial flats on mucks and shallow peat soils	Red maple, American elm, and black ash	Balsam poplar, *balsam fir*, yellow birch, white pine, tamarack, white-cedar, and basswood	

Geographical and Synecological Descriptions of Regions

In this section the natural range of balsam fir will be treated by adopting a provisional classification of forest regions. Those regions considered are the Boreal Region, the Great Lakes–St. Lawrence Region, and the Acadian–Appalachian Region. Ecotone areas and outliers will be considered together with their nearest associated major region. The subdivisions of these regions will be treated, for the most part, according to the classification systems of Rowe (1959) for Canada, and of Braun (1950) for the United States. These subdivisions are shown in Plate 2.

Plate 6 illustrates climatic conditions by using climographs for representative stations throughout the range of balsam fir. The climographs show the monthly mean temperature and average precipitation for individual months and for the year. The mean daily temperature over the balsam fir range varies from —23C to —4C (—10 to 25F) in January and from 13C to 20C (55 to 70F) in July. The frost-free season varies from 90 to 180 days. Mean annual precipitation ranges from 390 to 1400 mm (15 to 55 in.) and growing season precipitation from 150 to 620 mm (6 to 25 in.).

Plate 7 presents the distribution of the broad soil groups that are found within the range of balsam fir. Balsam fir seldom penetrates arctic soils. It occupies mainly podzols which are frequently intermingled with rock outcrops and peat deposits. In the western part of its range, gray-wooded soils predominate. In the eastern and central areas gray-brown podzolic, humic gley, and other soils are locally important.

THE BOREAL FOREST REGION

Balsam fir is a boreal species, although it has a more southern affinity than black spruce, white spruce, or tamarack. The eastern part of the Boreal Region is the area of optimum development, where the individual trees attain their greatest stature and age, basal area is highest, and proportionate volume is greater than in other areas.

The forest communities of the Boreal Region have been described very extensively, the geographic point of view predominating. The bulk of the information will be presented following the forest classification of Halliday (1937) and the forest region description of Rowe (1959) as the major sources. Additional details will be taken from Hills' (1960a) geographical model of Ontario. The forest classification by Linteau (1955) gives a deep insight into the synecological conditions of the northeastern section of the Boreal Region.

Climatic and edaphic conditions. According to the Köppen (1936) classification, most of the balsam fir range in the Boreal Region belongs to the "Dfc" climatic region, where D = middle latitude humid climates with cold winter; temperature of the coldest month less than 3C (26.6F); f = precipitation throughout the year; c = cool, short summers where temperatures above 10C (50F) occur for only 1–3 months.

Using the climatic descriptions of Thornthwaite (1941), the climate of the eastern part of the Boreal Region is humid to superhumid with abundant moisture at all seasons; temperatures are cool temperate-plus. In the central and northwestern parts, the climate is subhumid to subhumid-plus, with moisture generally deficient at all seasons; temperatures are classed as cool temperate to cool temperate-plus. The potential evapotranspiration computed according to Thornthwaite (1948) and given in the Canadian forest classification (Rowe, 1959) ranges from 380 to 530 mm (15 to 21 in.) per year. In the western parts of the balsam fir range precipitation becomes somewhat limiting, but in general temperature is the dominant climatic element in the Boreal Region.

Mean monthly temperatures and precipitation for characteristic stations are presented in Plate 6. Additional climatic data on the Boreal Region are given by Halliday (1937), Canada Geographical Branch (1957), Rowe (1959), and other sources, and are summarized as follows: the daily mean temperature ranges from —23C to —4C (—10 to 25F) in January, and from 13C to 18C (55 to 65F) in July. The frost-free season extends 90–160 days. The number of degree days above 6C (42F) ranges from 1000 to 2500. Bright sunshine prevails, on the average, from 1400 hours per year on the coast of Newfoundland to 7000 hours per year in Alberta. Precipitation shows a marked decline from east to west and is paralleled by an equally marked decline in the abundance of balsam fir. Annual precipitation equals or exceeds 1400 mm (55 in.) in southeastern Newfoundland, but declines rapidly to 560 mm (22 in.) in the north and to 380 mm (15 in.) in the west. Precipitation during the growing season varies from 230 to 460 mm (9 to 18 in.).

Newfoundland belongs to the Appalachian physiographic region. Anticosti Island in the Gulf of St. Lawrence is a part of the Great Lakes–St. Lawrence Lowlands. The west coasts of Hudson Bay and James Bay are part of the Hudson Bay Lowlands. Most of the balsam fir range in the Boreal Region is located on the Canadian Shield, the western part of the range lying within the Interior Plains physiographic region.

The geomorphology of Newfoundland is complex, with Precambrian and Paleozoic rocks intermingled. The rock underlying most of the balsam fir area within the Boreal Region is of Precambrian origin. Paleozoic sediments predominate in the Gaspé Section (B-2, according to the Canadian forest classification — see Plate 2), in the Hudson Bay Lowlands Section (B-5), and in the Manitoba Lowlands Section (B-15). Mezozoic sedimentaries predominate in the Mixed-Wood Section (B-18a), Hay River Section (B-18b), and the Lower Foothills Section (B-19a).

Nearly all the region was glaciated and covered with varying thicknesses of drift, glacial lake deposits, and marine deposits. There are large areas with exposed rocks. Podzol soils predominate, especially in the eastern part of the region. A wide area of peatlands with alluvial and brown wooded soils extends south and west from James Bay. Gray wooded soils and peat prevail in the Northern Clay Belt Section (B-4). Gray wooded soils mixed with podzols, rock outcrops, and peats dominate the remaining western part of the region. An area of gray wooded soils and peat is located north of Lake Winnipeg. West of this lake lies an area with high lime soils interspersed with peat. The soils over the wide range of the Boreal Region where balsam fir occurs are shown in Plate 7.

Geographic survey of forest communities. Balsam fir is a dominant species in Newfoundland, forming dense stands in the Northern Peninsula Section (B-29) and growing together with black spruce in very open and patchy stands in the Newfoundland–Labrador Barrens Section (B-31). It is found on the better soils

and accounts for 70 per cent of the volume of all conifers in the Avalon Section
(B-30). There, the species grows well early in life but soon slows down. Trees at
maturity seldom exceed 40 feet in height (Wilton, 1956; Rowe, 1959). Balsam fir
tends to invade black spruce stands after the latter are cut. In most cases this is
only a temporary stage in the secondary succession, balsam fir being followed
by the reinvading but slowly establishing black spruce (Ellis, 1960).

Balsam fir grows together with white spruce, black spruce, and paper birch
in mixed stands of low stature on wind-swept Anticosti Island (Section B-28c).
There is a virtual absence of shrubs even in open stands. Burned areas reproduce
easily with spruce, but the large deer population eliminates balsam fir seedlings
and saplings so that the latter seldom reach maturity (Rowe, 1959).

Black spruce, white spruce, and tamarack are the primary species in the Forest-
Tundra Section (B-32). Balsam fir occurs only rarely in the Ungava area. The
tundra barrens are intermingled with patches of stunted forest. Pines are not
present, and paper birch, aspen, and balsam poplar are infrequent.

Hustich (1949) investigated the forest geography of the Labrador Peninsula.
He prepared distribution ranges for the major species, an outline for major zonation of vegetation, and a forest type classification. An arrangement of these types
in moisture-nutrient coordinate axes indicating the position of balsam fir is shown
in the upper right diagram of Plate 5. Hare (1950) presented a somewhat different classification and used vegetational designation as a substitute for Hustich's
edaphic terms for major type groups. A Boreal-mixed forest ecotone zone was
added to the zonal divisions of the Boreal Forest in Labrador and Ungava. Wilton
(1959) worked out the local variations for forest types in the Grand Lake and Lake
Melville areas of Labrador (Section B-12). This section includes outlying forest
areas in valleys with better soil development and supporting richer forest vegetation than the surrounding subarctic vegetation of the adjacent uplands (Rowe,
1959).

The Northeastern Transition Section (B-13a) and the Fort George Section (B-
13b) are similar to Sections B-31 and B-32 described above, and to Section B-5 (see
below). These areas represent patches of forest mixed with barren land, giving
them a subarctic appearance. Jack pine is present in Section B-13a and becomes
more abundant in Section B-13b. Outposts of northern white-cedar, black ash, and
American elm appear in Section B-5 (Rowe, 1959). Hustich (1955) described the
forest communities in the Moose River area, Ontario (Section B-5). Balsam fir
appears on the uplands, mostly with black spruce, and shows poor development,
having slow growth and an asymmetrical form. The species attains greater height
on river banks, where it grows together with white spruce, aspen, paper birch,
and balsam poplar (Halliday, 1937; Rowe, 1959).

Balsam fir is of considerable importance in the Gaspé Section (B-2), where it
appears in both mixed and pure stands. It occurs together with aspen, balsam
poplar, paper birch, and white spruce on upland till sites and on alluvial soils in
the Gouin Section (B-3). Black spruce dominates large areas of the Northern Clay
Belt Section (B-4), where on better drained sites it is admixed with balsam fir.
In the East James Bay Section (B-6) balsam fir occurs mainly in valleys (Rowe,
1959).

Sections of the Boreal Region bordering the Great Lakes–St. Lawrence Region are penetrated frequently by white and red pine, sugar maple, yellow birch, and white-cedar. Balsam fir continues to occupy favorable sites in the border areas although its importance decreases, particularly to the west.

Admixed with white spruce, trembling aspen, balsam poplar, and paper birch, balsam fir occurs on the better drained sites in the lowland sections (B-14, B-15, B-21) found on the silty bed of glacial Lake Agassiz in Manitoba. The situation is similar in the Mixed-Wood Section (B-18a) on the upland soils derived from mixed Cretaceous deposits. The importance of balsam fir as a major component of the forest vegetation declines rapidly west of Manitoba. Though found in the mixed-wood forests on the best soils in Section 22a (mostly in Saskatchewan), it occurs very sparingly in Section B-20, and only in the southern portions of Sections B-22b and B-23a (northwest Saskatchewan and Alberta). The species occurs together with subalpine fir in the major body of the Lower Foothills Section (B-19a) in Alberta (Rowe, 1959).

Raup (1946) described seven forest types in the Athabaska–Great Slave Lake area, one of which, balsam fir–white spruce, contains considerable quantities of balsam fir. This type, the most mesophytic in the area, is limited to the floodplains of the Athabaska and Clearwater rivers in the southern part of the region (Section B-23a).

At the western border of the balsam fir range there is a transition from prairie to forest vegetation. This transition zone is also a meeting ground of Atlantic and Pacific floras (Moss, 1953). Balsam fir occurs there in mixed stands, associated with white spruce, aspen, balsam poplar, and paper birch on flat areas and gentle slopes.

The Ontario model by Hills. The general features of the forest geographical model of Ontario by Hills (1960a) discussed previously are illustrated in Plate 4. Balsam fir does not occur in the Hudson Bay site region (1E), the humid northwestern part of Ontario where closed forest stands are limited to the warmer ecoclimates. In the James Bay (2E) region the species occurs on fresh * sites having normal ecoclimates, and is associated with black spruce, white spruce, and aspen. It is able to persist on wet soils together with balsam poplar and paper birch in the Lake Abitibi (3E) region. In the Lake Timagami (4E) region the tree also occupies the drier soils with colder ecoclimates, and its associated species include white-cedar, white pine, and red pine in addition to the others mentioned above. The widest range of sites is in the Georgian Bay (5E) region, where it occupies all surfaces ranging from fresh soils with normal ecoclimates to wet soils and cold ecoclimates. On wet soils with normal ecoclimates associated species are red maple and black ash; on wet soils with cold ecoclimates, tamarack. Accompanied by sugar maple, yellow birch, hemlock, white pine, and white spruce in the most mesic conditions, balsam fir cannot compete with southern species for the most mesic sites in the Lakes Simcoe-Rideau (6E) region, where it is less abundant and occurs on fresh and wet soils with colder ecoclimates together with white and black spruce and tamarack. Under normal ecoclimates balsam fir can persist only on wet sites associated with hemlock, yellow birch, elms, and white-cedar. Rare in

* Apparently meaning moist.

the southernmost site region of Lake Erie and Ontario (7E), it occurs there on the wettest and coldest sites together with black and white spruce, sugar maple, red maple, and yellow birch.

The regions described above all belong to the eastern (E) series. The western (W) series is characterized by increased aridity which results in reduction of suitable sites and changes in associated species. The outstanding features shown by balsam fir in Ontario are (1) the progressive shift in its adaptability from warm dry sites in the north to cool wet sites in the south, and (2) its wide range of adaptability over many sites in central Ontario in contrast to its narrow range of adaptability on few sites at the northern and southern ends of the province.

Forest types by Linteau. Linteau (1955) prepared a forest site classification for the northeastern Coniferous Section which, following Rowe (1959), is now subdivided into the Laurentide-Onatchiway (B-1a) and Chibougamau Natashquan (B-1b) sections. The classification applies directly to an area of 75,000 square miles in central Quebec, although experience shows that it can be successfully applied over an even wider area. Abbreviated descriptions of forest types are presented in Table 5. Characteristic combinations of species of the ground vegetation are used as type names. A diagram showing the location of the types in edaphic coordinates is presented in Figure 6 (diagram 2).

Following the lead of Cajander (1923), Linteau attempted to solve simul-

TABLE 5. CLASSIFICATION OF FOREST TYPES (SITE TYPES) IN THE
NORTHEASTERN CONIFEROUS SECTION (B-1) OF THE
BOREAL FOREST REGION OF CANADA

Landforms	Topography	Soil Texture, Structure, and Consistence	Soil Profile and Mois- ture Regime	Species Composition in Mature Stands		Silvicultural Notes
				Major	Minor	
DRYOPTERIS-OXALIS (Dry-O T) Site Class I						
Marginal moraine; ground moraine	Moderately steep lower slope	Loose sandy loam (A and B) over compact sand (C); single-grained structure	Iron humus podzol; moist	*Balsam fir;* or paper birch and *balsam fir*	Paper birch and white spruce	Trees firm-rooted, suited to selection cutting. Advance reproduction thin.
CORNUS-MAIANTHEMUM (Co-Ma T) Site Class I						
River terrace	Flat to moderately steep	Mellow silt loam (A and B) over deep mellow loam (C); granular structure	Normal iron podzol; moist	*Balsam fir,* paper birch and white spruce; or paper birch and black spruce; or white spruce, trembling aspen, and *balsam fir*	Black spruce	Trees firm-rooted, suited to selection cutting. Good reproduction under shade.

Source: after Linteau, 1955.

TABLE 5. Continued

Landforms	Topography	Soil Texture, Structure, and Consistence	Soil Profile and Moisture Regime	Species Composition in Mature Stands		Silvicultural Notes
				Major	Minor	
Lacustrine plain	Upper slope near lake	Mellow sandy clay loam (A and B) over sandy loam (B$_2$ and B$_3$) over deep loose coarse sand (C); A and B of granular structure	Moist			

PTERIDIUM-ARALIA (Pte-Ar T) Site Class I

Landforms	Topography	Soil Texture, Structure, and Consistence	Soil Profile and Moisture Regime	Major	Minor	Silvicultural Notes
Marginal moraine	Moderately steep middle slopes	Loose sandy loam (A and B) over moderately compact loamy fine sand (C); A and B of granular structure	Iron humus podzol; moist	Black spruce and paper birch; or black spruce	Paper birch and *balsam fir* (mostly in the understory)	Trees firm-rooted, suited to selection cutting. Good reproduction under shade.

HYLOCOMIUM-OXALIS (H-O T) Site Class II

Landforms	Topography	Soil Texture, Structure, and Consistence	Soil Profile and Moisture Regime	Major	Minor	Silvicultural Notes
Marginal moraine	Moderately steep middle slopes	Loose rocky sandy loam over compact subsoil; single-grained structure	Iron humus podzol; very moist	*Balsam fir* and white spruce; or *balsam fir*	Paper birch, black spruce, and white spruce	Trees shallow-rooted. Reproduction easy.

CALLIERGON-OXALIS (Cal-O T) Site Class II

Landforms	Topography	Soil Texture, Structure, and Consistence	Soil Profile and Moisture Regime	Major	Minor	Silvicultural Notes
River terrace	Flat	Moderately compact very fine sandy loam over loose subsoil; single-grained structure	Iron humus podzol; very moist	*Balsam fir* and white spruce	Paper birch, white spruce, and black spruce	Trees firm-rooted, suited to selection cutting. Reproduction easy.

HYLOCOMIUM-CORNUS (H-Co T) Site Class II

Landforms	Topography	Soil Texture, Structure, and Consistence	Soil Profile and Moisture Regime	Major	Minor	Silvicultural Notes
Marginal moraine	Gentle lower slopes	Loose stony sandy loam over sand; single-grained structure	Iron humus podzol; very moist	*Balsam fir* and black spruce; or *balsam fir* and white spruce	Paper birch and black spruce	Trees shallow-rooted. Reproduction easy.

HYPNUM-HYLOCOMIUM (H-H T) Site Class II

Landforms	Topography	Soil Texture, Structure, and Consistence	Soil Profile and Moisture Regime	Major	Minor	Silvicultural Notes
Marginal moraine	Gentle lower slopes	Loose sandy loam (A and B) over moderately compact loamy fine sand; single-grained structure	Iron humus podzol; very moist	*Balsam fir* and black spruce or white spruce	Paper birch	Selection cutting possible. Spruce difficult to reproduce.

TABLE 5. Continued

Landforms	Topography	Soil Texture, Structure, and Consistence	Soil Profile and Moisture Regime	Species Composition in Mature Stands Major	Species Composition in Mature Stands Minor	Silvicultural Notes
Lacustrine plain	Gentle lower slopes near lake	Loose loamy sand (A and B) over compact sand (C); single-grained structure	Very moist			

HYPNUM-CORNUS (Hyp-Co T) Site Class II

Landforms	Topography	Soil Texture, Structure, and Consistence	Soil Profile and Moisture Regime	Major	Minor	Silvicultural Notes
Marginal moraine	Gentle lower slopes	Loose A₂ over indurated B sandy loam over slightly compact coarse sand	Normal iron podzol; fresh	Black spruce and *balsam fir*; or black spruce; or white spruce and *balsam fir*	White spruce, paper birch, trembling aspen, and *balsam fir*	Suited to selection cutting. Advance reproduction abundant.
River terrace	Flat	Loose A₂ sandy loam over slightly compact fine sandy loam (B₁) over compact B₂ and C very fine sandy loam; single-grained structure	Iron humus podzol; fresh			
Pitted outwash	Gentle lower slopes	Loose A₂ sandy loam over indurated sandy loam (B) over slightly coarse sand; single-grained structure	Normal iron podzol; fresh			

HYPNUM (Hyp T) Site Class II

Landforms	Topography	Soil Texture, Structure, and Consistence	Soil Profile and Moisture Regime	Major	Minor	Silvicultural Notes
Marine beach	Flat	Sand; B is cemented and C slightly hard; single-grained structure	Normal iron podzol; fresh	*Balsam fir* and black spruce; or black spruce	Paper birch, *balsam fir,* trembling aspen, and jack pine	Trees shallow rooted. Clearcutting favors *balsam fir.*
Level outwash	Flat plain	Loose A₂ and B₁ and moderately compact B₂ sandy loam over deep C very compact sand; single-grained structure	Iron humus podzol; fresh			
Marginal moraine	Moderate upper slopes	Loose A₂ loamy fine sand over B and C compact loam;	Iron humus podzol; fresh			

TABLE 5. Continued

Landforms	Topography	Soil Texture, Structure, and Consistence	Soil Profile and Moisture Regime	Species Composition in Mature Stands		Silvicultural Notes
				Major	Minor	
		single-grained A_1 and B_1; massive B_2 and C				
Ground moraine	Gentle lower slopes near valley bottoms	Hard clay loam; slightly to very compact from surface to subsoil; massive	Iron humus podzol; fresh			

CALLIERGON-LEDUM (Cal-Le T) Site Class III

Landforms	Topography	Soil Texture, Structure, and Consistence	Soil Profile and Moisture Regime	Major	Minor	Silvicultural Notes
Level outwash	Flat	Loose or crumbly silt loam (A_2 and B_2) over moderately compact coarse sand; crumb structure B_2 to single-grained B_3 and C	Somewhat dry	Black spruce and jack pine; or black spruce	*Balsam fir* and jack pine	Shallow-rooted open stands. Advance reproduction entirely black spruce layers. Clearcut.

CALLIERGON-VACCINIUM (Cal-Va T) Site Class II

Landforms	Topography	Soil Texture, Structure, and Consistence	Soil Profile and Moisture Regime	Major	Minor	Silvicultural Notes
River terrace	Gentle lower slopes	Plastic sandy loam (B_2) over loose loamy fine sand (C); single-grained structure	Normal iron podzol; fresh	Black spruce; or black spruce and jack pine	*Balsam fir*	Suited to partial cutting. Advance reproduction present.

KALMIA-VACCINIUM (Ka-Va T) Site Class III

Landforms	Topography	Soil Texture, Structure, and Consistence	Soil Profile and Moisture Regime	Major	Minor	Silvicultural Notes
River terrace	Flat	Loose gravelly coarse sand; single-grained structure	Normal iron podzol; somewhat dry	Jack pine and black spruce; or black spruce	Paper birch, aspen, and *balsam fir* may be present but are negligible	

CLADONIA-LEDUM (Cla-Le T) Site Class III

Landforms	Topography	Soil Texture, Structure, and Consistence	Soil Profile and Moisture Regime	Major	Minor	Silvicultural Notes
River terrace	Flat	Loose sand (A_2) over compact sand (B_2) and coarse sand (C); single-grained structure	Normal iron podzol; somewhat dry	Jack pine and black spruce		No *balsam fir*.

CLADONIA-VACCINIUM (Cla-Va T) Site Class III

Landforms	Topography	Soil Texture, Structure, and Consistence	Soil Profile and Moisture Regime	Major	Minor	Silvicultural Notes
River terrace	Flat	Loose sand (A_2 and B_2) over	Normal iron podzol;	Black spruce and		No *balsam fir*.

TABLE 5. Continued

Landforms	Topography	Soil Texture, Structure, and Consistence	Soil Profile and Moisture Regime	Species Composition in Mature Stands		Silvicultural Notes
				Major	Minor	
		loose gravelly coarse sand; single-grained structure	somewhat dry	jack pine		

HYPNUM-KALMIA (Hyp-Ka T) Site Class III

Landforms	Topography	Soil Texture, Structure, and Consistence	Soil Profile and Moisture Regime	Major	Minor	Silvicultural Notes
Level outwash	Flat, lakeshore	Cemented (B) sandy loam over compact coarse sand (C); single-grained structure	Iron humus podzol; somewhat wet	Black spruce	Jack pine, paper birch, tamarack, and *balsam fir* may occur	Suited best to partial cutting, avoiding large openings. Reproduction difficult. Advance reproduction almost entirely black spruce layers.
Marginal moraine	Gentle middle slope	Hard (B$_1$) sandy loam to loose (B$_2$ and B$_3$) and slightly compact (C) sandy loam; single-grained structure	Iron humus podzol; somewhat wet			
Lacustrine plain	Gentle lower slope	Fine sand (A and B), slightly hard and compact, over coarse compact sand (C); single-grained structure	Iron humus podzol; somewhat wet			

KALMIA-LEDUM (Ka-Le T) Site Class III

Landforms	Topography	Soil Texture, Structure, and Consistence	Soil Profile and Moisture Regime	Major	Minor	Silvicultural Notes
Marginal moraine	Flat	Loose bouldery sandy loam (A$_2$ and B$_1$) over hard (B$_3$) and compact (C) loamy coarse sand; single-grained structure	Somewhat wet	Black spruce	Tamarack and *balsam fir*	Shallow-rooted open stands. Reproduction primarily black spruce layers with little *balsam fir*. Clearcut.
Lacustrine plain	Flat	Plastic (A$_2$) sandy loam over compact (B and C) silt loam; massive	Somewhat wet			

SPHAGNUM-RUBUS (Sph-Ru T) Site Class IV

Landforms	Topography	Soil Texture, Structure, and Consistence	Soil Profile and Moisture Regime	Major	Minor	Silvicultural Notes
Lakeshore	Flat, swamp	Hydro-mor (organic) soil over compact sand	Humus podzol; wet	Black spruce	*Balsam fir*	Shallow-rooted open stands. Clearcut.

TABLE 5. Continued

Landforms	Topography	Soil Texture, Structure, and Consistence	Soil Profile and Mois- ture Regime	Species Composition in Mature Stands		Silvicultural Notes
				Major	Minor	
		SPHAGNUM-LEDUM (Sph-Le T) Site Class IV				
Level outwash	Flat, swamp	Sticky (A$_2$) to hard (B$_2$) sandy loam over compact gravelly coarse sand; massive (A and G) shot (B$_2$)	Humus podzol; wet	Black spruce		No *balsam fir*.

taneously both of the major problems in forest type classification, i.e., (1) that of establishing productivity classes, and (2) that of developing management units which would respond similarly to a given treatment. Spruce-fir site indexes in the area varied from 15 to 55 feet at 50 years of age. The range of site indexes was sub-divided into four 10-foot site classes. Forest site classification allowed the determination of site productivity within class limits in 88 per cent of the cases. The somewhat greater minimum site index (29 ft) on xeric sites as compared with the minimum site index (15 ft) on bog sites was attributed to the absence of low-nutrient sands in the xeric area studied. The classification reaffirms the adaptability of balsam fir to the best (mesic) sites in the Boreal Region. Balsam fir is absent from the nutrient deficient Cladonia-Ledum and Cladonia-Vaccinium forest types.

To illustrate the optimum conditions for balsam fir in the northeastern coniferous section of the Boreal Forest Region, more detailed information is presented with respect to the Dryopteris-Oxalis (Dry-O T) forest type as described by Linteau (1955). The type occurs on the lower slopes of moderately steep marginal or ground moraines. The soil is loose sandy loam over compact sand having single-grained structure forming a moist, iron-humus podzol. The water table lies about 25 inches below the surface and moves downward as the growing season advances. Litter and humus layers are thin. The average depth of the A$_0$ and A$_2$ horizons is 2.2 and 1.8 inches, respectively. The silt-clay content averages 30 per cent in the B-horizon. The pH of the humus layer is 4.90 $\pm$ 0.60. The water-holding capacity of the B and C horizons is 50–60 per cent. The soil nutrient values are: total nitrogen in the humus layer, 1.84 $\pm$ 0.22 per cent; calcium, 13.0 $\pm$ 3.8 m.e. / 100g; exchangeable potassium, 1.02 $\pm$ 0.12 m.e. / 100g; and phosphorus, 51 $\pm$ 10 ppm. As compared with other forest types in the region, the Dryopteris-Oxalis type is characterized by thin A$_0$ and A$_2$ horizons, moderate silt-clay content, high pH, high water-holding capacity, very high proportion of nitrogen, and moderate amounts of other nutrients.

The forest stands reach the highest site index (51 ft at 50 years) of all forest types in the area. At 90 years the heights of dominants are 70 feet, which is lower than the maximum height of 90 feet recorded for trees in Section B-2. Balsam fir is the main species; some paper birch and white spruce may be admixed. Advance reproduction of balsam fir is sparse.

Shrubs are rare, including only *Alnus rugosa, Amelanchier bartramiana, Sambucus pubens, Sorbus americana,* and *Viburnum cassinoides.* Of dwarf shrubs only *Moneses uniflora* is listed. Of the herbs *Oxalis montana* is constantly (80–100 per cent) present, *Cornus canadensis* often (40–60 per cent) present, and *Clintonia borealis, Maianthemum canadense,* and *Solidago macrophylla* seldom (20–40 per cent) present. *Actea rubra, Aralia nudicaulis, Aster acuminatus, A. lowrianus, A. macrophyllus, Coptis groenlandica, Rubus canadensis, Smilacina trifolia, Streptopus roseus, Trientalis borealis,* and *Viola incognita* are rare (1–20 per cent).

Of the pteridophytes *Dryopteris spinulosa* is constantly present; *Dryopteris disjuncta, D. noveboracensis, D. phegopteris, Lycopodium annotinum, L. clavatum,* and *Pteridium latiusculum* are rare.

The mosses *Calliergon schreberi* and *Dicranum scoparium* are often present, *Plagiothecium denticulatum* and *Thuidium delicatulum* are seldom present, while *Dicranum undulatum, Hylocomium splendens, Hypnum crista-castrensis, Sphagnum capillaceum,* and *S. palustre* are rare.

The site, among the best to be found in the region, is not widely distributed, occurring mainly in the Laurentides Park. The trees are firmly rooted and lend themselves well to selection cutting.

Forest succession. Forest succession has been given little study in the Boreal Region. Dansereau and Segadas-Vianna (1952) studied primary bog succession in northern Quebec and Ontario. The arboreal stage of succession starts with tamarack, which is gradually replaced by lowland black spruce, and later by balsam fir or by white-cedar terminating in an upland black spruce climax. The balsam fir subclimax ("quasi-climax" of the authors) can persist for a long time. The final black spruce climax stage was questioned by Oosting (1956), who considered that the bog succession in this region will end with a white spruce–balsam fir climax. Actually this seems to be only one of a number of possibilities. The development depends on the initial potential which can further develop, stabilize, or regress, depending on the systematically acting forces and occasional interferences that result over time. The diagram by Linteau (Fig. 6) indicates the possible forest community changes resulting from variations in drainage conditions and related changes in nutrient status. The prevailing trend is paludification.

Secondary succession has been studied more frequently, with the practical view usually stressed. Although the spruce-fir forest is more stable in the eastern sector of the Boreal Region, even there continued undisturbed conditions do not necessarily lead to a balanced, all-aged selection forest as observed by Baskerville (1960) in the Green River watershed in New Brunswick (Section B-2). Mac-Lean (1960) investigated secondary succession in Sections B-4, B-8, and B-9 in Ontario. Balsam fir was the only species to improve its relative position in slowly disintegrating overmature aspen-birch-spruce-fir stands. In such stands abundant herbaceous vegetation, shrubs, and hardwood sprouts develop, particularly on moist, nutrient-rich soils. In addition to the severe competition afforded by the lush regrowth, balsam fir may be adversely affected by browsing, which increases its susceptibility to disease. The situation is similar after partial cutting. Earlier cutting may result in better conifer reproduction.

Fires create favorable conditions for the reproduction of black spruce, aspen, paper birch, and sometimes for white spruce; but fires are generally detrimental to balsam fir (MacLean, 1960). Balsam fir seed is released in October after the fire season has ended. The following spring fires may destroy the seed present in the forest floor, as well as the seedlings and saplings. Greater resistance to fire damage and a more extended seed-release period help the restoration of spruces, particularly black spruce, in the fire-ravaged areas. The aspen-birch-spruce-fir cover type which develops after fires varies in composition depending on soils and other conditions. Light surface fires do not consume all accumulated organic matter, but stimulate growth of shrubs and hardwood suckers. More severe fires eliminate competition and expose mineral soil which may favor conifer reproduction. Hustich (1949) observed balsam fir spreading after fires and logging in the southern part of the Labrador Peninsula. Similar reports have also been made from other areas. The particular conditions need specific study.

Plochmann (1956) investigated secondary succession in northwestern Alberta and concluded that after disturbance succession leads to a spruce-poplar-fir climax type which has cyclic development. The optimum stage, a many-storied structure of white spruce (about 80 per cent by volume), paper birch, balsam fir, balsam poplar and some aspen, is followed by a decadent one-storied stage of somewhat more spruce and balsam poplar. The reproduction stage consists of alternating grass and shrub, hardwood, and spruce phases. Plochmann's conclusion was strongly criticized by Rowe (1961), who suggested that the white spruce–poplar–fir forest is a product of disturbance in northwestern Alberta. Climatic conditions have greatly restricted the role of balsam fir in this area, with the white spruce stands unable to reproduce without disturbance. Decadent forests do not of themselves become rejuvenated, but rather tend to remain open, unhealthy, ragged, and frequently brush-filled. Rejuvenation of these forests can take place only through such natural disturbances as fire, flood, windfall, or through the cultural activities of man.

Biotic factors play an important role in forest succession, at least locally. The effects of spruce budworm attack have been subjected to much discussion, but reports are contradictory. These are discussed at some length in Chapter 5. European spruce sawfly and birch dieback have contributed to an increased proportion of balsam fir in the forests of the Gaspé Peninsula (Blais, 1961). Deer are destroying balsam fir reproduction on Anticosti Island (Rowe, 1959). Deer and elk have browsed heavily on balsam fir, thus eliminating some of the competition of white spruce in Riding Mountain Park, Manitoba (Rowe, 1955).

THE GREAT LAKES–ST. LAWRENCE FOREST REGION

The discussion will treat the Great Lakes–St. Lawrence Region as outlined by Rowe (1959) in Canada, the western portion of the Northern Forest Region as defined by the Society of American Foresters (1954), and the Great Lakes–St. Lawrence Division of the Hemlock–White Pine–Northern Hardwood Region as interpreted by Braun (1950). A few small locations that lie outside the regional boundaries have also been included.

Forest communities in this region have been investigated rather intensively, primarily from the viewpoint of synecology with emphasis on the analysis of vegetation. Most of the studies have a local character. Systematic studies are available from the Lake Edward Experimental Forest, Quebec, initiated by Heimburger, and from Wisconsin, where forest type classification has been examined by Wilde, and the vegetation continuum has been developed by Curtis.

The diversity of the region is well demonstrated in the study of the vegetation of the rugged Batchawana Bay area (Section L-10) in Ontario by Hosie (1937). Hosie described eleven forest communities belonging to the deciduous, transitional, and northern forest formations, balsam fir being present in all three. The study also described the various types of forest succession and indicated the role played by balsam fir.

It is rather difficult to summarize the large number of local studies yielding some information on balsam fir but primarily having other objectives. For this discussion, the local studies will be grouped into those concerned with dry-mesic communities, mesic communities, and lowland communities. The systematic studies will be discussed first, after a general consideration of climatic and edaphic conditions of the whole region.

Climatic and edaphic conditions. According to the Köppen (1936) classification, the area belongs to the "Dfb" climatic region where: D = humid microthermal climate; temperature of coldest month less than 3C (26.6F); f = precipitation through the year; b = short warm summers, mean temperature above 10C (50F) at least 4 months.

In addition to the information presented in Plate 6, the following details are available: The mean daily temperature ranges from −18C to −4C (0 to 25F) in January, and from 16C to 21C (60 to 70F) in July. The frost-free season extends for 90–180 days. The number of degree days above 6C (42F) varies from 2500 to 4000. Bright sunshine, on the average, lasts for 1800–2200 hours per year. Annual precipitation ranges from 1070 mm (42 in.) in some eastern areas to 510 mm (20 in.) in northwestern Minnesota. The amount of precipitation during the growing season varies from 150 to 380 mm (6 to 15 in.), (U.S. Dep. Agric., 1941; Can. Geogr. Branch, 1957). The values for evapotranspiration according to Thornthwaite (1948) are from 480 to 610 mm (19 to 24 in.) increasing southward (data from Rowe 1939, Thornthwaite and Mather, 1955). In this region precipitation is seen as playing a more important role than temperature in effecting forest boundaries (see Livingston, 1921). In the Boreal Region, temperature is the more important element.

The region is mostly located on the Canadian Shield within the Superior Province of the Laurentian Upland. The province is characterized as a submaturely dissected, recently glaciated peneplain on crystalline rocks of complex structure, Precambrian in origin. Also within the region is the Central Lowlands (Great Lakes–St. Lawrence Lowlands) physiographic province, underlain by limestones and limy shales of Paleozoic origin. The range of balsam fir includes the Eastern Lake, Western Lake, and Wisconsin Driftless sections of the province. The Eastern Lake Section is characterized by maturely dissected glaciated cuestas and lowlands, moraines, plains, and lacustrine lakes. The Western Lake Section is a young glaciated plain with moraines, lakes, and lacustrine deposits. The Wisconsin Driftless Section is a maturely dissected plateau and lowland. It also includes the margins of old eroded drift (Fenneman, 1938; Loomis, 1938; Can. Geogr. Branch, 1957).

Podzol soils interspersed with rock outcrops and peat prevail in sections L-7, L-4a and b, L-9, and L-10. Typical podzols dominate in sections L-5 and L-6. Section L-8 is on gray wooded soils with dark gleysolic soils and peat. Sections L-4b, c, and d have podzols with brown podzolic soils and dark gleysolic soils. Dark gleysolic soils occur with podzol soils in sections L-2 and L-3. Gray wooded soils occupy section L-4e and parts of sections L-10 and L-11 (Rowe, 1959).

In the U.S. portion of the Great Lakes–St. Lawrence Region, true podzols are restricted to the Upper Peninsula and northern part of the Lower Peninsula in Michigan, and to northern Wisconsin. Podzols are rare in northern Minnesota. According to the distribution of spruce-fir types (U.S. Forest Service, 1949) and a soil map prepared by the North Central Agricultural Experiment Stations (1960), balsam fir occurs mostly on rocky and stony land, on shallow peat and muck soils, and on humic-gley soils. Muck soils and humic-gley soils are distributed together with podzols and brown podzolic soils in the bed of glacial Lake Agassiz. The greatest amount of rock outcrop occurs in northeastern Minnesota. Gray wooded soils occur predominantly in central Minnesota and to a lesser extent in Wisconsin and Michigan. The southern range of balsam fir is dominated by gray-brown podzolic soils although the most southern outposts are on humic-gley, organic soils, and regosols.

Forest types of the Lake Edward Experimental Forest. The Lake Edward Experimental Forest in Quebec (Section L-4a) is one of the most intensively investigated areas in the region. The types were studied originally by Heimburger in 1936. Subsequent reports have been published by Sisam (1938a,b), Heimburger (1941), and Ray (1941, 1956). Abbreviated type descriptions are presented below. Detailed vegetation analysis of the principal forest types is illustrated in Figure 7.

The Cornus type (Co T) which, according to Ray (1956), could be called softwood type or red spruce type, occurs on rocky slopes and at the borders of lakes and swamps. The humus layer is a pronounced mor, very acid, and rather low in lime. Compared with the other forest types, the A-horizon has a rather high lime content. In the B-horizon, pockets of hardpan are beginning to form. The soil is an iron podzol. The forest stand is composed chiefly of red spruce, balsam fir, white-cedar, paper birch, red maple, and poorly developed yellow birch. Numerous stumps indicate the former dominance of white pine.

The Oxalis-Cornus type (O-Co T), the softwood-hardwood or spruce–balsam fir–yellow birch type, occupies lower slopes and flats. The soil is developed on sandy glacial till and is deep, sandy, clearly podzolized, fairly well drained, and moist. The humus layer, a moderately acid to very acid mor, is rather low in lime. Balsam fir and yellow birch predominate. Balsam fir reaches its optimum development in this type, which also includes red spruce, paper birch, white-cedar, and red maple. This type very closely resembles the Oxalis-Cornus type of the Adirondacks, the main difference being the much greater abundance of mountain maple and balsam fir and the slightly poorer development of yellow birch in the former.

The Viburnum-Oxalis type (Vi-O T), the hardwood-softwood or spruce–yellow birch–maple type, occupies rather deep layers of glacial till mainly on upper

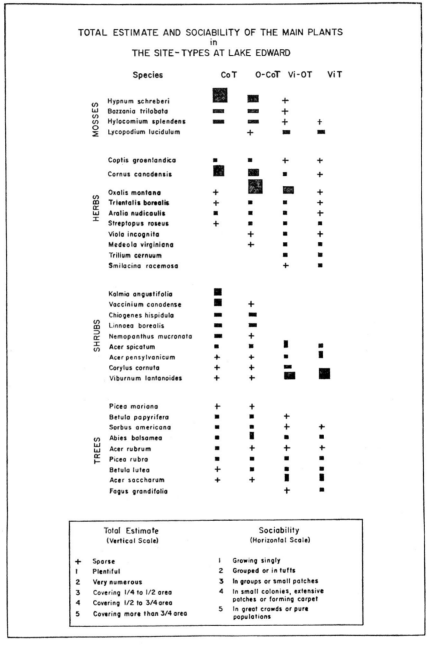

Figure 7. Analysis of vegetation of the principal forest types in the Lake Edward Experimental Forest, Quebec, as presented by Ray (1956), and based upon preceding studies by Heimburger and Sisam. Co T = Cornus type; O-Co T = Oxalis-Cornus type; Vi-O T = Viburnum-Oxalis type; Vi T = Viburnum type. For example, in the Cornus type the total estimate and sociability of *Hypnum schreberi* are both 5; of *Abies balsamea,* both 1. (Courtesy of the Department of Forestry, Canada.)

slopes and some hilltops. The soil profile is intermediate between an iron podzol and a brown podzol soil. The humus layer is usually a mor, but is sometimes a mull, and is moderately to strongly acid and low in lime. In the lowest layer of the B-horizon there is often a clay hardpan. The forest stand is composed chiefly of hardwoods including yellow birch, sugar maple, red maple, and some beech mixed with red spruce and balsam fir. Red spruce is more abundant than balsam fir. Leaf litter is thick. This type very much resembles the Viburnum-Oxalis type of higher elevations in the Adirondacks.

The Viburnum type (Vi T), a hardwood type of maple-beech, is found on hilltops and slopes, on sandy loams much like those on which the Viburnum-Oxalis type is found. The soil profile often approaches that of a brown forest soil. The humus layer, usually a mull but sometimes a mor, is moderately to strongly acid and low in lime. There is no hardpan in the B-horizon. The forest stand is composed largely of good yellow birch, beech, and sugar maple, with an admixture of good quality red spruce. The proportion of red maple, balsam fir, and paper birch is insignificant. Balsam fir is often defective. There is a moderately thick undergrowth of *Viburnum alnifolium* and *Acer pensylvanicum*. *Acer spicatum* is less frequent. The type corresponds to the poorer aspects of the Viburnum type in the Adirondacks.

The Kalmia-Ledum type (K-L T) is a black spruce swamp or black spruce muskeg type with some balsam fir and white pine on the interspersed morainic humps. The soil is sphagnum peat with a layer of decomposed sedge peat underneath and is about 90 cm deep, underlain by coarse lake sand. The ground vegetation is commonly *Kalmia angustifolia* and *Ledum groenlandicum* growing over a mat of *Pleurozium* (*Hypnum, Calliergon*) *schreberi* and *Sphagnum acutifolium*. Other plants occurring most frequently are *Vaccinium myrtilloides*, *Gaultheria hispidula*, and *Carex trisperma*. On gravelly humps are *Cornus canadensis, Coptis groenlandica, Dalibarda repens*, and the mosses *Hylocomium splendens* and *Dicranum undulatum*.

The Sphagnum-Oxalis type (S-O T), another swamp type, is found on shallow alluvial deposits. The soil is a shallow layer of dark, somewhat greasy, mucky peat on silty loam with a swamp profile. The stand is composed mostly of *Alnus rugosa* with occasional scattered balsam fir, white spruce, paper birch, red maple, and *Sorbus americana*. *Sphagnum acutifolium* and *Oxalis montana* are the chief ground-cover species. Included also are *Cornus canadensis, Maianthemum canadense, Clintonia borealis, Dalibarda repens, Aster macrophyllus, Carex trisperma, Thelypteris spinulosa* var. *intermedia, Pyrola secunda*, and *Aster acuminatus*.

Forest types in Wisconsin. Wilde has investigated forest types in the Lake States area since 1932. The most important study, with respect to balsam fir, was made in Wisconsin (Wilde et al., 1949). Balsam fir does not occur in the forest-prairie transition zone, is of little importance on immature soils, but occurs quite generally in forest types in the podzol area. It is most abundant in Clintonia-Lycopodium (hemlock–yellow birch on loam podzols), Vaccinium-Cornus-Rubus (white pine–red pine on gley-podzolic soils), Oxalis-Coptis (white-cedar–white spruce–balsam fir or white-cedar–tamarack on gley-podzol clays or woody peat, respectively) forest types. Balsam fir is of less importance in the Gaultheria-Maianthemum (white pine–red pine on podzolic sands), Smilacina-Polygonatum (sugar maple–basswood–red oak–white pine on podzolic loams), Rubus parviflorus (white pine–white-cedar on strongly podzolized clays), Galium-Equisetum (sugar maple–basswood–elm–yellow birch on podzolized gley loams) and Urtica-Thalictrum (elm–black ash–white-cedar on lacustrine muck or stream bottom soils) forest types.

Plate 5 (lower right diagram) presents an organization of the forest type system outlined by Wilde into an edaphic field. The importance of balsam fir in each type is shown.

The vegetation continuum in Wisconsin. Beginning with Gleason's (1926, 1939) individualistic concept of the plant association, the Wisconsin school, under the leadership of Curtis, developed the concept of the vegetation continuum and new methods for field work and for the analysis of vegetation. The vegetation of the state has been for many years the subject of systematic study, culminating in a major monograph, *The Vegetation of Wisconsin,* by Curtis (1959). The methods developed have also been applied to the analysis of vegetation in other areas (Maycock and Curtis, 1960; Loucks, 1962a; and others).

The basic principle underlying the continuum concept is the orderly and systematic transition of vegetation from one community to another as the ecological factors change. By definition, the continuum principle rejects the concept of discrete plant associations and abrupt changes in the vegetation. However, the Wisconsin summary does describe the major community types and presents keys for their identification, thus in part restricting the concept of continuum developed earlier. Balsam fir is present in six community types. Abbreviated descriptions of these types are presented in Table 6.

The original tabular descriptions include extensive species lists, structural analyses of typical forest stands, and other statistical data about vegetation, information on distribution of community types, climatic and edaphic information, and references to more detailed investigations.

The data are analyzed with respect to a moisture coordinate axis and three-dimensional similarities of species composition. Ecographs or diagrams showing multivariate distribution have been prepared for balsam fir and other major species. The method and materials presented by the Wisconsin school with respect to the important ecological factors will be treated in more detail in Chapter 3.

Dry-mesic communities. Information on balsam fir in dry-mesic communities growing on shallow soils usually near bodies of water has been presented by a number of workers. One of the important early studies was made on Isle Royale, Lake Superior, by Cooper (1913).

The communities investigated were developed on Precambrian rocks. Climax stands were characterized by a thinly-scattered white pine overstory with white spruce, balsam fir, and paper birch forming the main stand. Balsam fir of all age classes was present, though much affected by windfall and fungi. The successful establishment and growth of young white spruce and paper birch were largely dependent on openings in the canopy formed by windfall. An undergrowth of *Taxus canadensis* was prominent. Herbaceous cover was generally sparse except for *Cornus canadensis, Trientalis borealis, Linnea borealis, Mitella nuda, Aralia*

TABLE 6. SOME CHARACTERISTICS OF FOREST COMMUNITIES
WITH BALSAM FIR IN WISCONSIN [*]

Soils Occupied	Importance Values of Balsam Fir and Major Dominants		Constancy of Balsam Fir and Presence of Prevalent Ground Layer	
	Species	Value	Species	Per Cent
NORTHERN MESIC FOREST				
Gray-brown	Abies balsamea	3	Abies balsamea	30
Brown podzolic	Acer saccharum	107	Maianthemum canadense	85
	Tsuga canadensis	79	Polygonatum pubescens	82
	Fagus grandifolia	40	Streptopus roseus	73
	Betula alleghaniensis	29	Aralia nudicaulis	71
NORTHERN DRY-MESIC FOREST				
Podzol	Abies balsamea	6	Abies balsamea	36
Gray-brown podzolic	Pinus strobus	75	Maianthemum canadense	100
	Acer rubrum	37	Aralia nudicaulis	98
	Quercus borealis	36	Aster macrophyllus	90
	Betula papyrifera	30	Pteridium aquilinum	88
NORTHERN DRY FOREST				
Podzol	Abies balsamea	1	Abies balsamea	11
Gray-brown podzolic	Pinus banksiana	65	Vaccinium angustifolium	92
	Pinus resinosa	48	Maianthemum canadense	89
	Pinus strobus	43	Pteridium aquilinum	87
	Quercus ellipsoidalis	37	Rubus pubescens	82
NORTHERN WET FOREST				
Azonal peat	Abies balsamea	24	Abies balsamea	31
	Picea mariana	139	Carex trisperma	94
	Larix laricina	56	Ledum groenlandicum	90
	Thuja occidentalis	45	Smilacina trifolia	85
	Pinus banksiana	14	Gaultheria hispidula	83
NORTHERN WET-MESIC FOREST				
Azonal peat	Abies balsamea	45	Abies balsamea	95
	Thuja occidentalis	91	Aralia nudicaulis	96
	Tsuga canadensis	40	Maianthemum canadense	96
	Betula alleghaniensis	34	Dryopteris austriaca	92
	Fraxinus nigra	27	Rubus pubescens	92
BOREAL FOREST				
Gray wooded	Abies balsamea	69	Abies balsamea	100
Podzol	Pinus strobus	34	Galium triflorum	100
	Thuja occidentalis	32	Maianthemum canadense	100
	Betula papyrifera	26	Cornus canadensis	97
	Picea glauca	25	Lonicera canadensis	97

[*] Importance values are the sums of relative frequency, relative density, and relative dominance of individual species. Relative values of a species frequency, density, and dominance are expressed as percentages of the total frequency, density (number of individuals), and dominance (coverage) of all the species in a community. Species presence is the percentage of occurrence of individual species in a number of stands, each whole stand being considered as a separate sample unit. Species constancy is considered here as the percentage of occurrence of individual species in a number of stands based on measured samples of 80 trees per stand.

Source: after Curtis, 1959.

nudicaulis, and *Coptis groenlandica* in the openings. The most important mosses were *Pleurozium schreberi, Hylocomium proliferum, Hypnum crista-castrensis*, and *Hylocomium triquetrum*.

Balsam fir was found growing vigorously on shallow soils formed on rocks containing anorthosite and diabase on Beaver Island in Lake Superior by Lakela (1948). Associated species were *Pinus strobus, Picea glauca, P. mariana, Thuja occidentalis, Betula papyrifera, Populus deltoides, Alnus crispa, Cornus stolonifera, C. baileyi, Amelanchier sanguinea*, and *Sambucus pubens*. On the more exposed cliffs, the following shrubs were present: *Rosa acicularis, Diervilla lonicera, Spirea alba, Rubus strigosus, Ribes* sp., *Vaccinium* sp., *Arctostaphylos uva-ursi., Ledum groenlandicum, Chamaedaphne calyculata, Potentilla fruticosa*, and *Salix bebbiana*. Herbs included *Lycopodium annotinum, Cornus canadensis, Viola incognita, Dryopteris spinulosa*, and others.

Waring (1959) investigated dry-mesic communities in the Precambrian rock outcrop area of northeastern Minnesota. He recognized three forest types with increasing soil depth, moisture, and nutrient regimes. Balsam fir was present on all three types, but did not succeed in developing a distinct understory in the most xeric type. There it was more frequently accompanied by black spruce, while white spruce increased with mesification.

Balsam fir is also an important component of forest communities growing on shallow soils underlain by limestone on Mackinac Island in Lake Michigan (Potzger, 1941). Associated species were white spruce, white-cedar, paper birch, white pine, and, to a lesser extent, sugar maple, beech, red oak, and ironwood.

Balsam fir and white spruce form impressive stands on limestone outcrops in Wisconsin's Door County Peninsula, and in Bruce Peninsula, Ontario (Maycock and Curtis, 1960). Potzger and Wales (1950) described spruce and balsam fir growing on shallow soils in the Quetico-Superior area.

Dry-mesic communities with sufficient moisture for balsam fir and species of similar requirements can be also found on the shallow soils of talus slopes. In the southern parts of the balsam fir range they need to be north-facing and have access to spring water.

The Apple River Canyon, St. Croix County, Wisconsin, is a meeting ground of over 90 extraneous species derived from floras centered in all directions away from the area, and reaching here the boundaries of their ranges. Balsam fir occurs on the cool, moist talus slopes on the north-facing side of the canyon, where it is associated with *Pinus strobus, Betula papyrifera, Tilia americana, Populus grandidentata, Ulmus americana, Thuja occidentalis, Juglans cinerea, Ostrya virginiana, Betula alleghaniensis, Acer rubrum, A. negundo, Taxus canadensis*, and a large number of other woody and herbaceous species (Russell, 1953).

The outlying balsam fir communities in Winneshiek and Allamakee counties in the driftless area of northeastern Iowa are located on north-facing slopes of exposed Galena-Platteville limestones which are noted for their sinks, caves, and big springs. Balsam fir is accompanied by white pine, *Taxus canadensis*, and a number of other plants normally growing in more northern areas (Conard, 1939). At Wykoff, Fillmore County, southeastern Minnesota, the situation is similar. Balsam fir is associated with such northerly species as *Rubus pubescens, Rhamnus*

alnifolia, Cornus canadensis, Pyrola secunda, P. virens, and *Mertensia paniculata.*
At this locality balsam fir maintains itself by layering (Moyle, 1946).

Specific conditions of moisture supply and the cooling effect of north-facing
slopes have contributed to maintain balsam fir and ecologically related species in
these southerly outlying areas. Further north and also within the range of balsam
fir similar topographic situations may include niches of arctic or alpine conditions
along with the characteristic vegetation of such areas. Such conditions were
described by Butters and Abbe (1953) at the shore of Lake Superior in Cook
County, Minnesota, and by Soper and Maycock (1963) at Old Woman Bay,
Algoma District (48°47′N, 84°54′W) on the east shore of Lake Superior in Ontario. According to these authors, groups of arctic and alpine plants have been
reported from all parts of the Lake Superior shore and on a number of islands
in the lake. The arctic effect is created by the north and northeast exposures of
rocky slopes aided by the cooling effects of ponds and sprays from the lake. The
rock-barren tundra described by Soper and Maycock (1963) at Old Woman Bay
consists of a characteristic cover of a brilliant orange-colored lichen, *Caloplaca
elegans* (Link.) Th. Fr., a thick carpet of mosses, a sparse cover of vascular plants
numbering about 40 species, and a few small trees and saplings in sheltered places.
The surrounding boreal forest consists of *Thuja occidentalis, Picea glauca, Betula
papyrifera, Pyrus decora,* and *Abies balsamea.* Tree saplings and a few *Betula alleghaniensis* are present, and there is an abundance of *Alnus crispa.*

Although occurrences of balsam fir on shallow soils in dry-mesic communities
as described above are not typical for the species, they do offer some interesting
opportunities for research and may have some specific practical value.

Mesic communities. As indicated in Plate 5 (lower left diagram), the adaptability of balsam fir to mesic sites is more limited in the Great Lakes–St. Lawrence
Region than in the Boreal and to some extent the Acadian–Appalachian regions.
Under the more southerly climatic conditions balsam fir is unable to compete
with tolerant hardwood species for the more nutrient enriched sites.

The presence of balsam fir in mesic communities described by Gates (1926),
Kittredge (1938), and Martin (1959) is indicated in the diagrams in Figure 5
and will be discussed together with additional material in the section dealing
with succession.

In addition to systematic studies of the mesic communities described above,
brief mention should be made of two other studies. Lee (1924) and Kell (1938)
described, from the polyclimax viewpoint, spruce-fir-birch climax communities
occurring on rather coarse-textured soils but having adequate water supply at
Itasca Park, Minnesota. Besides balsam fir, white spruce, and paper birch, the
associated species include aspen, red pine, white pine, ash, red maple, and
American elm. Pines were not able to reproduce in undisturbed conditions. Kell
(1938) considered that spruce and balsam fir could not form intermingling associations with maple and basswood.

Grant (1934) described balsam fir–basswood communities on heavy clay soils
as the local climatic climax community in Itasca County, Minnesota, about 100
miles east of Itasca Park. Associated trees included *Abies balsamea, Quercus
borealis* var. *maxima, Acer saccharum, Betula alleghaniensis, Picea glauca, Ostrya*

virginiana, and *Betula papyrifera.* Shrubs present were *Corylus cornuta, Cornus alternifolia, Salix bebbiana, Prunus nigra, Alnus rugosa, Acer spicatum, Crataegus rotundifolia, Amelanchier bartramiana, Ribes triste, Viburnum trilobum, Dirca palustris,* and *Rubus pubescens.* The pteridophytes found were *Pteridium aquilinum, Osmunda claytoniana, Athyrium filix-femina,* and *Lycopodium annotinum.* This community, however, cannot be considered as a climatic climax for the area, since sugar maple–basswood can attain much better development on rich but well-aerated soils, and balsam fir does not succeed under those conditions (Bakuzis et al., 1962).

Lowland communities. There is a large variety of lowland communities in which balsam fir is an associated species. Such communities have been described by Stallard (1929), Kell (1938), Conway (1949), Buell and Niering (1953), Clausen (1957), Christensen et al. (1959), Hartley (1960), and others. A few examples taken from these studies will be examined more closely.

Kell (1938) described a black ash–American elm–balsam fir community, which she regarded as a climax association on peat at Itasca State Park. Other tree species in this community were *Betula papyrifera, Populus tremuloides, P. balsamifera, Quercus macrocarpa,* and *Picea mariana.* Balsam fir was the third ranking species, paper birch was unimportant, and green ash and red maple appeared occasionally. There were no indications that upland species were invading the peat association.

Buell and Niering (1953) also studied the plant communities at Itasca State Park. In a sloping bog with areas of different moisture content the higher better-drained area was dominated by balsam fir about 50 feet in height. Associated with the fir in order of their importance as mature trees were black and white spruces, the latter up to 145 years old and averaging 60 feet in height, and paper birch and tamarack. Reproduction of both fir and spruce was present. Principal shrubs were *Acer spicatum, Alnus rugosa,* and *Cornus stolonifera.* The main herbs were *Aralia nudicaulis, Mitella nuda, Cornus canadensis, Rubus pubescens, Thalictrum dioicum, Caltha palustris,* and 67 minor species. Principal mosses were *Mnium* sp., *Thuidium delicatulum, Sphagnum* sp., *Pleurozium schreberi,* and *Hypnum crista-castrensis.*

The tamarack-dominated lower and wetter zone contained an admixture of balsam fir, black spruce, and white spruce. The tree and shrub canopy was open. Important shrubs were *Alnus rugosa, Cornus stolonifera, Acer spicatum, Rhamnus alnifolia, Ribes americanum, R. triste,* and *Prunus virginiana.* Important herbs were *Aralia nudicaulis, Caltha palustris, Mitella nuda, Galium triflorum, Thalictrum dioicum, Rubus pubescens, Cornus canadensis, Calamagrostis canadensis, Impatiens capensis, Viola pallens,* and *Carex* sp.

Hartley (1960) described outlying bog communities at Tamarack Creek, Trempealeau County, and at the LaCrosse River, in the Wisconsin Driftless area. Although balsam fir was reported, the bogs were occupied chiefly by three species of trees: *Larix laricina, Betula alleghaniensis,* and *Acer rubrum,* 11 shrubs, 5 ferns, and over 20 species of herbs. The list indicates a much richer flora than is found in tamarack bogs farther to the north. This community is in spectacular contrast to one only 50 miles southwest where balsam fir and other boreal species occur on north-facing talus slopes.

Forest succession. Succession has been investigated intensively and for a long time in the Great Lakes–St. Lawrence Region. As discussed above, the primary succession diagrams actually are models of the regional edaphic fields. Figures 5, 6, and Plate 5 show the place of balsam fir in the "primary succession" in different forest regions.

One of the earliest outlines of primary succession in the region was prepared by Cooper (1913) for Isle Royale, Lake Superior. Succession terminates in a balsam fir–paper birch–white spruce climax association that also contains scattered white pine. Bergman and Stallard (1916) and Stallard (1929) considered white pine–red pine as the climax association of northern Minnesota, although the presence of maple and oak communities as representatives of beech-maple and oak-hickory climax associations was also recognized. Grant (1929, 1934) investigated succession in central Minnesota and concluded that the balsam fir–basswood community is the local climax association. As indicated in Plate 5 the basswood–sugar maple association without balsam fir occupies the more mesic sites in that area. The combination of sugar maple–basswood and balsam fir–spruce–birch communities in a single transitional climax association at Itasca Park, Minnesota was rejected by Kell (1938). Buell and associates (Buell and Gordon, 1945; Buell, 1956; Buell and Martin, 1961) also confirmed the presence of separate communities existing side by side in this area. More intensive soil studies in the area in the future may provide for an explanation of the existence of different communities side by side on seemingly similar sites.

The primary succession diagram (Fig. 5) by Kittredge (1938) was prepared for an aspen study in northern Minnesota and Wisconsin. For some reason it does not show balsam fir in wet situations, where it occurs rather frequently in that area. Kittredge recognized that the region investigated is transitional between maple-basswood and spruce-fir climaxes.

Primary succession in the transition zone between the northern coniferous and central deciduous forests was studied by Gates (1926) at Douglas Lake, Cheboygan County, Michigan. The suggested pattern of succession is shown as diagram 2 in Figure 5.

Balsam fir appears only in the hydrosere in this diagram. The strong trend toward mesification and soil enrichment shown in the vertical axis is the dominating theme of this and many other "primary succession" diagrams.

Martin (1959) prepared a primary succession diagram (Fig. 5) for Algonquin Park, Ontario (Section L-4b, Plate 2). The diagram points out the poorly defined relationship existing between balsam fir and hemlock. In the climax stands, hemlock accounts for 72 per cent and balsam fir for only 4 per cent of the basal area. However, there was very little hemlock reproduction anywhere, while balsam fir was reproducing abundantly in openings.

Martin's diagram differs from the other primary succession schemes as it shows such pioneering species as *Populus* sp., *Betula papyrifera*, and to some extent also *Acer rubrum* at early stages of the succession at intermediate moisture conditions. On the primary succession scale (or edaphic coordinates) this represents a short period in the development of new surfaces, which should not be confused with other situations where these species occur frequently after secondary disturbances.

As indicated by a study of the development of vegetation on spoil banks in central Minnesota (Leisman, 1957), balsam fir may appear together with birch and aspen and some jack pine and red pine on lean spoil banks 50 years after deposition.

Darlington (1931) described the climax hemlock community in the Porcupine Mountains of Michigan. Associated species included sugar maple, yellow birch, basswood, white-cedar, white pine, balsam fir, and white spruce. Mature balsam fir was rare and occurred only in openings. Seedlings, though abundant, were not able to compete successfully with the seedlings of more tolerant species.

Large openings in mature stands are favorable for balsam fir reproduction in competition with more tolerant species, as observed in the Lake Edward Experimental Forest (Ray, 1956). When balsam fir is associated with less tolerant species such as white pine, red pine, white spruce, and paper birch, the latter benefit from larger openings, especially those resulting from uprooted trees. Decadent balsam fir is frequently blown down without affecting the pine or birch overstory. This provides a chance for the subclimax species to persist in the community.

In the absence of disturbance the spruce-fir communities, under certain conditions, are invaded by tolerant hardwoods in the Keweenaw Peninsula, northern Michigan (Maycock, 1961). On the other hand, after severe disturbance such as fire or logging, spruce-fir may occur as a pioneer community on wet soils having warm ecoclimates in central Ontario (Section L-4d), as shown by Hills and Brown (1955). On such sites without disturbance, black ash and red maple, sometimes admixed with elm and yellow birch, prevail. A reappearance of the balsam fir understory after fires was described by MacLean (1949) and Jarvis (1960) in the Goulais watershed area of Ontario (Section L-10). In general, secondary succession following fire recapitulates certain phases attributed to primary succession. These phases may last a shorter or longer time depending on local conditions. Grant (1929) investigated a burn succession in central Minnesota. Balsam fir was present in limited quantity (about 1 per cent by number) in the pioneer stages of all three succession series: the spruce swamp, the hardwood, and the pine.

Fires destroyed large areas of forests in the Great Lakes region even before the beginning of extensive logging operations in the area (Graham, 1941; Brown and Curtis, 1952; Potzger and Wales, 1950). That logging operations, often followed by fire, have drastically changed the natural vegetation over vast areas has been pointed out repeatedly by many authors. Kittredge and Gevorkiantz (1929) estimated that in the Lake States alone, 21 million acres of aspen forest could be attributed to these disturbances. Heinselman (1954) estimated that 1,430,000 acres of aspen cover type would revert to spruce-fir within one rotation.

Fires have assisted in the spread of jack pine onto mesic sites which are later invaded by balsam fir reproduction (Stott et al., 1942; Lake States For. Exp. Sta., 1942). Hansen (1946) found that jack pine stands in Minnesota are readily invaded by red pine and balsam fir and to a lesser degree by paper birch, white pine, and other species. On good sites balsam fir invasion is handicapped by hardwoods.

Plate 8 illustrates the competitive vigor of balsam fir in northeastern and north-central Minnesota. In the northeastern counties the growth of balsam fir relative to its present growing stock far exceeds that of other species. Toward the periphery of its range in central Minnesota, the growth rate of advance reproduction is more in balance with the amount of growing stock in merchantable or mature stands.

THE ACADIAN-APPALACHIAN FOREST REGION

This region is a provisional combination of the Acadian Forest Region of Canada (Rowe, 1959), and the Northern Appalachian Highland Division of the Hemlock–White Pine–Northern Hardwoods Region (Braun, 1950) in the United States. Placing the boundary of the Great Lakes–St. Lawrence Region along the International Border is not a satisfactory solution. According to Braun (1950), sections L-5 and L-6 of Halliday (1937), should be included in the Appalachian Highland Division. The description given by Rowe (1959) indicates that sections L-4a and L-5 (see Plate 2) are closely related to the Acadian Region in Canada. The combination of Appalachian and Acadian areas into one region as given here is primarily one of convenience in discussing community and geographical relationships in balsam fir.

Some of the oldest geographical and synecological studies in the region were carried out by Graves (1899), who developed approaches in New York that were somewhat similar to those of his European contemporaries, particularly Morozov and Cajander. Hawley and Hawes began their investigations in 1912. Important studies of the vegetation in this area were made by Bray (1930), Heimburger (1934), and Westveld (1952, 1953), Westveld, chmn. (1956). More recently a number of Canadian authors have been active, particularly Halliday (1937), Place (1955), and Loucks (1956, 1962b). The spruce-fir types were the first to attract the interest of investigators and have been the most thoroughly investigated.

Climatic and edaphic conditions. According to the Köppen (1936) classification, the area lies within the same broad climatic zone as the Great Lakes–St. Lawrence Region. The Dfb classification is discussed above in connection with that region. The southernmost localities where balsam fir is present are located in the Dfb Climatic Region. This region is described as a mesothermal humid climate; the temperature of the warmest month is below 22C (71.6F).

The mean daily temperature ranges from −12 to −4C (10 to 25F) in January and from 13 to 21C (55 to 70F) in July. The frost-free season varies from 90 to 180 days. Mean annual precipitation ranges from 890 to 1400 mm (35 to 55 in.), and warm season precipitation from 380 to 640 mm (15 to 25 in.), U.S. Dep. Agric. (1941), Can. Geogr. Branch (1957). The values for evapotranspiration according to Thornthwaite (1948) are from 510 to 560 mm (20 to 22 in.), increasing southward. Precipitation is not limiting to forest distribution and is generally favorable to high forest productivity. Temperature is limiting to forest distribution in some small alpine areas and in restricting growth in adjoining subalpine zones.

Some characteristics of a subalpine zone appear starting at 460 m (1500 ft) elevation on Cape Breton Island (Loucks, 1962b) in the northern section A-6 (Rowe, 1959). Alpine regions occur on Mt. Katahdin in Maine, the Green Mountains in Vermont, the White Mountains in New Hampshire, and the Adirondacks in New York (Antevs, 1932).

The area belongs to the major physiographic division of the Appalachian Highlands. The bedrock consists of Paleozoic sedimentaries, mainly Carboniferous, with some older rocks, mainly of Cambrian and Silurian age. The sedimentary rocks are frequently interrupted by Precambrian intrusives, mainly granites and gneisses. Much of the area is above 610 m (2000 ft) in elevation. The Catskills and Green Mountains rise above 1200 m (4000 ft), the Adirondacks above 1500 m (5000 ft), and the White Mountains above 1800 m (6000 ft). The area where balsam fir occurs was entirely glaciated except for a southern extension of its range in Pennsylvania (Fenneman, 1938; Loomis, 1938; Can. Geogr. Branch, 1957; and other sources).

Developed podzol soils occur in the northern part of the region and in the Adirondacks. Brown podzolic soils occur in the southern parts of New England but are rare in New York. Most of the soils south of the Adirondacks in New York are weakly podzolized, being predominantly gray-brown podzolics (U.S. Dep. Agric., 1938; Cline, 1953; Can. Geogr. Branch, 1957).

Early studies of spruce-fir types. Graves (1899) studied the spruce-fir types in the Adirondacks. Although he recognized the importance of detailed description, Graves considered only four major forest types, i.e., swampland, spruce flats, hardwood land, and spruce slopes. Hawley and Hawes (1912) added two temporary types, birch-poplar, and old-field spruce, both of which would revert to one of the permanent types if left undisturbed.

Descriptions of spruce-fir types by Zon (1914), Westveld (1931), and Donahue (1940) agree substantially with those of the earlier workers. As representative of these studies, Zon's description (1914) is taken as the basis for the treatment given here.

The swamp type (Zon, 1914) or the spruce swamp type (Westveld, 1931; Donahue, 1940) occurs on low, poorly drained soils. Balsam fir often forms pure stands but more commonly is mixed with black spruce, red spruce, white-cedar, and tamarack. Balsam fir stands are characteristically composed of small, slender trees that are relatively free of diseases, especially butt rot, and climatic injuries. Root growth begins on swampland about five weeks later than on slopes or dry flats. Other associated species, according to Westveld (1931) are hemlock, white pine, black ash, red maple, paper birch, and yellow birch. About 70 per cent of the herbs are mosses, 10 per cent are ferns and fern allies, and 20 per cent are flowering plants (Zon, 1914).

The flat or spruce flat type is intermediate with respect to moisture condition between the swamp and hardwood types. The spruce flat type occurs on the low swells adjoining wet swamps, on the gentle lower ridges, and on knolls. It is fairly well drained. The fern moss (*Hylocomium proliferum*) is the principal ground cover species. Balsam fir grows rapidly, attains a moderately large diameter and height, and has a straight, clear bole. The tree is subject to butt rot. Balsam fir frequently forms pure stands. In mixed stands the common associated species are red spruce, yellow birch, and red maple, the latter two of small size and relatively unimportant. Westveld (1931) also listed paper birch, hemlock, white-cedar, and scattered white spruce and white pine as additional associates. The upper limit of the spruce flat type is marked by the appearance of beech and sugar maple. According to Zon (1914), mosses form 60 per cent of the ground cover while ferns and flowering plants account for 5 per cent and 20 per cent, respectively. The type is of considerable commercial importance and occurs extensively on level areas.

The hardwood slope or spruce-hardwood type is well drained. The soils frequently contain large quantities of rock. Balsam fir never occurs in pure stands, being mixed with red spruce and hardwoods. Of the latter, the most important are yellow birch, red maple, sugar maple, and beech. Westveld (1931) also listed white pine and hemlock as occurring occasionally. He subdivided the hardwood slope type into two subtypes, the yellow birch–spruce subtype border-ing the spruce flat, and the sugar maple–spruce subtype bordering the northern hardwood type. Red spruce is well developed but less abundant than in the other types. Balsam fir, unless shaded severely, reaches its best development in this type. The species is particularly susceptible to butt rot and large trees are often culls. Hardwood litter predominates on the forest floor. Mosses cover 5 per cent of the ground surface. Ferns and fern allies are common, covering 30 per cent of the ground. Flowering plants account for 25 per cent of the ground cover. The type is of high commercial value and covers large areas in the region.

The mountain-top or spruce slope type occurs at elevations above 700–920 m (2500–3000 ft). These slopes are usually steep, rocky, and thin-soiled. Balsam fir often forms pure stands, but over-all spruce predominates (Westveld, 1931). Yellow birch and paper birch are also present. The chief ground cover is com-posed of *Sphagnum* mosses. Timber yields are fairly high except near the upper limits. At timberline balsam firs and black spruce form dwarf communities, and are the last tree species to persist before giving way to alpine vegetation.

The New England classification of spruce-fir types. In 1953 Westveld published a new classification of the spruce-fir types of eastern North America. He reclas-sified the forest cover types of the Society of American Foresters (1940) into three major groups: the spruce-fir group, the spruce-fir-hardwoods group, and the hardwoods-spruce-fir group, emphasizing the difference between temporary and permanent (climax) forest types. Westveld recognized red spruce, white spruce, and black spruce phytogeographic zones and described the ground vegetation equivalents for each type.

The zonation was further developed by Westveld, chmn. (1956). The New England states were divided into six vegetation zones: Spruce–Fir–Northern Hard-woods, Northern Hardwoods–Hemlock–White Pine, Transition Hardwoods–White Pine–Hemlock, Central Hardwoods–Hemlock–White Pine, Central Hardwoods–Hemlock, and Pitch Pine–Oak zones. The first four are in the New England Section (9G; Braun, 1950) and the latter two in the Glaciated Section (4E) of the Oak-Chestnut Region. Balsam fir plays an important role in the first zone, which is further subdivided into two intermingled subzones: Spruce–Fir–Intolerant Hard-woods and Tolerant Hardwoods–Spruce–Fir. Westveld, chmn. (1956) listed the forest cover types recognized by the Society of American Foresters (1954) according to their successional status in each zone. The same cover type could appear in different natural vegetation zones and subzones.

Forest types of the Adirondacks. Heimburger (1934) recognized three series of forest types in his studies in the Adirondacks. The series represented climatic variations and to some extent differences in moisture conditions.

The subalpine series includes the upper mountain slopes, usually above 980 m (3200 ft), where the forest is composed largely of red spruce in its lower reaches, and of balsam fir with an admixture of paper birch, mountain ash, and occasionally white spruce and stunted yellow birch at higher elevations.

The western series occupies the major part of the Adirondacks on a variety of sites ranging from poor, sandy, gravelly, outwash plains to fertile deep glacial till on lower mountain slopes. The elevation ranges from 450 to 980 m (1500 to 3200 ft) above sea level.

The eastern series occupies valleys of the extreme eastern part of the Adirondacks. The climate is drier and more continental. The soils contain more soluble lime. Soils vary from gravelly sands of outwash plains to fine silty lake sands and glacial till. They are less podzolized than in the corresponding types of the western series, and the humus is much thinner. The elevation ranges from 240 to 750 m (800 to 2500 ft) above sea level. The forest usually consists of pines and hemlock. Many plants of the western series are lacking, or if present, occupy different sites. Abbreviated descriptions of the forest types are presented in Table 7. In addition to these, Heimburger also prepared a diagram (see diagram 1 of Figure 6) relating the forest types in a genetic classification following the principles introduced by Sukachev. Heimburger (1934) described 22 forest types in the Adirondacks and noted a transitional type, Oxalis-Dicranum (O-D T), which is more widely distributed in the White Mountains of New Hampshire.

Studies of alpine vegetation. Except for the systematic study of subalpine forest

TABLE 7. FOREST TYPES OF THE ADIRONDACKS, NEW YORK

Type Name and Abbreviation	Parent Material	Soil Profile	Humus	Tree Species
SUBALPINE SERIES				
Hylocomium-Cornus H-Co T	Gravelly sand or bedrock at high elevations	Strong podzol, poorly drained	Mor, very acid* low lime†	*Balsam fir*, black spruce, and mountain paper birch
Oxalis-Hylocomium O-H T	Usually glacial till	Humus podzol, sometimes tending toward brown forest soils	Mor or "alpine humus," or occasionally fibrous mor. Mod. to extr. acid and low in lime	*Balsam fir*, red spruce, mountain ash, and mountain paper birch

*Slightly acid indicates pH greater than 5.5; moderately acid – pH 4.5–5.5; very acid – pH 3.5–4.5; extremely acid – pH below 3.5.

† Lime content below 0.5% is considered very low; 0.5–1.0% lime content is low.

Source: after Heimburger, 1934.

TABLE 7. Continued

Type Name and Abbreviation	Parent Material	Soil Profile	Humus	Tree Species
Aspidium-Oxalis Asp-O T	Thin glacial till on bedrock, at foot of some steep slopes	In places with good drainage and on wet rich soils with seeping ground water. Frequently more a brown forest soil than a podzol	Mor or alpine humus. Mod. to very acid and low in lime	*Balsam fir* (serious rot), paper birch, red spruce, mountain ash, and yellow birch

WESTERN SERIES (SOFTWOOD AND MIXED-WOOD TYPES)

Type Name and Abbreviation	Parent Material	Soil Profile	Humus	Tree Species
Vaccinium-Gaultheria V-G T	Gravelly outwash plains, eskers, and alluvial sands	Iron podzol, often with a hardpan	Mor, mod. to very acid and low in lime	Red pine, white pine, paper birch, aspen, black spruce, tamarack, and scattered *balsam fir* and red spruce
Cornus-Maianthemum Co-Ma T	Gravelly outwash plains and alluvial sands (better than V-G T)	Iron podzol, sometimes with a hardpan	Mor, mod. to very acid and low in lime	Red pine, white pine, *balsam fir*, red spruce, paper birch, aspen, and red maple
Hylocomium H T	Better-drained lake sands, flat deltas, and old flat beach lines with a boulder pavement	Iron podzol, often with a hardpan	Thick mor, very acid and with low lime	Black spruce, red spruce, white pine, tamarack, *balsam fir*, paper birch, yellow birch, aspen, and red maple
Bazania-Oxalis Ba-O T	Eskers, ledges, or escarpments, somewhat drier than surrounding area	Podzol with a rusty and somewhat compact B-horizon	Mor, very to extr. acid and low in lime	Hemlock, red spruce, *balsam fir*, red maple, white pine, and occasional poor yellow birch and beech
Oxalis-Cornus O-Co T	Gravelly upper margins of flats, swamp margins, poorly-drained till, low deltas, dry till and gravelly escarpments	Humus or iron podzol with occasional hardpan	Mor, mod. to extr. acid and low in lime	Red spruce, yellow birch, *balsam fir*, red maple, paper birch, beech, and hemlock; occasionally a white pine overstory
Viburnum-Oxalis Vi-O T	Unmodified glacial till	Podzolic soil with occasional claypan	Mor, mod. to extr. acid, and with more lime than O-Co T	Red spruce, sugar maple, beech, yellow birch, red maple, *balsam fir*, white pine, and hemlock

WESTERN SERIES (HARDWOOD TYPES)

Type Name and Abbreviation	Parent Material	Soil Profile	Humus	Tree Species
Viburnum Vi T	Glacial till. Drier than Vi-O T	Weakly podzolized brown forest soil with occasional claypan	Mor to mull, mod. to very acid, with more lime than Vi-O T	Beech, sugar maple, yellow birch, red spruce, hemlock, and occasional *balsam fir*

TABLE 7. Continued

Type Name and Abbreviation	Parent Material	Soil Profile	Humus	Tree Species
Oakesia Oa T	Grenville limestone outcrops	Brown forest soil	Mor to mull, weakly to mod. acid, with more lime than Vi T	Beech, sugar maple, white ash, white elm, bass-wood, red spruce, and yellow birch; *balsam fir* not mentioned
Arisaema Aris T	Well-drained sandy glacial till or talus	Brown forest soil	Mull, very acid to neutral, with lime variable	Sugar maple, beech, white ash, white elm, bass-wood, black cherry, yellow birch, and a few small rotten red spruce, with scattered *balsam fir* and hornbeam in the understory

EASTERN SERIES (SOFTWOOD AND MIXED-WOOD TYPES)

Type Name and Abbreviation	Parent Material	Soil Profile	Humus	Tree Species
Vaccinium-Myrica V-M T	Coarse outwash sands	Iron podzol with no hardpan	Thin mor, mod. acid and with lowest lime in the series	Red pine, white pine, red oak, aspen, *balsam fir*, paper birch, red maple, white spruce, pitch pine, and jack pine
Cornus-Pyrola Co-Py T	Coarse silty sand	Podzolic soils	Mor, weakly to mod. acid, with lime above avg. for the series	White pine, *balsam fir*, red pine, paper birch, red spruce, red maple, beech, and sugar maple
Maianthemum-Carex Ma-Ca T	Lake sand and glacial till	Podzolic soils	Mor, mod. to very acid, with low lime	White pine, hemlock, red spruce, *balsam fir*, red maple, paper birch, and red oak
Aster-Gaultheria As-G T	Dry rocky glacial till	Podzolic soils	Mor, weakly to mod. acid, with lime above avg. for the series	White pine, beech, aspen, paper birch, and red maple; *balsam fir* not mentioned
Aster-Corylus As-Cor T	Well-drained lake sands, shallow glacial till and lime-stone outcrops	Podzolic soils and podzols	Mor, weakly to mod. acid, with lime avg. for the series	White pine, sugar maple, aspen, paper birch, *balsam fir*, red maple, and black cherry

EASTERN SERIES (HARDWOOD TYPES)

Type Name and Abbreviation	Parent Material	Soil Profile	Humus	Tree Species
Aster-Aralia As-Ar T	Shallow glacial till	Brown forest soil	Mull, weakly to mod. acid, with lime avg. for the series	Sugar maple, aspen, red maple, striped maple, paper birch, and *balsam fir*
Maianthemum-Viburnum	Glacial till	Weakly podzolic soils	Mor, nearly a mull, weakly to	White pine, beech, hemlock, sugar

TABLE 7. Continued

Type Name and Abbreviation	Parent Material	Soil Profile	Humus	Tree Species
Ma-Vi T			strongly acid, with low lime	maple, yellow birch, and red spruce (rare); *balsam fir* not mentioned
Aralia-Oryzopsis Ar-Or T	Glacial terraces, eskers, and on rocky glacial till on hills	Weakly podzolized brown forest soil	Mull approaching a mor, nearly neutral, with high lime	Red oak, basswood, white ash, sugar maple, and hornbeam; *balsam fir* not mentioned
Dicentra Di T	Moist rich glacial till	Brown forest soil, sometimes slightly podzolized	Mull, mod. acid to neutral, with lime variable	Sugar maple, basswood, black cherry, beech, white elm, and white ash, occasionally with a white pine overstory; *balsam fir* reproduction present
Mitella Mi T	Very rich moist glacial till	Brown forest soil	Mull, mod. to weakly acid, with high lime	Sugar maple, basswood, white ash, and hemlock; *balsam fir* not mentioned

types by Heimburger (1934) very few investigations have been made of balsam fir growing under subalpine and alpine conditions. A major exception is the work of Antevs (1932) on the alpine vegetation of the Presidential Range in the White Mountains of New Hampshire. This range consists of a narrow ridge extending for about 23 km (14 miles) in a north-south direction culminating in the summit of Mt. Washington at 1917 m (6288 ft). Precipitation at the summit averages 2000 mm (80 in.) annually. Mean temperature for winter (November to April) is —11.3C (11.7F), spring (May to June) is 4.2C (39.7F), summer (July to August) is 8.4C (47.1F) and fall (September to October) is 1.8C (35.2F). The prevailing winds are from the west and northwest and frequently reach 100 mph.

Forest trees of merchantable size, primarily red spruce, extend to about 1280 m (4200 ft) elevation. At this level scrub forest begins with the main species, balsam fir, associated with paper birch and mountain ash. At higher elevations the trees become shorter and the proportion of black spruce increases. The scrub forest, only 2–3 m (6–10 ft) tall reaches its upper limit at 1460–1580 m (4800–5200 ft). Above these limits a few solitary balsam fir, black spruce, birch, and a few other woody species persist in sheltered spots. The upper limit of the scrub forest depends on wind, temperature, snow cover, and soil — in that order of importance. The average life span of balsam fir and black spruce ranges from 60 to 100 years.

The alpine flora of the White Mountains includes 63 species of arctic plants and 126 species of boreal plants. The vegetation was classified into alpine heaths, heath meadows, bogs, bog meadows, and scrub thickets. These formations were divided into communities or community groups. Balsam fir was noted as present in all formations but not all communities. Balsam fir and black spruce form the scrub thickets in sheltered situations in the alpine zone. The undergrowth of these thickets is fairly uniform. The most important constituents in an approximate order of frequency are sphagnums and other mosses, reindeer lichens, *Cornus canadensis, Coptis trifolia, Solidago macrophylla, Oxalis montana, Trientalis borealis, Linnea borealis, Lycopodium annotinum* var. *pungens, Vaccinium* sp., *Ledum groenlandicum, Veratrum viride, Geum peckii, Clintonia borealis, Maianthemum canadense, Spiraea latifolia* var. *septentrionalis,* and *Betula papyrifera* var. *minor.* In several places the vegetation consists mainly of carpets of ground-hugging balsam fir, black spruce, willow or *Vaccinium* which grow and migrate leeward of prevailing winds by developing roots under branches and dying off on the windward end.

Canadian investigations. Canadian studies have produced several geographic descriptions of forest communities (Halliday, 1937; Rowe, 1959; Loucks, 1962b). The forests of the Acadian Region, as outlined by Rowe (1959), or the Canadian part of the Acadian-Appalachian Region, consist of red spruce associated with balsam fir, yellow birch, and sugar maple, with some red pine, white pine, beech, and hemlock. Black and white spruce, red oak, American elm, black ash, red maple, paper birch, gray birch, and balsam poplar are widely distributed. The presence of red spruce and gray birch and the limited appearance of jack pine and northern white-cedar is characteristic. Balsam fir is distributed over all of the Acadian Region, but is especially prominent on Cape Breton Island (Sections A-6 and A-7) and in the East Atlantic Shore Section (A-5b).

The forest classification developed for the Maritime Provinces (Loucks, 1962b) includes the Acadian Region and parts of the Boreal and Great Lakes–St. Lawrence regions as recognized by Rowe (1959). Loucks recognized seven forest zones, 11 ecoregions and a greater number of forest districts. Balsam fir appears as the most characteristic species in the Fir-Pine-Birch Zone, and is second only to black spruce in the Spruce Taiga Zone. Balsam fir is the second most important species in the Spruce-Fir Coast Zone and in the Red Spruce–Hemlock–Pine Zone. In the Sugar Maple–Yellow Birch–Fir Zone balsam fir ranks third in importance.

Considerable additional work of a more detailed synecological nature has been done, but much of the information has never been published except as progress reports of limited circulation. Place (1955) described a forest classification scheme based on the work of Long (1952) which resembles an approach used by Pogrebniak (1955). Long recognized four moisture levels and three associations, i.e., red maple, red spruce–balsam fir, and black spruce that correspond to three nutrient levels.

Loucks (1956) developed a system of forest type classification based on moisture, nutrient, and ecoclimate axes for the Acadian Experimental Forest. A diagram of this classification is shown in Plate 5. Balsam fir penetrates into the most xeric type, pine-spruce-Cladonia, but is typically absent from three wet types: spruce-larch-Carex, black spruce–Sphagnum, and black spruce–Chamaedaphne. The requirements of black spruce are similar to the requirements of red spruce, but white spruce is limited to the four nutrient-enriched types.

Loucks (1960, 1962a) prepared synthetic scalars for the evaluation of moisture, nutrient, and local ecoclimate (heat) coordinates from edaphic and various other measurements. He also presented an analysis of the forest vegetation of New Brunswick following the techniques of the ordination method (Curtis, 1959). The ecological models developed by Loucks will be further considered in Chapter 3.

A number of interesting new Canadian investigations dealing with forest succession in the Acadian-Appalachian Forest Region will be completed in the near future.

Forest succession. Succession has been little studied in this area. The general trend is probably toward paludification or bog development as has been observed in Europe and in Alaska where similar climatic conditions exist. Impoverishment of soils due to leaching may explain the aggressiveness of balsam fir and spruce as observed by Potzger (1953b) and others. The diagrams presented in Plate 5 could also be considered as illustrations of primary succession patterns but without the sweeping trends toward any terminal community.

Revegetation of landslides in the White Mountains of New Hampshire was studied by Flaccus (1959), who classified landslide habitats as bare cliffs and ledges, rock debris or slide talus deposits, areas of erosion in glacial till, residual areas, and areas of deposit. Balsam fir is capable of seeding-in immediately after landslides in the crevices of cliffs and ledges wherever a small amount of soil exists. Paper birch and red spruce are frequent co-pioneer species on such areas. Balsam fir is absent from unstable talus deposits at the foot of ledges where rubble continually accumulates. Balsam fir and red spruce are sparse on the upper and middle eroded deposits of glacial till which are steep and unstable. Further down on more gentle slopes, balsam fir and red spruce appear following the pioneering paper birch, yellow birch, aspen, pin cherry, and other species. On the residual areas left after landslides, drainage and exposure are increased, and trees have little chance to survive. Flaccus studied more intensively the revegetation of deposit areas with chronosequences extending for 72 years. Herb and shrub stages are of comparatively short duration, and paper birch, yellow birch, pin cherry, and aspen come in immediately.

Balsam fir appears as a pioneer species after severe disturbance in beech–yellow birch stands in Cape Breton, Section A-7 (Fuller, 1919; Nichols, 1935). Balsam fir is accompanied by white spruce, rather than red spruce, which is not common to this area. Balsam fir is found only occasionally in climax stands of beech–sugar maple–yellow birch.

Many other aspects of succession have been studied in this region, especially those resulting from fire, logging, and land abandonment. Since these have primarily a bearing on forest management problems, they will be discussed in Chapters 6 and 7.

Conclusions

1. Information on the geographical and synecological distribution of balsam fir is abundant but highly diversified because of the variety of natural conditions under which the species is found, and because of the many different methods and viewpoints by which the species has been studied.

2. Mapping of ranges, organization of existing data, and geographic listings of associated species are the usual methods adopted in plant geographic studies.

These, however, are not only insufficient for making analyses by modern statistical methods but are unsatisfactory for explaining the fundamental problems of species distribution.

3. Synecological models have been developed for describing and analyzing plant communities by diagrams illustrating primary succession, genetic relationships, vegetational and environmental gradients, and continua. More complex models embodying bivariate fields and multidimensional coordinate spaces are being developed. These latter will serve as a pool for storing vast amounts of information and provide a basis for more fundamental generalizations. The methods developed thus far are still rather crude and need further improvement and refinement before specific problems relating to balsam fir and other species can be investigated.

4. The boundaries of the balsam fir range are not yet well established in all the areas where the species occurs. Particularly in need of investigation are those localities in which balsam fir comes in close contact with subalpine fir and Fraser fir. Geographic distribution of varieties is also little known and needs additional study.

5. Considerable success has been made in the geographic classification of forest regions and forest associations in Canada. Halliday's work in 1937 can be considered a historic landmark. The seemingly conflicting views have contributed to the emergence of better knowledge and toward new concepts in forest synecology. Since balsam fir is widely distributed in Canada, both geographical and synecological investigations have contributed much valuable information about this species.

6. Balsam fir is only locally important in the United States, although it is present in areas that have been very intensively investigated. Most studies have been conducted by individual workers or small groups working in limited areas. For effective work, large-scale systematic studies carried out over a term of years by a team of specialists are needed. The most important recent contributions to forest geographical and synecological problems, with respect to the distribution of balsam fir in the United States, have been carried out by Braun in her studies on the Deciduous Forest and by Westveld and his associates in the mixed hardwood-conifer forests of the northeastern United States. More of these kinds of investigation are needed.

7. The synecological studies have shown that balsam fir and white spruce are the most demanding species in the Boreal Forest Region. Balsam fir reaches its best development on sites slightly more moist than mesic. In the Great Lakes–St. Lawrence Forest Region, the species is largely precluded from sites better than intermediate in nutrients, and from xeric and hydric extremes. In the Acadian-Appalachian Forest Region, it extends into richer sites but normally cannot compete successfully with more mesic hardwoods on the best sites. In this region distribution is strongly influenced by mountainous topography and elevation.

8. Away from the Boreal Region, balsam fir retains its boreal character as expressed in its tendency to occupy moist and cool sites, e.g., swamp borders, heavy clay soils, mountain tops, north-facing slopes, and shaded understories in pine and hardwood stands.

9. The problems of forest succession have intrigued many investigators. Studies of succession have shown that balsam fir is very responsive to a changing environment and to competition from associated species. Studies in vegetation dynamics, both in undisturbed natural stands and in disturbed areas, need to be continued for a proper evaluation of the environmental factors that are most important and as a guide in forest management. Improved methods in studying forest succession are much to be desired.

3

ECOLOGICAL FACTORS

Most of the available information on the relationships of balsam fir to the different ecological factors is of a plant geographical or synecological character. The response of balsam fir to various factors in different regions has been observed primarily in mixed stands. Even in pure stands balsam fir competes with understory vegetation. It is well known that species under conditions of competition respond quite differently to environmental factors than when growing in isolation. No reports were found on trenching experiments or other attempts to isolate balsam fir from the competition of other species. The few physiological studies that have been made are discussed in Chapter 1. Unfortunately the methods of solving factorial problems developed by Jenny since 1941 have not been applied to balsam fir problems except for the chronosequences (Leisman, 1957; Flaccus, 1952) which were discussed in Chapter 2.

Those aspects of recently-developed ecological models in which balsam fir is involved will be discussed first. These models are based on a physiological classification of the important factor complexes — moisture, nutrients, heat, and light — which are the components of ecosystem matter-energy budgets.

Further discussion will generally follow the more conventional approach which makes use of a classification of climatic, edaphic, and biotic factors. Both systems of classification are complementary in that they can be used for both theoretical and applied purposes. Man's cultural activity, a biotic factor, will be considered as a secondary measure that acts upon primary climatic, edaphic, and biotic conditions, and will be discussed in later chapters.

Information on climatic, edaphic, and certain biotic factors as they apply to balsam fir is scanty. On the other hand, the literature dealing with the relationships of balsam fir to fungi and insects, two important biotic factors, is very abundant. The latter, therefore, will be treated in separate chapters. In this chapter wildlife will be considered as a special biotic factor.

Ecological Models

The concept of ecosystem occupies the central position in contemporary ecological thought. It has not only established a strong link between physiology and

84

ecology but has also influenced biogeography and related branches of knowledge. Ecosystem patterns are rapidly advancing to a prominent position among biological models.

In Chapter 2 the models which were developed from primary succession schemes and genetic forest classification patterns were modified to account for the essential components of the ecosystem matter-energy budget. These components have been primarily the moisture and nutrient regimes. The models can be further extended to include heat and light as additional coordinate axes in their basic structure.

Several workers have contributed to the initial phases of model constructions. Insofar as their studies relate to balsam fir, their data are reviewed in appropriate places in this monograph. The discussion which follows will be limited to those models in which the principles of model construction are illustrated.

It should be pointed out that models constructed from data based on distribution of plants and their responses to habitat conditions are merely first approximations to the needs of science. There is an inherent hazard that circular reasoning will result when data of this sort are relied upon. Independent measures of effective physical factors and knowledge of feedback mechanisms are really needed as basic information for model construction.

WISCONSIN MODELS

Several types of models have been developed by the Wisconsin ecological school under the leadership of Curtis and his associates. These are characterized by the construction of a multidimensional space based on vegetation similarity indexes. Curtis (1959) calculated the similarity indexes according to the formula

$$C = \frac{2w}{a+b}$$

where C is the similarity index, a and b are the sums of the quantitative species data in stands A and B, and w is the sum of the data common to both stands. For tree species the importance values (for definition see Table 6) were used in the formula; for other species frequency percentages were employed. In presentation of the data the scales selected were inversely proportional to the similarity indexes. The similarity indexes are ordinated along several axes, referred to as first, second, and third axis. Stands which are most dissimilar are placed at the two ends of the first axis; and the locations of all other stands are placed in a relative position on this axis with respect to the two most dissimilar stands. After the first (horizontal) ordination, there are many stands located at the same position on the first axis which are dissimilar among themselves. A second choice is then made of the two most dissimilar stands among those ranked equal on the first axis. These two stands then become the end points for the second (vertical) axis. When all the stands are realigned in accordance with the second axis, their position is represented by points on a plane. The third axis is determined in a similar manner, and, for its presentation requires a space model. The three bivariate combinations can be shown separately on a plane. The entire ordination space can be subsequently related to various environmental gradients by using geometrical transformations.

Curtis' (1959) three axes are moisture, temperature (heat), and brightness (light). These were used to portray Wisconsin plant communities. Illustrations taken from his study are shown in Plate 9. For easier comparison with other models, the orientation of some of the axes has been changed.

The upper two diagrams in Plate 9 present the location of Wisconsin plant communities in moisture-heat and moisture-light planes with axes oriented toward a relative scale of increasing intensity for each factor. Communities having balsam fir as an associated species are shown situated in the wet-cold and wet-dark positions of the plane. When importance values are added to the diagrams by means of contour lines (lower left diagram), the position and importance of balsam fir in each community is made more clear. As indicated, balsam fir reaches its greatest importance value in the boreal forest community. The lower right diagram examines the position of balsam fir relative to other species in the boreal-hardwood transition forests of the Great Lakes area, based on the work of Maycock (1957). To be interpreted, the vegetation similarity axes must be transformed into environmental factor axes; however, the more or less central position of balsam fir is clearly shown. Abbreviated descriptions of Wisconsin community types which include balsam fir are given in Tables 6 and 15.

Another model showing the position of the species in Wisconsin plant communities was derived from the data presented by Wilde et al. (1949) and is shown in Plate 5.

SASKATCHEWAN-MANITOBA MODELS

By a choice of indicators, Rowe (1956) emphasized the need for evaluating moisture regime, nutrients, and radiant energy in relation to forest stands. He first prepared a moisture-preference scale having five classes for individual undergrowth species in the Mixed-Wood Section (B-18 according to Halliday, 1937). Moisture coordinate values for communities (VMI = vegetation moisture index) were computed on the basis of species presence. Ubiquitous and rare species were excluded, and progressively greater weights were assigned to the species of wetter sites. Balsam fir was present in the three wettest classes.

As stated by Rowe (1956), the principal factor controlling the structure of undergrowth is light, for although temperature, root competition, litter accumulation, and soil characteristics exert a subsidiary effect, their main influence is on composition.

Light conditions were characterized by the relative cover-abundance of different physiognomic groups of forest undergrowth plants such as tall, medium, and low shrubs; herbs and grasses; and lichens and mosses. A diagram was presented for evaluating different combinations corresponding to a gradual change of overstory species from aspen-poplar to spruce. The effects of light on admixtures of balsam fir with other species were not shown directly, but could be obtained by an analogous approach.

Rowe suggested that a nutrient-factor axis could be established in a similar way, although the problem was considered more difficult. The geographical distribution of plants or phenological data was suggested as the basis for a heat-factor axis.

NEW BRUNSWICK MODELS

An early model (developed by Loucks, 1956) dealt with the edaphic field of the Acadia Experimental Forest, New Brunswick (see Plate 5). Loucks (1962) later described a method of model construction based on synthetic environmental scalars. Scalars are unified single expressions of different kinds of related measurements. Three scalars were developed, i.e., a moisture regime scalar incorporating three major factors, a nutrient scalar based on six factors, and a heat (originally called local climate) scalar based on four major factors. The scalars can be of different ranks, e.g., the moisture regime scalar is built upon a "run-off" scalar based on slope steepness and position, soil water-holding capacity, and depth to the prevailing water table. Different components in a synthetic scalar may be weighted in accordance with their known importance as shown by previous studies or by local observations. Scalars are developed for regional use.

Nutrient regime is defined "in terms of those soil factors which contribute to soil fertility, and which at their optimum favor the greatest production of what is believed to be the most demanding indigenous species of trees, shrubs, and herbaceous plants." The synthetic nutrient scalar is developed separately for mineral soils (organic layer up to 18 in. thick) and organic soils (organic layer over 18 in.). Factors incorporated into these scalars are humus content of the A-horizon, depth of solum, silt and clay content, depth of decomposition, run-off accumulation, and others.

The synthetic heat scalar is determined from daytime radiation by slope and aspect, sensible heat on east and west slopes, night temperature, and soil water-holding capacity.

Loucks (1962) also contrasted his environmental scalars with the phytosociological (similarity index) axes of the Wisconsin school by using the plant communities of northwestern New Brunswick for the comparison. In general the two sets of axes were found to form diagonals with each other.

Plate 10 shows the distribution of the importance values of balsam fir, approximately by quartiles, in phytosociological and environmental ordination. The drawing is a simplification and approximation of the original drawing by Loucks (1962). The first axis ordinates 63 sampled areas arranged in a continuum from a sugar maple (left) to a black spruce (right) stand. The second axis represents an expansion of the midportion of the first axis, in a place where balsam fir communities are dominant. The continuum represents a series of communities from a balsam fir stand at the low end to a white-cedar stand at the upper end. The third axis represents an expansion of the far-left portion of the first axis, where balsam fir is less prominent. The series of communities represented by this continuum ranges from a sugar maple stand at the low end to a balsam fir–yellow birch stand at the upper end.

The bivariate distributions indicate a lower left-central position for balsam fir, a situation not unlike that shown in the lower right diagram of Plate 9.

In moisture-nutrient and moisture-heat scalar coordinates (lower set of diagrams), balsam fir occurs over a wide range of conditions, most abundantly in the central portions of the moisture and heat axes and somewhat above center on the nutrient axis. The species is replaced by black spruce at low nutrients under the greatest variety of moisture positions, by white-cedar and white spruce in high nutrient-mesic and mesic-wet moisture positions, and by sugar maple in high heat–high nutrient positions. Other forest communities are more intermediate.

Loucks (1962) also noted that balsam fir occurs primarily on soils where no water table occurs within the major rooting zone. There is a significant increase of balsam fir basal area growth with increasing nutrient-scalar values, and the highest importance values for the species occur at intermediate heat scalar magnitudes.

MINNESOTA MODELS

Minnesota models originate from a reconnaissance study made of 356 forest communities throughout Minnesota in 1957 and from an intensive study of 55 forest communities in the Central Minnesota Pine Section in 1960 and 1961 (Bakuzis, 1959, 1961). The first studies were made using models of synecological coordinates, but other coordinate systems based on soil organic-mineral and soil N-P-K (nitrogen-phosphorus-potassium) elements have also been used. The method used in the development of synecological coordinates resembles the techniques used in the construction of ecological models by Rowe (1956), Ellenberg (1950), and Curtis (1955), and is basically patterned after the models developed by Pogrebniak (1930, 1955) and others. In contrast to the main approach used by the Wisconsin school (Curtis, 1959), the synecological coordinate method does not attempt to construct a multidimensional vegetation space from similarity indexes of species composition directly, but from the beginning total similarity is partitioned into similarity of occurrence along individual gradients of moisture, nutrients, heat, and light. These four factor complexes are the major synecological axes. The term "synecological" is used to emphasize the point that all observations concerning the occurrence of individual species are made under conditions of competition with other species. Competition results in species response that is quite different from that when competition is absent. Those species occurring predominantly at the lowest intensity of an environmental factor complex receive the lowest coordinate magnitude (value) of 1; the highest intensities are graded as 5, and intermediate coordinates are 2, 3, or 4 depending on their relative position of occurrence along the gradient. This procedure is repeated independently for each of the four factor complexes. Thus in synecological coordinates, a species is assigned four values, corresponding to its relative position with all other species in its requirements for moisture, nutrients, heat, and light.

Provisional assessment of coordinate values for individual species was obtained from previous studies or observations by many workers, mostly botanists and foresters. Among such studies perhaps the best suited for this purpose are the well-known tolerance tables for forest trees (e.g., Baker, 1950), which classify species into groups, or rank them sequentially in relation to changing intensity of environmental factors. The coordinate values first assigned are provisional and are later adjusted and refined for each species according to its occurrence in different forest communities within a region. Tables 8, 10, 14, and 20 present synecological coordinates for balsam fir and other tree species in Minnesota.

In the use of synecological coordinates an assumption is made that a community coordinate represents the averages of the individual coordinates of all species present. No weightings are used, and no ubiquitous species or rare species are excluded. All relationships in the models are investigated with respect to community coordinates which, as statistical averages, show a central tendency as compared with the original coordinates of individual species. The ecological interpretation of this statistical tendency is that species are evaluated with respect to microsites while communities are interpreted in terms of macrosites. A scatter diagram of all community coordinates represented in a geographic area provides for the regional synecological field when two synecological axes are used, or it provides for a space when more axes are represented. Although at the start each species has only one coordinate value for each axis, it may occur in different combinations with other species in communities with widely scattered community coordinates. The distribution pattern of the coordinates for all communities in which a species is present is the ecograph for that species. Ecographs may show not only the range of the field or the space where a species is present, but may also indicate the frequency of occurrence or give other quantitative data. A comparison of ecographs of different species provides information on their possible interrelationships. When ecographs of major species are superimposed in moisture-nutrient axes, diagrams similar to the primary succession diagrams worked out by the Clementsian school are obtained. This provides additional evidence of the edaphic nature of the successional schemes. Direct measurements of the environment in physiographic terms can be plotted into synecological coordinate systems and their patterns compared with the patterns of different characteristics of the vegetation. The distribution of balsam fir in synecological coordinates in Minnesota forests is shown in Plate 11.

The four coordinate axes: moisture, nutrients, heat, and light, provide 6 bivariate and 12 trivariate combinations. The third dimension cannot be fully represented on a plane, but its average values can be plotted with the aid of contour lines. The outlines represent the total synecological space of Minnesota forests, and the shaded areas indicate the space where balsam fir is present. The outlines reflect some basic relationships of the structure of forest ecosystems in Minnesota. All outlines with moisture as one of the coordinate axes have a characteristic triangular form as a result of the merging of upland and lowland populations. There is a very strong positive correlation between community nutrient and heat coordinates. The outline of the moisture-light field is an inverted triangle. The two remaining fields are approximately elliptical. There is rather strong negative correlation between community nutrient and light coordinates. The relationships between community heat and light coordinates show a slight negative correlation. The ecographs for individual species (see Bakuzis, 1959) indicate that some species are unable to become shade tolerant in moving southward. This weakens the correlation within the Minnesota forest vegetation as a whole with respect to these two coordinate axes. These kinds of relationships have been observed over a wide range of geographical conditions by many workers (Boyko, 1947; Walter, 1947, 1955; Pogrebniak, 1955; and others).

Balsam fir is distributed over a wide range of conditions in Minnesota forests. The four-dimensional ecosystem space is not shown because of presentation difficulties, but in this type of space only paper birch among Minnesota species shows wider occurrence (Bakuzis, 1959). In the coordinates shown in Plate 11, balsam fir is most restricted along the heat axis. This may be due to its inability to compete with tolerant hardwoods, under increasing heat, although balsam fir increases somewhat in shade tolerance under these conditions. The closely related nutrient axis imposes similar restrictions on balsam fir. The wide range of the species along the moisture axis is explained by the presence of balsam fir seedlings on dry and wet sites. However, as indicated in Plate 16 balsam fir cannot become successfully established along this entire range of moisture conditions, at least not in the Central Pine Section of Minnesota. The position of balsam fir on the light axis is also shown.

The tridimensional series of diagrams in Plate 11 shows the averages of the third coordinate when the other two are known. Variability of the third dimension was not computed, but there was nearly a functional relationship among the magnitudes of the three coordinates. Coordinate patterns for the third dimension in balsam fir communities do not greatly differ from the third dimension patterns for all Minnesota forest communities. When the diagrams are examined in pairs of identical axes, it appears that the contour lines for the third axis may run parallel, perpendicular, or sloping. These diagrams demonstrate the intricacies and yet the great regularity of the interrelationships within forest ecosystems. Close study of various species' models, of models showing specific characteristics of the environment, and of models from other regions, reveals that they contain a great amount of information in condensed form which can lend itself to the development of new and interesting hypotheses.

Plate 12 presents some information obtained from vegetation and soil analyses of 55 forest communities in the Central Pine Section of Minnesota. The edaphic conditions will be treated in more detail in another section of this chapter, and other related problems will be referred to in appropriate places in this monograph (e.g., Plate 16). Plate 12 illustrates some of the characteristics of forest ecosystems which can be analyzed with the aid of models of this type.

The models shown in Plate 13 are based on 30 sample areas in the upland forests of the Central Pine Section of Minnesota. The models illustrate various types of biological information available for balsam fir. The first group of diagrams shows the distribution of measurable basal area of balsam fir, i.e., of trees from 1-inch dbh and over, and its reproduction in six bivariate combinations of the four coordinate axes. The diagrams indicate those conditions which are critical for balsam fir reproduction to become established.

The lower two sets of diagrams introduce the application of two soil triangular coordinate systems in the study of the same area. Both systems are constructed on the results of soil analyses for the upper 6-inch (15 cm) horizon. The organic-mineral coordinate system resembles the well-known soil texture triangle. The three axes are silt plus clay, sand plus gravel (up to 2 mm diam), and organic content. All percentages add up to 100 thus enabling their use directly in the construction of a triangular coordinate system. Although the site complex is located eccentrically, the field covered by the 30 sample areas is large enough to counteract this drawback. To conform with the soil N-P-K coordinate system, the second diagram of the middle set shows the data transformed by a simple arithmetical manipulation in order to centralize the field. The diagrams show that balsam fir reproduction on soils having a large percentage of sand and gravel seldom becomes established, but point out that establishment is readily accomplished on soils with a high percentage of organic material and silt and clay.

In N-P-K coordinates the proportionate parts do not add up to 100 per cent, hence it was necessary to apply transformations in order to compute coordinates and to centralize the field. The establishment of balsam fir stands is apparently related to accumulation of nitrogen in the topsoil. Total nitrogen less than 0.2 per cent by analyses (or 30 per cent in the transformed coordinates) is associated only with transient reproduction. The inconclusive results shown for balsam fir by P and K values probably indicate that even the lowest amounts of potassium (50 ppm) and phosphorus (10 ppm) in the soils examined are not limiting to the species establishment.

Climatic Factors

The wide diversity of climatic conditions within the range of balsam fir is shown in the regional descriptions presented in Chapter 2. Recently efforts have also been made to correlate climatic data with growth of forests (Paterson, 1961; Lemieux, 1961), but the data available for balsam fir were insufficient to apply the Paterson or similar indexes.

Information on the interrelationship of balsam fir and climatic factors is primarily of two kinds. One approach ranks balsam fir among other species in its tolerance of different factors. The other kind of information deals primarily with the destructive effects of frost, heat, drought, flooding, hail, snow, wind, and fire. The discussion which follows treats the major individual climatic factors or factor-complexes individually. Those discussed are heat, light, precipitation, wind, and fire.

HEAT

In the wide range of balsam fir distribution, the frost-free season varies from 80 to 180 days (from data by Ward et al., 1938). Halliday (1937) described the frost-free season as less than 100 days in the Boreal Region, 120 days in the Acadian (Atlantic) Region, and 150 days in the Great Lakes–St. Lawrence Region. In the area of optimum growth it is 110 days.

Mean annual temperature within the range of balsam fir varies from 25 to 45F, and in the optimum growth area it is 35 to 40F. Mean January temperature ranges from −15 to 34F, while in the optimum area the range is from 0 to 10F. The corresponding mean temperatures for July are from 57 to 72F and from 60 to 65F (Ward et al., 1938; U.S.D.A., 1941; Can. Geogr. Br., 1957; see also the climographs for balsam fir, Plate 6). Of interest is the fact that the mean annual maximum temperature throughout the range of balsam fir is about the same, i.e., 86F, but the mean annual minimum temperature varies greatly, i.e., from −49 to 5F.

From a study of the geographic distribution of 13 major forest cover types in the United States, Woodward (1917) concluded that spruce-fir types occur in areas having the shortest growing seasons.

According to Hutchinson (1918), the following list indicates the order of tolerance to low temperatures of some eastern species in Ontario, beginning with the most tolerant.

Picea mariana	*Pinus strobus*
Picea glauca	*Acer saccharum*
Pinus banksiana	*Fagus americana*
ABIES BALSAMEA	*Quercus rubra*
Populus balsamifera	*Carya amara* (*C.*
Betula papyrifera	*cordiformis*)
Thuja occidentalis	*Castanea dentata*
Tsuga canadensis	*Juglans cinerea*

The position of balsam fir among other species in Minnesota forests as indicated by the synecological coordinates for heat is shown in Table 8 (Bakuzis, 1959), higher values denoting warmer conditions. The heat coordinates presented here are average expressions of macroclimatic heat effects as modified by microclimatic conditions, on trees growing in forest stands in competition with other species.

TABLE 8. SYNECOLOGICAL COORDINATES OF HEAT ON A RELATIVE SCALE
FROM 1 TO 5 INDICATING THE PREVAILING OCCURRENCE
OF TREE SPECIES IN MINNESOTA FORESTS

Tree Species	Heat Coordinates	Tree Species	Heat Coordinates
Larix laricina	1.0	*Acer saccharum*	3.5
Picea mariana	1.0	*Quercus rubra*	3.5
Thuja occidentalis	1.3	*Quercus macrocarpa*	3.8
ABIES BALSAMEA	1.4	*Tilia americana*	3.8
Picea glauca	1.5	*Ulmus americana*	3.9
Betula papyrifera	1.7	*Ostrya virginiana*	4.3
Pinus banksiana	1.9	*Fraxinus pennsylvanica*	4.4
Pinus strobus	2.0	*Quercus alba*	4.6
Populus balsamifera	2.1	*Carya cordiformis*	4.8
Populus tremuloides	2.1	*Acer negundo*	5.0
Pinus resinosa	2.1	*Populus deltoides*	5.0
Betula alleghaniensis	2.4	*Ulmus thomasii*	5.0
Acer rubrum	2.6	*Juniperus virginiana*	5.0
Fraxinus nigra	2.6	*Juglans nigra*	5.0
Quercus ellipsoidalis	3.0	*Celtis occidentalis*	5.0
Populus grandidentata	3.1	*Acer saccharinum*	5.0

Source: Bakuzis, 1959.

A comparison with the table of Hutchinson (1918) shows that, although two different regions are involved, the over-all agreement is quite good. There are 10 species common to both tables, and only one, *Thuja occidentalis* is in disagreement by more than three ranks. The rank correlation coefficient between the two sets of data is: $R = +0.67$, significant at the 5 per cent level. The ranks for balsam fir differ by only one.

Although late frost damage of leaders and new lateral shoots — particularly of young balsam fir in open areas — has been widely reported, the damage is considered light or moderate. Frost damage can be recognized by browning of the needles or by a slight displacement of the leaders from the growth shoot of the previous season. Plate 14 illustrates several degrees of frost injury sustained during a May freeze in Minnesota (Hansen and Rees, 1947).

Bonner (1941) reported that climatic injuries in the Clay Belt (Section B-4) of Ontario are rather unimportant, but the injury that is noticed is especially serious in the advance growth in cutover areas. Smerlis (1961) reported slight damage by late frost in Laurentides Park, Quebec. Light to moderate and occasionally also severe late-frost damage to balsam fir reproduction was reported from localities in Nova Scotia, New Brunswick, and Quebec (Can. For. Ent. Path. Branch, 1960–62). In Wisconsin, Wilde (1953) described early-fall frost damage in open places. The effect of winter frost in producing lesions in balsam fir leading to decay of the sapwood, and in producing such injuries as frost crack, cupshake, and starshake was reported by Pomerleau (1944).

Arnold (1958) studied frost damage to 2-year-old balsam fir seedlings in a New Hampshire forest nursery following alternate freezing and thawing. The roots of the seedlings were snapped off at cleavage lines in the light loamy sand soil at a depth of 1.5 inches.

Balsam fir is very susceptible to sunscald in the Clay Belt area of Ontario (Millar, 1936). Sunscald was also reported from Wisconsin by Wilde (1953).

LIGHT

Maps in the *Atlas of Canada* (Can. Geogr. Branch, 1957) indicate that in southeastern Canada where balsam fir reaches its greatest intensity of distribution the total annual number of bright sunshine hours ranges between 1400 and 1800. Distribution of balsam fir, however, appears to be particularly affected by precipitation and temperature since it does not follow similar light conditions in central Canada. In the United States the greatest concentration occurs in New England, where bright sunshine during summer months (June-August) does not exceed 9 hours daily, and distribution is practically limited to areas having less than 10 hours of sunshine daily.

Rating the shade tolerance of forest trees is an old practice. The oldest such rating involving balsam fir was prepared by Zon and Graves in 1911. They rated balsam fir as the most tolerant species of the forests of eastern North America. Zon (1914) rated red spruce and hemlock more tolerant but northern white-cedar less tolerant than balsam fir. In Hutchinson's rating (1918) for Ontario, balsam fir was considered less shade tolerant than *Fagus americana, Tsuga canadensis, Acer saccharum, Tilia americana, Sorbus americana, Betula alleghaniensis, Ulmus americana, Picea mariana,* and *Quercus rubra.* It was rated more tolerant than *Pinus strobus, Betula papyrifera,* and *Populus tremuloides.* Baker's rating (1950) placed balsam fir second only to hemlock among all eastern conifers. Because hardwoods were rated separately, comparisons of balsam fir with the very shade-tolerant American beech and sugar maple were not made. Of course these ratings depend on the criteria used and on the other environmental conditions under which the species are observed. Graham (1954a) attempted to systematize the most important criteria and to introduce a pattern for scoring. His tolerance scores for trees in Iron County, Michigan are reproduced in Table 9.

The position of balsam fir among other species in Minnesota forests according to average synecological light coordinates is shown in Table 10, with low values indicating greater tolerance.

As given for other synecological coordinates, the magnitudes are adjusted for Minnesota conditions. The variability in terms of community coordinates is shown in Plate 11 for balsam fir. Fifteen common species are listed in Tables 9 and 10. The rank correlation coefficient, R, between the two sequences was $+0.74$, significant at the 1 per cent level. Disagreement about balsam fir for five ranks possibly indicates a regional difference for this species. The correlation between the ranks of ten species common to both the Minnesota list and that of Hutchinson was $+0.92$, which is highly significant. The ranks of balsam fir differed by only one.

Since trenching experiments have not involved balsam fir, and other ways of excluding competition effects have not been investigated, shade tolerance remains as a poorly understood complex of interactions. Place (1955) argues that if tolerance means the ability of a species to survive intense shade and root competition and its capacity to respond to release, then red spruce seedlings, followed by balsam fir, black spruce, and white spruce, should be considered the most tolerant under New Brunswick conditions. The same order of tolerance has also been reported from the northeastern United States (Westveld, 1953). As early as 1914 Zon observed that balsam fir seedlings endure heavy shade better during their first five years than in later life.

It is known that shade restricts the growth of *Abies alba* (Zentgraf, 1949), and this may be true of other species of firs. By adversely affecting competing vegetation, and by modifying the effects of frost and drought, shade may also favor growth of a species. Shaded seedlings are affected less by the balsam woolly aphid (Balch, 1952). Light plays a complex indirect role in budworm attacks as discussed in Chapter 5.

Lee (1924) presented a sequence of plant communities from Itasca Park, Minnesota, by ranking them according to decreasing light intensities under their canopies. Mature jack pine stands showed the highest light intensities followed by mature red pine, mature white pine, young red pine, young jack pine, sugar maple–basswood, and finally by spruce-fir, which had the lowest light intensity.

Mean light intensity at different levels above the forest floor in four forest types in the Adirondacks of New York was reported by Donahue (1940) and is shown in Figure 8. The type-sequence from hardwood to spruce swamp shows increasing light intensity in forest stands with increasing soil moisture. This is essentially the same relationship indicated in Plate 11 by the moisture-light coordinates.

PRECIPITATION

Distribution of mean annual precipitation in the natural range of balsam fir, as shown in Plate 6 by climographs, varies from 20 inches in the northwest to 40 inches in the east. Halliday (1937) described mean annual precipitation in Canada as varying from 15 to 30 inches in the Boreal Forest Region, 25 to 45 inches in the Great Lakes–St. Lawrence Region, and 40 to 55 inches in the Acadian Region (Can. Geogr. Branch, 1957). The minimum amount of annual precipitation tolerated by balsam fir was reported by Zon (1914) to be 25 inches, while in Maine, its

TABLE 9. SHADE TOLERANCE OF FOREST TREES IN IRON COUNTY, MICHIGAN, ON A RELATIVE SCALE FROM 0 TO 10

Species	Tolerance Score
Tolerant	
Tsuga canadensis	10.0
ABIES BALSAMEA	9.8
Acer saccharum	9.7
Tilia americana	8.2
High mid-tolerant	
Picea glauca	6.8
Picea mariana	6.4
Betula alleghaniensis	6.3
Acer rubrum	5.9
Thuja occidentalis	5.0
Pinus strobus	4.4
Low mid-tolerant	
Prunus serotina	2.4
Fraxinus nigra	2.4
Pinus resinosa	2.4
Intolerant	
Pinus banksiana	1.8
Betula papyrifera	1.0
Larix laricina	0.8
Populus tremuloides	0.7

Source: Graham, 1954a.

TABLE 10. SYNECOLOGICAL COORDINATES OF LIGHT ON A RELATIVE SCALE FROM 1 TO 5 INDICATING THE PREVAILING OCCURRENCE OF TREE SPECIES IN MINNESOTA FOREST COMMUNITIES

Tree Species	Light Coordinates	Tree Species	Light Coordinates
Acer saccharum	1.0	*Juniperus virginiana*	3.0
Carya cordiformis	1.0	*Acer rubrum*	3.4
Ostrya virginiana	1.1	*Populus grandidentata*	3.4
Thuja occidentalis	1.2	*Acer negundo*	3.4
Quercus alba	1.3	*Quercus macrocarpa*	3.5
Tilia americana	1.5	*Populus balsamifera*	3.5
Celtis occidentalis	1.6	*Picea mariana*	3.5
Betula alleghaniensis	1.7	*Fraxinus pennsylvanica*	3.6
Fraxinus nigra	1.8	*Populus deltoides*	3.9
Juglans nigra	2.0	*Populus tremuloides*	4.2
ABIES BALSAMEA	2.0	*Acer saccharinum*	4.5
Ulmus americana	2.0	*Pinus resinosa*	4.5
Picea glauca	2.3	*Betula papyrifera*	5.0
Ulmus thomasii	2.6	*Pinus banksiana*	5.0
Quercus rubra	2.7	*Quercus ellipsoidalis*	5.0
Pinus strobus	2.8	*Larix laricina*	5.0

Source: Bakuzis, 1959.

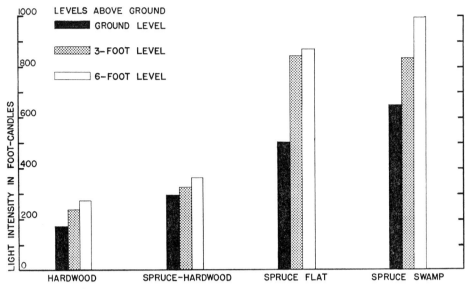

Figure 8. Mean light intensity in foot-candles at 0, 3, and 6 feet above the forest floor by main
forest types in the Adirondacks, New York. 1 Lux = 0.0929 foot-candles.
(Redrawn with permission from Donahue, 1940.)

optimum growth area in the United States, the mean is 43 inches. Under optimum
conditions in Canada, mean annual precipitation is 30 to 40 inches.

There is very little information available about the influence of balsam fir stands
on the distribution of precipitation received. Trimble (1959) investigated the interception of annual rainfall and snow by different forest cover types in Pennsylvania and in the White Mountains of New Hampshire. The results are shown in
Table 11.

Net interception of snow differs only slightly from net interception of rain.
Spruce and spruce-fir cover types show the greatest percentage of interception of

TABLE 11. AVERAGE ANNUAL RAINFALL AND SNOW INTERCEPTION (IN PER-
CENTAGES) IN DENSE STANDS AT AGE OF MAXIMUM INTERCEPTION BY FOREST
COVER TYPES IN PENNSYLVANIA AND THE WHITE MOUNTAIN
AREA IN NEW HAMPSHIRE

| | Data for Rainfall in Pennsylvania | | | | | | Net Snow Interception in New Hampshire |
| | Gross Interception | | Stemflow | | Net Interception | | |
Forest Cover Type	With Leaves	Without Leaves	With Leaves	Without Leaves	With Leaves	Without Leaves	
Northern hardwoods	20	17	5	10	15	7	10
Aspen-birch	15	12	5	8	10	4	7
Spruce, spruce-fir	35	—	3	—	32	—	35
White pine	30	—	4	—	26	—	25
Hemlock	30	—	2	—	28	—	25
Red pine	32	—	3	—	29	—	30

Source: Trimble, 1959.

precipitation. Lee (1924) found that evaporation from soil under balsam fir in northern Minnesota accounted for only 8 per cent of the total precipitation. This compares closely with the 10 per cent given by Zentgraf (1949) for *Abies alba* in Germany.

Effects of summer drought and winter drying of balsam fir have been frequently reported. Winter drying seems to be more widespread. The effect of drought on balsam fir planted in shelterbelts in the prairie region of western and southwestern Minnesota has been very serious (Deters and Schmitz, 1936). Following the drought of 1934 and preceding years, 30 per cent of balsam fir died. Balsam fir planted in Nebraska towns has also been seriously affected by drought (Pool, 1939).

An interesting observation on the effect of drought on needle length during the drought year of 1936 was made by Schantz-Hansen and Joranson (1939) at the Cloquet Forest Experiment Station, Minnesota. Compared with the preceding year, the needle length of white pine, red pine, and jack pine decreased significantly. No effect was observed on black spruce, and an increased length was noted for balsam fir. The authors explained that balsam fir completes its foliage growth earlier than the other species and drought effects were not made evident.

Winter drying, which is both a temperature and a moisture problem, occurred in the Adirondacks in 1947–1948 and was described by Curry and Church (1952). In descending order, needle-browning and defoliation occurred in the following species: *Picea rubens, Tsuga canadensis, Pinus strobus,* and *Abies balsamea. Tsuga canadensis* and *Pinus strobus* recovered rapidly, but *Picea rubens* and *Abies balsamea* showed permanent damage and some trees died as a result of extensive bud mortality. Balsam fir in the upper midwestern states showed only a slight browning of needles during the same winter and recovered satisfactorily (Stoeckeler and Rudolf, 1949).

Reeks (1958) described the combined effect of drying and alternating mild weather and frost in early spring in Manitoba as resulting in rapid browning of conifers. Balsam fir was most severely affected.

The Canada Forest Entomology and Pathology Branch (1962) reported winter drying of balsam fir in Nova Scotia and New Brunswick. Smerlis (1961) observed occasional drying in balsam fir stands younger than 35 years of age in Laurentides Park, Quebec. Needles were most seriously affected.

The Canada Forest Biology Division (1953, 1955) reported extensive hail damage in the Quetico Provincial Park, Ontario, in 1951. Severe mortality occurred in 1952 and continued into 1953 as cankers developed on the injured trees. The rating of major species in increasing order of susceptibility was given as follows: jack pine, red pine, white pine, poplar, aspen, spruce, balsam fir, and birch. Severe injury to balsam fir reproduction by hail was reported from the Port Arthur District, Ontario, in 1954. Damage on a 30-square-mile area caused the death of 50 per cent of the branches on balsam fir. Thomas (1956) reported on hail damage in the Lake Nipigon area of Ontario, where all species were affected. Balsam fir reproduction was damaged most severely.

The Canada Forest Entomology and Pathology Branch (1961) reported heavy hail damage in Quebec in stands where balsam fir predominated. Balsam fir was the most seriously affected species. Hail damage caused numerous wounds which often resulted in death of the leaders. Black spruce was the least affected species while paper birch was in an intermediate position. Apparently the contradiction in the reports of the relative susceptibility of balsam fir to hail damage can be attributed to different stand conditions in the areas investigated.

Glaze damage by ice storms has also been frequently reported. Serious damage occurred in parts of New Brunswick, Nova Scotia, and Prince Edward Island in January 1956. Of the species affected balsam fir ranked third behind aspen and tamarack in sustaining serious injury. Damage resulted in broken tops, lost cone crops, and the creation of abundant entrance points for trunk rots (Can. For. Biol. Div., 1957). Ice storm damage on the plateaus of Cape Breton Island in March 1960 was responsible for many broken tops in dominant and codominant balsam fir (Can. For. Ent. Path. Branch, 1960). The main part of the balsam fir range lies within the area of greatest occurrence of heavy glaze. However, balsam fir and spruce with their very straight and strong central trunks and small flexible branches are relatively resistant to glaze damage (Lemon, 1961).

In November 1958 an ice storm caused severe damage to some species in forest stands on a portion of the Sandilands Forest Reserve in southeastern Manitoba. Damage was characterized by the bending and breaking of tree stems. Young stands suffered the most bending damage, but pole stands had a higher percentage of breakage. Older stands were practically undamaged. Jack pine was most severely affected, followed by white-cedar and black spruce. Damage to balsam poplar, trembling aspen, white spruce, balsam fir, paper birch, green ash and tamarack was minor (Cayford and Haig, 1961).

The Lake States Forest Experiment Station (1939) reported snow-breakage in balsam fir up to 4.5 inches dbh in northern Wisconsin.

WIND

Relative susceptibility to windthrow and wind-breakage as affected by site conditions, former cutting practices, size, age, and stand composition has been given frequent attention during recent years. Such specific problems as the relationship between air mass movements and development of insect epidemics have also gained attention.

Zon (1914) stated that balsam fir is easily uprooted or broken by wind because its root system is shallow and its bole is brittle. The answers given to a questionnaire on the occurrence of blowdown of pulpwood stands in eastern Canada showed blowdown to be most frequent in narrow valleys, at lake ends, on ridges, on south or southeast-facing slopes, and on shallow (6–18 in.) soils in mature stands with spruce and balsam fir the dominant species. Balsam fir with butt rot was most susceptible, with most damage occurring from August to October after heavy rains (Weetman, 1957). A dense, compact, impermeable zone 12–18 inches below the surface of the mineral soil was found to be a contributing factor in the susceptibility to windthrow of balsam fir in spruce-fir stands in Maine (McLintock, 1958).

An intensive survey of windfall was made in the New York Adirondacks by Behre (1921) in 1916. His observations showed that balsam fir stands were severely affected by windthrow and breakage. Large trees were more often broken whereas small trees were more often uprooted. Breakage was greater on the drier sites but windthrow was greater on wet sites. Balsam fir was damaged more on exposed hardwood sites than in spruce types (evidently spruce-flat) or lowlands.

Schantz-Hansen (1937) reported a storm with winds up to 60 mph following rain in the Cloquet Experimental Forest in Minnesota. Little damage was done to the upland stands, but in an all-aged, uncut spruce-balsam fir stand, 29 per cent of the spruce and 27 per cent of the balsam fir were windthrown. According to Cheyney (1942), more wind damage occurs on wet sites than on drier sites.

Loss due to wind on upland sites in the Patricia District of Ontario was 26 per cent as compared with 18 per cent on lowlands (Seed, 1951). It was explained that trees on the upland were smaller and had one-sided root systems. The basal area ratio of windthrow to wind-breakage was 3:1 in spruce and 1:1 in balsam fir. Breakage occurred in a zone extending from the ground to 15 feet up the bole and was more noticeable both on small suppressed trees and on large trees.

Greater damage on ridgetops than in swampy areas following a storm in May 1953 at Itasca State Park, Minnesota, was reported by Lundgren (1954). On the ridgetop only one of the 58 original balsam firs remained standing. By contrast 45 out of 63 paper birches survived. Uprooting was the major cause of balsam fir loss. Most of the uprooted trees had root rots and most of the broken trees were heavily infected with butt rots. Paper birch was not in leaf at the time of the storm. Although lack of leaves allows the hardwoods to survive heavy wind storms, damage to associated conifers is often increased when the other trees are leafless (Bowman, 1944).

Stoeckeler and Arbogast (1955) reported on wind loss on the Argonne Experimental Forest in Wisconsin during a severe and extensive storm on October 10, 1949, when wind velocities averaged 50 to 75 mph with gusts up to 102 mph. Effects were evaluated on a cover-type basis. In the hemlock-hardwood type, hemlock was most seriously damaged and balsam fir was next, followed by white-cedar, white spruce, and white pine. In a second-growth hardwood type the damage was confined mostly to conifers; 58 per cent of the economic loss was balsam fir and 26 per cent was hemlock. In the balsam fir type, trees larger than 8 inches dbh and older than 50 years were reported as highly susceptible to windthrow. Only 42 per cent of the windthrown balsam fir showed external evidence of defect.

The effect of cutting practices on wind damage was studied by McLintock (1954). Following partial cutting on five areas in Maine and northern New Hampshire, wind losses were 0.12 cords of merchantable spruce and balsam fir per acre per year over a study period of 3.5 years. Wind damage increased sharply when more than 20 per cent of the total basal area was cut. More than 60 per cent of the damage to balsam fir was breakage. Two-thirds of the damage to spruce was uprooting. Large firs and dominant spruces were especially vulnerable. By retaining 75 per cent of the basal area and taking other precautions, McLintock (1954) estimated that losses could be reduced from 0.12 to 0.05 cords per acre per year.

Arguing against exaggeration of wind damage, Kelly and Place (1950), reported that with proper selection only 1 per cent of the basal area was lost in 12 years, although up to 70 per cent of the basal area of the original stand was removed by cutting.

Hurricane damage in 1954 in the New England states and Canada stimulated more careful study of the effects of wind and storms. Grisez (1955) reported on September hurricane damage at Penobscot Experimental Forest in Maine prior to leaf fall. Wind velocity of 40 mph with gusts up to 62 mph was accompanied by a 7-inch rainfall. Of the 750 cords blown down, 70 per cent was balsam fir. In stands with light cuts (15–35 per cent by volume) the loss was 0.18 cords per acre. In heavily cut stands (50–90 per cent), the loss was 0.37 cords per acre. Damage was also considerable (0.25 cords per acre) in uncut stands, but this was attributed in part to large defective trees.

A survey in eastern Canada indicated that once a wind has achieved hurricane velocity (75 mph) on the Beaufort scale, nearly all mature trees in pulpwood stands will be blown down. With a somewhat lower wind velocity (about 60 mph) the overmature and many mature trees will be blown down if in a weakened condition or situated in an exposed area. Management can be effective in protecting against damage from winds below velocities of 60 mph (Weetman, 1957).

There seems to be a rather widely held opinion that high winds cause breakage of feeder rootlets as a result of swaying of the tree trunks and resulting tension. Redmond (1954) excavated two balsam fir trees 6–8 inches dbh in September, 1954, two or three days after Hurricane Edna struck Nova Scotia. Gusts up to 80 mph were accompanied by 0.5 inches of rain. The trees were growing on sandy loam, and no damage to rootlets of the balsam fir was found.

Indirect information is available on reproduction and seed dispersal as well as on the spread of insects as a result of wind movements. It would seem that a broad survey of wind effects similar in extent to reproduction surveys could furnish leads for more detailed future research.

FIRE

Fire is unique among the factors in the degree to which it has been influenced by man's activities. Although a high proportion of fires in the balsam fir region are still started by man, fire is far less frequent now and covers less area annually than before the arrival of European man, due to his intensified efforts in the past half century to prevent and control wild fires. The reduction in fire frequency has been favorable for extension of coverage by balsam fir. Very recently foresters have begun to use controlled burning as a tool to modify the forest environment, as primitive man (Stewart, 1951) and early prospectors of northern Minnesota (Winchell, 1879; Hall, 1880) did formerly on a much wider scale and in a less controlled manner. However, since fire can be started by lightning and spread by wind, especially under conditions of low atmospheric humidity and in areas of southwesterly exposure and of coarse-textured substrata, it is in part climatically and in part edaphically dependent. Thus fire cannot be considered as a purely cultural activity used by man to modify the climatic, edaphic, and biotic conditions of his living space.

Effects of fire on secondary succession were discussed in Chapter 2. Practical aspects of the use and control of fire in the forest will be treated in Chapters 6 and 7 in connection with reproduction and stand development.

In the United States approximately 3 per cent of all forest fires are caused by lightning but with great variation between regions in the incidence of fires of such origin. In some years 5 per cent of the forest fires in Maine have been lightning-caused (Hawley and Stickel, 1948), although normally it is much less. The four-year averages of the percentage of number of fires set by lightning range from 0.7 to 1.4 in the eastern United States (U.S. forest fire statistical data, compiled by Davis, 1959). In Europe, spruce and fir are the species most frequently struck by lightning (Walter, 1951), probably as a result of their specific morphological character, and especially because their acid sap and preference for moist substrata provide an excellent electrical conductor upward from the water table, whence comes the initial great discharge.

Highly susceptible to fire damage (Zon, 1914), balsam fir is rated the least resistant species in the northeastern states by Starker (1934). In laboratory tests the bark proved to be more heat resistant than the bark of beech, but less resistant than that of hemlock (Stickel, 1941).

Balsam fir sheds its bark more quickly after fire than other species. Three years after a fire in eastern Canada the remaining merchantable volume loss amounted to only 3.4 per cent where burning was severe, 15.4 per cent where moderate, but 42.6 per cent where light, because of the different rate of progress of associated sapwood decay and wind-breakage (Skolko, 1947).

Diebold (1941) studied the effect of fire on the depth of forest floor in the New York Adirondacks. On spruce-fir slopes he found that fire had reduced the thickness of the duff on the forest floor from 14 to 2 inches. Severe to moderate erosion following fires was observed on 30,000 acres in the Adirondacks. As contrasted with fire, logging reduced the depth of the forest floor very little.

Fire-fighting methods in the spruce-fir cover type have been set forth in some detail by the Northeastern Forest Fire Protection Commission (U.S. Forest Service, 1958). Descriptions of specific fire conditions in this cover type were also included. Fire in the spruce-fir cover type may be ground, surface, or crown. In periods of drought it is likely to be a combination of all three. Tree crowns and duff, together with slash, provide the most highly inflammable combination of fuels. Under such conditions fire spreads at a high rate. Heavy smoke ahead of the fire often makes direct attack impossible. When burning material becomes buried under the surface, subsequent fires may occur as surface fuels are again ignited.

Edaphic Factors

In evaluating the effects of edaphic factors a more or less general soil classification is of particular importance. Although local classifications can be developed using only a few edaphic characteristics found to be essential to growth on the basis of intensive local investigation and accumulation of experience, these seldom have validity for application elsewhere. Classification is the framework into which can be fitted a great number of uncontrolled factors in routine research or practice. General classification systems are extremely difficult to prepare, however, because of their over-all complexity and the fact that they must serve different purposes. The views of foresters on soil classification differ markedly from the views of agronomists, and the latter viewpoint has dominated the approach to soil classification used in the United States. The interests of forestry are best served by an integrated ecosystem approach. The general difficulties involved in the problem of soil classification have also contributed to the present unsatisfactory conditions in our knowledge of the relationships between balsam fir and various edaphic factors. Moreover, edaphic conditions within the wide range of balsam fir are very diversified. Information on the species and the soil in which it grows is scanty and is based primarily upon small-scale research projects in areas where the species is often only secondary in importance.

Much of the natural range of balsam fir is on the Canadian Shield, which consists chiefly of metamorphic and igneous rocks with sedimentary rocks in certain areas. The Canadian Shield is closely related geologically to the Appalachian Highlands, although the history of the latter region is more complex. These two physiographic regions support the greatest area of balsam fir. To a lesser extent the Great Lakes–St. Lawrence Lowlands and the Western Interior Lowlands contribute to the range of the species.

With a few local exceptions the entire present-day range of balsam fir was glaciated. Both the Keewatin and Labrador glaciation centers were located on the Canadian Shield.

Balsam fir grows on soils underlain by Precambrian gneiss, granite, schist, anorthosite, diabase, slate, sandstone, and Silurian, Ordovician, and Cambrian limestones. These formations are covered by layers of residual, glacial, and organic deposits of varying thicknesses (Halliday, 1937).

Balsam fir grows on heavy clay soils (Grant, 1934), loams and sandy loams (Westveld, 1949), gley loams (Wilde et al., 1949), sandy glacial till (Heimburger, 1941), lake sand (Sisam, 1938a,b), rock outcrops and rocky slopes (Zon, 1914; Schantz-Hansen, 1923, 1934), fine and coarse woody peats (Wilde, 1933), and on other organic soils.

The soil profiles on which balsam fir occurs range from rendzinas and brown forest soils through all grades of podzolization to strong ortstein podzols, gleys, and organic profiles. However, a glance at the soils map (Plate 7) shows that to the west, away from the podzols and brown podzolic soils and their associated gleysolic and humic gley soils, the intensity of balsam fir decreases sharply. Characteristic of the latter group are the widely distributed gray-wooded soils.

The optimum edaphic conditions available for balsam fir change from region to region. In the Boreal Region the white spruce–balsam fir type occurs on the optimum sites of the region (see Plate 5); in the Great Lakes–St. Lawrence Region, the best site occupied by balsam fir is the white spruce–balsam fir–paper birch type (No. 36a according to the Society of American Foresters, 1954); and in the Acadian-Appalachian Region it is the red spruce type (No. 32a). In the two latter regions the best site which balsam fir can occupy remains below the regional optimum. In the Lake Edward Forest in Quebec (Section L-4a of Rowe, 1959) balsam fir shows its best production and reproduction in relation to other species in the Oxalis-Cornus type, which develops on sandy glacial till and has a podzol profile (Heimburger, 1941). In Minnesota the balsam fir–basswood community on heavy clay soils described in Itasca County by Grant (1934) could probably be considered as the edaphic optimum available for balsam fir in this part of the Great Lakes–St. Lawrence Region.

PHYSICAL PROPERTIES OF SOIL

Soil texture is a frequently used characteristic in the study of growth relationships. Donahue (1940) performed analyses for the four major forest types in the New York Adirondacks. Figure 9 gives the characteristic soil profiles and the prevailing species composition.

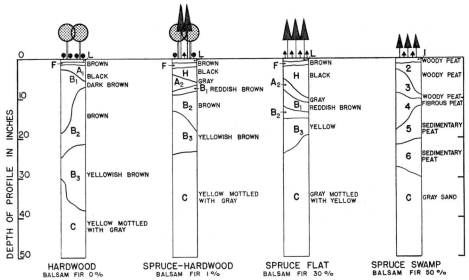

Figure 9. Soil profiles of major spruce-fir forest types in the Adirondacks, New York. (Redrawn with permission from Donahue, 1940.)

On a volume basis balsam fir was distributed approximately 1 per cent in the spruce-hardwood type, 30 per cent in the spruce flat type, and 50 per cent in the spruce swamp type. Table 12 presents the soil texture data. Rock outcrops and boulders occupied 17.1 per cent of the total soil volume in the spruce-hardwood type, 13.4 per cent in the hardwood type, and 7.7 per cent in the spruce flat type. There was almost no difference in percentage of clay among the hardwood, spruce-hardwood, and spruce flat types, although the A and B horizons of the spruce-hardwood type had a slightly higher silt plus clay content than the hardwood type. The pH values ranged from 3.12 to 5.77, with acidity for various forest types increasing in the following order: hardwood, spruce-hardwood, spruce swamp, and spruce flat. Total height at maturity increased in the following order: spruce swamp, spruce flat, spruce-hardwood, and hardwood. The number of roots per square foot of rock-free soil surface in the vertical plane increased in the same order. Roots were excavated to the full depth of rooting, 3–5 feet deep. In all types the greatest concentration of roots was in the F-layer. The productivity figures for these types are given in Chapter 9; synecological descriptions are in Chapter 2.

Plate 12 presents some characteristics of the edaphic field or the distribution of 55 forest communities in moisture-nutrient axes for the Central Minnesota Pine Section. This area has a wide range of moisture and nutrient conditions as compared with other Minnesota forest sections. The Central Minnesota Pine Section is located between Itasca State Park, Red Lake, Grand Rapids, and Brainerd, an area occupied predominantly by the gray-wooded Nebish-Rockwood, and the brown-podzolic Menahga soils. Diagram 9 shows the silt plus clay percentage of the upper soil horizon and its general but wave-like increase as moisture intensity and nutrient level increase. The thickness of the organic layer increases with moisture. Silt plus clay in stands with a measurable basal area of balsam fir ranged from 3 per cent to 85 per cent. Thickness of the organic layer varied from 1 inch to 12 feet. The highest groundwater level observed was at one foot below the surface.

TABLE 12. ANALYSIS OF FOREST SOILS BY TEXTURE CLASS
PERCENTAGES IN THE ADIRONDACKS, NEW YORK

Soil Horizon	Coarse Sand	Fine Sand	Silt	Clay
HARDWOOD (BALSAM FIR 0%)				
A_1 ....41.5	42.1	13.6	2.8
B_1 ....51.7	32.8	12.5	3.0
B_2 ....39.7	41.6	15.3	3.4
B_3 ....59.5	25.5	11.8	3.2
C_1 ....71.6	15.2	10.6	2.6
SPRUCE-HARDWOOD (BALSAM FIR 1%)				
A_2 ....48.6	30.9	17.0	3.5
B_1 ....36.1	44.3	15.7	3.9
B_2 ....50.0	29.1	16.5	4.4
B_3 ....56.2	22.2	18.6	3.0
C_1 ....70.3	20.6	7.4	1.7
SPRUCE FLAT (BALSAM FIR 30%)				
A_2 ....33.6	37.7	25.2	3.5
B_1 ....30.4	52.3	14.3	3.0
B_2 ....72.0	18.0	7.0	3.0
B_3 ....47.0	37.5	11.8	3.7
C_1 ....82.7	7.7	8.6	1.0

Source: Donahue, 1940.

Kell (1938), working in Itasca Park, Minnesota, suggested that the occurrence of different species is closely correlated with the moisture equivalent of the soil. She found an average moisture equivalent of 8.0 (ranging from 2.3 to 25.1) for balsam fir and 12.0 (ranging from 6.0 to 22.7) for basswood. Since measurements of moisture equivalents exclude the effect of the soil organic constituents, the values bear a direct relationship to soil texture. Her samples were taken from each decimeter of mineral soil depth from 10 to 60 cm. Averages from five samples were considered as representative of the soil profile.

Lee (1924) investigated the physical properties of soil under different forest cover types in Itasca Park, Minnesota. Information contained in Table 13 is taken from study, which does not distinguish between temperature differences due to cover characteristics and those due to intrinsic soil characteristics.

Lieth and Quellette (1962) studied soil respiration in some forest communities in the Boreal Forest Region in the Gaspé Peninsula, Quebec. In a mature stand of the Abietum balsameae association growing on loamy soil without an understory, but with a thick cover of forest litter, a soil respiration rate of 65 mg CO_2/m^2/hour was found. In a young stand with a rich understory but with otherwise similar conditions, the rate was 81 mg CO_2/m^2/hour. Measurements were made in July and August 1960 during which the daily average temperature was about 13–15C and monthly precipitation was about 30 mm. The values obtained are characteristic of low productivity, as compared with values obtained by similar methods in the temperate zone of central Europe.

Temporary flooding and stagnant water affect many other factors and may cause complicated changes in vegetation. The composite effect of temporary natural lake shore flooding was studied by Ahlgren and Hansen (1957) in northern Minnesota. Balsam fir 1–2 feet tall were submerged for varying periods up to 48 days. They concluded that the growth rate of the terminal shoots, ability of foliage to endure and recover from submergence, and tree mortality of balsam fir and

TABLE 13. AVERAGE SOIL TEMPERATURE IN DEGREES C, FOR
SEPTEMBER 3–16, 1922, IN ITASCA STATE PARK, MINNESOTA

Forest cover	Depths of Reading (Ft)			
	0.5	1.0	2.0	3.0
Pinus banksiana	14.31	14.09	14.16	14.20
Pinus resinosa	14.28	14.10	14.14	14.16
Acer saccharum	14.10	14.01	14.08	14.09
Picea-Abies	13.82	13.78	13.80	13.81

Source: Lee, 1924.

black spruce were affected less than white spruce, white pine, or red pine. However, aspen, balsam poplar, black ash, red maple, bur oak, and basswood had greater tolerance to flooding than any conifer. Earlier, Gates and Woolett (1926) studied the more prolonged effects of beaver dam building in Carp Creek at Douglas Lake, Cheboygan County, Michigan. Ninety-five per cent of all balsam fir and 71 per cent of all white-cedar were dead or dying. Severe damage was also noted in red maple although American elm and black ash withstood the flooding well.

Averell and McGrew (1929) investigated swamp forests in northern Minnesota and concluded that balsam fir was not a dominant species. Furthermore, they found it only where peat was less than three feet deep. Even after drainage of stagnant sphagnum bogs, increased leader growth of balsam fir was found to be very slow as compared with black spruce (Satterlund and Graham, 1957).

In the Central Pine Section of Minnesota discussed previously the greatest depth of peat at which a species was able to reach a dominant or codominant status was as follows: tamarack, 18 feet; black spruce, 15 feet; balsam fir, 10 feet; black ash, 5 feet; white-cedar, 1 foot; American elm, 6 inches. On the average balsam fir codominance, however, was limited by a 9-foot depth of peat. Balsam fir reproduction was found on peat up to 19 feet deep. These generalizations are strongly influenced by limited data. For example, white-cedar has been found on bogs with 7 feet of peat in northern Minnesota and on 20 feet of peat in the Cedar Creek Preserve in east-central Minnesota. According to Pierce (1953) deficiency in oxygen and low values of redox potential do not prevent reproduction of black spruce, white-cedar, balsam fir, and speckled alder but are mainly associated with slow growth. Steward (1925) listed the swamp species beginning with those capable of enduring poorly drained sites in the Upper Peninsula of Michigan as follows: *Larix laricina, Picea mariana, Thuja occidentalis, Abies balsamea, Fraxinus nigra, Acer rubrum, Pinus strobus, Populus tremuloides,* and *Betula papyrifera.*

Physical properties of soil affect all major constituents of the ecosystem matter-energy balance: moisture, nutrients, heat, and light. Heat and light act upon the vegetation primarily through its aerial parts and are commonly considered to be climatic factors. Although precipitation is a major climatic effect, moisture differences within a climatic region are expressed entirely (except for interception) through the rooting zone of vegetation, so that moisture then becomes an edaphic factor. In contrast to nutrients, however, its action is primarily of a physical nature. Therefore the relationship of species to the moisture regime is considered in this section as relating to the physical properties of soils.

Relative moisture requirements or tolerances of prevailing species have been presented by many authors on local bases. Most descriptions of a species contain at least some relative evaluation of its relationships to moisture conditions. The Society of American Foresters (1954) cover types are organized according to the occurrence of type dominants in a sequence of increasing moisture intensity. Table 14 shows the position of balsam fir among other species according to their average moisture synecological coordinates in Minnesota, higher values denoting moister conditions.

TABLE 14. SYNECOLOGICAL COORDINATES OF MOISTURE ON A RELATIVE SCALE FROM 1 TO 5 INDICATING THE PREVAILING OCCURRENCE OF TREE SPECIES IN MINNESOTA FOREST COMMUNITIES

Tree Species	Moisture Coordinates	Tree Species	Moisture Coordinates
Juniperus virginiana	1.0	*Populus deltoides*	2.9
Quercus ellipsoidalis	1.0	*Betula papyrifera*	2.9
Pinus banksiana	1.0	*Ulmus thomasii*	2.9
Pinus resinosa	1.1	*Fraxinus pennsylvanica*	3.0
Populus grandidentata	1.3	*Acer saccharinum*	3.2
Quercus rubra	1.5	*Acer saccharum*	3.2
Quercus macrocarpa	1.5	*Celtis occidentalis*	3.3
Juglans nigra	1.6	*Acer negundo*	3.3
Pinus strobus	1.7	*Ulmus americana*	3.5
Acer rubrum	1.9	ABIES BALSAMEA	3.6
Populus tremuloides	2.0	*Populus balsamifera*	3.7
Quercus alba	2.1	*Betula alleghaniensis*	3.7
Ostrya virginiana	2.4	*Fraxinus nigra*	3.9
Tilia americana	2.5	*Thuja occidentalis*	4.2
Carya cordiformis	2.7	*Picea mariana*	4.5
Picea glauca	2.8	*Larix laricina*	5.0

Source: Bakuzis, 1959.

Of 32 major forest species in Minnesota, only 6 occurred on the average at greater moisture intensity than balsam fir. There are 16 species common to both Table 14 and the Society of American Foresters (1954) forest cover-type moisture sequence. The rank correlation coefficient between these two sequences is +0.83 and significant at the 1 per cent level. Balsam fir ranked twelfth in both compared sequences.

CHEMICAL PROPERTIES OF SOIL

The difficulties in evaluating chemical soil analyses in relation to tree growth are great, and such information is not abundant. For balsam fir, even illustrative data are scarce. Some insight with respect to the average composition of the A_1 soil layer in Wisconsin community types including balsam fir can be obtained from Table 15 which contains data derived from Curtis (1959). Other information on Wisconsin forest community types is given in Table 6 and Plate 9.

TABLE 15. SUMMARY OF SOIL ANALYSIS FOR WISCONSIN PLANT
COMMUNITY TYPES INCLUDING BALSAM FIR AS AN
ASSOCIATED SPECIES

Community type	% Water Retaining Capacity	pH	Available Nutrients (ppm)			
			Ca	K	P	NH$_4$
Northern dry forest 120		4.9	1255	94	?	?
Northern dry-mesic forest ...127		5.2	1985	77	?	?
Northern mesic forest 247		5.6	4215	93	?	?
Northern wet-mesic forest ..495		5.5	3780	84	9	18
Northern wet forest 670		4.7	755	109	14	23
Wisconsin boreal forest 223		5.1	2785	150	35	18

Source: Curtis, 1959.

According to soil analysis data for upland forests of the Minnesota Pine Section, total nitrogen content varied from 0.08 per cent to 0.27 per cent, the amount of available phosphorus ranged from 14 to 40 ppm, and the potassium content varied from 30 to 130 ppm in the A_1 soil layer in those forest communities having measurable balsam fir basal area. Nitrogen content was the highest, and phosphorus and potassium were the lowest in the top 6-inch layer of that portion of the upland forest with balsam fir (see Plate 13).

Balsam fir has been observed growing on soils with a great range of soil acidity. The Oxalis-Cornus forest type in the Lake Edward area (Section L-4a, Rowe, 1959) is known to have the best development of balsam fir as compared with other species. The pH in the F-layer was 4.9; and in the H-layer, as low as 4.2 (Heimburger, 1941). Wilde (1953, 1954) found that balsam fir in Wisconsin was growing well on strongly acid soils (pH of 4.6), and was surviving on soils with pH values less than 4.0. Table 15 presents additional information on pH values in Wisconsin forest communities where balsam fir was present. Averell and McGrew (1929) found the species in northern Minnesota growing in drained bogs on peat 0.7–1.0 feet thick with pH ranging from 6.5 to 6.7. (Some pH-values for forests in central Minnesota are given in Plate 12.) Optimum growth occurs on soils where pH of the upper organic layer is between 6.5 and 7.0.

Balsam fir becomes more suppressed by other species with diminishing pH values (compare Plates 12 and 16). The species was reported on Mackinac Island growing on limestone rocks with a pH of 6.0 to 8.0 (Potzger, 1941). These figures demonstrate the great adaptability of balsam fir to a wide range of soil acidity conditions.

A pH value of 3.9 was reported (Melin, 1930) for freshly fallen balsam fir needles collected at Mount Desert Island, Maine. Maki (1950) found a pH for fresh leaves of 4.55 (4.41–4.71) in northern Minnesota. Lutz and Chandler (1951) compiled information from several sources giving the range of pH as 3.0 to 4.9 for freshly fallen needles of this species.

The deposition of balsam fir litter was studied by Chandler (1944). Some results are given in Tables 16 and 17. Table 16 can be considered as an illustration. For more general conclusions about the seasonal pattern of litter deposition or the amount deposited more data would be necessary. Table 17 gives a comparison with the litter production of other species.

The pattern of forest litter distribution (Plate 12) within the edaphic field in Minnesota indicates that forest stands with balsam fir do not accumulate large amounts of foliage litter but tend rather to accumulate dead wood (slash).

Chandler (1944) further investigated the mineral nutrient content of freshly fallen needles for several species in the New York Adirondacks, including a 25-year-old balsam fir. The trees were growing on Becket sandy loam, an acid podzol developed from syenites, granites, and gneisses (Table 18).

When compared with other data for the United States compiled by Lutz and Chandler (1951), it appears that many hardwood species return to the soil relatively greater amounts of calcium than does balsam fir, but that many conifers return less. Balsam fir, however, exceeds most species in the nitrogen content of its freshly fallen litter. The amount of ash in balsam fir needles, 3.08 per cent, according to Melin (1930), is comparatively low, especially when compared with hardwood species. Return of nutrients in pounds per acre according to information provided by Chandler (1944) is shown in Table 19.

Decomposition rates for litter from different tree species were reported by Melin (1930). He attributed the low decomposition rate of *Abies balsamea* and *Populus grandidentata* to the high content of lignin in these species. His data also showed only 0.83 per cent nitrogen in balsam fir needles, and he considered that such low nitrogen content also retards decomposition. The data by Chandler (1944) showed 1.25 per cent nitrogen based on dry weight in freshly fallen needles. Maki (1950) found 1.23 per cent nitrogen in fresh needles and 1.33 per cent in dead needles.

The relative position of balsam fir according to its occurrence at different nutrient levels appears in the primary succession diagrams worked out for different parts of the species range (see Chapter 2). Other information is scattered through many sources, particularly in botanical literature. Plate 13 shows the distribution of balsam fir and its reproduction in soil N-P-K coordinates. Table 20 shows the relative position of balsam fir among 33 Minnesota forest species according to their

TABLE 16. OVEN-DRY WEIGHT OF FRESH BALSAM FIR LITTER
IN THE ADIRONDACKS, NEW YORK

Period	Lbs per Acre
May 1 to July 1, 1942	198.0
July 1 to August 20, 1942	90.5
August 20 to October 3, 1942	1125.0
October 3 to May 27, 1943	1157.0
Total annual deposition	2570.5

Source: Chandler, 1944.

TABLE 17. ANNUAL DEPOSITION OF LITTER IN THE
ADIRONDACKS, NEW YORK

Species	Lbs per Acre Oven-Dry
Balsam fir	2570.5
Norway spruce	3660.6
Red spruce	1670.1
White pine	2730.4
Red pine	3367.4
White-cedar	2009.5

Source: Chandler, 1944.

TABLE 18. MINERAL NUTRIENT CONTENTS (IN PERCENTAGES) OF
OVEN-DRY WEIGHT OF FRESH LITTER FOR DIFFERENT SPECIES
IN THE ADIRONDACKS, NEW YORK

Species	Ca	Mg	K	P	N
Balsam fir	1.12	0.16	0.12	0.09	1.25
Norway spruce	1.96	0.23	0.39	0.09	1.02
Red spruce	0.79	0.20	0.35	0.10	0.89
White pine	0.60	0.16	0.18	0.05	1.14
Red pine	0.58	0.18	0.35	0.07	0.69
White-cedar	2.16	0.15	0.25	0.04	0.60

Source: Chandler, 1944.

TABLE 19. AMOUNTS OF NUTRIENTS IN POUNDS PER ACRE
RETURNED TO THE SOIL ANNUALLY BY DIFFERENT TREE
SPECIES IN THE ADIRONDACKS, NEW YORK

Species	Ca	Mg	K	P	N
Balsam fir	28.8	4.1	3.1	2.3	32.1
Norway spruce	65.4	7.7	13.2	3.0	34.4
Red spruce	13.2	3.3	5.8	1.7	14.9
White pine	16.4	4.4	4.9	1.4	31.2
Red pine	19.5	6.3	11.8	2.4	23.2
White-cedar	43.3	3.0	5.0	0.8	12.1

Source: Chandler, 1944.

TABLE 20. SYNECOLOGICAL COORDINATES OF NUTRIENTS ON A RELATIVE SCALE
FROM 1 to 5 INDICATING THE PREVAILING OCCURRENCE OF TREE
SPECIES IN MINNESOTA FOREST COMMUNITIES

Tree Species	Nutrient Coordinates	Tree Species	Nutrient Coordinates
Larix laricina	1.0	*Fraxinus nigra*	3.5
Pinus banksiana	1.1	*Quercus rubra*	4.2
Picea mariana	1.2	*Betula alleghaniensis*	5.0
Pinus resinosa	1.6	*Juglans nigra*	5.0
Quercus ellipsoidalis	1.8	*Fraxinus pennsylvanica*	5.0
ABIES BALSAMEA	2.0	*Tilia americana*	5.0
Betula papyrifera	2.1	*Ulmus americana*	5.0
Pinus strobus	2.1	*Populus deltoides*	5.0
Picea glauca	2.2	*Ostrya virginiana*	5.0
Thuja occidentalis	2.3	*Acer negundo*	5.0
Populus tremuloides	2.3	*Acer saccharinum*	5.0
Acer rubrum	2.3	*Acer saccharum*	5.0
Juniperus virginiana	2.6	*Quercus alba*	5.0
Populus balsamifera	2.7	*Ulmus thomasii*	5.0
Populus grandidentata	3.0	*Carya cordiformis*	5.0
Quercus macrocarpa	3.2	*Celtis occidentalis*	5.0

Source: Bakuzis, 1959.

average occurrence at different intensities of available nutrients on a relative scale.
Values increase with nutrient levels.

The Deciduous Forest Formation is rather poorly represented in Minnesota,
which probably explains the failure to differentiate among species at the highest
nutrient levels. Balsam fir ranks sixth from the oligotrophic end of the scale in
Minnesota forests, although in the Boreal Forest Region it is among species occupying the richest sites. As with other ranked sequences, the variability should be
evaluated from species ecographs (see Plate 11).

Wildlife: A Biotic Factor

The interacting ecological relationships of forests and their vertebrate inhabitants, both animals and birds, are often not recognized and are seldom fully understood. The forest with its trees, associated shrubs, vines, herbs, fungi, and other
kinds of plants, its lower animals, and its own special microclimate furnishes the
general matrix environment of its wildlife inhabitants. Wildlife, while seldom an
obviously major ecological factor affecting the forest, does influence, sometimes
in a controlling way, the development of the vegetation. Under certain conditions
wildlife may be of considerable importance in influencing plant succession and
the species which become the dominants of the permanent vegetation.

This review has been expedited by the monographic publications which have
been prepared for several wildlife species. Of special interest are *North American
Moose* (Peterson, 1955), *The Ruffed Grouse* (Edminster, 1947), *The Ruffed
Grouse, Life History, Propagation, Management* (Bump et al., 1947), and *The
Deer of North America* (Taylor, 1956).

Examination of range maps of animals and birds of Canada reveals many which
overlie the range of balsam fir but few which approximately coincide (Can. Geogr.

Branch, 1957: plates 39, 41, 42, 43). This review is restricted to the few which appear to have the closest affinity to the northern coniferous forest and have some specific meaning for balsam fir.

Carnivorous animals such as the lynx, marten, fisher, wolverine, wolf, and others associated with the northern coniferous forest are largely omitted in this discussion. These animals eat little plant food; their ecological requirements relate to the vegetation only indirectly through their prey. It is possible that some of these carnivores — those which travel or find their prey in tree tops — may have a stronger dependency on vegetation of a certain type, although little information is available on this aspect of animal ecology.

GENERAL ECOLOGICAL AND TAXONOMIC PROBLEMS

Many of the problems with which the animal ecologist deals are held in common with plant ecologists. This is an obvious result of the general dependent relationship which animals have within the plant community. As a consequence of the primary role of vegetation and of the fact that community classification has been predominantly the work of plant rather than animal ecologists (Dice, 1952), such classifications are not always the most useful in dealing with problems of animal ecology.

There are few references giving systematic consideration to the distribution of animals among the major plant geographic areas of the continent. Merriam (1898) described Hudsonian and Canadian life zones including the characteristic animals. Shelford (1945, 1963) delineates the biomes of North America and describes the Boreal Coniferous Forest biome including dominant and influent plants and animals. Wolverine, lynx, red squirrel, and certain races of snowshoe rabbit are reported present throughout nearly the entire biome. Woodland caribou was historically a permeant dominant present in all faciations in the central area of this transcontinental forest, but balsam fir is not included in a listing of its foods. Moose is reported as inhabiting spruce–balsam fir climax as well as subclimax stands. Shelford notes that both the caribou and moose tend to maintain subclimax conditions by their feeding preferences. Subclimax types are probably optimum for these species. Spruce grouse, northern flying squirrel, and red squirrel are listed as minor permeant influents having preference for the spruce–balsam fir climax. Shelford also describes the southern ecotone between the boreal spruce–balsam fir and the deciduous forest as a red pine–moose faciation having the same dominants and permeant influents as the boreal forest with few species of animals characteristic of it.

Problems peculiar to animal ecology and resulting in added complexity result from the mobility of the animal organisms. In his book *Natural Communities*, Dice (1952) points out the value of "community interspersion" to those species whose daily and seasonal mobility takes them to several different kinds of communities. A related principle is involved in the habitat preferences of many animals for "edge cover" (Leopold, 1933).

In the use of scientific names dependence has been placed on such comprehensive listings as those by Miller and Kellogg (1955); the American Ornithologists' Union (1957); as well as on several monographic studies of individual species or genera. Mammalian names as given by Miller and Kellogg and bird names as listed by the American Ornithologists' Union are used.

Animal taxonomists have been diligent in recognizing taxonomic categories of subspecific ranks for the mammals in general. These subspecies units are largely of the nature of geographic races. However, a considerable amount of published data dealing with management problems, forest and wildlife interrelationships, and related nontaxonomic information fails to recognize the existence of these subspecies.

MOOSE AND WOODLAND CARIBOU

These animals are considered together because their ranges are generally contiguous, because they have been influenced — largely in dissimilar ways — by common historical developments, and because the sources of published information are often the same for both species.

Peterson (1955) recognized four subspecies of moose in North America. Of these, the eastern moose (*Alces alces americana* (Clinton)) and the northwestern moose (*Alces alces andersoni* Peterson) cover a range which coincides reasonably well with the general limits of balsam fir distribution except for its southern extension along the Appalachian Mountains and for western Canada where the moose covers a more extensive range than does balsam fir.

The summer food of moose is reported to be largely the leaves of broad-leaved trees and shrubs, herbs, and aquatic plants (Murie, 1934; Peterson, 1955). In winter, balsam fir becomes an extremely important constituent of the diet as reported from Isle Royale in Lake Superior (Murie, 1934; Aldous and Krefting, 1946; Krefting, 1951) and elsewhere (Cahalane, 1947). Even greater use of this species was reported from Newfoundland (Pimlott, 1953), New Brunswick (Wright, 1952), and Maine (Dyer, 1948), where balsam fir is more strongly dominant in the forest stands. Moose stomachs collected in early fall at Isle Royale (Krefting, 1951) also contained balsam fir. Moose browse on trees and shrubs by nipping twigs and stripping leaves. They also chew the bark of some species. Only a few cases of balsam fir bark being eaten were reported; aspen, mountain ash, and red maple bark are listed as highly preferred. Newsom (1937) notes the use of balsam fir as "rubbing trees" in Quebec. This is probably done to loosen and help cast the horns.

The role of fire in improving moose range on Isle Royale has been discussed by Aldous and Krefting (1946) and Krefting (1951). Hosley (1949) recommended scattered cutting in old stands rather than burning as being more effective and less destructive of the shallow organic soil layers.

According to Pimlott (1953), the degree to which balsam fir reproduction will be browsed by moose depends to some extent on the availability of large timber nearby to provide suitable escape and winter cover. This suggests avoidance of selection or group selection silvicultural methods where high moose populations exist on forest management areas.

In 1946 a browse survey on Isle Royale (Aldous and Krefting, 1946) showed that balsam fir ranked third in importance and formed 13.6 per cent of the food eaten. The heavy moose population that built up subsequent to the 1936 fire was reported to be keeping balsam fir reproduction down to snow-level heights. In 1948 balsam fir assumed the position of first importance in moose diet, making up 21.8 per cent (Krefting, 1951). By 1950 the preferred balsam fir was being replaced by formerly unbrowsed white and black spruce, and the proportion of balsam fir in the moose diet had dropped to 5 per cent. Excessive browsing had resulted in killing 600 balsam fir trees per acre in some areas or in stunting growth so that trees of 30 years or older were only three feet tall. Any future conversion to a more pure forest of balsam fir was felt to be impossible as long as a large moose population was present.

Pimlott (1955, 1961) reported balsam fir, paper birch, and aspen as key browse species for moose in Newfoundland. He observed that as the moose population increased in density the utilization of balsam fir increased greatly until it became the most important food for the moose.

Hosley (1949) cited 1847 land survey data based on witness tree counts which showed that balsam fir made up 40 per cent of the trees, ranking first in number among all trees on Isle Royale. This was before the migration of moose to the island in 1912. He concurred that moose browsing greatly reduced the proportion of balsam fir in favor of spruce.

To some extent the range of woodland caribou (*Rangifer caribou caribou* (Gmelin)) involves balsam fir forest types. Pimlott (1953) cited information from Newfoundland indicating that these animals are most common where forest cover is present on less than a fourth of the area, the remaining portion being covered with large tracts of muskeg, barrens, and water. The species partially overlaps the range of moose, but its distribution generally lies further north. In northern Minnesota and in Canada generally several authors (Hickie, undated; de Vos and Peterson, 1951; and Peterson, 1955) recorded a succession in ungulate dominance over the past 100 years from caribou to moose to deer. Fires and logging from 1880 to 1910 were considered to be responsible for this change. Few records giving the extent that caribou injures or consumes balsam fir were found in the literature. Cringan (1957) reported that on the Slate Islands in Lake Superior, balsam fir made up 32.3 per cent of the available browse but that it constituted only 1.1 per cent of the food eaten by the caribou. He concluded that balsam fir was unimportant in caribou diet. This may change in the future, however, as both caribou and moose have been reported in the last decade in increasing numbers at the southern boundaries of their ranges.

WHITE-TAILED DEER

Of the 30 subspecies in the white-tailed deer complex only the northern woodland white-tailed deer (*Odocoileus virginianus borealis* Miller) has extensive coincidence of range with balsam fir. Numerous authors report that the white-tailed deer was originally found considerably south of the present northern limits of its distribution. Jenkins and Bartlett (1959) indicated that before 1950 only the southern part of Lower Michigan was good deer range. There the interspersion of hardwood forest, swamps, and prairie openings provided ideal habitat. Trippensee (1948) also described the white-tail as originally an inhabitant of the hardwood forest. In northern New England and southeastern Canada where the virgin forests were mostly spruce-fir, deer were practically unknown until logging encouraged the development of mixed coniferous-hardwood forests. A similar pattern of movement northward following settlement and logging was reported from Wisconsin by Dahlberg and Guettinger (1956) and from Minnesota by Erickson et al. (1961).

The food habits of white-tailed deer have been widely studied and, as might be expected of an animal so extensively distributed, reports on its food preferences are highly divergent. There is essential agreement, however, that balsam fir ranks low. Together with spruce, balsam fir is avoided if better foods are available and, except where deer populations are excessive and other browse has been depleted, damage is usually moderate. No reports were found describing the use of balsam fir for food by deer other than by browsing the needles and twig tips.

Howard (1937) reported moderate browsing of balsam fir by deer at Wilderness State Park in Michigan under conditions of critical food shortage. The trees were not browsed heavily enough to have established a browse line as was the case with white-cedar. Jenkins and Bartlett (1959), also in Michigan, called balsam fir a "stuffing" food of little value to maintaining good health, and noted it as a constituent of swamp conifer types eaten when other conifers were scarce. Erickson et al. (1961) listed 26 species of Minnesota plants commonly used for deer browse. These were placed in three preference categories: good, medium, and poor. Balsam fir was placed in the medium preference group along with such species as aspen, hazel, jack pine, and basswood. Graham (1954b) reported that in the western part of the Upper Peninsula of Michigan deer browsing in northern hardwoods has curtailed yellow birch, white-cedar, and hemlock, favoring the less desirable balsam fir and sugar maple.

Davenport (1939) conducted controlled feeding experiments and analyzed balsam fir browse. He reported a protein content of 8.25 per cent and confirmed the low rating of the species in maintaining deer health. This conclusion was reported by others (Lake States Forest Experiment Station, 1938). The protein content of balsam fir browse has been shown to be greater than that of white-cedar, a far better deer browse species. Chemical analyses of browse alone do not explain why balsam fir is a food poor for deer but excellent for moose.

Gill (1957) described balsam fir as having medium or occasionally high palatability and as being more resistant to browsing than white-cedar in Maine deer yards. He noted its value for deer food as fair in combination with other foods and indicated that as a sole diet it will not maintain deer.

On the basis of late winter and early spring observations in New York, Spiker (1933) rated balsam fir at the bottom of the food preference list. However, Pearce (1937), also in New York, found deer to have browsed on 89 per cent of the balsam fir but only 0.5 per cent of the red spruce in deer yards in the spruce-flat type. In these same yards yellow birch and red maple were almost totally browsed.

In Wisconsin, Stoeckeler et al. (1957) concluded that a 6–8 year period of rather low deer population is needed to permit successful reproduction in second-growth hardwood-hemlock stands. Balsam fir seedlings were more abundant and larger within fenced enclosures than where available to deer. In another Wisconsin report (Lake States Forest Experiment Station, 1959) red maple and aspen competing with balsam fir and spruce on the Argonne Experimental Forest were being eliminated by snowshoe hares and deer while balsam fir and spruce were browsed only very lightly. Swift (1948) found that deer damage to balsam fir seedlings was greatest in northwestern Wisconsin, an area where deer attain their greatest general abundance.

A somewhat heavier use of balsam fir by deer is reported from northern Minnesota where that tree species is more abundant than elsewhere in the Lake States. Aldous (1948) reported 62 per cent of the balsam fir on his plots were browsed. Krefting and Stenlund (1951) noted a general increase in browsing of balsam fir from 1940 to 1949 on five wintering areas in northern Minnesota. By 1949 this species made up one-third of the amount of browse consumed by deer. On the Tamarac National Wildlife Refuge, Krefting et al. (1955a) noted an increase in browsing on balsam fir from "occasional" in 1942 to the extent that there was a marked browse line in 1946. This coincided with a population build-up resulting from complete protection.

Heavy use of balsam fir by deer in Riding Mountain Park, Manitoba, was noted by Rowe (1955), to a degree that mortality of seedlings and saplings could occur. However, both deer and elk were referred to, and a separate evaluation for deer was not reported.

Rowe (1959) also noted that on Anticosti Island in the mouth of the St. Lawrence River a large resident deer herd had eliminated the understory of shrubs, including balsam fir young growth.

Observations on experimental cuttings in a mixed-wood stand in the Pasquia Hills, Saskatchewan, from 1924 to 1956, indicated that balsam fir seedlings failed to develop into saplings. This was attributed to excessive browsing (Waldron, 1957).

While balsam fir has not been widely planted in Europe and there are few references to it in connection with wildlife, Zederbauer (1919) reported it to be heavily browsed by deer in Austria.

Seasonal variation in the extent to which balsam fir is used has been given limited study. Aldous (1948) found that in Minnesota it formed 13.4 per cent of the contents of deer stomachs in the fall and ranked second in quantity of food eaten. By late winter balsam fir increased to 42.7 per cent of the food eaten, thus taking first rank among all species. In a stomach count, Aldous (see Erickson et al., 1961) reported balsam fir present in 86 per cent of 21 deer shot in Minnesota in fall 1936 and in 80 per cent of 51 deer found dead from starvation or accident in winter 1937. Although balsam fir was the most heavily used species, the authors concluded it was a starvation food since many deer were found that had their stomachs filled with its buds and twigs (Aldous, 1942). Similar results in Wisconsin were reported by Dahlberg and Guettinger (1956). Stomach analyses in the fall showed 11.5 per cent balsam fir by volume and 36 per cent by occurrence in all deer stomachs. This compared with 43 per cent by volume eaten and 85 per cent by occurrence in winter deer taken from starvation areas.

While deer do not appear to use balsam fir effectively for food, there is much evidence showing the importance of the species as deer cover. Studies in the Lake States indicate that large areas of potential hardwood winter range cannot be used by deer because of lack of coniferous cover (Taylor, 1956). In Upper Michigan (Bartlett, 1950), winter cover can be a mixture of swamp conifers and lowland and upland hardwoods. Balsam fir is a constituent of these winter cover types. Krefting (1941) found balsam fir to be the most abundant species in the understory of two major deer yards along the north shore of Lake Superior in Minnesota. Erickson et al. (1961) list balsam fir as occurring in a high proportion of the deer yards in central and northern Minnesota.

In the central Adirondacks, winter deer yards are usually in the lower level spruce slopes and spruce-balsam fir swamps (Webb, 1948). The northern hardwoods appear to be avoided. This suggests that the winter yarding areas are selected primarily for cover rather than food values since these types rate very low as food sources.

An interesting experiment in planting large-sized trees in small tracts for winter cover for deer and grouse was reported by Krefting (1959). Balsam fir as well as white and black spruce (6–8 ft) were transplanted from nearby areas into patches 1100–2000 square feet in size in aspen stands and older thickets lacking in winter cover. After 24 years, survival varied from 36 per cent to 85 per cent, and the author indicated that deer, ruffed grouse, varying hare, red squirrels, and moose had made use of the plantings as protective cover. They were particularly attractive to deer. Balsam fir appeared to be a better species for this purpose than either black or white spruce. Because it sheds its lower branches earlier, it provides more accessible winter cover. In addition it is preferred to the spruces as a deer food. To improve the cover in some Minnesota deer yards biologists of the Minnesota Department of Conservation planted balsam fir, white spruce, jack pine, and red pine in three areas in central Minnesota in 1957. The browse in these areas was considered abundant (Erickson et al., 1961).

Numerous studies have been made on such topics as season of cutting, degree of cutting, orientation and configuration of cutting areas, the use of fire, bulldozing to stimulate sprout growth, and other techniques. It is not within the scope of this discussion to review these numerous reports since many are unrelated to balsam fir. It should be noted, however, that balsam fir is highly tolerant to such herbicides as 2,4-D and 2,4,5-T (Leinfelder and Hansen, 1954). This permits the use of aerial applications of 2,4-D as described by Krefting, et al. (1955b, 1956) to stimulate browse production of many hardwood trees and shrubs without injury to balsam fir and other conifers.

HARES

The American varying hare (*Lepus americanus* Erxleben), more popularly known as the snowshoe rabbit, is the characteristic lagomorph of the northern coniferous forest of North America (Cook and Robeson, 1945). Four subspecies are recognized (Miller and Kellogg, 1955) as occurring within the general range of balsam fir. Two subspecies of the European hare (*Lepus europaeus* Pallas) have also been introduced and are reported to be rather well established in southern Ontario, Michigan, New York, and Connecticut.

It is of interest to note that the Virginia varying hare (*Lepus americanus virginianus* Harlan) has an extension of its range in the mountains of West Virginia and

Virginia which is not continuous with its range in Maine and New England, there being a gap of at least 100 miles between the two ranges (Brooks, 1955). This coincides roughly with the mountain extension of the spruce-balsam fir cover types.

Various habitat requirements have been reported for the species. Grange (1932, 1949) stressed the importance of "underbrush" and the need for a considerable portion of either aspen or balsam fir, or both, in the range. Dell (1952) stated that best success with restocking in New York was achieved in areas with many-aged blocks of spruces and pines interwoven with open lanes. Severaid (1942) reported that in Maine the hare favors softwood swamps and balsam fir thickets in the younger age classes. MacLulich (1937) indicated that in Canada coniferous swamps, willow-alder swamps, and poplar-birch second growth on logged areas and burns constitute the most widely-used habitats. In eastern Ontario cedar and spruce swamps are preferred. According to Trippensee (1948), hares are seldom found in cover not having sufficient density to shield from detection from above, and conifers must be present. He cited Seton as reporting that the chief abundance of hares is in the Canadian life zone and that they are less plentiful in the Hudsonian Zone. Brooks (1955) associated the Virginia subspecies with areas now or formerly having red spruce, while Kellogg (1939) suggested that this subspecies may have occurred in the high elevation spruce-fir forests as far as eastern Tennessee and western North Carolina.

Damage to trees by the varying hare was reported (Leopold, 1947) as being more localized in space and time than that resulting from deer browsing. Damage occurs in areas having brushy cover in the winter months and at intervals of time during which hare populations fluctuate widely. Conifers are important as food species as well as cover. Most writers noted that pines and spruces rather than balsam fir make highly favored food (Martin et al., 1951; Van Dersal, 1938; Cook and Robeson, 1945; Grange, 1932; Lake States Forest Experiment Station, 1936). Some variation in reported use of balsam fir as food has been noted. Cook and Robeson indicated that in cover containing an abundance of balsam fir seedlings the hares occasionally nibble the balsam fir but that a single bite suffices. In artificial feeding tests (Severaid, 1942) some individuals stripped the needles of small balsam fir trees while others ignored them.

It appears that this animal is not generally a serious deterrent to the regeneration of balsam fir within its main range of occurrence. On the other hand, balsam fir is important as one of the conifers providing necessary protective cover for the hare.

MICE

A mouse census taken in a spruce–balsam fir–hemlock stand on the Penobscot Experimental Forest in Maine by Grisez (1954) showed populations varying from 7 to 16 per acre. Nearly all were red-backed voles (*Clethrionomys gapperi* (Vigors)). There were also several deer mice (*Peromyscus maniculatus maniculatus* (Wagner)) and one meadow vole (*Microtus pennsylvanicus pennsylvanicus* (Ord)).

Rodent population studies on the University Forest in Maine have focused attention on the population trends over time (Quick, 1954, 1955). One of the plots studied was in a spruce-balsam fir forest cover type. Red-backed voles were most common in 1953 and deer mice were next with 13 and 9 animals per acre. No other species were reported that year. There was a sharp decrease in 1954 to 5 red-backed voles and no deer mice. The absence of deer mice was interpreted to be a sampling error because in 1955 there were 2 deer mice per acre. Shrews were almost totally absent in the population records for the spruce-balsam fir plot. The author felt that population changes were not dependent on cutting which took place in the type.

Small mammal populations in New Brunswick and in the Gaspé Peninsula were reported by Morris (1955). The red-backed voles predominated in all coniferous and mixed-wood cover types. Deer mice, cinerous shrews (*Sorex cinereus* Kerr), and short-tailed shrews (*Blarina brevicauda* (Say)) were also common. At the peak of the seasonal cycle the combined population in different stands varied from 0.6 to 15.0 per acre. Morris also reported a direct relationship between population density and amount of protective cover on the ground. Slash from partial cuttings was most favorable.

Field mouse damage to plantings in New York was studied by Littlefield et al. (1946). Damage was confined to feeding on the phloem from the roots up to the snow line during the winter. They rated balsam fir as "slightly favored" by the mice, Scotch pine and Douglas fir being the only species in the "highly favored" class and white spruce the only species rated "not favored." This intermediate rating was based on observations that about 10 per cent of the planted trees were attacked and less than 5 per cent were girdled.

Krefting (1956) observed heavy damage to both balsam fir and red pine in Minnesota. Mortality by mouse girdling was about 25 per cent the first year after planting and about 50 per cent after the second winter.

The influence of mammals and birds in retarding natural and artificial reseeding of coniferous forests in the United States was discussed by Smith and Aldous (1947). These authors listed 44 species of mammals and 37 species of birds as coniferous seed eaters. Because of their wide habitat and distribution, mice perhaps affect the seed supply most. The white-footed mouse is most important.

A study by Abbott and Hart (1960) on the Penobscot Experimental Forest in Maine revealed that mice and voles did not eat balsam fir seed if white spruce seed was available. Laboratory experiments by Abbott (1962) with white-footed mice, red-backed voles, and meadow voles confirmed their definite distaste for balsam fir seed when white pine, red pine, hemlock, red spruce, and white spruce seeds were available. It is interesting to speculate on the possible influence of this selective feeding in favoring balsam fir over spruce in the forest regeneration.

Hamilton and Cook (1940) and Hamilton (1941) point out the great abundance of small mammals above, on, and under the forest floor and emphasize their ecological and economic significance. Small mammals often number 50–200 or more per acre, although the summer breeding population seldom exceeds three pairs per acre. Small mammals are 8–35 times as numerous as birds in the same habitat, and their activities extend throughout the year. Deer mice alone eat 30 per cent of their weight or about six grams of food daily, and the total small mammal population may consume 0.85–1.2 pounds of food per acre daily.

RED SQUIRRELS

Balch (1942) described red squirrel (*Tamiasciurus hudsonicus hudsonicus* (Erxleben)) damage in the Maritime Provinces of Canada. Apparently the squirrels prefer to feed on the male flower buds of balsam fir, which occur near the tips of the twigs. By late winter large numbers of both spruce and balsam fir shoots about two inches long have accumulated beneath the trees from red squirrel cuttings begun late in summer and continued through fall and winter. Flower buds are preferred to shoot buds. Another type of injury consists in the cutting of leaders and sometimes laterals from small balsam fir and spruce. Often some of the large buds are eaten. This type of feeding apparently takes place in late winter and spring when other foods are scarce. Unusual amounts of such types of injury may be evidence of an abundance of male flowers in the approaching seed year.

Pulling (1924) suggested that in Canada and northeastern United States small mammals, especially red squirrels, may somewhat influence the establishment of balsam fir at the expense of spruce. It was pointed out that balsam fir seed is released early when there is an abundance of other food available for the squirrels and other rodents. Spruce seed is released later and is used by these animals to a greater extent.

Martin et al. (1951) noted that red squirrels were using balsam fir seed, bark, and wood for food, although chipmunks and white-footed mice were found eating only the seeds.

BEAVER

Little actual data are available on the extent to which beaver (*Castor canadensis* Kuhl) affect balsam fir. Vesall et al. (1948) found balsam fir present on only 2 per cent of 1635 acres flooded by beaver in Koochiching County, Minnesota, black spruce being the main species affected. In the Penobscot Experimental Forest in Maine, Hart (1956) reported beaver to have caused, by flooding, 14 per cent of all mortality on spruce-fir stands studied over a seven-year period.

In Michigan, Gates and Woollett (1920) reported 95 per cent mortality of balsam fir from flooding. Balsam fir was also cut by the beaver and may have been used in dam building. Krefting (1963) listed balsam fir with other species cut by beaver for food or dam building.

Patrick and Webb (1954) noted that in some beaver colonies in the Adirondacks the beaver are reluctant to cross through balsam fir, spruce, or cedar fringes to obtain aspen for their food. A number of contradictory reports of balsam fir used as food by beaver have been made. Johnson (1927) observed beaver-cut balsam fir in the Adirondacks, but there was no evidence of its use for food. Van Dersal (1938) reported its use as a building material but never as food. However, Martin et al. (1951) listed balsam fir seed, bark, and wood as minor components in the beaver diet.

PORCUPINE

While there has been considerable study of the damage done by porcupine (*Erethizon dorsatum* (Linnaeus)), little published information relative to balsam fir could be found. Reeks (1942) studied porcupine damage in the Maritime Prov-

inces. He found the food preference in New Brunswick to vary in the following decreasing order: spruce, pine, larch, balsam fir, hemlock, birch, beech, maple, aspen.
In Nova Scotia it appears to be spruce, birch, hemlock, balsam fir, beech, larch,
maple, and pine. In the mixed forests of northeastern Wisconsin, Krefting et al.
(1962) found a somewhat different preference pattern. Tamarack, yellow birch,
sugar maple, and eastern hemlock were severely attacked. Pines were somewhat
less susceptible, and little or no damage was recorded for balsam fir, white-cedar,
and paper birch.

Bark is the main source of winter food, while herbaceous growth constitutes
the main diet in spring and summer. Among balsam fir, spruce, and larch, the
entire stem of young trees was frequently stripped, while on other species the
bark feeding was more apt to be in patches (Reeks, 1942).

The use of balsam fir for food by porcupine was also recorded by Van Dersal
(1938) and Martin et al. (1951).

BEAR

Few published reports of bear (*Euarctos americanus americanus* Pallas) damage to balsam fir could be found. In Maine, Zeedyk (1957) found that bear girdled
balsam fir, red spruce, and white-cedar in pole and timber-sized trees. Balsam fir
seemed to be preferred to spruce, and locally the damage could be quite severe.
Bears also lick the sap and gum from the inner bark.

On the Lake Edwards Forest Experimental Area in Quebec, Ray (1941) reported very serious mortality to balsam fir by bears clawing off the bark to obtain
the juices beneath. Over 90 per cent loss was sustained on one sample plot where
the growth of the balsam fir had been greatly stimulated by the girdling of suppress-
ing hardwoods. The damage was only slight on the adjacent plot trees which had
not been released from hardwood competition.

Etheridge and Morin (1961) reported on decay associated with bear damage in
the Gaspé Peninsula, Quebec. They found that basal scars 11–15 years old had
induced decay from *Stereum chailletii* and *S. sanguinolentum*, extending upward
an average of 4.5 feet in the trees.

GROUSE

Of the North American birds commonly called "grouse," two species are birds
of forest habitat with ranges lying appreciably within the general range of balsam
fir. These include three subspecies of the spruce grouse (*Canachites canadensis*
(Linnaeus)) and the ruffed grouse (*Bonasa umbellus* (Linnaeus)), with a number of subspecies variously recognized by different authorities.

The spruce grouse, with its three subspecies — *C. canachites canachites* (Linnaeus), *C. canachites canace* (Linnaeus), and *C. canachites osgoodi* (Bishop) —
appears to be strongly associated with northern climax conifers, including spruce-
balsam fir in its various associations. Bent (1932) calls it a bird of the wilderness,
largely dependent on buds, tips, and needles of spruce, balsam fir, and larch for
its winter diet. Martin et al. (1951) listed balsam fir as making up 5–10 per cent of
the fall and winter diet. Grange (1948) also related the adverse effect of settlement
on the species in Wisconsin where it is now rare, the few individuals remaining

being confined to the immediate vicinity of undisturbed coniferous forest. From the limited amount of study which has been given by wildlife biologists to the spruce grouse, it is not clear to what extent it is specifically dependent on the spruce-fir Boreal Forest as opposed to other conifers, nor can it be determined whether its elimination from much of its more southern range is a result of man's activity as a predator or his general disturbance of cover conditions through logging and settlement.

Bump et al. (1947) described the ruffed grouse as characteristically transitional and boreal in distribution, with its optimum range in southern Canada and the northern United States. Considering all its subspecies, however, the total range extends considerably south of these optimum areas (American Ornithologists' Union, 1957). Edminster (1947) described three types of grouse range based on the dominant plant associations and land use patterns in northern New York and New England, and the northern and middle Appalachians. In the former area spruce-balsam fir types are the "most indicative" and northern hardwoods second. The spruce-fir type is not important for grouse in the Appalachians.

A résumé of the diet reports (Bent, 1932; Bump et al., 1947; Edminster, 1947; Grange, 1948; Martin et al., 1951) for ruffed grouse indicates that balsam fir is not an important item in its diet. However, it is important in providing fall, winter, and early spring shelter. King (1937) mentioned balsam fir thickets as forming desirable late-winter and early-spring cover if the thickets are near openings and near such shrubs as bearberry (*Arctostaphylos uva-ursi*), the fruit of which is used as food when uncovered by snow melting in the spring. Balsam fir could probably be replaced by spruce and cedar for this function. Magnus (1949) and Marshall and Winsness (1953) studied the ruffed grouse habitats in the Cloquet Forest in Minnesota. In general it appears from these studies that in the spring ruffed grouse prefer stands with an admixture of balsam fir more than in other seasons.

OTHER BIRDS

It is recognized that the great number and variety of small, forest-inhabiting birds are of some significance to the forest and certainly of interest and importance in themselves. The ecological significance of songbirds in balsam fir forests has not been studied systematically, however, and only the few reports available are discussed here.

It is possible that seed-eating birds may play a role of some significance in adversely affecting regeneration of certain tree species. In some cases seed germination may be assisted by the passage of seeds through birds (Krefting and Roe, 1949). Insect eaters may be of help in maintaining populations of certain insects at endemic levels. It is also possible that in some situations birds may act as carriers of spores of destructive forest pathogens. Not much has been done to focus attention on these problems in relation to balsam fir.

An extensive and useful summary of the published literature dealing with North American wildlife including birds and the plant materials which they use has been prepared by Martin et al. (1951). The firs have been grouped together and rated as moderately important to a number of birds, their evergreen foliage being particularly useful for winter cover. Of the songbirds listed, four are com-

monly found within the range of balsam fir and have it listed as an item in
their diet. These birds together with the estimated proportion of balsam fir in their
diet are as follows: boreal chickadee (*Parus hudsonicus* Forster) 2–5 per cent,
red crossbill (*Loxia curvirostra* Linnaeus) ½–2 per cent, white-winged crossbill
(*Loxia leucoptera* Gmelin) 2–5 per cent, and yellow-bellied sapsucker (*Sphyrapicus varius* (Linnaeus)) ½–2 per cent. In view of the relatively small use made of
balsam fir it seems doubtful that the small bird population could exercise any significant effect on balsam fir regeneration.

Kendeigh (1947) made a study of bird populations during a spruce budworm
outbreak in the Superior section of the Boreal Forest (B-9, in Halliday, 1937). He
noted that an application of one pound of DDT in an oil spray in late May and
June to check the outbreak produced a small immediate mortality of birds, but
did not affect the size of the total breeding population nor the success of raising
young. Principal forest species included an overstory of paper birch, jack pine, and
white spruce, with an understory of balsam fir, black spruce, and white spruce.
The breeding population was at least 319 pairs per 100 acres in the forest and at
least 169 pairs in the cutover area. The spruce budworm was very abundant and
constituted the chief food for several species, especially warblers, vireos, kinglets,
and chickadees.

An analysis of bird populations in relation to forest succession in Algonquin
Provincial Park, Ontario, was made by Martin (1960). He reported that fairly
distinct bird communities could be recognized for bogs, the spruce–balsam fir
Boreal Forest, the Hemlock Forest, and the Deciduous Forest. The Boreal Forest
had 232 territorial males per 100 acres, including 32 different species. This was a
greater variety of species than in any of the other vegetational communities, but in
total numbers there were more birds per 100 acres in the bog and hemlock communities. It is interesting to note that studies in the spruce forest (Piceetum
myrtillosum) in the Leningrad area of Russia revealed a maximum of about 300
birds per 100 acres. Population densities were greatest in thicket stage stands, but
the number of species was highest in mature stands (Pospelov, 1957, as cited in
For. Abstracts, 19(3): 331).

Conclusions

1. The available information on balsam fir and its relationships to climatic and
edaphic factors do not cover the field systematically. Basic information from physiological and ecological experiments is lacking. Study of the effects of competition
on the relationships of the species to its physical environment is needed to properly assess species adaptation to environmental conditions.

2. Further development of ecosystem models will aid in the solution of complex
interrelationships. The available models involving balsam fir primarily present
static relationships and do not include feedback systems. Relative estimates should
be replaced by direct measurements, and techniques for integrating diversified
measurements already available need to be developed to produce information of
greater precision and validity.

3. Balsam fir withstands very low winter temperatures, but its reproduction
may be hindered occasionally by early frosts and winter drying. It grows well in

areas having low light intensities but abundant precipitation. Whether balsam fir is affected more or less adversely than other species by snow breakage, hail, and glaze damage is not known, but it is more subject to fire damage than a number of other species.

4. Wind, particularly in effecting blowdown and tree breakage is of basic concern in the silviculture of balsam fir. A large scale survey of wind effects would be desirable to furnish information basic to developing preventative silvicultural measures.

5. For practical evaluation of the complexity of ecological factors, regional forest classification systems need to be developed. Forest cover and forest environments should be treated concurrently.

6. The site index, especially of shade tolerant species like balsam fir, is much affected by species competition, and great care must be exercised in its use to evaluate the potential productivity of the physical environment. Soil-based site index curves, currently popular in other forest regions, should be explored for their possible usefulness in species of the Boreal Forest.

7. Few data from soil analyses are available which can be related to the growth of balsam fir. Especially lacking is information from low-lying swamp and peat areas. Few data are available from peat analyses. Soil acidity data indicate that balsam fir can grow over the whole range of pH values found in forest soils, but there are few data available giving the optimum pH conditions for growth.

8. Balsam fir annually deposits a moderate amount of litter that has a rather high acid reaction and a low ash content. The latter, however, is moderately high in calcium and high in nitrogen content. Additional data on litter content and productivity may be useful in evaluation of sites for balsam fir.

9. Studies in wildlife management and applied animal ecology frequently fail to consider the subspecies of the animal involved. Studies at the subspecific level may reveal variations in environmental affinities within the species and enable wildlife managers to more adequately assess the potential of habitat improvements.

10. Because of the great dependency of many wildlife species on "edge cover" and interspersed vegetational patterns, the silviculturist with a knowledge of these ecological requirements can exercise considerable control over the productivity of the forest for the wildlife resource. At the same time he should be able to so manipulate the forest that excessive damage by wildlife is avoided or at least much reduced.

11. Wildlife biologists, like foresters, are primarily concerned with the basic problems of harmonizing organisms with their environments, and with manipulating the latter to obtain the optimum productivity in terms of the organisms. Because of these common interests, wildlife biologists, like foresters, have need of synecological classifications which will identify ecosystems in relationship to animals and plants. The extent to which the same classification can serve the needs of foresters and wildlife biologists must, of course, be a matter of investigation.

12. From the standpoint of wildlife, much remains to be learned about the relative productivity of spruce–balsam fir stands handled on a partial cutting versus a clear-cutting basis with respect to the ways in which the different habitat needs for such species as deer, moose, and ruffed grouse are met.

4

MICROBIOLOGY

THIS chapter reviews an important segment of biotic factors involved in the ecology of balsam fir. It considers all microorganisms directly and indirectly associated with its reproduction and growth from the seedling stage to maturity, with its decomposition in the forest, and with the deterioration of its wood products.

Early studies were primarily concerned with the taxonomy of the microorganisms and the gross effects produced in the host. Several decades ago principal emphasis was placed on the pathogenic fungi and the economic losses they produced. More recently, the physiology of the microorganisms and their ecologic relationships to each other and to their host have been given much needed attention. Canadian investigators in particular have led the way in studies aimed at solving some of the complex microbiological problems.

Microorganisms Involved

Saprophytic and parasitic fungi constitute the majority of microorganisms that are included in the microbiology of balsam fir, and most of this chapter is concerned with them. However, the literature contains occasional references to viruses, bacteria, slime molds, algae, and lichens. The little attention given to this miscellaneous group of microorganisms is a result of lack of evidence that they are responsible for economic losses. Future microbiological studies should give greater emphasis to these microorganisms and should attempt to elucidate the roles they play in the reproduction, growth, and deterioration of balsam fir. There is reason to believe that they may prove to be far more important than presently suspected, and that the ecologic, symbiotic, and pathological relationships with which viruses, bacteria, slime molds, algae, and lichens are associated may be far more complex than those for fungi. The information available on these miscellaneous microorganisms is reviewed prior to presenting the abundant literature on the fungi.

MISCELLANEOUS MICROORGANISMS

No virus disease of balsam fir has been reported. However, the occurrence of virus diseases of the spruce budworm (see Chapter 5), the most important insect enemy of balsam fir, indicates that viruses cannot be dismissed as part of the microbiological complex of this species.

Bacteria are important in humus decomposition and soil development. Bacterial diseases of the spruce budworm are very important in the control of this insect. Bacteria are found in areas of high moisture content in balsam fir heartwood and appear to be closely associated with its development as first noted by Schrenk (1905). Observations on the occurrence of so-called wetwood and bacteria in balsam fir have recently been summarized by Hartley et al. (1961).

Etheridge and Morin (1962) studied the occurrence and development of wetwood in balsam fir. Through a detailed examination made by sectioning 126 trees, they determined the location of wetwood or high-moisture-content zones within trees, the relationship of these moist areas to branch stubs and defects, and the moisture content of affected wood. They observed that most of the high-moisture-content zones were located around dead branch stubs, in the heartwood of butts, and around injuries. They attributed most of the wetwood to water moving into the bole of the tree through dead branch stubs. Such movement would appear highly unlikely and further study may show that the association of wetwood with dead branch stubs relates more to the functioning of these stubs as entrance courts for microorganisms than for the entrance of water. Moisture contents as high as 250 per cent (based on oven-dry weight) were found in wetwood; moisture contents as low as 54 per cent were encountered in dry heartwood; and moisture contents averaging 93–94 per cent appeared to be common for normal heartwood. The investigators did not attempt to determine the existence of microorganisms in the wetwood zones. The evidence for such a common association of microorganisms with wetwood in hardwoods (Hartley et al., 1961) suggests that a similar situation probably prevails in balsam fir.

During the past twenty years studies of antibiotics and of competition among microorganisms inhabiting the same environment suggest that bacteria of the soil, bark, and dead branches may be more important than presently suspected in preventing infection by parasitic fungi. Nissen (1956) reported that Actinomycetes, the fungal bacteria, had an inhibiting effect on root rot of Norway spruce in Denmark caused by Fomes annosus.* Since all forest soils abound in various species of Actinomycetes, their possible effect on balsam fir rot must be considered.

Algae play an important role in forest soils. It has been suggested that they may help supply oxygen in swampy soils. The limitations of balsam fir on peat soils have been discussed in Chapter 3, but no special attention has been paid to the algal flora by investigators. This is an area requiring additional research.

Lichens commonly grow as epiphytes on balsam fir and are known to interfere with gas exchange. The influence of lichens on rough bark formation and canker development has not been studied in balsam fir.

A similar relationship exists with respect to the occurrence of slime molds (Myxomycetes). The slime molds associated with balsam fir have been listed by Moore (1906, 1913), Farlow and Seymour (1888), Seymour (1929), and Weh-

*In order to save space the authority for the accepted Latin names of fungi is not given in the text. For complete designations and synonyms, see Appendix I.

meyer (1950), but no specific information on their role in balsam fir ecology or physiology was reported. The list of slime molds, containing nine species, is given in Appendix I.

FUNGI ASSOCIATED WITH BALSAM FIR

The variety and abundance of parasitic and saprophytic fungi found on balsam fir is indicated by the species list presented in Appendix I. Although this list of 262 species may not be complete, the majority of fungi described as causing diseases of living trees or growing on dead needles, twigs, bark, and wood have been included. It is evident from this list that fungi are more cosmopolitan than higher plants. Many of the fungi associated with balsam fir in North America are associated with silver fir in Europe and with other tree species throughout the world (Saccardo, 1882–1931; Oudemans, 1919; and others).

The list shown in Appendix I contains 164 species of Basidiomycetes, 66 species of Ascomycetes and 32 species of Fungi Imperfecti. No Phycomycetes are included in the list, possibly because none of the authors considered nursery-seedling diseases, most of which are caused by this class of fungi.

Like all tree species, balsam fir is subject to attack by fungi from the seedling stage to maturity. Although it has not been documented, it is quite likely that the early disappearance of a large number of newly established, succulent balsam fir seedlings is due to attack by damping-off fungi. In the sapling stage balsam fir is subject to needle rusts, needle casts, cankers, and root rots. When the tree becomes merchantable, heart rot fungi are of primary concern. Since both heartwood and sapwood of balsam fir lack decay resistance, insect-killed trees, wind-thrown trees, and wood products are subject to early and rapid deterioration by a large number of decay fungi.

Unresolved taxonomic problems make it difficult to correlate and integrate the information on the microbiology of balsam fir contributed by different authors over a period of many years. Appendix I provides a general summary of the information on the identification and synonomy of the fungi associated with balsam fir. Only a few of these can be discussed in the text.

FUNGI INVOLVED IN SYMBIOTIC AND OTHER RELATIONSHIPS

Fungi are not only involved in the various diseases of balsam fir and in the decay of its wood, but are also frequently associated in complex interrelationships with each other and with their host. The literature contains a considerable number of references to such relationships.

Mycorrhizal formations on the roots of balsam fir by various fungi have been studied. Moore (1922) and McDougall (1922) reported on an ectotrophic mycorrhiza caused by a species of *Cortinarius*. Kelley (1950) noted that the genus *Abies* has not been studied intensively to determine the fungi responsible for the abundant mycorrhizal formations that are almost universally present. He produced a picture showing the "Hartig net" of an unidentified mycorrhizal formation on *Abies balsamea*. Redmond (1957) noted the presence of mycorrhizae on balsam fir roots in New Brunswick. It is suggested in the literature that mycorrhizal fungi may at times be identical with those later causing root rot and other root injury.

Although saprophytic fungi are generally associated with the decomposition of dead plants, a number of complex interrelationships between saprophytic and parasitic fungi have been described in the literature on balsam fir.

Darker (1932) pointed out that a large number of saprophytic fungi invade balsam fir needles after infection by Hypodermataceae (see Phacidiaceae and Hysteriaceae in Appendix I), and prevent the parasitic fungi from fruiting and producing inoculum for further infection. Thus, *Bifusella faullii* and *Hypodermella mirabilis* are followed by *Leptosphaeria* sp., *Lophodermium autumnale*, and *Stegopezizella* (*Phacidium*) *balsameae*; and *Lophodermium piceae* is followed by *Cladosporium herbarum*. According to Wagener and Davidson (1954), stumps of cut trees may be attacked by *Fomes annosus* and the fungus from such infected stumps may enter surrounding living trees through root connections. There is also evidence to indicate that when such saprophytic fungi as *Trichoderma* sp. and *Peniophora gigantea* develop in stumps before attacks by *Fomes annosus* they may exclude the latter.

Although *Coryne sarcoides* has been found in balsam fir wood, it does not of itself cause significant damage; but some strains apparently produce substances inhibitory to the wood-destroying fungi *Coniophora puteana* and *Polyporus tomentosus* (Etheridge and Carmichael, 1957). Krstic (1956) reported that the metabolic products of *Penicillium glaucum* retard the growth of the wood-rotting fungi *Coniophora cerebella* (*C. puteana*) and *Stereum purpureum*. It has also been shown that germination of *Polyporus schweinitzii* spores is retarded by *Trichoderma viridis*. Jackson (1947) described *Trichomonascus mycophagus* as a parasitic fungus on *Corticium confluens*, a decay fungus growing on the bark of balsam fir.

The abundant saprophytic fungi responsible for the decay of branches and natural pruning may well play an added role, i.e., that of reducing or possibly stimulating infection by heart rot fungi. Recent studies indicate that such saprophytic fungi living on dead branches may influence infection by *Stereum sanguinolentum*, the fungus causing red top rot of balsam fir. Schaeffer (1926) pointed out that branch infection by *Valsa* sp. and other saprophytic fungi together with crowding within the stand may play an important part in natural pruning. According to Long (1924), another saprophytic or weakly-parasitic fungus, *Cenangium abietis*, which also occurs on balsam fir, plays a role in the pruning of ponderosa pine. This fungus is known to cause a twig blight of balsam fir, but to what extent it is harmful or beneficial is not known.

According to Pomerleau and Etheridge (1961), Etheridge and Pomerleau (1961), and Etheridge (1962), *Kirschteiniella* (*Amphisphaeria*) *thujina* is one of the most common fungi on dead branches of balsam fir. It is a blue-stain fungus which was noted earlier by Kaufert (1935) as being present in the heartwood of 10 per cent of all overmature balsam fir in the Lake States, in some cases developing around the margins of heart rot. It was reported by Crowell (1940a,b) in Quebec, and by Christensen and Kaufert (1942) in the heartwood of northern white-cedar in Minnesota. Laboratory tests by Pomerleau and Etheridge (1961) showed that *Kirschteiniella thujina* can retard the development of decay by *Stereum sanguinolentum*. Whether it can prevent or reduce decay in living trees has not been determined.

These examples of the interrelationships of saprophytic and parasitic fungi emphasize the need for considering the entire fungal flora of balsam fir in any consideration of the pathology of this species.

Seedling and Sapling Diseases

It is rather difficult to separate the diseases caused by pathogenic fungi according to age classes of the host plant or by the affected parts. In seedling and sapling stages, foliage and bark diseases predominate, although root and stem diseases also may cause serious injury.

YOUNG SEEDLING DISEASES

There are very few reports on seedling diseases of balsam fir because the latter is not widely grown in forest nurseries, and little attention has been paid to diseases of naturally occurring seedlings.

According to Boyce (1961) seedlings of the genus *Abies* are attacked by damping-off fungi. However, traditional damping-off fungi of the genera *Pythium, Phytophthora, Fusarium, Rhizoctonia,* and others have yet to be reported on balsam fir. *Fusoma parasiticum* has been reported as causing damping-off of balsam fir in German forest nurseries. When compared with other species, however, balsam fir proved to be rather resistant to attack by this fungus (Manshard, 1927). *Pestalotia hartigii* was found associated with stem girdling of 2–4 year old seedlings of balsam fir, but there is suspicion that the real cause of this injury was high soil temperatures and that the fungus was only saprophytic (Boyce, 1961; Guba, 1961).

"Snow mold," caused by *Phacidium infestans,* has been described by Faull (1929a,b) as present on balsam fir seedlings during certain winters of heavy snowfall and late melting. Injury usually is limited to browning and death of needles and occasionally to death of terminal buds. Under nursery conditions, control of snow mold can be accomplished by spraying with lime-sulfur (Hubert, 1931; Boyce, 1948; Baxter, 1952a).

The presence of *Phacidium infestans* and two related species, *P. balsameae* and *P. abietinellum* on the needles of balsam fir has been reported in U.S. and Canadian forest disease surveys. The rather extensive occurrence of *Phacidium infestans* on the needles of balsam fir reproduction in Laurentides Park, Quebec, was investigated by the Canadian Forest Biology Division in 1956. Examination of 2617 trees 9–55 years of age revealed *Phacidium infestans* fruiting on the needles of 25 per cent of the trees; *Lophodermium* sp. on 28 per cent, *Adelopus balsamicola* (*A. nudus*) on 11 per cent; *Trichosphaeria parasitica* (*Acanthostigma parasiticum*) on 7 per cent; *Hypodermella nervata* on 5 per cent; *Peridermium balsameum* (*Milesia kriegeriana*) on 1 per cent; and *Bifusella faullii* on less than 1 per cent. Smerlis (1961) reported that *Phacidium infestans* caused severe defoliation of balsam fir 6–15 years old in all crown classes.

Reid and Cain (1962) described a new genus, *Nothophacidium,* and considered that *Phacidium infestans, P. balsameae,* and *P. abietinellum* found on balsam fir belong to the common species, *Nothophacidium abietinellum.* Additional work will be needed to corroborate this finding.

NEEDLE CASTS AND BLIGHTS

As a result of the action of these fungi, portions of needles, whole needles, groups of needles, and twigs may be killed, and even the stem may be invaded and stimulated to form witches' brooms (Darker, 1932). The most important fungi causing these diseases belong to the Sphaeriaceae (genus *Rehmiellopsis*), Hysteriaceae (genera *Hypodermella* and *Lophodermium*), and Phacidiaceae (genera *Bifusella* and *Phacidium*).

Rehmiellopsis sp. causes tip blight of balsam fir. The earliest symptoms appear as yellowish-pink spots on needles of the current season's growth when the bud scales begin to drop off. The young twig usually continues to develop, but before reaching maturity it becomes brown, shriveled, and brittle. Needles turn dark reddish-brown and finally gray, adhering for one to two seasons. Small black fruiting bodies appear about six weeks after infection, but the spores do not reach maturity until the following spring when new growth is developing. No conidial stage has been observed (Waterman, 1945).

Species of *Hypodermella* cause needle casts of balsam fir. Several species may be present on the same needle. *Hypodermella nervata* has long, dark-colored pycnidia occupying the needle groove. *Hypodermella mirabilis* appears on the sides of the groove. They attack whole needles and may destroy a year's growth (Darker, 1932; Boyce, 1948, 1952).

Species of *Lophodermium* are well known on Scotch pine, and several important studies on these needle-cast fungi have been made. Spaulding (1912) reported that *Lophodermium nervisequum* caused serious mortality of balsam fir in the Adirondacks. The identity of this species was questioned by Darker (1932), who described several new *Lophodermium* species occurring on balsam fir. *Lophodermium lacerum* and *L. piceae* were reported causing slight damage in stands of all ages up to 55 years in Laurentides Park, Quebec (Smerlis, 1961).

Bifusella faullii, considered the most serious of the needle casts on balsam fir, attacks whole needles and can destroy the current growth. Although most injurious to juvenile plants, it also attacks older trees in dense stands (Darker, 1932). Ascospores infect the needles in July. The following spring the needles change color to light brown and later to buff. Pycnidia appear on the upper surfaces of the needles. The ascospores are discharged the following July when the needles are in their third growing season. In the Laurentides Park study, *Bifusella faullii* was found only in stands younger than 25 years of age.

Potebniamyces balsamicola, an Ascomycete associated with twig and branch blight of balsam fir, was described for the first time by Smerlis (1962). Previously considered an unidentified species of *Phacidiella* (Smerlis, 1961), it is the principal fungus associated with dying twigs and branches of balsam fir in central and eastern Quebec and is present also in northeastern New Brunswick. Stands younger than 15 years of age are attacked, especially dominant and codominant trees, but the damage caused is not severe. Only infrequently have the organisms causing needle casts and blights been reported to cause serious damage. In the majority of cases these diseases are limited to needles of the lower branches, particularly those covered by snow. They may be responsible for some loss of needles and twigs and small reduction of growth, but this rarely reaches serious or economic proportions. Most of these diseases can be classed with the large group of endemic diseases constantly present in the forest, which rarely if ever develop to epidemic or serious loss conditions and for which there are at present no practical or necessary control measures. In some cases these diseases may even be beneficial, speeding up mortality in overdense, stagnant sapling stands.

RUSTS

The rust fungi attacking balsam fir cause loss of needles, deformation of branches and stems, stimulation of the formation of witches' brooms, and occasionally death of individual trees. Needle rusts have been observed in near epidemic proportions, but even under these conditions the only measurable effect appears to be some loss of growth. This condition is similar to that found in black spruce on which needle rusts may also occasionally become epidemic.

Balsam fir is attacked by a number of heteroecious rusts. Primary infection occurs through the current year's needles. The pycnia appear as small, round, slightly raised, orange or yellowish spots usually on the undersurface of the needles. The orange-yellow or white aecia show up on the undersurfaces in the spring following the pycnidial stage. Position and size of pycnia, color of aecia, time of development, and age of the needles are diagnostic characteristics. Not all of the spore stages and alternate hosts are known (Hubert, 1931; Boyce, 1948; Baxter, 1952a). Kamei (1940) noted that *Milesia* rusts are not selective for specific species of *Abies*.

A rust said to be caused by *Caeoma arcticum* of the family Pucciniaceae (Seymour, 1929) is listed, but it may only be an unclarified synonym for some better known rust. Considerable disagreement still exists with respect to the classification of rusts belonging to the Melampsoraceae.

Melampsoraceae rusts having their pycnial and aecial stages on balsam fir are represented by the following genera: *Calyptospora* (uredia on *Vaccinium* sp., telia unknown), *Hyalopsora* (uredia and telia on *Phegopteris* sp. and *Dryopteris* sp.), *Melampsora* (uredia and telia on *Salix* sp. and *Populus* sp.), *Melampsorella* (uredia and telia on *Cerastium* sp. and *Stellaria* sp.), *Milesia* (uredia and telia on ferns), *Uredinopsis* (uredia and telia on ferns), and *Pucciniastrum* (uredia and telia on *Epilobium* sp.).

Melampsorella caryophyllacearum causes witches' brooms on balsam fir. Witches' brooms are also caused by *Milesia polypodophila* (Hunter, 1927, 1948; Hubert, 1931; Boyce, 1948; Baxter, 1952). Species of *Melampsorella* are considered the primary cause of witches' broom formation on balsam fir in Minnesota.

The U.S. Disease Survey also lists *Gloeosporium balsameae*, belonging to family Melanconiaceae of the Fungi Imperfecti (Weiss and O'Brien, 1953), as a rust of balsam fir.

CANKERS

Cankers caused by fungi are generally less common on softwoods than on hardwoods; however, there are quite a few reports of cankers on balsam fir. Canker fungi cause necrosis of cortical tissues which, when combined with the formation of callous tissue by the host, results in the production of cankers. In addition to reduction of growth of individual trees, cankers may serve as entrance points for wood-rotting fungi. When cankers occur on twigs and small branches, the entire twig or branch may be killed; such cankers are frequently called twig blights. The fungi usually enter through wounds caused by atmospheric conditions and insects.

In Ontario, according to Quirke and Hord (1955), cankers and twig blight occur most commonly on reproduction and immature trees along stand borders and on mature trees growing on shallow dry soils. The first symptom usually is a reddening of the needles, which may adhere for several years before falling. Branches attacked become constricted. On larger trees elliptic cankers may develop on the trunk at the base of invaded branches. A year after cankers appear, the characteristic flask-shaped fruiting bodies form on the dead bark.

Until recently the fungi considered responsible for most of the cankers of balsam fir were *Nectria* (*Creonectria*) *cucurbitula* and *N.* (*Thyronectria*) *balsamea*. The first fungus has also been reported from Norway (Robak, 1951). Roll-Hansen (1962) linked *N. cucurbitula* sensu Wollenweber, to *Cephalosporium*, the latter its conidial stage. Nectria canker on balsam fir planted in Norway was identified by him as *N. cucurbitula* var. *macrospora* and its conidial stage, *Cylindrocarpon cylindroides*.

Christensen (1937) and Christensen and Hodson (1954) have described a stem canker caused by *Cephalosporium album*. Never very prevalent and rarely causing death, it is an inconspicuous canker on suppressed trees. The cankers are oval or elliptic, the dead bark is slightly sunken, and resin exudes from the surface. Several outer rings of wood beneath the canker are normally stained brown. Roll-Hansen (1962), referring to the work of Christensen and Hodson and to specimens sent him from Canada, questioned whether the fungus involved was identical with his described *Cephalosporium* stage of *Nectria* "*cucurbitula.*"

A canker caused by *Fusicoccum abietinum* was observed in southeastern Canada, particularly on Cape Breton Island, by the Canada Forest Biology Division (1956–1958). The description of this species (Hubert, 1931) indicates its close relationship to the *Nectria* sp. cankers reported in Ontario (Quirke and Hord, 1955). More recently, *Fusicoccum* (*Phoma, Phomopsis*) *abietinum* has been reported rather widely in New Brunswick and Nova Scotia (Can. For. Ent. Path. Br., 1961).

Balsam fir dieback which started in Ontario in 1954 was most commonly associated with *Thyronectria* (*Nectria*) *balsamea*, *Dermea* (*Cenangium*) *balsamea*, and *Valsa* (*Cytospora*) *abietis* (Raymond and Reid, 1957, 1961). *Thyronectria* and *Valsa* have been reported in the United States by Weiss and O'Brien (1953).

The occurrence of *Aleurodiscus amorphus* in the northeastern United States, with occasional killing of sapling balsam fir, has been reported by Hansbrough (1934) and Boyce (1948).

In the taxonomic reorganization of the family Tryblidiaceae by Seaver (1951), the former *Scleroderris* and *Bothrodiscus* cankers of balsam fir are now recognized as *Godronia abietina* and *Godronia abietis* cankers. According to Boyce (1948), the *Godronia* cankers develop during the dormant season and are limited primarily to the bark, causing little actual damage.

Other canker fungi on balsam fir are *Tympanis abietina* and *T. truncatula* (Groves, 1952); *Lachnella agassizii, L. arida, L. hahniana,* and *L. resinaria* (Seaver, 1951). These cankers do not cause much damage. According to Anderson (1902), they cause bark to roughen and later to scale off.

In the Lake States none of the balsam fir cankers occur frequently enough to constitute a serious disease problem. Together with the rusts and leaf-cast fungi, the canker fungi produce the type of scattered endemic losses that go largely unnoticed and for which there is no control except the elimination of infected trees in logging and stand improvement operations. The recent report on the more abundant occurrence of different cankers in northeastern United States and Canada indicates that under certain conditions cankers on balsam fir may become a serious problem in balsam fir management.

Decay

The losses caused by the decay or wood-rotting fungi exceed those caused by all other disease organisms attacking forest trees (Peace, 1939). This relationship is even more striking for balsam fir. Decay in living trees is largely limited to heartwood, with the sapwood occasionally invaded in overmature trees as heart rots reach advanced stages. Decay of sapwood in living trees is normally limited to wounds. Insect-killed trees, wind-thrown trees, and wood products are attacked by a large number of fungi that cause sap rots and are capable of extending into the heartwood. Balsam fir heartwood possesses little or no natural durability and is thus subject to attack by both sap rot and heart rot fungi in harvest products and in dead trees.

Decay fungi are commonly designated according to the part of the tree affected: root, butt, and top rotting fungi. They may also be designated according to the type of decay produced, i.e., brown rots, white rots, white pocket rots, white stringy rots, and others. The kind of rot that results depends on enzymes produced by the causal fungus and the mode of action of these enzymes. Thus, brown rot fungi hydrolyze the polysaccharides and cellulose in preference to, or more rapidly than, lignin (Cartwright and Findlay, 1944). The fungi causing white rots attack both lignin and cellulose, and those causing white pocket or stringy rots attack all wood constituents but localize their activity.

Although the majority of wood-destroying fungi belong to the Basidiomycetes, some Ascomycetes and a few Fungi Imperfecti have been shown to be capable of causing decay (Ainsworth and Bisby, 1954).

DECAY OF LIVING TREES

As previously indicated, the preponderance of decay in living trees is limited to the heartwood and is caused by heart rot fungi. Exceptions to this rule occur when sap rot fungi invade the wood around wounds caused by mechanical damage and canker fungi. A further exception may be the invasion of roots of suppressed trees and broken roots of all trees by a host of organisms, probably including many sap rot fungi.

Fungi responsible for heart rots. Early work on balsam fir heart rots indicated that four fungi, *Stereum sanguinolentum, Poria subacida, Polyporus balsameus,* and *Polyporus schweinitzii* were largely responsible for the majority of the losses (McCallum, 1928; Spaulding et al., 1932; and Kaufert, 1935). *S. sanguinolentum* was reported as causing most of the red top rot; *P. subacida,* as causing most of the white stringy rot of butts and roots; *P. balsameus,* as causing some of the brown cubical butt rot; and *P. schweinitzii,* as responsible for part of the brown cubical butt and root rot. Some of the early identifications were based on observation of the type of rot. The observations were then compared with decay descriptions known to be associated with the several species of fungi. In some cases sporophores associated with the decay served as a guide. In other cases isolations were made in which the unknown cultures were matched with known cultures. Most of the early workers using decay cultures as a means of identification frequently encountered organisms that did not match the known cultures (Kaufert, 1935; Spaulding et al., 1932). In addition, the cultural characteristics of such fungi as *P. subacida* are not sufficiently distinctive to permit positive identification on gross characteristics alone. Because of these complications, many early workers attributed to *P. subacida* far more decay than it is now known to induce. A similar situation holds for the brown butt rots, although in many cases the reports of a number of unidentifiable cultures may account for the fact that this type of decay is caused by several fungi. However, the early conclusion that *Stereum sanguinolentum* is the chief organism responsible for red top rot is still accepted today.

Very helpful in the identification of the fungi responsible for heart rot of balsam fir have been the cultural characteristic studies of wood-rotting fungi by Fritz (1923) and Nobles (1948). Fritz showed the difficulty of identifying *P. subacida* by means of cultural characteristics.

More recently, studies of fungi responsible for balsam fir heart decays, in Canada by White (1951) and Basham et al. (1953), and in the United States by Prielipp (1957), have clarified the situation with respect to the fungi responsible for different types of heart rot and have considerably increased the list of organisms involved.

A cultural study should be made to determine the fungi responsible for the white stringy butt rot in Minnesota and Wisconsin, most of which was attributed to *P. subacida* by Kaufert (1935). The conclusion that *P. subacida* is responsible for only a part of this type of decay in Canada (Basham et al., 1953) and Upper Michigan (Prielipp, 1957) indicates that the same situation probably prevails in Minnesota.

Basham et al. (1953) determined the frequency of heart rot fungi among balsam fir from material collected in the United States and Canada. The results of this extensive study are shown in Table 21.

Prielipp (1957), in his study of balsam fir heart rot in Upper Michigan, found that 249 of the 484 trees studied had decay. He reported as follows on the identity of the causal organisms:

"Red heart rot was caused by *Stereum sanguinolentum* in 90 per cent of the top rot infections. White stringy butt rot was caused by five fungi with *Poria subacida, Armillaria mellea,* and *Odontia bicolor* accounting for 30 of the 39 white stringy rots sampled. Twenty-three cultures of brown cubical rot were made. Of this number, *Poria cocos* and *Polyporus balsameus* each accounted for 30 per cent of the infections while 22 per cent was caused by *Coniophora puteana* and 17 per cent by *Merulius americanus.*"

TABLE 21. FREQUENCY WITH WHICH DIFFERENT FUNGI WERE ISOLATED FROM DECAYED HEARTWOOD OF BALSAM FIR IN EASTERN NORTH AMERICA

Fungus	Type of Decay in Order of Frequency of Occurrence	Number	Percentage
Stereum sanguinolentum	Red heart rot, red butt rot	312	35.3
Corticium galactinum	White stringy butt rot, red butt rot, red heart rot	234	26.5
Odontia bicolor	White stringy butt rot, red butt rot	72	8.2
Coniophora puteana	Brown cubical butt rot	56	6.3
Polyporus balsameus	Brown cubical butt rot	55	6.2
Poria subacida	White stringy butt rot	45	5.1
Armillaria mellea	White stringy butt rot	45	5.1
Omphalia campanella	White stringy butt rot	17	1.9
Merulius himantioides	Brown cubical butt rot	17	1.9
Stereum chailletii	White stringy butt rot, red heart rot, red butt rot	8	0.9
Fomes pinicola	Brown cubical heart rot, brown cubical butt rot	6	0.7
Fomes pini	White pocket heart rot	4	0.5
Polyporus abietinus	White stringy butt rot	3	0.4
Polyporus circinatus	White stringy butt rot	2	0.2
Trechispora raduloides	Red heart rot	2	0.2
Polyporus volvatus	Red heart rot, red butt rot	2	0.2
Polyporus schweinitzii	Brown cubical butt rot	1	0.1
Polyporus guttulatus	Brown cubical butt rot	1	0.1
Polyporus resinosus	Brown cubical butt rot	1	0.1
Peniophora gigantea	Red heart rot	1	0.1
Total		884	100.0

Source: Basham et al., 1953.

Although these studies have greatly extended the list of fungi responsible for balsam fir heart rots, their principal contribution has been to show that the important white stringy butt rots previously attributed primarily to *Poria subacida* are caused by *Corticium galactinum, Odontia bicolor, Omphalia (Xeromphalina) campanella,* and *Armillaria mellea* in addition to *P. subacida*; that the brown cubical rots formerly attributed primarily to *P. balsameus* are due to *Coniophora puteana, Poria cocos,* and *Merulius* sp. in addition to *P. balsameus*; and that the red top rot is largely due to *S. sanguinolentum* as reported by earlier workers.

In a 1956 study in Quebec by the Canada Forest Biology Division, 1215 of 1822 balsam fir trees sampled were found to be infected with heart rot. The organisms present were distributed as follows: *S. sanguinolentum,* 34.8 per cent; *Corticium galactinum,* 32.9 per cent; *Poria subacida,* 9.6 per cent; *Coniophora puteana,* 5.2 per cent; and *Fomes pini,* 0.8 per cent.

According to Basham et al. (1953), the nine principal heart rot fungi of balsam fir are *Stereum sanguinolentum,* causing almost all upper trunk rot; *Corticium galactinum, Odontia bicolor, Poria subacida, Armillaria mellea,* and *Omphalia campanella* associated with most of the white stringy butt rots; and *Coniophora puteana, Polyporus balsameus,* and *Merulius himantioides* (*M. americanus*) accounting for most of the brown butt rots. The results of Prielipp's (1957) studies indicate that, in the United States, *Poria cocos* should be added to this list. *Poria cocos* was first noted on balsam fir by Baxter (1947). This species occurs on many hardwoods and conifers and is even found on corn (Wolf, 1926).

Smerlis (1961), investigating the pathological conditions in 6–55-year-old balsam fir stands in the Hylocomium-Oxalis forest type in Laurentides Park, Quebec, made 182 isolations of trunk rots and 133 isolations of root and butt rots. Of the top or trunk rot cultures, 76 per cent were *Stereum sanguinolentum,* 4 per cent were *Corticium laeve,* about 10 per cent were *Stereum chailletii,* and 19 per cent were not identified. Of the root- and butt-rot cultures, *Corticium galactinum* made up 41 per cent, *Coniophora puteana* 10 per cent, *Corticium fuscostratum* 5 per cent, *Armillaria mellea* 2 per cent, and *Odontia bicolor, Omphalia campanella, Poria asiatica, Stereum chailletii, S. purpureum,* and *S. sanguinolentum* each about 1 per cent. The remaining fungi were not identified.

Prevalence of heart rot. All research workers who have studied the occurrence of heart rot in balsam fir have reported on the prevalence of decay and have attempted to relate the latter to site, age, and other factors. Since prevalence of heart rot in all tree species is recognized to be directly correlated with age, this variable must be considered in connection with all reports of frequency of occurrence. Other factors such as site may enter the picture, but age is by far the most important.

McCallum (1928), Kaufert (1935), Spaulding and Hansbrough (1944), Morris et al. (1955), and Prielipp (1957) presented data indicating the relationship of age to infection. This information is summarized in Table 22.

Although there are differences in the results reported, these may be due to type of stands sampled, regional variation in the species, and other factors. There is general agreement that infection occurs in some trees before they reach 40 years of age and that the amount of infection increases rapidly thereafter. At 100 years of age most trees are infected with heart rot.

The above authors agree that infection of balsam fir by different fungi occurs at different ages. Although there is some difference in the infection age by different species of butt rot fungi, the most striking difference is between the infection age of butt and top rot fungi. The results reported by Spaulding and Hansbrough (1944) and Prielipp (1957) given in Table 23 illustrate this difference.

It is evident from these data that infection by butt rot fungi occurs earlier than that by top rot fungi, and that infection percentages for butt rot are much higher. As will be shown later, in the older age classes there is greater cull loss from top rot than from butt rot even though the percentage of infection is lower. Much loss by wind breakage results because infection by butt rot fungi occurs at an early age and is found in a high percentage of trees at age 70 and over.

TABLE 22. INCIDENCE OF HEART ROT AT DIFFERENT AGES IN BALSAM FIR
IN THE UNITED STATES AND CANADA

| Age (Years) | Percentage of Trees Infected | | | | |
	Quebec[a]	Minnesota and Wisconsin[b]	Northeastern United States[c]	Upper Michigan[d]	New Brunswick[e]
41–50	0	17	20	30	8
51–60	5	39	40	43	35
61–70	9	56	55	56	55
71–80 37		69	62	67	68
81–90 37		79	68	75	77
91–100 59		86	78	81	83
101–110 69		92	80	83	88
111–120 71		95	87	87	91

Sources: [a]McCallum, 1928; [b]Kaufert, 1935; [c]Spaulding and Hansbrough, 1944; [d]Prielipp, 1957; [e]Morris et al., 1955.

TABLE 23. PERCENTAGE OF BALSAM FIR TREES INFECTED BY
BUTT ROT AND TOP ROT FUNGI COMPARED IN RELATION
TO AGE OF TREES

| Age (Years) | Butt Rot Fungi | | Top Rot Fungi | |
	Northeastern United States[a]	Upper Michigan[b]	Northeastern United States[a]	Upper Michigan[b]
41–50 15		28	1	2
51–60 35		43	5	8
61–70 45		50	10	12
71–80 55		70	13	17
81–90 60		73	17	22
91–100 66		70	22	30
101–110 71		70	28	39

Sources: [a] Spaulding and Hansbrough, 1944; [b] Prielipp, 1957.

TABLE 24. PERCENTAGE OF BALSAM FIR TREES WITH VARIOUS ROTS AND
PERCENTAGE OF TOTAL WOOD VOLUME DECAYED IN VARIOUS AGE
CLASSES AT LAURENTIDES PARK, QUEBEC

Age Class (Years)	No. of Trees Examined [*]	Rot	Trunk Rots	Root and Butt Rots	Wood Volume Decayed
6–15	3800	5.3	0.4	5.1	0
16–25	1650	10.4	1.8	10.4	0.1
26–35	660	29.5	10.8	34.0	1.2
36–45	480	34.2	12.2	28.8	0.9
46–55	180	46.7	12.6	43.3	1.4

Source: Smerlis, 1961.
[*] Approximate, original data in two groupings.

The presence of trunk and root or butt rots in immature balsam fir stands is shown in Table 24. The volume of decay in these immature stands is small. It should also be pointed out that the investigations by Smerlis (1961) were limited to the Hylocomium-Oxalis forest type in Laurentides Park. The early development of root and butt rot as compared to trunk rot is supported by these investigations.

The information available on the relationship of infection and development of cull to site quality and other factors is not very conclusive. Zon (1914) found that butt rot of balsam fir admixed with hardwoods was more common on slopes than on lowlands. Kaufert (1935) found that 66 per cent of balsam fir on ridges had butt rot as compared with 51 per cent on lowlands. Top rot appeared to be unaffected by these broad site differences. Spaulding and Hansbrough (1944) stated: "butt and top rot is independent of character of site and is approximately equal in both dry and wet situations." Prielipp (1957) reported that in upper Michigan infection by butt rot fungi occurs later and develops more slowly in lowland balsam fir than in upland balsam fir, and that consequently the rotation age for the former can be increased. According to Heimburger and McCallum (1940a,b), *Poria subacida* butt rot was prevalent in balsam fir growing on a wide range of site conditions. The fungus, *Polyporus balsameus* occurred on 61.2 per cent of plots taken on slopes and on 42.2 per cent of lowland plots.

The same forest types were used as the basis for determining incidence of different decays of balsam fir by Basham et al. (1953), and results are given in Table 25.

TABLE 25. FREQUENCY OF OCCURRENCE OF VARIOUS ROTS IN
BALSAM FIR IN TWO FOREST TYPES OF
EASTERN NORTH AMERICA

Type of Observation	Mixed Slopes		Softwood Flats	
	Number	Per Cent	Number	Per Cent
Trees examined	1004	100.0	901	100.0
Trees with decay	720	71.7	580	64.4
Trees with trunk rot	364	36.4	371	41.2
Trees with butt rot	687	68.4	409	45.3
Trees with white butt rot	519	51.7	360	39.9
Trees with brown butt rot	153	15.2	24	2.7
Trees with red butt rot	15	1.5	25	2.8

Source: Basham et al., 1953.

Pomerleau (1958) found significant differences in the proportion of decayed to sound wood and in volume increment of balsam fir on different sites in Quebec. Balsam fir on Dryopteris-Oxalis forest types produced 14.5 per cent more gross merchantable volume, and 24.7 per cent more net merchantable volume than balsam fir in a Hylocomium-Oxalis forest type.

Basham et al. (1953) further reported that *Poria subacida* occurs more frequently in balsam fir on mixed wood slopes, whereas *Armillaria mellea* develops best in balsam fir on softwood flats. These findings, together with those provided by workers in the Lake States and northeastern United States, and the more recent work of Davidson (1957) and Pomerleau (1958) indicate the need for additional research on the relationship of site and other growing conditions to the incidence of heart rot fungi.

The influence of site on the occurrence of balsam fir cull in Ontario is related by Morawski et al. (1958), who found that up to about 100 years of age there was little difference in the degree of cull between dry and moist sites and that trees on both sites had less cull than those on fresh* sites. His report showed that after 100 years cull was least in balsam fir growing on moist sites, intermediate on fresh* sites, and highest on dry sites. These results agree quite well with the reports of Kaufert (1935) and Prielipp (1957), who reported a higher percentage of infections and somewhat more cull in upland balsam fir than on moist slopes.

In general, it can be concluded that the existing relationships of site conditions to the frequency of infection and the amount of cull are not striking and are possibly not of great commercial importance. Their existence would indicate nonetheless that more intensive study, with greater attention paid to the many variables in site quality, together with teamwork by pathologists and ecologists, may in the future uncover more positive proof that relationships do exist.

The information available on the effect of tree vigor on the incidence of decay is as lacking in clear-cut relationships as that of decay to site conditions. Kaufert (1935) reported that while fast-growing trees contained a greater volume of cull than slow-growing trees, the percentage of cull was about the same. Spaulding and Hansbrough (1944) developed the same conclusion. These authors noted that suppression of balsam fir by hardwoods favors the development of top rot. In a study made in New Brunswick, Morris (1948) failed to find any relationship between tree vigor and percentage of decay or cull.

Amount of decay and cull and rotation age. The primary purpose of most investigators who have studied heart rots of balsam fir has been to determine the percentage of decay and cull, and to relate these to age in an attempt to establish pathologic cutting cycles. As might be expected, there is considerable variation in the results reported, although there is much less difference concerning recommendations on pathologic rotation for this species.

The following variables in decay and cull measurements account for the differences in results reported by investigators:

(a) Regional variation of balsam fir, difference in growing conditions, and possible variation in the prevalence of various heart rot fungi are important. Balsam fir in southeastern Canada is known (see Chapter 8) to grow to larger sizes and to produce heavier volumes per acre than in other areas.

(b) Correction of age determinations on trees infected with butt rot is subject to large errors. It is possible that the difference in amounts of decay reported for balsam fir by Kaufert (1935) and Prielipp (1957) is partially due to this variable. Kaufert corrected the age of trees having butt rot by applying age-height data collected on saplings, a practice followed by most investigators. The recognized slow growth of balsam fir seedlings and saplings under dense overstories (the probable origin of most of the trees sampled by Kaufert) might have resulted in considerable under-estimation of their true ages.

(c) Culling practices vary with region and with time. Adoption of the prevailing

* Intermediate between dry and moist.

culling practices, the procedure followed in most studies, readily explains the difference in results reported in Table 26. When studies were made by Kaufert (1935), balsam fir pulpwood was not in demand. Material from the older trees containing some heart rot, which became marketable in later years, had no market at that time and was culled.

In view of these variables it is remarkable that there is so much agreement among the results of different investigators. The proportion of trees with decay up to the age of 70 years is of particular interest in this respect.

TABLE 26. A COMPARISON OF CULL AND NET MERCHANTABLE VOLUME OF BALSAM FIR IN THE UNITED STATES AND CANADA

	Percentage of Cull					Net Merchantable Volume in Cubic Feet per Tree				
Age (Years)	On-tario [a]	North-eastern United States [b]	Minne-sota and Wis-consin [c]	Upper Michi-gan [d]	New Bruns-wick [e]	On-tario [a]	North-eastern United States [b]	Minne-sota and Wis-consin [c]	Upper Michi-gan [d]	New Bruns-wick [e]
40	0	2	0	0	0	1.1	0.6	1.1	2.5	1.0
50	2	6	6	7	5	1.7	2.3	2.0	4.2	2.0
60	4	10	9	12	16	2.3	4.3	3.0	5.9	2.7
70	6	14	16	16	20	3.0	6.4	3.9	8.0	3.6
80	8	17	27	18	25	3.5	8.5	4.4	10.0	4.5
90	11	20	39	23	29	4.0	10.4	4.7	10.8	5.5
100	13	23	48	29	33	4.3	12.3	4.5	10.6	6.4
110	17	26	56	36	36	4.4	14.1	4.1	9.9	7.3
120	21	30	65	50	40	4.5	15.4	3.4	7.9	8.1
130	25	34	74	..	45	4.5	16.6	2.6	..	8.7
140	29	40	83	..	49	4.4	17.1	1.7	..	9.3

Sources: [a] Morawski et al., 1958; [b] Spaulding and Hansbrough, 1944; [c] Kaufert, 1935; [d] Prielipp, 1957; [e] Morris et al., 1955.

The results of studies by Spaulding and Hansbrough (1944) and by Prielipp (1957) (see Table 27) are quite typical of those reported for the amount of cull caused by butt rot and top rot fungi. Butt rot develops at an earlier age, infects a higher percentage of trees, and causes more cull than top rot up to age 80–90. At that age, top rot and associated cull increase rapidly, and in overmature trees far exceed the cull caused by butt rot.

It should be pointed out that top rot fungi are not necessarily more important in the total pathology of mature and overmature stands than are butt rot fungi. Weak-ening of trees by butt rot fungi and subsequent losses through windthrow are not reflected in cull figures.

Other studies on decay and cull in balsam fir have been made by Pomerleau (1948), Davidson (1951), and Gevorkiantz and Olsen (1950). In the latter study cull was determined on a stand basis. The results agree with those of other in-vestigators.

The pathologic rotation age of tree species is not as significant today as 25 years ago. Rotation age is determined by many factors, and few stands of pulpwood species such as balsam fir will be grown this long. In the less accessible areas of the United States and in much of Canada, however, management and utilization have not yet reached the stage where pathologic rotation can be ignored. There is very close agreement that 70–80 years is the approximate age for pathologic rotation for this species. This is the conclusion of most investigators, whose results were obtained on individual trees, as well as that of Gevorkiantz and Olsen (1950), who worked on a stand basis. In discussing pathologic rotation, Prielipp (1957) stated: "In Upper Michigan the age of stand break-up varies with soil moisture, and occurs at approximately the following ages: 70 years for uplands, 80 years on transition, and 90 years in swamps. On upland sites decay enters at 30–35 years, and the quality rotation is set at 45–50 years. The quality rotation age is 55–60 years for transition areas and 65–70 years for swamps. . . . Balsam fir stands that have been suppressed severely, subjected to mechanical injury from logging or climatological factors, or exposed to fire should be closely watched These stands will ultimately have excessive cull and should be handled accordingly."

TABLE 27. PERCENTAGE OF MERCHANTABLE VOLUME OF
BALSAM FIR CULLED FOR DIFFERENT AGE CLASSES
IN UPPER MICHIGAN [*]

Age (Years)	Butt	Top	Total
50	5	2	7
60	8	4	12
70	9	7	16
80	10	8	18
90	11	12	23
100	12	17	29
110	14	22	36
120	17	33	50

Source: Prielipp, 1957.
[*] Based on 476 trees.

Entrance courts. Earlier investigators agreed that heart rot fungi of balsam fir enter the trees through dead branches, branch stubs, wounds, and roots. Infection usually does not begin until heartwood formation is underway (Lorenz and Christensen, 1937). As early as 1914, Zon reported that trees developed rot within 5–7 years after wounding.

Wounds on balsam fir are caused by logging, fire, frost, wind, glaze, insects, wildlife, rock pressures on roots, and other factors. While most wounded trees develop decay soon after injury, visible wounds are found on only part of the trees with heart rot. Kaufert (1935) found that of 1170 trees examined, 18 per cent had wounds, but heart rot of some type was present in 59 per cent of the trees. Almost all wounded trees (95 per cent) had heart rot, but only one-third of the heart rot could be detected from the presence of wounds on the bole. Spaulding and Hansbrough (1944) found that only 12 per cent of the heart rot was detectable from the appearance of trunk wounds. The relationship between external defects and heart rot of balsam fir was further indicated by Prielipp (1952) and is given in Table 28.

The occurrence of both top rot caused primarily by *Stereum sanguinolentum* and butt rot caused by a number of fungi in a high percentage of trees showing no external indication of decay makes the problem of heart rot detection extremely difficult. While trees with broken tops and top wounds of considerable age are very likely to contain top rot, even more trees without external wound symptoms contain important amounts of top rot.

TABLE 28. PERCENTAGE OF BALSAM FIR TREES HAVING ROTS
IN UPPER MICHIGAN

Bole Characteristics	Top Rot	Butt Rot	Total Rot
No visible defect	3	28	29
Damaged roots	17	83	92
Woodpecker holes	30	81	89
Branch stubs, lower bole	17	62	73
Branch stubs, upper bole	33	46	63
Cracks	22	64	73
Mechanical injury	30	59	69
Flat-topped trees	20	54	66

Source: Prielipp, 1952.

Trees with butt wounds are rather likely to contain considerable decay, but the same may be true of trees without visible wounds. Most investigators agree that the majority of butt rot fungi enter through broken roots. Evidence of broken roots is present on most balsam firs from saplings to mature trees. Root breakage probably occurs because of swaying, frost heaving, and other forces. Because of the susceptibility of balsam fir to attack by heart rot fungi, logging damage is more important with this species than with such associated conifers as black and white spruce. Trimble (1942) reported that after a partial cut in a balsam fir stand, 9 per cent of the remaining trees were injured and 3 per cent were killed during logging.

Root injuries in balsam fir stands 6–55 years old were investigated by Smerlis (1961) in Laurentides Park, Quebec. His results are summarized in Table 29.

The frequency of injury by the root weevil (*Hylobius pinicola* (Couper)) is surprisingly high, and its association with root and butt rot fungi is statistically significant. Dominant, codominant, and intermediate tree root systems are mostly affected.

Intensive studies of the entrance courts of *Stereum sanguinolentum* have been undertaken in Canada. Faull and Mounce (1924) expressed the opinion that *S.*

TABLE 29. PERCENTAGE OF BALSAM FIR TREES WITH ROOT INJURIES OR
DISEASE IN LAURENTIDES PARK, QUEBEC

Age Class (Years)	Number of Trees Examined	Injured or Diseased Root Systems	Root Weevil Injuries	Mechanical Injuries	Unknown Injuries
6–15	3697	14.9	5.7	3.5	6.4
16–25	1390	48.0	22.5	4.8	25.6
26–35	567	73.6	34.7	13.2	38.1
36–45	399	71.7	40.9	11.8	26.3
46–55	157	68.8	43.3	10.8	25.5

Source: Smerlis, 1961.

sanguinolentum infects primarily through dead branches or branch stubs and rarely fruits on living trees. Davidson and Newell (1962) dissected 370 branches and stubs of balsam fir from Nova Scotia and New Brunswick. Red rot, caused by S. *sanguinolentum* was present in 96 branches of which 70, or 19 per cent, had received infection from the outside. Of these, 44 had wounds, 7 had dead and broken branch terminals and 6 had dead or broken secondary branches. The entrance courts for red rot in the remaining 13 branches could not be traced. Only 14 per cent of the small branches (0.1–1.0 in. diam) had decay, whereas 35 per cent of larger branches and an average of 19 per cent of all branches contained decay. Artificial inoculation was successful on 69 per cent of all living branches and on only 8 per cent of dead branches. These results indicate that large wounded branches of balsam fir provide more suitable infection courts for *Stereum sanguinolentum* than do dead branches and branch stubs. This is an important observation and contrary to commonly held views.

The studies by Etheridge (1962) confirmed the observations made by Davidson and Newell (1962). Injured leaders and branches were found to be equally important as infection courts, each being associated with about one-third of all infections. The remaining infection was due to either broken tops or trunk wounds. This study also indicated that uninjured branches and leaders of balsam fir do not normally serve as infection courts for *Stereum sanguinolentum*. Slowly dying branches and leaders rarely serve as points of entry for the fungus. Studies are now underway to determine the factors associated with slowly dying wood that make it unsuitable for this fungus, and also to investigate further the agents causing the different infection courts.

A recent publication by Davidson and Etheridge (1963), in which they review the earlier work by Davidson and Newell (1962) and by Etheridge (1962), indicates that fresh wounds are the primary infection court for S. *sanguinolentum*. Aging of wounds apparently increases resistance to infection. These workers also indicate that dead branches and limbs are not the important infection courts they were earlier thought to be.

DECAY OF INJURED, DEAD, AND DOWN TREES, AND WOOD PRODUCTS

Balsam fir trees killed by fungi, insects, and other agents, are subject to rapid deterioration by many fungi, primarily sap rotters. Windthrown trees vary in their rate of deterioration because they may remain alive for several years.

Windthrow and fire effects. The deterioration of windthrown balsam fir in Newfoundland was investigated by Davidson (1952) and by Davidson and Newell (1953). Trees studied had been windthrown 3–4 years earlier. Incipient heart rot comprised 1.2 per cent of the merchantable volume. Incipient sap rot was present in 98 per cent of the trees and 34 per cent of the volume. Advanced sap rot was present in 22 per cent of the trees and 1.0 per cent of the merchantable volume. Most of the sap rot was due to *Stereum chailletii*. Since windthrown balsam fir having only the incipient stage of decay is usable in sulfite pulping, the decay losses after 3–4 years are not high. Presence of incipient decay, however, causes sinkage in water transport of pulpwood which can result in large losses.

This study in Newfoundland was continued by Stillwell (1959). An additional

175 dead trees 4–8 years of age were examined. Incipient sapwood decay decreased from 32 per cent to 13 per cent of the total gross merchantable volume from the fourth to the eighth year, while the volume of advanced sapwood decay increased from 2 per cent in the fifth to 45 per cent in the eighth year. All trees examined in the seventh and eighth years following death contained 0.5–0.8 inches of decayed sapwood. *Stereum sanguinolentum* and *S. chailletii* were the two main fungi associated with incipient decay. Advanced decay was caused chiefly by *Lenzites saepiaria* and *Polyporus abietinus*.

According to Stickel and Marcott (1936) 45 per cent of fire-scarred trees in northeastern United States became infected within three years. Skolko (1947), in his study of balsam fir deterioration in fire-injured and fire-killed stands, indicated that decay losses are correlated with the severity of the burn. Five years after a severe fire only 1 per cent of merchantable volume was lost because of sapwood decay. The losses following lighter fires were much larger. Three years after moderate fire the losses amounted to 13.3 per cent, while after a light fire 27.7 per cent of merchantable volume was lost. Decay also contributed indirectly to the breakage losses, which were negligible three years after severe fire, but which amounted to 2.1 per cent after moderate fire and to 14.5 per cent after light fire. Stillwell (1958) noted that in some instances salvage operations were profitable many years after fire killing of balsam fir, provided that the product produced permitted the use of fire-killed and charred wood.

Insect relationships. The complex interrelationships between the insects and fungi attacking balsam fir were mentioned earlier, particularly the importance of injuries caused by *Hylobius pinicola* to decay of immature trees. The "gout disease" following attack by *Adelges piceae* (Ratz.) may be caused by *Creonectria (Nectria) cucurbitula* (Raymond and Reid, 1961). The same authors also noted the association between canker-causing fungi and blight, involving the fungi *Thyronectria (Nectria) balsamea*, *Dermea balsamea*, and *Valsa abietis*, and the sawyer beetles (*Monochamus* spp.). Stillwell (1960) noted the association between woodwasps and *Stereum chailletii* after trees were weakened by the balsam woolly aphid.

Most of the studies of the interrelationships of insects and fungi have been concerned with the spruce budworm and decay. Peirson (1923) observed that *Armillaria mellea* commonly invaded budworm-killed balsam fir and spruce. According to Basham et al. (1953), Schierbeck in 1922 prepared a list of fungi in order of importance in the decay of budworm-killed balsam fir: *Armillaria mellea*, *Fomes pinicola*, *Polyporus abietinus*, *Stereum sanguinolentum*, *Polyporus balsameus*, *Poria subacida*, and *Lenzites saepiaria*. Stillwell and Redmond (1956) and Redmond (1957) noted that only 3 per cent of the butt rot of budworm-killed trees could be attributed to budworm attack. Rankin (1920) and McCallum (1925) also indicated the lack of association between budworm damage and occurrence of butt rot. During a recent budworm epidemic Stillwell and Redmond (1956) observed rootlet mortality up to 75 per cent in trees defoliated for three consecutive years. The death and decay of these rootlets was not considered responsible for the entrance of heart rot fungi. Stillwell (1956) noted that budworm damage may serve as entrance points for top rot fungi. All balsam fir trees surviving the budworm epidemics of 1912 and 1920 in New Brunswick, on which the branches had been killed to a diameter of 0.5 inches, contained rot 20 years later (Stillwell, 1956). No decay was found in the base of dead tips if they were less than five years old or less than 0.5 inches in diameter.

Stillwell (1960) attempted to determine the reasons for the earlier deterioration of trees severely weakened by spruce budworm attacks as compared to windthrown trees. He noted that wood-destroying fungi penetrated the sapwood of budworm-killed balsam fir to an average depth of 0.4 inches within one year, whereas two years after windthrow decay was negligible. His observations in the Kedgwick watershed of New Brunswick showed that infection with *Stereum chailletii* and *S. sanguinolentum* may have resulted from attack by siricids or woodwasps. The woodwasps were identified as *Xeris spectrum* (L.) and *Sirex cyaneus* F. A possible symbiosis between siricids and fungi has previously been discussed (Cartwright, 1938, and others).

Rapid deterioration of balsam fir killed by the eastern hemlock looper (*Lambdina fiscellaria fiscellaria* (Guen.)) at North River, Nova Scotia, was also attributed to siricid-borne infection (Stillwell, 1962).

Basham (1951a,b), in his studies of budworm-killed balsam fir near Sault Ste. Marie, Ontario, showed that the first sap-rotting fungus to invade was *Stereum chailletii*; this was soon followed by *Polyporus abietinus*, a common white rot fungus. In the 182 trees sectioned, 92 per cent of the decay found was attributed to these species. The remaining 8 per cent was owing to the brown rot fungi, *Fomes pinicola* and *Coniophora puteana*. The percentage of recoverable volume declined from 88 per cent one year after death to 37 per cent after five years.

Basham et al. (1953) showed by inoculation studies that *P. abietinus* could develop on budworm-killed balsam fir without the previous invasion by *S. chailletii*. Continuing his studies of budworm-killed trees, Basham (1959) isolated from discolored cambium of recently killed trees a yeast fungus, *Ceratocystis (Ophiostoma) bicolor*. The principal species isolated from stained sapwood was *S. chailletii* and an unidentified hyphomycete. One year after death of the trees, *P. abietinus* was the only fungus consistently isolated from decayed sapwood. There was no evidence of one fungal species inhibiting or paving the way for another. The author attributes the succession of fungi to changes in the wood, particularly to changes in moisture content.

It is evident from these studies that unless salvage operations can be initiated shortly after serious injury or death, deterioration by wood-rotting fungi will have proceeded to the point where losses will be severe. The same situation holds for unpeeled balsam fir pulpwood. It deteriorates rapidly, and storage for longer than one year may result in lower yields and higher bleaching costs. Peeled balsam fir pulpwood can be stored 2–3 years in woodyards with no greater loss than may occur in unpeeled pulpwood after one year.

Descriptions of the Principal Heart Rot Fungi

Stereum sanguinolentum A. & S. ex Fr. (red top rot, red heart rot, "sapin rouge"). The many investigators all agree that this fungus causes most of the balsam fir trunk rot. Infection occurs through branch and other wounds. Often the decay is confined to the upper portion, but it may extend throughout the trunk. The wood becomes reddish-brown in color and attains a very high moisture content (Stephenson, 1950). The decay occurs as a solid circular mass or in irregular patches. As the decay proceeds white mycelial sheets are formed in the wood. Incipient decay is pink or red in color. Advanced decay has a light brown color and is dry and friable in texture, with a tendency toward stringiness. Fruiting bodies (1–8 cm) are thin, somewhat soft, cream to grayish-brown, resupinate or partly bracket-like. The hymenium is fuscous and bleeds when wounded. Spores are oblong or cylindrical. Fruiting bodies occur on dead trees or slash (Faull and Mounce, 1924; Stephenson, 1950; Brooks, 1953). According to Baxter (1952a), it is also serious on pulpwood, and if inoculated it can grow on wood pulp (Kress et al., 1925).

Corticium galactinum (Fr.) Burt (white stringy butt rot). This species was found by Basham et al. (1953) to be responsible in eastern North America for 26.5 per cent of all decay infections, 40.3 per cent of all butt rots, and 54 per cent of all white stringy butt rots. It is difficult to distinguish *C. galactinum* from *Poria subacida*. This fungus is also barely distinguishable from *Corticium odoratum*. It was therefore overlooked until 1951, when it was reported by the Canada Forest Biology Division (1951), Davidson (1951), and White (1951). More accounts of this fungus may be expected in the future as it receives added study. The first description of *C. galactinum* was furnished by Fries (1851) in Denmark from material sent from South Carolina (cited in White, 1951). Schrenk (1902) described it (under the name *Thelephora galactina*) as a root disease of apple trees in West Virginia, Kentucky, Southern Illinois, Mississippi, Arkansas, and Oklahoma. It is now known to occur in North America from the west coast to Texas and eastern Canada, the West Indies, Europe, and Japan. Schrenk and Spaulding (1909) discovered it on oak roots in the Ozark Mountains. Burt (1926) found it on *Rubus* sp. and *Lychnis alba*. Cooley and Davidson (1940) and Cooley (1948) extended the list of hosts to additional genera from which it was previously reported, and to *Kalmia* sp. They also noted it living saprophytically on dead stumps. Long (1917) was the first to discover the fungus on *Pinus echinata* stumps in Arkansas. Kress et al. (1925) found it occasionally on stored pulpwood, and after inoculation it was also found growing on pulp. Bisby et al. (1938) listed the fungus as a species occurring on twigs of balsam fir. Marshall (1948) found *Corticium galactinum* on the roots of several pines, especially *Pinus strobus*. Infected pines became stunted and the foliage turned yellow. Marshall was not certain whether it was a primary or secondary pathogen. The manual prepared by Nobles in 1948 for the identification of wood-destroying fungi includes this species. White (1951) found it on roots of balsam fir saplings in northern Ontario. He further expressed the opinion that infection may have occurred through root grafts or through wounded roots and that it may damage other conifers and hardwoods.

Odontia bicolor (A. & S. ex Fr.) Bres. (white stringy butt rot). According to Basham et al. (1953), this is another important species previously mistaken for *Poria subacida*. In their survey, *O. bicolor* was present in 8.2 per cent of all decayed balsam fir, comprising 12 per cent of all butt rots and 17.3 per cent of all white stringy butt rots.

The morphology and biology of some species of *Odontia* have been described by Brown (1935). According to Nobles (1953), *Odontia bicolor* is widely distributed in Europe and in North America. It is an important cause of butt rot in various species of conifers and hardwoods. Decay is confined to heartwood of butts and roots. Infection takes place through the roots and advances to the basal parts of the trunk. The incipient stage is pink, red, pinkish-brown or reddish-brown. The advanced stage is white of the pitted, stringy, or feather type, with black flecks. In final stages of decay the wood may be completely destroyed, resulting in a hollow butt or trunk.

Coniophora puteana (Schum. ex Fr.) Karst. (brown cubical butt rot). According to Basham et al. (1953) this fungus ranks fourth in incidence comprising 6.3 per cent of all balsam fir rots and 9.8 per cent of all brown cubical rots in eastern North America. It was described as associated with balsam fir butt rot by Spaulding and Hansbrough (1944). Basham (1951a,b) found it causing limited sapwood rot in dead balsam fir. It is also important on building timbers and occurs on slash (Boyce, 1948; Baxter, 1952a).

Polyporus balsameus Pk. (brown cubical butt rot). This species accounts for 6.2 per cent of all decay, 9.6 per cent of butt rots, and 41.4 per cent of brown cubical butt rots (Basham et al., 1953). Kaufert (1935) stated that it was responsible for 5.7 per cent of the cull caused by butt rots. Spaulding and Hansbrough (1944) found that 13 per cent of all cull due to decay in balsam fir was caused by this fungus. The species was first reported in association with balsam fir by Peck (1878) and since then has been reported by many authors. Hubert (1931) and Boyce (1948) credited this fungus for rapid deterioration of budworm-killed balsam and spruce. The later reports by Basham (1951a,b), Basham et al. (1953), and Basham and Belyea (1960) did not mention it in this connection, but *Polyporus abietinus* was noted as the most important sapwood rot. In longitudinal section the early stage of decay is warm buff, buff-yellow to light buff in color, fading into the whitish color of the normal wood. In the later stage the decay color is light brown, and when the rot is crushed to a powder it becomes clay colored. The older rot at the base of infected trees sometimes has a hazel-brown or snuff-brown color and occurs as cubes. The sporophores are annual, fleshy, shelf-like and rigid when dry, up to three inches wide, pale brown on the upper surface which is also distinctly marked with concentric zones, and white on the undersurface. The sporophores commonly develop in overlapping clusters (Hubert, 1929, 1931; Boyce, 1948; Baxter, 1952a).

Poria subacida (Pk.) Sacc. (white stringy butt rot, spongy root rot, stringy butt rot, feather rot, yellow stringy rot). Ranking sixth in incidence among all balsam fir rots, *Poria subacida* comprises 5.1 per cent of all rots, 7.9 per cent of all butt rots, and 11.1 per cent of all white stringy butt rots (Basham et al., 1953). These figures indicate a far less important position for this fungus now than was previously reported. According to Kaufert (1935), it was responsible for 89.0 per cent of cull due to butt rots in the Lake States. Spaulding and Hansbrough (1944) credit it with 28 per cent of all cull in the northeastern United States. Prielipp (1957) in more recent studies found *P. subacida* in 13.7 per cent of the trees with rots. The difficulty of recognizing this species was pointed out by Fritz (1923). McCallum (1928) was not certain of his identification of this species. Recently, White

(1951), Davidson (1951), and Basham et al. (1953) have shown that *Corticium galactinum* and *Odontia bicolor* are responsible for a large part of the decay formerly attributed to *P. subacida*. Boyce (1948, 1961) and Baxter (1952) have also associated this species with deterioration of dead trees, contrary to the reports of McCallum (1928), Basham (1951a,b), and Basham et al. (1953). In the early decay stages the fungus causes slight darkening of the wood with faint outlines of pockets. Later, in the springwood, irregularly shaped pockets appear which soon run together forming masses of white fibers. The annual rings separate easily from each other. Finally, the heartwood becomes a soft spongy mass of water-soaked white fibers. The fruiting bodies are annual, crust-like, white to ivory-yellow or pinkish-buff, and are usually found on the underside of exposed roots or root crotches (Hubert, 1931; Boyce, 1948; Baxter, 1952a).

Armillaria mellea (Vahl. ex Fr.) Quél. (white stringy butt rot, shoestring rot). According to Basham et al. (1953) this species accounts for 5.1 per cent of all balsam fir rots, 7.9 per cent of all butt rots, and 11.1 per cent of white stringy butt rots. Seldom extending more than two feet above the ground, and therefore of much less importance than *Poria subacida*, it is generally considered a secondary fungus (Boyce, 1948; Baxter, 1952a; Christensen and Hodson, 1954). In the early stages of decay the wood appears water-soaked and later develops an indistinct purplish to brownish invasion zone. In the last stages the wood becomes soft and spongy, often stringy, and white or yellowish in color. Surrounding the whitish and yellowish areas of decay, or imbedded within them, are dark brown to black, narrow, thread-like zone lines, sometimes concentric. Both sapwood and heartwood are attacked. The fruiting bodies are borne in clusters at the base of the exposed roots. The caps are smooth, honey-colored or brownish. The black stems are shoestring-like rhizomorphs of two kinds, subcortical and subterranean.

Conclusions

1. Future investigations of the relationships between balsam fir and its microbiological complex must be based on physiological studies of both components, with emphasis on the ecosystem as a whole and its dynamic development. A number of recent Canadian studies on the decays of balsam fir illustrate the potentials of this type of research approach.

2. Added research is needed on the identification of fungi associated with balsam fir, their life histories, and their physiology.

3. *Corticium galactinum, Odontia bicolor*, and other fungi formerly overlooked or under-evaluated should receive added attention now that their importance is recognized. Silvicultural practices must take into account the life histories of these and other fungi.

4. Research on root development of balsam fir and associated forest vegetation is called for, with particular attention being given to entrance points for butt rot fungi and the interrelationships between them and the abundant soil microflora under different conditions.

5. There are already observations which justify detailed and fundamental work on the role of saprophytic fungi in pruning of balsam fir and the relationship of such pruning to infection by heart rotting and other fungi. The ecological succession

of wood-rotting fungi on budworm-injured, wind-thrown, and fire-killed balsam fir needs further study.

6. Research is necessary on the relationships between diseases of balsam fir and injuries caused by wildlife.

7. There is little information on the effect of injury to remaining trees during thinning or partial cuts. Although from some points of view balsam fir lends itself to all-aged management or the selection system, its extreme susceptibility to decay following injury may preclude any type of cutting system except clearcutting.

8. Balsam fir reproduction needs to be studied in order to determine whether its age can be ascertained from external appearance. The fact that this species can tolerate long periods of extreme suppression and that occasional trees 5–10 feet tall may be 50 years old is well known. Since the development of heart rot is directly correlated with age, and infection at age 50 regardless of diameter may already be quite high, should advance reproduction be left? It may well be that reproduction of this nature is not worth retaining. A technique or standard for recognizing such "old-age" reproduction might help to reduce a great deal of future loss from diseases.

9. There is need to adjust the pathological rotation of balsam fir to different climatic, edaphic, and stand conditions, taking into account stand histories and local economic factors. Such studies should be made throughout the range of balsam fir and are especially important in areas where the species plays an important role in forest management.

10. There is need for the establishment of sample plots in all parts of the balsam fir range on which detailed and continuous observations would be made to trace the development of fungi and insects and the deterioration they cause. Extension of the mycological studies to whole stands, using broad surveys and permanent plots within a framework of an advanced ecological forest classification system, would provide a much better chance to recognize those geographic regions, forest types, individual stands, and trees which are most seriously affected by diseases, and to work out management practices and technical measures for better control of fungi.

5

ENTOMOLOGY

INSECTS, like fungi, are among the biotic factors affecting trees and forest communities, and a consideration of their special relationships to balsam fir forms the subject of this chapter. Detailed information on insect taxonomy, morphology, population dynamics, and other specialized topics is not presented. However, some discussion of physiological processes inseparable from insect ecology has been included. Secondary processes initiated by insect activities are treated in the appropriate chapters; rots following budworm attack are discussed in Chapter 4, effects on reproduction and stand development in Chapters 6 and 7, and effects on the technical properties of wood in Chapter 9. Special consideration will be given to the spruce budworm and balsam woolly aphid because of their great economic importance. Increasing attention has been paid in recent years to the entire insect problem as it affects balsam fir. It is the special purpose of this chapter to review these developments.

Taxonomic and Ecological Problems

A number of related zoological problems are frequently referred to in forest entomology. Arthropods other than insects are occasionally discussed in entomological literature. It is only in recent years that systematic study of these organisms has been related to balsam fir. By contrast, lists of insects associated with balsam fir have been prepared and systematic surveys of insect populations have been conducted for a long time.

The ecological aspects of insect relationships have been of great interest and practical concern in forestry. Following the solution of major taxonomic problems and the completion of investigations concerning major life processes of insects, entomologists gave increased attention to broader insect-ecological relations. In these studies the interests of several disciplines have been involved. A classic example of this mutual approach is the concerted effort that is currently directed at unraveling the complex relationship that exists between the spruce budworm and its principal host, balsam fir.

TAXONOMIC SURVEY

In published information on balsam fir very few references relating to arthropods other than insects were found. Packard (1890) noted a mite (order Acarina) that was commonly found on fir. The mites would work at the base of leaves during summer and early fall causing the leaves to curl. Later they would spin a web in the leaf axils located at the base of shoots. The amount of damage done was slight. Jacot (1939) described the action of several saprophagous mites on fallen needles of balsam fir. These animals were among the microfauna found in the litter of spruce-fir forests. Haynes (1962) studied invertebrate predators on balsam fir foliage in northern New Brunswick and found 56 species of spiders.

Of the 600,000 to 700,000 species of insects that have been described, over 100 are known to be directly associated with balsam fir. Raizenne (1952) listed 34 species of Lepidoptera from Ontario. Craighead (1950) identified 25 species affecting balsam fir in the eastern United States. The report by Packard (1890) contained 29 taxa, but the specific identity of several was not established. Surveys mainly list insects of economic importance. The annual reports of the Canadian forest insect and disease survey (Can. For. Biol. Div., 1942–59) have listed the following insects as most common on balsam fir: *Acleris variana* (Fern.), *Adelges piceae* (Ratz.), *Choristoneura fumiferana* (Clem.), *Dioryctria reniculella* (Grt.), *Hemerocampa leucostigma* (J. E. Smith), *Lambdina fiscellaria fiscellaria* (Gn.), *Neodiprion abietis* (Harr.), and *Pleroneura borealis* Felt. More recently the Canadian survey (Can. For. Ent. Path. Br., 1960–61) has added *Dasyneura balsamicola* (Lint.), *Orgyia antiqua* L., and *Semiothisa granitata* (Gn.). Baldwin et al. (1952) considered spruce budworm, balsam woolly aphid, hemlock looper, carpenter ant, and gypsy moth as important pests of balsam fir in the northeastern United States. Roe (1950) surveyed balsam fir literature and listed spruce budworm, carpenter ant, false hemlock looper, balsam fir sawfly, and balsam woolly aphid as most important. In the Maritime Provinces (Morris, 1958) the following insects seriously affect balsam fir: spruce budworm, balsam woolly aphid, black-headed budworm, and eastern hemlock looper.

From a survey of the available literature a checklist of 122 species was compiled. This list together with synonyms is given in Appendix II. A synopsis by orders and families is presented in Table 30. The checklist includes only those insects that have been reported as found on balsam fir. Insect parasites and predators are not included.

GENERAL ECOLOGICAL RELATIONSHIPS

Insect epidemics may substantially alter the species composition of forest stands. In the last 30 years the volume of balsam fir has increased from 46 per cent to 80 per cent in the forests of the Gaspé Peninsula. This increase has been due primarily to the destruction of white spruce by the European spruce sawfly, and to birch dieback (Blais, 1961a). The spruce budworm kills balsam fir primarily. On the other hand, the insect often releases the frequently abundant advance reproduction of balsam fir, or promotes shrub development. As a result, forest composition may be greatly changed.

Dead trees add to the soil organic matter. In the decomposition of slash and litter, insects, in association with many different organisms, play an important role. Micro-environment is just as important in the ecology of insects as it is with fungi. Studies of the smallest units of ecosystems limited to niches and synusiae, or larger communities, are essential to the understanding of the many complicated interrelationships that exist between organisms and their environment.

TABLE 30. TAXONOMIC SYNOPSIS OF INSECTS
REPORTED ON BALSAM FIR

Order and Family	Number of Species	Order and Family	Number of Species
Hemiptera		Diptera	
Aphididae	2	Cecidomyiidae	1
Phylloxeridae	1	Coleoptera	
Diaspididae	1	Buprestidae	10
		Melandryidae	2
Lepidoptera		Cerambycidae	11
Gelechiidae	1	Curculionidae	6
Blastobasidae	1	Scolytidae	21
Tortricidae	5	Hymenoptera	
Pyralididae	2	Xyelidae	1
Geometridae	24	Diprionidae	2
Lasiocampidae	1	Siricidae	4
Sphingidae	1	Torymidae	1
Noctuidae	15	Formicidae	1
Arctiidae	3		
Lymantriidae	5	Total	122

Physiological characteristics of trees play an important role in their susceptibility to insect attack. Canadian studies of the spruce budworm have stimulated fundamental investigations in tree physiology (summarized in Chapter 1) which are now being conducted in Canada and the United States.

Insects require several years to build up populations in response to favorable conditions. This slow buildup makes possible the prediction of the size of insect populations. Sophisticated sampling designs which have been incorporated in mathematical models have not only been useful in the development of techniques and the logistics of insect control, but have also contributed to the general knowledge of the dynamics of insect populations (Morris, 1954, 1955, 1960; Morris et al., 1958; Waters, 1955; Watt, 1961; Morris, ed., 1963).

Opinions held by different workers reflect changing viewpoints on the causes of insect outbreaks. Entomologists first emphasized the dynamic balance existing between insects and their predators and parasites. Later workers laid stress on the climatic conditions leading to insect epidemics. More recently, the evaluation of host conditions was stressed. Recognition of the complexity of the problems involved led some workers to consider all aspects of insect outbreaks in a single model. In this regard, the rapidly developing general systems theory may in the future provide new and more useful models for understanding population dynamics, not only for insect outbreaks but for entire ecosystems.

Spruce Budworm

The spruce budworm, *Choristoneura fumiferana* (Clem.), is the most destructive insect of balsam fir. It also attacks several spruces and other conifers including Douglas fir, white fir, and larches. Economic effects of spruce budworm attacks include direct killing of trees, loss of increment, degradation of wood, and deterioration of dead, standing timber (Graham, 1956a). Indirect effects involve other organisms, environmental factors, and the economic structure of forest management.

Spruce budworm epidemics have stimulated much research. According to information compiled by Neatby (1955), investigations from 1909 to 1919 were chiefly limited to observations, and extensive studies did not begin until 1924 with the work of Swaine, Craighead, and Bailey. During the period 1924 to 1944 considerable survey data were accumulated. Heavy losses of timber in Ontario from 1935 to 1945 stimulated further research. The modern forest insect laboratory at Sault Ste. Marie was established in 1945 primarily to intensify spruce budworm research in southern Ontario. In eastern Canada, where ravages of the spruce budworm were felt from the late 1940's to the mid-1950's, a special program known as the Green River Project was set up in northern New Brunswick, primarily to study the insect and its effect on the spruce-fir forests. A comprehensive monograph on the dynamics of epidemic spruce budworm populations based on fifteen years' research by many workers associated with the Green River Project and a companion aerial spraying project has been published (Morris, ed., 1963). The spruce budworm has also received some research attention at forest laboratories in Manitoba, Alberta, and British Columbia, although in these areas the problem has never been so acute. In the United States, investigations on the spruce budworm have not been given such major consideration, although moderate increases in funds and personnel and expansion in facilities have been made in the Lake States and in the Northeast during the past decade.

HISTORICAL RECORDS

Historical reviews of spruce budworm outbreaks have been presented by several authors. In general, two kinds of information are presented: (1) direct reports beginning with casual notes and ending with data from modern surveys, and (2) indirect information from careful analysis of stems and stand structure by age and size classes.

From an analysis of spruce stems and a study of the distribution of balsam fir age classes, Peirson (1950) concluded that an outbreak of spruce budworm occurred in Maine about 1770. No written reports of this outbreak were found.

A spruce budworm epidemic of about 1770 was also reported by Greenbank (1956) in New Brunswick. He listed 1806, 1878, 1912, and 1949 as years of other budworm outbreaks in the same area. Analysis of the radial growth of affected trees indicated that each outbreak lasted approximately ten years.

Swaine et al. (1924), Brown et al. (1949), and Peirson (1950), referring primarily to letters first mentioned by Packard (1890), give the period from 1807 to 1818 as the first documented spruce budworm outbreak in North America. The outbreak was centered in Maine east of the Penobscot River.

Turner (1952) traced a spruce budworm outbreak in the Algoma District of Ontario to the period from 1832 to 1835. Blais (1954) investigated the Lac Seul area in northwestern Ontario and concluded that an epidemic occurred about 1865, lasted for seven years, and covered nearly ten thousand square miles. Since this area had never been logged, he concluded that budworm outbreaks are not necessarily preceded by logging disturbances.

A spruce budworm outbreak occurred between 1874 and 1880 in Maine, New Brunswick, and Quebec (Swaine et al., 1924; Brown et al., 1949; Peirson, 1950; Blais, 1961a). Graham (1956c) estimated that a budworm epidemic also occurred in the Upper Peninsula of Michigan about 1880.

Another cycle of budworm activity started between 1909 and 1913 and continued for a decade in New Brunswick, Quebec, Maine, and Minnesota (Peirson, 1923; Graham, 1923; Swaine et al., 1924; Graham and Orr, 1940; Brown et al., 1949; Peirson, 1950). The loss of timber in this epidemic was estimated as equivalent to 250 million cords of balsam fir and spruce.

Gryse (1944) showed graphically the areas of budworm outbreaks in Canada from 1909 to 1944. Webb et al. (1961) presented a cartographic history of outbreaks in eastern North America from 1949 to 1959. Further illustrations have appeared in the annual reports of the Canada forest insect and disease surveys (Can. For. Biol. Div., 1942–59; and Can. For. Ent. Path. Br., 1960–62). It is obvious from these records that the budworm in Canada is very dynamic, with activity subsiding in some districts and developing in others. These more recent general trends are reviewed separately by provinces.

In Ontario new outbreaks of budworm started in 1935, damage in the ensuing years being considerable. Elliot (1960) estimated that 17 million cords of spruce and fir were destroyed between 1937 and 1955 in the Lac Seul and Lake Nipigon areas. According to survey reports, 12,000 square miles were affected in 1952. High mortality in a balsam fir stand in Ontario is illustrated in Figure 10. Extensive damage has continued in recent years, although a decline had been forecast for 1962.

Spruce budworm has been an important pest since 1939 in Quebec. In 1949 the insect was the most serious forest pest in the province. By 1953 spruce budworm was affecting 4500 square miles. The highest population intensity was reached in the Gaspé in 1958, but since 1957 a general decline was observed in the remainder of the province. The insect was at a low level in 1960 and 1961.

An increase in budworm attack was reported from New Brunswick in 1948. The buildup was rapid, affecting 500 square miles in 1950 and increasing to over 16,000 square miles by 1956. A marked decline which began in 1958 was partially due to an intensive aerial spray program (Webb et al., 1959).

In Nova Scotia the mainland was free from budworm outbreak until 1950. Light defoliation has continued since that time up to 1961. Cape Breton Island was more seriously affected during this period.

Figure 10. Heavy mortality following severe spruce budworm infestation in a balsam fir stand
at Manitowik Lake, Ontario. (Reproduced from Turner, 1952.
Courtesy of the Department of Forestry, Canada.)

In Newfoundland budworm defoliation generally has not been serious or widespread. A slight increase was observed between 1959 and 1961.

The spruce budworm has been spreading slowly in Manitoba since 1944 when it was observed in the Spruce Woods Forest Reserve. It has been reported on balsam fir since 1951 near the Saskatchewan boundary in northern Manitoba and also in Saskatchewan. The outbreak in this latter area has continued during recent years, affecting 1000 square miles in 1961 (Elliot, 1962).

In the United States, budworm outbreaks have been most serious in Maine and Minnesota. The latest outbreak in Maine started in 1947 and by 1949 had covered four million acres (U.S. Bureau Ent., 1952). Waters and McIntyre (1954) reported a sharp decline in 1951 and 1952. This was followed by renewed outbreaks which reached a new high of three million acres by 1956 (Waters and Waterman, 1957). Since then there has been a decline. In 1961 about one million acres were affected (Bean and Waters, 1961). Outbreaks have also been reported in New York, Vermont, and New Hampshire, but budworm populations have generally remained at an endemic level.

Absent for two decades, spruce budworm was again found throughout spruce-fir types in the Lake States in 1954 (Beckwith and MacAloney, 1955). In 1956 moderate to heavy defoliation was observed on 419,000 acres in Minnesota (Bean and Graeber, 1957). The affected area increased to 960,000 acres by 1958. Budworm populations in Wisconsin and Michigan remained low during this period (Schmiege and Anderson, 1958, 1960; Anderson and Schmiege, 1959, 1961).

BIONOMICS

The scientific name of spruce budworm has been changed many times since it was first described in 1865 by Clemens (Freeman, 1958). A list of synonyms (see Appendix II) shows that the insect has received seven names grouped under five different genera. These changes reflect both increasing knowledge of the spruce budworm and the group to which it belongs as well as changing concepts from a genetic and evolutionary viewpoint. The presently accepted name, *Choristoneura fumiferana*, was proposed by Freeman in 1947. The spruce budworm is primarily associated with the Boreal Forest, but total distribution is considerably wider. Freeman (1958) gives its range as extending from Virginia northward to Labrador, westward across Canada and the northern United States to British Columbia, southward in the Cordilleran region to Arizona and California, and northward to Yukon.

For several decades entomologists have recognized a form of spruce budworm that feeds primarily on jack pine (Graham, 1928). Freeman (1953) distinguished between the two forms, *C. fumiferana*, spruce budworm, and *C. pinus* Free., jack pine budworm. The morphological differences between the larvae of the spruce and jack pine budworms were described by MacKay (1953) who differentiated them on the basis of head structure. Cox (1953) reported morphological differences between the adults and larvae of the two species. Campbell (1953) studied the morphological differences involving the pupae and egg clusters. The *C. fumiferana* ordinarily emerges and matures before *C. pinus*. Larvae of *C. pinus* have not been reported on balsam fir nor have larvae of *C. fumiferana* on pines. The unmated female adults tend to remain on their host trees. Smith (1954) concluded that interbreeding of the two species in nature is uncommon even in places where they are coexistent.

At the present, in Canada, *Choristoneura fumiferana* is recognized as a complex of three more or less distinct forms: (a) an eastern one-cycle form, (b) a western one-cycle form, and (c) a western two-cycle form. The one-cycle forms complete one generation in one year, while the two-cycle forms require two years (Miller, 1963a).

The spruce budworm *C. fumiferana* is very similar to *C. muriana* (Hb.), a European fir budworm, in its morphology, physiology, and natural enemies. The absence of spruce budworm outbreaks in Europe is due to differences in their respective host species, *Abies balsamea* and *A. alba*. There are also differences in behavior with respect to their penetration into buds and needles after hibernation (Franz, 1957).

The adult is typically a grayish colored moth with a wingspread of about 25 mm. Brownish forms are also rather frequent. The moths are active in the northeastern states in late June and July and in the Lake States and in New Brunswick during July and early August (Jaynes and Carolin, 1952; Miller, 1963a). They often occur in such large numbers that in flight they resemble a "cloud of grayish snowflakes whirling around the treetops" (Graham and Orr, 1940). Fully gravid females cannot fly until some oviposition has occurred (Wellington and Henson, 1947), but females carrying fewer eggs can fly before any oviposition takes place (Blais, 1953a). The flight period lasts for three weeks (Swaine et al., 1924).

Balsam fir and white spruce are the favored hosts for oviposition (Swaine et al., 1924; Jaynes and Carolin, 1952) in eastern Canada, the northeastern United States, and the Lake States. From the New Brunswick experiments, Greenbank (1963d) found no significant difference in oviposition preference between balsam fir and white spruce, but there was a definite host selection preference for these species in comparison with black spruce. Red spruce apparently is somewhat intermediate as a host species, being affected much less than white spruce or balsam fir but more so than black spruce.

For egg-laying the moths prefer tall, open-grown trees. This preference has been attributed by some workers (e.g., Blais, 1952) to the availability of large amounts of staminate flowers providing pollen, an important food for the second-instar larvae. Greenbank (1963c), however, emphasized the primary importance of warm, dry microclimates in the open-grown trees. These trees are, of course, also heavy pollen producers, but depending on phenological conditions, pollen may or may not be available as food for the second instars of the spruce budworm.

The females deposit about 175 small, flat, pale green eggs in overlapping patches on the needles. Other reports indicate 170–270 and 100–200 such eggs, while in New Brunswick (Miller, 1963a) the oviposition rate is 200 per female. The number of eggs per mass may vary from 10 to 100 but commonly averages about 20. The disc-shaped eggs are about 750 μ long, 500 μ wide, and 200 μ high (Stairs, 1960). Most of the eggs are laid on the foliage of the terminal twigs. More eggs are laid on the ventral surfaces of needles than on the dorsal surfaces (Hawley and Stickel, 1948; Blais, 1952; Neatby, 1955). They are laid in July or August and hatch in about ten days. The developing embryo goes through a sequence of changes which has been described in detail by Stairs (1960).

The newly hatched larvae, about two mm long, have pale, yellowish-green bodies with brown heads. They do no feeding. Depending on weather conditions first-instar larvae may move toward the branch tips (photopositive movement), after which they spin silk threads and launch or rapidly establish hibernacula (photonegative movement). Larvae launched on silk threads may be carried considerable distances by the wind (Mott, 1963a).

Laboratory experiments have shown that the preferred hibernation sites on mature trees are the flower bracts and on immature trees bud scales. To a lesser extent, needle axils, bark crevices, and lichens are used. Very few overwinter on the trunks of balsam fir. Choice of hibernation sites is often affected by weather conditions (Greenbank, 1963c). In these sites the larvae spin their small hibernacula, develop into second-instar larvae, and overwinter.

In diapause the larvae may withstand temperatures as low as —60F, but after they have begun feeding in the spring they are very susceptible to low temperatures (Atwood, 1945b). In northern New York larvae of the July 1946 hatch had suffered

75–84 per cent mortality during hibernation (Jaynes and Speers, 1949). According to Miller (1958) mortality in hibernacula accounts for about 15 per cent (9 to 22 per cent) of the budworm population in New Brunswick. Morris (1963c) concluded that survival of small larvae contributes significantly to variation in generation survival but varies considerably from place to place. In laboratory studies it was found that at 71F and long photoperiods some second-instar larvae did not enter diapause.

When spring air temperatures rise to 60F or above, the second-instar larvae emerge from hibernation. In New Brunswick this often takes place in early May (Miller, 1963a). Depending on weather conditions, the beginning date of rapid emergence may vary by as much as four weeks (McGugan, 1954). Laboratory studies have shown that 2.5C approximates the threshold temperature that terminates diapause. Taking this as a base, Bean (1961a) calculated from standard thermograph records the number of degree-hours that would establish the approximate date when the first larvae would emerge. In Minnesota this method was tested for four seasons, and an average error of only 1–2 days was found. Postemergence weather may markedly influence the rate of development (Rose and Blais, 1954, and others).

Several authors have investigated microclimatic effects on the development of spruce budworm larvae. Wellington (1948) found the internal temperature of flower buds to be 4–5F higher than in vegetative buds during cool spring weather. This may give the young larvae additional protection. Wellington (1950) further noted the difference in temperature between exposed and protected parts of balsam fir trees; within the silken tents spun by the larvae, inside temperatures may exceed outside temperatures by 8–13C. Shepherd (1958) studied temperatures of budworm larvae during a period of clear weather. Average temperature of larvae on reproduction under a spruce overstory exceeded average temperatures of the surrounding air by 2.3C. In a stand opening, the difference was 3.7C. The internal temperature of larvae also influences their photic orientation (Wellington et al., 1951) as shown by some experiments with polarized light.

The fall launching of first instars is the first major migration dispersal of young larvae. A second important larval dispersal takes place in May when second instars, in response to light, move toward the branch tips, spin silken threads, and are launched on wind currents to other parts of the tree, to other trees in the stand, or even to other stands (Miller, 1963a). Most mortality in young larvae is associated with these two major dispersal periods. Of these, spring dispersal losses are more variable than fall losses. In the New Brunswick study area the dispersal losses of small larvae averaged 64 per cent of the total hatch in the period 1952–1956 (Miller, 1958).

Greenbank (1963b) was unable to establish any relationship between small-larval survival and weather conditions due, in part at least, to the inability to separate the period into first-instar and second-instar survival. In addition, the time of emergence of larvae in the spring could not always be precisely determined. A large proportion of a hibernating population emerges from a given plot on the same day, and weather conditions on that day are the determining factors in whether or not dispersal will take place.

The feeding habits of spruce budworm larvae have elicited considerable attention with respect to their effect on the host and the development of the insect. Emergence usually takes place before foliar buds of the current season burst. The larval food supply in this interval consists largely of pollen produced in the male strobili. Second-instar larvae feeding on pollen may remain there up to the fourth instar before moving onto the new foliage. If male strobili are not available, the larvae either mine into the one- or two-year-old needles or into the expanding vegetative buds. Typically, only one old needle is mined and third-instar larvae begin feeding on newly opened vegetative buds by late May or early June (Bess, 1946; McGugan, 1954; Miller, 1963a).

The importance of staminate flowers in the larval diet has been investigated in both laboratory and field studies. Jaynes and Speers (1949) observed in laboratory experiments that pollen-fed larvae matured three weeks earlier than larvae fed on foliage alone. Blais and Thorsteinson (1948) observed 4–5 days' earlier development in larvae which fed on naturally-maturing pollen. Greenbank's (1963b) experiments in New Brunswick showed clearly that, given a choice, second-instar larvae will settle on foliage with staminate flowers more often than on nonflowering foliage. Branches having large clusters of male strobili carried a greater concentration of second instars than branches having few clusters. This preference was not, however, expressed equally well in every year. In years when emergence preceded the bursting of flower buds, more larvae were found feeding on needles than on staminate flowers. Greenbank emphasized that staminate flowers may not be used as food to the same extent every year.

Over a three-year period Greenbank (1963b) found that feeding on staminate strobili under field conditions promoted larval development by 3 to 7 days depending on weather conditions. However, the key factor in survival was total generation time, and this is influenced much more by weather conditions than by pollen supply. A wet, overcast spring does not promote fast larval development even when large supplies of pollen are available, and in such a spring maturity is reached 7 to 12 days later than in a warm, dry spring.

Ordinarily, larvae do not move about as long as the food supply lasts. When the foliage on one tip is consumed, they move to another nearby or spin down on silken threads to other branches. The feeding period lasts 3–4 weeks. Late-larval stages, particularly the sixth instar in late June and early July, accomplish most of the defoliation. At low and moderate population densities, larvae confine their feeding to current shoots. At high densities they are forced to feed on old foliage. Fifth and sixth instars can complete their development on old foliage, but this causes a marked reduction in the size of the pupae and the fecundity of the adults (Blais, 1952; Greenbank, 1963c). Laboratory experiments have also shown that larvae forced to feed on old foliage give rise to smaller pupae and less fecund adults. Adults raised on trees from which the current year's foliage was removed laid fewer eggs than adults raised on incompletely defoliated trees (Blais, 1953a; Miller, 1957).

Spruce budworm larvae feeding on black spruce develop more slowly and have a higher mortality rate than those feeding on white spruce and balsam fir. Black spruce buds open later, but when male strobili are abundant, the development of budworm larvae feeding upon them is similar to that of larvae feeding on balsam fir or white spruce. The late opening of black spruce buds explains the relative immunity of that species to severe budworm damage (Blais, 1957).

Young larvae have yellow-orange bodies, a blackish-brown head and a pale-brown prothoracic shield. Mature larvae have olive-brown bodies, a shining black head, and a brownish shield. Their length is about 25 mm. The body is marked with conspicuous whitish-yellow areas around the setae and by a laterally extended yellowish stripe (Jaynes and Carolin, 1952). There are six instars (McGugan, 1954). The width of frass can be used to gauge spruce budworm larval instars. Frass collected in traps can be used to calculate the percentage of each instar present. Neither the length nor the volume of the frass can serve for this purpose (Bean, 1960).

Mature larvae spin loose, silken webs on their feeding shoots. Within this web they transform to pupae. Pupae are about 12 mm long. Their color changes from pale yellowish-brown to dark reddish-brown with blackish transverse bands and spots. The adults emerge about ten days later, in midsummer (Brown et al., 1949; Peirson, 1950; Craighead, 1950).

PREDATORS AND PARASITES

Several decades ago birds were considered of great importance in the control of spruce budworm. The early studies were local in nature and the data obtained serve mainly as illustrations. Tothill (1923) estimated that birds effected 13 per cent control of spruce budworm during an outbreak in New Brunswick in 1918. Kendeigh (1947) concluded from an Ontario study that birds consumed about 4 per cent of all budworm larvae. George and Mitchell (1948) estimated the consumption by birds to be 3–7 per cent of budworm larvae at Lake Clear, New York, in 1946. During an outbreak in Maine, an examination of the stomachs of nearly 500 birds revealed that 20–40 per cent of the birds' diet consisted of budworm larvae (Mitchell, 1952; Dowden et al., 1953).

Morris et al. (1958) investigated the numerical response of birds and mammals to budworm populations during the decade 1947–1956 in the Green River area. The spruce budworm does not become attractive to most predators until the fourth instar. Most predation occurs in the fifth and sixth instars and in the pupal stage. Birds feed on larvae and pupae until late July, on moths in late July and August, and on egg masses and first instars after late August. The budworm is eaten by terrestrial mammals only when populations are very high. At such times, because of the lack of foliage, many larvae drop from the trees. The effect of birds and mammals on the budworm population during the outbreak years was negligible. Although the budworm population increased 8000 times, the population of the most responsive predator, the bay-breasted warbler (Dendroica castanea Wilson), increased only twelve times. However, when the population is at an endemic level of 1000 or fewer budworms per acre, birds may become important. During the ten-year period of observation rodents and insectivores showed no numerical response, but fluctuated independently of budworm density. However, populations of the red-backed vole (Clethrionomys gapperi Vigors) and the deer mouse (Peromyscus maniculatus Wagner) were much reduced from normal at the peak of the budworm cycles. This may have been associated with a shortage of tree seed resulting from budworm damage.

The importance of arachnids, particularly spiders, as predators of the spruce budworm has recently been investigated in New Brunswick by Loughton et al. (1963). As a group, spiders outnumber all other species of arthropods on the foliage of balsam fir trees. About 75,000 spiders per acre have been recorded in the Green River area. Of the two major groups, hunting spiders are considered more promising as insect-control agents than web-spinners. Serological tests from collections taken in the Fredericton area showed that 20 per cent of the spiders had fed on spruce budworm larvae in 1959 when the larval population was high and 8 per cent in 1960 when populations were low (Loughton et al., 1963). The data obtained were considered too few to give an accurate assessment of the role of spiders in budworm larval mortality, but at least they pointed to spiders as important predators.

Red mites have been observed feeding on the eggs of spruce budworm. Serological tests showed that 22 per cent of the mite populations sampled in the Fredericton area in 1959 had fed on spruce budworm eggs. In 1960 the figure was 24 per cent. Loughton et al. (1963) felt that this predation level may be of considerable significance at times in affecting mortality levels of the spruce budworm.

Although insect predators are of little importance in affecting spruce budworm control during the epidemic phase, in normal periods they may exert a greater influence. Interesting food relationships may develop between the larvae of spruce budworm and the larvae of spruce needle worm (*Dioryctria reniculella*) following late spring frosts. The latter may become predators of spruce budworm larvae and pupae (Barker and Fyfe, 1947; Warren, 1954).

Considerable work has been done on insect parasites of the spruce budworm (Coppel and Maw, 1954; Coppel and Smith, 1957; Dowden, 1962; Dowden and Carolin, 1947; Jaynes, 1954; Miller, 1960; Peirson, 1950; Raizenne, 1952; Thomson, 1957a,b, 1958). Wilkes et al. (1948) listed 45 species of parasites and predators; Brown (1952) listed 61 species; McGugan and Blais (1959) listed 75 species. Reeks and Forbes (1950) listed 21 species of native insect parasites of spruce budworm in the Maritime Provinces grouped as follows: Tachinidae 7, Chalcidae 2, Braconidae 3, Pteromalidae 1, Ichneumonoidae 8. Daviault (1950) listed the following insects as the most common and most active parasites of the spruce budworm in Quebec: *Apanteles fumiferanae* Vier., *Glypta fumiferanae* (Vier.), *Meteorus trachynotus* Vier., *Itoplectis conquisitor* (Say), *Phryxe pecosensis* T.T., and *Aplomyia caesar* (Ald.). Interest has been shown in introducing parasites from British Columbia and Colorado to eastern North American forests (Baird, 1947; Dowden, 1962). Search for parasites has also been made in Europe (Franz, 1952).

Limited data suggest that parasitism may at times be relatively high at endemic population levels. However, when climatic conditions in combination with favorable stand conditions combine to produce a favorable environment for population increase, parasites cannot prevent the population "release." In six generations the spruce budworm population can increase ten thousand times. At the same time, small larvae and moths are dispersed over wide areas. An increase in and dispersal of parasites does not follow this pattern. Only in the final phases of an outbreak after population density has already been greatly reduced by the action of other factors, can parasites kill an appreciable proportion of the larval population and combine with other factors to bring about the collapse of the epidemic (Miller, 1963b).

Several years ago viruses appeared as a promising control measure for spruce budworm (Graham, 1956a), and in 1949, Bergold reported on a polyhedral virus disease which showed some promise. More recently, however, there has been a tendency to regard virus and other diseases of spruce budworm as having low virulence (Neilson, 1963). According to Neilson three virus diseases have been found on spruce budworm in New Brunswick. The most prevalent, capsule virus or granulosis, has always been considered of questionable diagnosis. The nuclear and cytoplasmic polyhedroses present little difficulty in diagnosis but they are rare. Too little is yet known of these viruses to assess their role in controlling spruce budworm populations; at present they are not considered to have much importance.

Induced bacterial diseases, after early failures, have recently shown more promise (Prebble, 1961). Commercial preparations of several strains of *Bacillus thuringiensis* Berliner have gained world-wide attention as selective pathogens of the larvae of a number of Lepidoptera, including some forest species (Krieg, 1961). The first test against the spruce budworm, in which the *soto* variety was used, failed. More recently Angus et al. (1961) found that spruce budworm is susceptible to commercial preparations of *B. thuringiensis* var. *thuringiensis* under laboratory conditions. In Quebec, Smirnoff (1963) tested thuricide (*B. thuringiensis* in an inert filler) under both forest and laboratory conditions from 1959 to 1961. Mortality as high as 80 per cent was achieved under field conditions in 1960. Results were affected by temperature and precipitation. In the laboratory various strains of *B. thuringiensis* and *B. cereus* Frankland and Frankland were also tested with promising results. The availability of several strains of bacteria may prove helpful in controlling the budworm under different environmental conditions.

McLeod (1949) listed *Hirsutella* sp., *Empusa* sp., *Beauveria* sp., and *Tsaria* sp. as the most important fungus diseases of spruce budworm. Detailed studies are in progress on the ecological requirements of these species (MacLeod and Lougheed, 1956, and others).

In Ontario, Thomson (1955a,b; 1957a,b; 1958, 1960) reported on a microsporidian (protozoan), *Perezia fumiferana* Thom., infesting spruce budworm. In some instances 40 per cent of sixth-instar larvae were infected by the microsporidian. The organism produces a disease that slows the development of larval and pupal stages, and foreshortens the life of adult insects. All infected females transmit the disease to their offspring, and some transmission also occurs through the male parents. The effectiveness of microsporidians in reducing the life expectancy of budworm moths was questioned by Neilson (1963), at least insofar as his data from New Brunswick revealed. Neilson suggested, however, that the presence of a microsporidian may have had an adverse effect on fertility, a condition that Thomson was unable to show. Although he did not offer an explanation for these discrepancies, Neilson suggested that differences in laboratory methods and in budworm populations from the two areas may be partly responsible.

CONDITIONS ASSOCIATED WITH OUTBREAKS

A distinction should be made between the conditions which are prerequisite for initiation of an outbreak, and its consequent spread, and the conditions which result in high mortality of forest stands after an outbreak has reached epidemic proportions.

The comprehensive study on the dynamics of epidemic spruce budworm populations in New Brunswick, referred to earlier (Morris, ed., 1963), pointed to the climatic release of budworm populations following several years of dry sunny weather in combination with the availability of extensive, continuous, mature, and overmature forest stands with a high percentage of balsam fir. Macroclimatic conditions in combination with the microclimatic, such as exist in mature stands, are considered essential in creating epicenters of outbreaks. In addition to stand maturity, microclimatic effects are influenced by topography, stand composition, and stand density. A supply of preferred food for spruce budworm larvae is essential, but the significance of the availability of staminate strobili is de-emphasized in this study. Similarly, the view that the budworm "starves itself out" is not supported. The study also noted that man's action has not increased the frequency of spruce budworm outbreaks (Morris, 1963a,b,c).

Swaine et al. (1924) and Graham and Orr (1940) pointed out that budworm epidemics occur at intervals of 50–70 years. McLintock (1949) suggested a cycle of 30–35 years in the northeastern United States. After an epidemic has destroyed stands over large areas, a period of time is necessary for balsam fir to recover. Wellington et al. (1950) suggested that a high proportion of mature balsam fir trees in forest stands is a prerequisite for budworm outbreaks, but that relaxation of climatic control is the immediate cause of rapid buildup of populations. Their studies showed that three or four years before budworm outbreaks, the number of cyclonic centers passing annually through the affected areas decreases. Outbreaks occur during periods of decreasing or minimum storminess, when precipitation is below normal and summer drought is common. Spring and fall droughts are sometimes associated with spruce budworm outbreaks. Dry, sunny weather promotes larval development and the simultaneous development of staminate flowers. Open winters with widely fluctuating temperature in November and December may be detrimental to hibernating larvae. Henson (1951) studied the flight habits of budworm moths in northern Ontario. He concluded that the insects were carried by convective storms which precede cold fronts. The adults may be carried by winds for hundreds of miles.

The hypothesis that climate controls budworm outbreaks has been supported by a number of authors. Greenbank (1956, 1957) supported this view, based on analyses of available weather data preceding the 1912 and 1949 outbreaks in New Brunswick. Outbreaks have progressed in an easterly direction beginning near the Ontario-Quebec border in 1939 and ending in the Gaspé Peninsula in 1958 (Pilon and Blais, 1961). The same authors also emphasize the importance of prevailing winds in moth dispersal. Greenbank (1963a) recognized that the spread of an infestation with the prevailing wind has in some instances accounted for the rise of populations in neighboring and in uninfested areas. However, climatic analyses for different regions of Canada suggest that the incidence of favorable climatic periods may also progress from west to east, and this provides an alternative and preferable explanation to moth dispersal for the apparent spread of an infestation over distances up to 1000 miles. Blais (1954, 1960b) presented additional evidence from Ontario and Quebec that spruce budworm attacks are a phase in a natural cycle of events associated with maturing balsam fir stands or forests.

Turner (1952) attempted to test the effect of "topographical sites." Sites were recognized as ridgetop, upper slope, lower slope, dry flat, lower flat, and wet flat. It was not possible to establish a definite relationship between the budworm-caused mortality and site conditions. McLintock (1955) was also unsuccessful in establishing a trend relating defoliation by spruce budworm to site, since stand conditions on well-drained sites differed materially from those on poorly drained sites. Morris and Mott (1963) noted that trees on ridge tops show defoliation before those in valleys. Blais (1958b) produced evidence that trees growing on poor sites offer less resistance to budworm than trees growing on better sites.

In areas of Ontario where forests contained much aspen as well as balsam fir, heavy feeding by the forest tent caterpillar preceded spruce budworm outbreaks. A similar coincidence was observed in western Quebec and New Brunswick. Both epidemics, according to Wellington (1950), are initiated by meteorological events. Ghent (1958a) noted the increased flower production of balsam fir as a result of the defoliation of overstory aspen by the forest tent caterpillar.

Birch dieback may also be a contributing factor in the rapid buildup of spruce budworm. In central New Brunswick extensive areas have been affected by this disease since 1932, resulting in an increase in the relative dominance and exposure of balsam fir. There is no reason to believe, however, that birch dieback has been a contributing factor to the initiation of budworm outbreaks (Mott, 1963b).

In the discussion of the life history of the spruce budworm the important role that flowering of balsam fir plays in the development of the insect was pointed out. This relationship also has a great impact on the trees, as indicated by Bess (1945), Balch (1946b), and Blais (1958a). Interrelationships between flowering and vegetative growth of balsam fir were investigated by Morris (1948, 1951), Mott et al. (1957), and other authors (see Chapter 1). They noted that flowering shoots bear only one-half the vegetative foliage present on nonflowering shoots. Therefore, defoliation affects the flowering shoots more severely. Flowering is closely related to age of the trees, site, and stand conditions.

Hardwood overstories generally have a protective effect on softwood understories, but admixed hardwoods cannot protect dominant softwoods from spruce budworm attacks (Craighead, 1925). Turner (1952) showed that the influence of a hardwood overstory on budworm-caused mortality of balsam fir depends on the relative height of the trees. Where fir was present only as an understory, its mortality increased with increasing proportion of hardwoods. Where fir was codominant, the proportion of hardwoods did not affect its mortality.

Blais (1958a) noticed in his eleven-year study of permanent plots in the Lac Seul area of northwestern Ontario that the first trees to succumb to budworm defoliation were suppressed balsam fir growing under mature trees. This event occurred in the year after heavy defoliation. In two more years mortality began in the dominant stand, after which it progressed rapidly.

One of the earlier silvicultural measures recommended for the handling of vulnerable balsam fir stands was the development of open-grown stands through heavy thinnings or partial cuts to encourage the vigor of single trees. Such open-grown stands were thought to have the greatest budworm resistance. One of the reasons for this may be the high dispersal loss of small larvae in open and mixed stands compared with a relatively low loss in dense stands of preferred species (Morris and Mott, 1963). On the other hand, observations by Graham (1935) suggested that dense balsam fir stands may be more resistant than open-grown stands because of unfavorable microclimates for the development of spruce budworm. The findings of Turner (1952) corroborated this. In view of such important contributing factors as size and age of stand and stand structure, density alone does not appear to be a reliable indicator of stand vulnerability.

Susceptibility and vulnerability of balsam fir to spruce budworm attacks have been investigated in Ontario since 1937, but early attempts to correlate stand conditions to budworm attack failed (Atwood, 1945a). Westveld (1945, cited from McLintock, 1949) developed a formula for rating stand vulnerability in the northeastern United States, which included a knowledge of balsam fir volume per acre and basal area of the average tree, the latter considered as representing stand age. A vulnerability index was obtained by multiplying balsam fir gross volume per acre in cords by the basal area of the average tree in square inches. McLintock (1949) suggested that 20–99 should mean low vulnerability, 100–249, medium vulnerability, and 250 or more, high vulnerability. Other ratings could be assessed by considering the gross volume of balsam fir per acre, e.g., 4 cords or more indicating high hazard, 2 to 4 cords, medium hazard, and ½ to 2 cords, low hazard (McLintock, 1949).

Balch (1946a) considered as highly susceptible those areas in New Brunswick on which mature (over 60 years) balsam fir predominates (over half the volume of the stand). Stands 40 to 60 years of age were considered as moderately susceptible, and stands under 30 years of age, or where balsam fir was killed in previous outbreaks, were considered slightly susceptible. These hazard areas according to Balch (1946a) can be delineated on maps.

In the Lake States balsam fir frequently occurs in small blocks of 40 acres or less. Bean and Batzer (1956) suggested that under such conditions, risk rating should primarily be based on (1) age; (2) density expressed as volume per acre of balsam fir and white spruce; (3) proportion of balsam fir in the stand; (4) size of area; and (5) vigor of trees. Operability, accessibility, and other economic factors should be considered in determining priority for cutting high-risk stands in advance of approaching epidemics.

A hazard rating system devised by Graham (1956b) in Michigan is based upon age of balsam fir, species composition, and stand structure.

In Ontario, Turner (1952) attempted unsuccessfully to apply the systems developed by Westveld (1945) and Balch (1946a). He pointed out that forest conditions favorable for insect attack are not the same as conditions resulting in high mortality once an infestation has started. The quantity of balsam fir required to support an epidemic need not be high. Outbreaks may occur in areas where the merchantable volume is as low as three cords per acre.

EFFECTS OF OUTBREAKS

There is a lag between initial severe defoliation and a decline in diameter growth of balsam fir. Belyea (1952a) noticed that the first suppression of radial growth at breast height occurred two years after severe defoliation by the spruce budworm. Blais (1958a), using growth ratio techniques, concluded that the suppression of radial growth at breast height occurs from two to four years after defoliation depending upon its severity and duration. Mott et al. (1957) investigated the effect of budworm defoliation along the entire stem. The radial-growth maximum, which occurs between the third and seventh internode from the apex, was the first place where defoliation was noted. With continued defoliation the annual ring at that location disappeared. According to Turner (1952), reduction of growth at the base of the tree occurred one or two years later.

Swaine et al. (1924), investigating the mortality caused by the 1912–1920 budworm outbreak in New Brunswick, found that all rootlets and roots less than two mm in diameter were dead six months before the death of the tree. Redmond (1957) showed that many rootlets were killed on trees which survived outbreaks. Further studies (Redmond, 1959) showed that when 70 per cent of new foliage was destroyed, mortality of rootlets was in excess of 30 per cent. Rootlet mortality reached 75 per cent when all new foliage was destroyed. After defoliation stopped, new rootlets developed. These secondary rootlets were unusually long and had little associated mycorrhiza (Redmond, 1957). After severe defoliation, mature and over-mature trees were unable to develop new rootlets for four or five years, although they could produce new foliage before succumbing. Stillwell (1960) noted a three-year lag in rootlet response after foliage had recovered.

Tree mortality after spruce budworm attack usually occurred in the fifth year of severe defoliation in the Lake Nipigon and Lac Seul areas in northwestern Ontario (Blais, 1958a). In southwestern Quebec, McLintock (1955) reported that trees two-thirds or more defoliated died after five years. He estimated that recovery of the surviving trees would take six or seven years. During this time the growth, on the average, would be reduced by 75 per cent.

Many authors have noted that mature balsam fir stands are more seriously affected by spruce budworm defoliation than younger stands, but the effect is not limited to the former. Baskerville (1960) investigated a stand 40–50 years old which was defoliated annually from 1950 to 1957 and had lost 52 per cent of its basal area by mortality. The surviving trees were so weakened that numerous potential entrance courts for pathogenic fungi were found. At Sturgeon Lake, Ontario, Ghent (1958b) found that budworm defoliation also affected advance reproduction. In general, height growth of seedlings responded immediately when the peak of budworm populations passed and heavy overstory mortality began. Sometimes, however, damage to seedling leaders by the feeding of late-instar larvae nullified the growth response to seedling release.

It was once thought that spruce budworm, having preference for balsam fir, might serve to increase spruce in forest stand composition. Ghent et al. (1957) investigated this problem in northwestern Quebec and north-central and north-western Ontario. They noted an increase of formerly less-abundant shrub species after spruce budworm outbreaks. Balsam fir and spruce seedlings were equally affected by the overstory release, and their relative proportion remained unchanged for at least five years. The development of the stand after budworm outbreaks depends on local conditions, and either spruce or balsam fir may gain the advantage. Dense balsam fir thickets in the Green River area were traced to severe budworm outbreaks in 1913–1919 (Vincent, 1962). The management problems of these stands are discussed in Chapter 7.

POPULATION DYNAMICS AND SURVEYS

Forest insect surveys can be classified as serving the purposes of detection, reconnaissance, hazard appraisal, or damage (Orr, 1954). Either aerial or ground surveys may be used. Sometimes they can be combined to good advantage.

Use of aerial surveys is reported in a review paper on the use of aircraft in forest insect control by Balch, Webb, and Fettes (1955–56). Heller et al. (1952) noted that light (15 per cent) defoliation can be detected from aircraft traveling 90–100 mph at a height of 200 feet. Waters et al. (1958) specified an altitude of 500 feet for best detection work. In two and a half weeks involving 44 hours' flying time, these workers surveyed 10 million acres of which 4.5 million were defoliated. For quantitative appraisal of top-killing and mortality, aerial photographic methods, especially color photographs, proved more suitable than aerial inspection (Waters et al., 1958).

Morris (1950) described ground surveys for studying insect populations on large trees. He also stressed the advantages of sequential sampling techniques for rapid assessment of population changes (Morris, 1954). These latter methods, which utilize broad population classes, require prior knowledge of class limits and distribution patterns of insect populations. Forest insect survey data will generally fit one of the following distributions: binomial, negative binomial, Poisson, or normal (Waters, 1955).

Surveys should account for different stages of insect life history (Fettes, 1949). It is important to determine egg-mass and larval-feeding indexes both in bud-feeding and in later stages. There should also be defoliation indexes. In an egg-mass survey, Blais and Martineau (1956) considered infestation light if there were 99 eggs or less; medium, 100–199 eggs; and severe, 200 or more eggs per 100 square feet of foliage.

Morris et al. (1956) emphasized the value of phenological observations, especially as related to shoot elongation, as a means to assess the development of spruce budworm. Balsam fir shoot elongation has been used for timing spruce budworm spraying programs in the Lake States (Bean, 1961b).

In addition to population and defoliation surveys in spruce budworm studies, the preparation of life tables is of primary importance. Population is expressed in basic units (no. per branch surface) and in absolute units (no. per acre). Correct timing in sampling requires a knowledge of insect life history and population stability both in time and place. Life tables are developed for successive generations in different stand types. The important sources of mortality are related to stand factors, population density, and climate. The first life tables were prepared by Morris in 1949, and were further developed during ten years of intensive study in the Green River area (Morris, 1955; Morris and Miller, 1954; Morris et al., 1958).

A typical life table for the budworm contains four age intervals: (1) eggs to first-instar larvae; (2) first-instar larvae to third-instar larvae; (3) third-instar larvae to pupae; (4) pupae to adults. In addition to these age intervals, life tables include (a) sex ratio; (b) fecundity; and (c) adult mortality and dispersal (Miller 1963a).

Within a stand, under high population levels, sampling should be stratified according to position in the crown. Finer stratification may be necessary to accommodate certain host species or flowering conditions. Within strata, cluster sampling may be advisable. Samples can be drawn from the same tree during successive budworm generations (Morris, 1955).

Population dynamics need to be studied at both epidemic and endemic levels. It is possible that a low equilibrium density is determined largely by interactions between the budworm and its enemies; a high one, by interactions between the budworm and foliage supply (Morris, 1963a).

In the New Brunswick studies, data are available only on the epidemic phase. Intensive studies are currently in progress on the endemic phase of the budworm cycle.

Of the many variables tested in survival rate studies, population density and weather have had the greatest effect on temporal variation in survival, and the density of balsam fir has had the greatest effect on spatial variation. During the outbreak the mass dispersal of adults had important effects that could not be modeled mathematically. Morris (1963c) suggested the intensive study on a very few plots of the age-specific survival and mortality factors that determine temporal changes in populations, and the extensive study on widely different types of plots of the factors and processes that determine spatial differences in survival.

The key-factor analysis suggests that changes in population density, though affected by many variables, are mainly determined by a few. This analysis requires only one population fix in each generation, and its density at the beginning of the large larval (third-instar) period is the logical choice. Third-instar larvae have larger density than pupae but can be sampled with fewer errors than eggs, small larvae, or pupae. The budworm populations fluctuate between outbreaks at such low densities that the measurement of age-specific survival rates becomes impractical. By refining this model — paying special attention to the effects of parasites and predators, and to the influence of stand microclimates on the mean density of budworm populations — the population dynamics of spruce budworm at low levels can be clarified (Morris, 1963c).

Miller and Macdonald (1961) suggested more specific changes in sampling design for low budworm density levels. Intensive four-level sampling from a cluster of ten trees is replaced by single-branch sampling from a large number of trees. Budworm distribution within crowns is assessed only on some plots. One population fix, preferably the third-instar larval stage, is obtained per year. They also suggested that various indirect sampling methods be used under low population conditions.

Applied control against spruce budworm may have different objectives such as prevention of outbreaks, suppression of outbreaks, and protection of forest stands. Applied control cannot be discussed with much confidence until the dynamics of both endemic and epidemic budworm populations are understood (Morris, 1963a,c). The prevention of outbreaks by the use of parasites, predators, diseases, and by the use of silvicultural practices has so far not been demonstrated to be feasible (Morris, 1963c). Attempts to suppress incipient budworm outbreaks have only recently been successful, but further trials are needed (Blais, 1960a, 1961b, 1963). Protection of forest stands has been gradually developed and has achieved a considerable measure of success (Morris, 1963c).

Protective measures against forest insects were begun several decades ago. Arsenical dusts were disseminated from aircraft against the spruce budworm and hemlock looper in the late 1920's. Results with the hemlock looper were encouraging, but the protected feeding habits of the budworm made the latter much more difficult to control (Nordin and McGugan, 1962).

Since the development of DDT, aerial spraying as a direct measure against spruce budworm and other forest insects has received a great deal of attention. According to Webb (1958) the first experimental DDT spraying in Canada was carried out on 93,000 acres in the Lake Nipigon area during 1945–1946. The first aerial spraying trials proved more effective on insects other than the spruce budworm because of the latter's concealed feeding habits. Early reports were discouraging (Anonymous, 1945). Difficulties in timing, improper flying techniques, and drifting of spray caused variations in mortality to range from 20 to 99 per cent. In the early applications, large dosages of DDT were used and the costs were high. From an initial rate of 3–5 pounds per acre, subsequent dosages have been reduced to ½ pound or less. This reduction was accomplished by doubling the spacing between flights, and by flying in pairs (Webb, 1958). Per acre costs have dropped from about three dollars in 1947 to one dollar in 1960. DDT can be effective if sprayed in dosages as low as ¼ pound in ½ gallon of water providing that the spray deposits 10 or more drops per cm² over the target area (Fettes, 1960).

In Canada, areas considered for spraying are usually selected on the probability that an additional year's attack might seriously threaten the lives of the trees. This policy is based on the assumption that natural controls might exert a regulating effect and thereby minimize unnecessary spraying.

Spraying early in the season results in lower budworm kill because the larvae are concealed in needles or buds. Early spraying, however, does reduce the amount of defoliation of the current year's growth. Reasonably good control is obtained when treatments are applied during the fourth and early fifth instars. After several severe defoliations, trees should be sprayed earlier in the season to preserve as much of the remaining foliage as possible (Blais and Martineau, 1960).

Important but less obvious effects of spraying extend beyond the treated generation and into the second and third post-spray generations. Severe population reduction takes place in the seven- to ten-day period following application. In the New Brunswick experiments in the first post-spray generation the survival of small larvae was considerably lower than that of the large larvae. In the second generation the survival of small larvae was high but that of large larvae tended to decrease (Macdonald and Webb, 1963). There was no evidence that spraying prolonged the outbreak, nor was the parasite complex affected adversely by the spraying.

Experience has shown that short-term protection is feasible. Webb (1955) suggests spraying at three-year intervals until the harvest cut, or until the outbreak subsides.

Other insecticides have been compared with DDT. In a spray chamber test, DDT was compared with BHC, aldrin, dieldrin, and endrin. Of these, DDT proved to be the most effective against spruce budworm (Secrest and Thornton, 1959). Since insect populations may develop resistance to certain insecticides (Randall, 1963), work on new chemicals should be continued. In general the use of DDT in spruce budworm control is considered a success (Blais, 1956; Daviault, 1954; Webb, 1954, 1956; Balch et al., 1955–56).

The use of predators and parasites to control the spruce budworm is still in the experimental stage. Hopes are placed in finding new and more effective ones than those known at present. The bacterium *Bacillus thuringiensis* and possibly others may offer an alternative to DDT for direct control. An alternative which does not have serious effects upon aquatic fauna is much desired (Prebble, 1961).

The effectiveness of silvicultural methods used to combat the spruce budworm are difficult to assess. Hart (1956) reported a planned experiment in Maine, but the budworm did not appear in the experimental area. A summary of management measures which have been recommended as possible control measures by numerous authors (Balch, 1953a; Gryse, 1944; MacAloney et al., 1947; McLintock, 1947; Westveld, 1946, 1954) include the following: (a) use a short rotation for balsam fir; (b) increase the proportion of spruce and other species; (c) maintain good growth and vigor; (d) avoid large pure stands of mature and overmature balsam fir.

It is important that forest insect control be approached with a thorough knowledge of the insect and its host.

Balsam Woolly Aphid

Balsam woolly aphid, *Adelges piceae* (Ratz.), an introduced pest, is the second most important insect attacking balsam fir in the Maritime Provinces of Canada and the northeastern United States. Its life cycle is complicated and presents interesting theoretical as well as practical problems. Annand (1928) clarified the systematics of several closely related aphids in a monograph on the Adelginae of North America. Lately, however, it appears that the taxonomic position of balsam woolly aphid is still unsettled. The differences between *Adelges* (*Dreyfusia*) *piceae* and *A.* (*D.*) *nüsslini* C.B. are not readily apparent. Eichhorn (1957) and Merker (1962) have investigated a third species, *A.* (*D.*) *merkeri* Eichh., which occurs on *Abies alba*, and is closely related to the other two. *A. merkeri* also attacks balsam fir (Balch, 1962). The use of different generic names for these aphids appears to follow national practices. German authors generally use *Dreyfusia*, Canadian authors prefer

Adelges, and many workers in the United States use *Chermes*. Since most references to the balsam woolly aphid have been supplied by Canadian authors, *Adelges* has been adopted as the generic name in this report.

The balsam woolly aphid was introduced from Europe and was first reported in North America from Maine by Hopkins in 1908 (MacAloney, 1935). Kotinsky (1916) noted its occurrence in Maine and New Hampshire. The aphid has since spread into Vermont, Massachusetts, and New York (MacAloney, 1935).

Balch (1934) noted the presence of balsam woolly aphid in Nova Scotia, Prince Edward Island, southern New Brunswick, and parts of New England. In Newfoundland it has been present since 1940 or earlier (Balch, 1952). In the latter province the aphid has made substantial progress and is considered the most important insect pest of balsam fir. From 185 square miles in 1949, the insect spread and by 1957 had infested 2400 square miles (Frost, 1958). A recent estimate (Can. For. Ent. Path. Br., 1961) gives the affected area as covering 3000 square miles.

At present the insect appears to be spreading rather slowly. However, its occurrence has caused much concern in the Maritime Provinces. Low winter temperatures periodically reduce the population level sharply. This occurred in the Maritime Provinces (Can. For. Biol. Div., 1958) and in northeastern United States (Waters and Mook, 1958) in 1957. Striking mortality of balsam fir caused by the balsam woolly aphid has occurred at medium elevations on Cape Breton Island, where trees of all ages are dead, dying, or much deformed by a swelling known as "gout" (Balch, 1956). Both *A. piceae* and *A. nüsslini* have been introduced to the Pacific coastal region where they attack a number of western species.

There are many references to the balsam woolly aphid. Particularly valuable studies have been made by Balch (1952) and Varty (1956).

LIFE HISTORY

The life cycle of Adelginae is typically complicated. In a complete life cycle there may be as many as five forms and two hosts. The primary host is commonly a *Picea* sp., whereas species of *Larix*, *Pseudotsuga*, *Tsuga*, *Pinus*, or *Abies* are usually secondary hosts (Annand, 1928).

Balch (1952) investigated the life cycle of *Adelges piceae* in eastern North America. The insect appears only on *Abies balsamea*, and then only as a parthenogenetic generation of the exsulis form. The exsulis form occurs in two series: sistens and progrediens. The sistens, or false stem mother, has a winter resting period in one of its generations (the hiemosistens) and a summer diapause in the other generation (aestivosistens). The progrediens generations have no resting period. The progrediens may develop from the first-laid eggs of the hiemosistens as winged progrediens (progrediens alata), or wingless progrediens (progrediens aptera). The winged progrediens is sterile. The wingless progrediens produces offspring which become aestivosistens. The insect depends for its multiplication on two generations of sistentes that feed on bark, twigs, or new shoots, but not on needles. The winged and wingless progredientes are rare, and their influence on population trends is negligible. Progredientes feed on needles, inserting their shorter stylets into stomata, while sistentes insert their longer stylets into bark, and by injecting saliva cause abnormal growth of bark and wood which on smaller branches appears as "gout."

The insect overwinters as neosistens (first instar). To overwinter satisfactorily the nymphs must have inserted their stylets into the bark. Kloft (1957) found that the protein content of bark is of particular importance for the development of aphids. All stages from egg to adult may enter winter, but the latter can withstand temperatures only as low as 3F. In very cold winters the neosistentes on balsam fir trunks below the snow line are the only ones that survive (Balch, 1952). Low temperature, —15 to —30F, is the best natural control for this insect (Brower, 1947).

Initiation of feeding may vary considerably in the spring. Feeding may begin even before bud-swelling. On the same tree a difference of twenty days may occur between first feeding dates of the individual larvae. Most of the adult hiemosistentes appear in late May. Egg laying begins within two or three days and continues for about five weeks. Incubation lasts for about twelve days. The eggs hatch and give rise to nymph aestivosistentes that, after a few hours, settle on the bark, insert their stylets, and begin feeding. Within two or three days they produce a flocculent waxy covering and enter diapause. The hibernation period lasts three to eight weeks. Most aestivosistentes commence their development in August. These latter insects, all females, lay their eggs from mid-August to October beneath a waxy covering deposited on the bark. The offspring, appearing largely in the fall, are all hiemosistentes. In spring some progredientes may develop. The waxen wool which covers the larvae and adults appears in two distinct periods, May and August. The insect is generally more numerous in the second period (Balch, 1952).

MacAloney (1935) stated that the balsam woolly aphid may cause damage in northeastern United States at elevations up to 3000 feet. MacAloney (1935) and Balch (1952) could not establish any site preference shown by the species. Brower (1947), however, reported that poorly drained sites are usually the centers from which the insects spread. As aids in identification of these sites, he presented lists of indicator plants.

Dominant, open-standing trees are the first to become infested, and from them the aphid spreads mainly on air currents to other trees and stands (Balch, 1952). Infestation of the trunk is usually heaviest below a height of 12–15 feet (Mac-Aloney, 1935). Saliva of the insect, possibly through its auxin content (Balch et al., 1964), causes formation of a type of compression wood, "Rotholz," in the sapwood. Pockets of Rotholz make it possible to date earlier attacks (Balch, 1952). Heavy attacks on the stem may kill a tree in 2–3 years (Balch, 1953b). Infestation of twigs and smaller branches causes them to swell, most noticeably in the tops of trees. Branchlets are thickened and irregularly twisted, giving rise to the condition known as gout. The top of the crown is often flattened or bent, sometimes umbrella-shaped and, in typical cases, dead. The death of trees may be caused by inhibition of bud growth resulting in gradual starvation through lack of new foliage, and by compression wood interfering with conduction in the xylem (Balch, 1952). Mortality due to *Adelges piceae* attack is considerable (Brower, 1947; Balch, 1952). Balsam woolly aphid may also be responsible for attacks by secondary insects, and provides entrance points for trunk rots.

Kloft (1955) called attention to interesting differences in reaction by *Abies alba* and *Abies balsamea* to *Adelges piceae* attack. Aphids feed on cells of phellogen, phelloderm, wood ray parenchyma, and annual ring tracheids. Proteins and polysacharides are removed. After some time these cells collapse, causing pathological bark formation in *Abies alba*. This deprives the woolly aphid of its food supply and the attack ceases. The effect on the trees is minor. However, in the case of *Abies balsamea*, swelling continues in the twigs and branches and the attacks continue. Eventually the trees are so weakened that they frequently die.

From ten years' observations in the Wind River Arboretum in Washington, Fraser fir (*Abies fraseri*) was found to be even more susceptible to attack by the balsam woolly aphid than balsam fir (Silen and Woike, 1959).

Clark and Brown (1957, 1958, 1959, 1960) and Brown and Clark (1956, 1959, 1960) have studied predators of the balsam woolly aphid in Canada since 1947. Since 1951, 17 predatory species (13 beetles and 4 flies) have been introduced from Europe, India, Pakistan, Japan, and Australia. Of these, *Laricobius erichsonii* Rosen, appears most promising. *Neoleucopis obscura* (Hal.), established in New Brunswick as early as 1933, later was found in Nova Scotia and Prince Edward Island. This insect has spread well but has not achieved good aphid control, presumably because of its specific life cycle. There are many reports of predators (Brown, 1947; Balch, 1952, 1956; Prebble and Bier, 1954; Waters, 1954; Carroll and Bryant, 1960). Difficulties arise in maintaining predator populations during periods of decline in the host species.

Balch (1952) noted that a pink fungus was often found on dead individuals of the second to fourth instars of the balsam woolly aphid. He could not establish the fungus as the primary cause of death.

CONTROL

Low winter temperatures have exercised the most effective control. Some control can also be attributed to predators. Control by chemicals under forest conditions or through forest management practices has so far proved of little value.

Nicotine sprays (Brower, 1957) and lime sulfur sprays (MacAloney, 1952) have proved ineffective. Balch (1952) noted that an effective spray would kill the dormant-stage larvae since they are present on the trees at all times. Dormant larvae and eggs are resistant to nicotine sulfate in concentrations of 1:400. At other stages the insects can be killed, but over 50 per cent of the adults may survive because of the protection afforded by the waxen "wool" that covers their bodies. Miscible oils at a concentration of 1:25, applied thoroughly, will kill at all stages, but as many as 5 per cent may survive because of protection given by lichens or by the accumulation of "wool" in bark crevices. The best time to spray is early April, just before molting commences. Under forest conditions it might be possible to apply basal sprays following cold winters which would kill those insects not protected by a snow cover (Balch, 1952).

Kotschy (1960) reported on a series of chemicals used to combat *Adelges nüsslini* in the Austrian Alps near Vienna. Jahn (1961) reported on similar measures in the Tyrol. The use of simple smoke cartridges to control *A. nüsslini* was demonstrated at the Thirteenth Congress of the International Union of Forest

Research Organizations in the forests of the Tyrol, Austria. Results of these new chemical control measures will be published shortly.

For management control measures, Balch (1952) recommended short cutting cycles for balsam fir, short rotations, favoring spruce in selective cuttings, and maintenance of full stocking in immature stands. Since initial infestations begin on large trees, their removal makes it possible to check outbreaks in early stages. Cutting or girdling decadent, unmerchantable trees would also be advisable. Aerial photography has been used successfully in surveys of damage caused by the balsam woolly aphid and for planning salvage operations (Pope, 1957). Frost (1958) suggested a system of cutting mixed conifer stands in Newfoundland whereby 10–20 black spruce seed trees per acre would be left. He also advocated prescribed burning as a management measure in stands threatened by the balsam woolly aphid.

Other Associated Insects

Except for the spruce budworm and balsam woolly aphid, the insects associated with balsam fir have received relatively little attention. In considering their role in the ecosystem, however, they may be of considerable importance. For example, the relationships between certain insects of the Siricidae and the spreading of decay organisms is well documented. *Sirex cyaneus* Fab., *Urocerus albicornis* (Fab.), (Belyea, 1952b; Felt, 1906), and *Xeris spectrum* L. (Stillwell, 1960) were found on weakened and dying balsam fir trees. As noted in Chapter 4, Cartwright (1938) found these insects to be vectors of the destructive balsam fir trunk rot, *Stereum sanguinolentum* (Smerlis, 1961).

Although the distinction between primary and secondary pests is rather arbitrary, it is useful to consider bark beetles, ambrosia beetles, and woodborers under the category of secondary insects. The primary insects will be discussed separately.

PRIMARY INSECTS

Aside from the spruce budworm and the balsam woolly aphid, the major primary insects are the black-headed budworm, hemlock looper, balsam fir sawfly, balsam twig aphid, balsam gall midge, balsam fir seed chalcid, and various species of weevils. These are discussed separately below. In addition both the gypsy moth, *Porthetria dispar* (L.) (Baldwin et al., 1952; Anderson, 1960), and the European spruce sawfly, *Diprion hercyniae* (Htg.) (Balch, 1942) are listed as attacking balsam fir. Presumably these unlikely associations occurred under starvation conditions. Two or more undescribed species of bud-miners, *Argyresthia* spp. (Lepidoptera, Yponomeutidae), have been noted for several years on balsam fir in Ontario. Since these insects kill terminal buds of side branches, heavy infestation could perhaps affect the growth and form of balsam fir trees (Lindquist and Sippel, 1961).

The black-headed budworm: Acleris variana (Fern.). This species often causes damage to balsam fir resembling that of the spruce budworm. Feeding is largely restricted to the newly developed needles. A heavily infested tree will appear brownish because of the many webs of partially eaten needles that are produced by pupating larvae. Defoliation by the black-headed budworm frequently accompanies spruce budworm outbreaks (McCambridge and Downing, 1960). The black-headed budworm is found in the northeastern states, and in Canada from the Gaspé to Alaska (Craighead, 1950). Typically grayish, with greatly variable markings of orange, white, and black, the moth has a wingspread of about 16 mm. The full-grown larva is about 13 mm in length and has a brownish head with a bright green body. In the earlier instars the head is black. The eggs overwinter, hatching when the buds break in spring. Pupation takes place in July, and the moths emerge in late July and August (Balch, 1932; Doane et al., 1936; Craighead, 1950). Outbreaks of this species have been reported from Cape Breton Island in 1929 and in the Maritime Provinces in 1947–1951. Mortality was light. A brief review of these outbreaks is given by Miller (1962).

Hemlock looper: Lambdina fiscellaria fiscellaria (Gn.). Although hemlock is preferred, balsam fir is one of the principal host trees. Damage consists of defoliation by the larvae, which feed on both old and new needles. In severe infestations trees may die after one complete defoliation. Although a serious pest of balsam fir in the Maritime Provinces (Morris, 1958), the insect is not reported to occur on balsam fir in the Lake States. The moth is creamy to grayish-brown with a wingspread of about 30 mm. It emerges in late August and September. Eggs are laid on needles, twigs, and in bark crevices, where they overwinter. Larvae hatch early in June and, when full grown, are about 30 mm long and yellowish-green to gray in color. Pupation takes place in August (Doane et al., 1936; Craighead, 1950; Schaffner, 1952). Control may be accomplished by DDT sprays.

Balsam fir sawfly: Neodiprion abietis (Harr.). Balsam fir is the preferred host for this insect, although it may also feed on spruce and larch. Damage consists of defoliation by the larvae. The balsam fir sawfly has been known since the latter part of the nineteenth century (Packard, 1890; Felt, 1906). In recent years the sawfly has been frequently reported in Canada, especially from Newfoundland, where it attacks small trees (Can. For. Biol. Div., 1950–59). Locally, the insect causes much damage in the Maritime Provinces. Peterson (1947) reported the balsam fir sawfly in Alberta, Saskatchewan, and Manitoba. In the latter province, serious damage was reported in 1959 (Can. For. Ent. Path. Br., 1960).

The adult sawflies emerge in July and August. Eggs are deposited in slits cut in the needles. The eggs overwinter and young larvae hatch in late May or June. The full-grown larva, about 18 mm long, has a black head and a greenish body marked with dark longitudinal stripes. Reddish-brown cocoons are spun among the foliage or in the litter. Pupation takes place in late July (Craighead, 1950). An intensive life history study is now in progress in Newfoundland (Carroll, 1962).

Balsam twig aphid: Mindarus abietinus Koch. This insect is distributed from New England to the Pacific Coast and attacks several species of spruce and fir. The aphid feeds on young new shoots causing curl and disfiguration of the needles and shoots. However, the injury is seldom severe (Craighead, 1950). The twig aphid has been identified on the windshields of airplanes at elevations from as high as 500 to 5000 feet (Balch, 1950).

Balsam gall midge: Dasyneura balsamicola (Lint.). The balsam gall midge is found on balsam fir and Fraser fir (Craighead, 1950). Larvae produce oval swellings near the bases of needles, sometimes causing severe needle cast at the end of the growing season that may impair the quality of Christmas trees (Giese and Benjamin, 1959). Abundant on balsam fir in Maine (Waters and Mook, 1958) and in the Adirondacks of New York (Craighead, 1950), the insect is also found in Wisconsin (Giese and Benjamin, 1959) and is widespread in Canada. Considerable damage to balsam fir was reported from Manitoba in 1955 and from Ontario in 1959 and 1960 (Can. For. Biol. Div., 1956–58; Can. For. Ent. Path. Br., 1960–61). The larvae feed throughout the summer and overwinter in the ground. Pupation occurs and adults emerge in spring (Giese and Benjamin, 1959).

Balsam fir seed chalcid: Megastigmus specularis Walley. Damage by this insect consists entirely of seed destruction by larvae. According to Hedlin (1956), the balsam fir seed chalcid was first reported from Canada in 1928 by Walley (1932). It is also known from Minnesota. Craighead (1950) noted that the life histories of the various *Megastigmus* species are similar. Adults emerge in the spring from infested seed that has fallen to the ground. Emergence occurs for over a month, and the adults live nearly two weeks. Oviposition commences about June 15 and larvae appear 9–10 days later. Eggs are deposited within the seed embryo. Larvae pass through the first four instars rapidly, but more than a month is required between the fifth and sixth. Each insect confines its feeding to the embryo and cotyledons of a single seed. The pupal stage lasts slightly more than a month (Hedlin, 1956). Huge quantities of seed may be destroyed in a severe infestation.

Weevils: Hylobius spp. Although *Pissodes dubius* Rand. is a secondary weevil attacking dead and dying balsam fir (Swaine et al., 1924), there are several species of *Hylobius* weevils which attack living trees. *Hylobius pales* (Hbst.), the pales weevil, has been reported attacking balsam fir (Carter, 1916). A single beetle can kill a 2- to 3-year-old seedling in a few days. *Hylobius pinicola* (Couper), the root weevil, has been reported attacking balsam fir roots in stands 6–55 years of age in Quebec (Can. For. Biol. Div., 1957). In the stands examined, 6 per cent to 77 per cent of the tree root systems were injured. The weevil is credited with providing infection points for root and butt rots. This insect has recently been reported from Quebec (Smerlis, 1957, 1961). Trees attacked by species of *Hylobius* were subsequently infected by 33 species of fungi. Of all *Hylobius*-attacked trees, 60 per cent had fungal infection. This insect has caused damage to balsam fir for at least 20 years in Laurentides Park, Quebec.

Doane et al. (1936) reported that most of the Hylobini in North America live as scavengers in the stumps and dead trunks of coniferous trees. Normally these insects are considered beneficial. According to Craighead (1950) *Hylobius pales* is a robust weevil 7–10 mm in length with a dark to reddish-brown body marked irregularly with hairs. The beetles overwinter in the litter. Depending on the locality, they become active sometime between April and June, and feed on the tender bark of sapling twigs or at the base of seedlings. The white eggs are laid about July 1 in the inner bark of freshly cut logs or in large roots of fresh stumps. The white, footless larvae hatch in about two weeks. Mature larvae pupate toward the end of August, the new adults emerging in September to begin feeding. It is at this time that severe damage occurs on young trees.

Lists of secondary insects following a spruce budworm outbreak were presented by Belyea (1948, 1952b). Mortality began in the fifth year after such an attack. The following insects were present only on dead trees: *Trypodendron bivittatum* (Kby.), *Serropalpus substriatus* Hald., *Sirex cyaneus*, *Sirex* sp. (noctilio group), *Urocerus albicornis*. Other species were very scarce, and their mode of action could not be definitely established. These included *Monochamus marmorator* Kby., *Pissodes dubius*, and *Tetropium cinnamopterum* Kby. Only *Pityokteines sparsus* Lec. and *Monochamus scutellatus* (Say) were attacking living trees, but observations indicated that they could not cause mortality in balsam fir.

MacAloney (1935) listed the insects associated with balsam woolly aphid as including *Pissodes dubius, Monochamus marmorator, M. scutellatus, Serropalpus barbatus* (Schall.), *Xylotrechus undulatus* Say, *Phymatodes dimidiatus* Kby., and *Camponotus herculeanus* L.

Mott (1954) studied the secondary insects in chemically debarked trees in the Green River watershed. Insects attacking dead and dying balsam fir trees were *Pityokteines sparsus, Polygraphus rufipennis* Kby., *Monochamus scutellatus, Trypodendron bivittatum.*

The bark beetles have little effect on the technical properties of wood. However, the ambrosia beetles, *T. bivittatum, Gnathotrichus materiarius* (Fitch), and the woodborer *M. scutellatus* do cause extensive damage and should be taken into consideration. Ninety-five per cent of all Cerambycidae attacking balsam fir pulpwood belong to the genus *Monochamus*. These beetles may cause up to 5 per cent volume loss and degrade the lumber (Wilson, 1961a,b). They usually carry the spores of several fungi causing decay of balsam fir (Raymond and Reid, 1961).

The secondary insects associated with balsam fir are represented by bark beetles, ambrosia beetles, and woodborers or sawyers. From each group one representative species will be described.

Balsam fir bark beetle: Pityokteines sparsus Lec. This is a distinctly secondary insect which feeds in the phloem. The beetle breeds in slash, but it is also found on the boles and limbs of dying or dead balsam fir trees (Graham, 1923). Some adults, after laying eggs in June, may attack fresh trees and establish a second brood. The latter then overwinters in the larval stage (Belyea, 1952b).

Balsam fir sawyer: Monochamus marmorator Kby. This woodborer is typical of many species of *Monochamus* (Felt, 1924). It is a large (15–30 mm), elongated cylindrical beetle. Eggs are laid in early spring, mostly in dead or felled trees, but also in green balsam fir. The larva bores beneath the bark and remains there for 40–60 days, then enters the wood and hibernates. A pupal cell is made the following spring or early summer. Later the new adult gnaws a round emergence hole and escapes (Craighead, 1950).

Ambrosia beetle: Trypodendron bivittatum (Kby.). This short, stout, shiny bronze-colored beetle, about 3 mm long, attacks balsam fir and many other conifers. The beetles bore tunnels that often penetrate deeply into the wood. Adults and larvae feed on symbiotic, ambrosia-producing fungi that grow in the tunnels (Doane et al., 1936).

CONTROL

Control measures can be prescribed only when the conditions favoring insect attack are related to the physiological state of the trees. Until recently very few physiological studies pertaining to balsam fir had been made.

Healthy stands, sanitation in the woods, and rapid transport of materials from the forest are desirable measures. Little fundamental study has been made of slash as a factor of the environment. Slash affects not only insects, but also diseases (e.g., *Stereum sanguinolentum* sporophores are exclusively on slash), fire hazard, and reproduction.

Debarking logs may be important in preventing attack by bark beetles and woodborers, but this procedure cannot prevent damage by all species of ambrosia beetles. Benzene hexachloride sprays can be used to protect cut-and-piled wood (Craighead, 1950; Wilson, 1961b). Wettable DDT and DDT in oils have also proved successful (Blais, 1953b). Piling in shade may also reduce the attacks of woodborers (Wilson, 1960).

It is beyond the scope of this monograph to review the scattered literature of the many additional species reported on balsam fir. Further information can be obtained by consulting the sources directly (Appendix II).

Conclusions

1. Insects constitute an important part of balsam fir ecology and management. As do fungi, insects require close physiological contact with their hosts. The biochemical and biophysical aspects of this interaction have not been given sufficient fundamental study.

2. Although major attention has been given to the spruce budworm and the balsam woolly aphid as the principal pests, research on other important insects should continue.

3. Spruce budworm outbreaks have stimulated fruitful basic research in many branches of biological science. These studies have also benefited the forest industries by pointing toward protective forest management practices and more effective control measures.

4. Research on stand vulnerability to spruce budworm attacks should be continued on the basis of very thorough site and tree classification. It appears that once a budworm epidemic has developed and gained momentum, the effects generated are such as to obliterate any site and stand differences that may have been responsible for the outbreak.

5. Technological progress in the development of less expensive chemical spraying methods should not hinder the search for other controls, particularly those of a biological nature.

6. Spruce budworm and balsam woolly aphid studies have shown that dominant trees are subject to more intensive attacks than are suppressed trees. This should be recognized in the development of measures for balsam fir stands to resist attacks by these insects.

6

REPRODUCTION

BALSAM fir silviculture is considered in this chapter and in the two following, Chapter 7 on Stand Development and Chapter 8 on Growth and Yield, with the latter having a strong mensurational character. Although most of the information on the fundamentals underlying silviculture has been discussed in preceding chapters, little attention was given to the influence of man's action on the forest. Some authors consider the practice of silviculture as belonging to the group of biotic factors while others consider it separately under cultural factors. In this presentation, man's activities in manipulating the forest are seen as a superimposed function modifying the primary climatic, edaphic, and biotic factors, and to a small but increasing extent guiding the species' genetic development.

While consideration of man's actions inevitably involves problems of forest policy and forest economics, very little has been published on the economic and policy problems specific to balsam fir. Therefore no separate treatment can be given to these areas, and discussion will be oriented largely along biological and technical lines.

Reproduction is a dynamic process affected by the inherited characteristics of the species, by the natural climatic, edaphic, and biotic factors, and by modifications imposed by man. Theoretically, conditions are never exactly repetitious, nor may differences always have practical significance. Forests are managed in units which are supposed to respond homogeneously to treatments. Permanent (climax, in the sense of the polyclimax point of view) forest types possess biologically equivalent ecological conditions. Stands of similar composition by species, age structure, density, origin, and development, belonging to the same permanent forest type, are the closest approximations to homogeneous units for treatment. It may be necessary to recognize subtypes, to consider synusiae (unions, ecological niches, layer communities), and to study spatial distribution and population dynamics in order to define optimal silvicultural practices for forest reproduction, protection against fungi and insects, and improvement of growth for balsam fir.

The basic knowledge accumulated about balsam fir discussed in previous chapters, as well as its application to specific problems, has not yet been molded into a clear, comprehensive system. Nevertheless, the reproduction studies that have been pursued for many years, though not always meeting present standards, provide information that may be helpful in planning future work.

Seed and Seedbeds

The botanical characteristics including flowering habits, seed-bearing maturity, and periodicity of seed crops, that relate to seed production and seed dispersal are discussed in Chapter 1. This section reviews seed crop and seedbed information related to the germination and establishment of seedlings.

SEED CHARACTERISTICS AND PROCUREMENT

Balsam fir cones are suitable for collection for only a short period of time. Seeds mature about the same time as those of white spruce. In Ontario this is after mid-August (Carman, 1953). Collection should be made just prior to natural seed dispersion. Collection by climbing is not feasible. Because tree felling when the cones are fully mature results in loss of cones by shattering, cutting must be done before cone scales separate from the central axis. Since at this time cones are somewhat immature, the collection procedure sometimes results in low germination and poor storage quality of the seed.

Balsam fir cones are considered ripe for harvest when their moisture content is down to 60 per cent as compared with 50 per cent for white pine and Douglas fir, and 40 per cent for Scotch pine, Norway spruce, and European larch (Rohmeder, 1960). Collectors should not include cones picked from slashings made early in the season. The cones must be spread in shallow layers not more than 2½ inches thick and turned as necessary to prevent molding. They open and release seeds without heating. Seed coats are soft, and dewinging requires considerable care (Roe, 1948).

One bushel (35.2 liters) of balsam fir cones weighs about 35 pounds. This amount yields 37–42 ounces (1000–1200 g) of cleaned seed averaging 60,000 (range 30,000 to 94,000) seeds per pound. The purity of commercial seed is reported to be 90 per cent and soundness is about 50 per cent (U.S. Forest Service, 1948). Heit and Eliason (1940) provide similar figures.

The moisture content at which balsam fir should be stored is not definitely known. For comparison, that of *Abies grandis* is about 6 per cent, and it seems likely that balsam fir should be stored at a similarly low moisture content. That balsam fir seeds occasionally show embryo dormancy, which can be overcome by stratifying the seed in moist sand at 41F, was determined in an experiment summarized in Table 31 (Roe, 1948). Further work indicated that results approach an optimum with 90 days of stratification. Heit and Eliason (1940) noted improvement of germination after 30 days of stratification at 2–4C. They also recommended fall seeding.

Stratification in moist vermiculite provided significant improvement over polyethylene bag storage in promoting early germination of balsam fir seed (Kozlowski, 1960). Stratification and storage were carried out for 10 weeks at 5C. Germination after 35 days was 41.9 per cent for vermiculite stratified seed, 26.6 per cent for seed kept in moist polyethylene bags, and 26.7 per cent for seed kept in a dry refrigerator. Germination after 70 days reached 42.0, 27.0, and 32.9 per cent respectively.

Balsam fir and other conifers were stratified in moist sand at 32–38F for periods of 3, 14, and 27 months. Thirty-day germination tests of the balsam fir seed gave germination percentages of 42, 50, and 0.5 per cent respectively (MacGillivray, 1955). From these results it was suggested that in cold swamps a few balsam fir seeds might remain viable in the forest floor for two or three years.

Rohmeder (1951) suggested that seed dormancy could be attributed to the inhibiting effect of turpentine on germination. He also advised a 42-day germination test using the Copenhagen tank germinator (Rohmeder, 1938). A 60-day laboratory germination test was recommended by Heit and Eliason (1940).

TABLE 31. PERCENTAGE OF BALSAM FIR GERMINATING IN A
60-DAY PERIOD UNDER DIFFERENT PRETREATMENTS AND
GERMINATION TEMPERATURES

	Germination Temperature (F)		
Pretreatment	68–86	50–77	50
None	3.6	0.6	0.2
Stratified 30 days at 50F	24.4	1.1	0.2
Stratified 30 days at 41F	34.4	5.2	0.6
Stratified 60 days at 41F	47.0	3.8	0.0
Watersoaked 2 days at 45F	14.6	0.9	0.2

Source: Roe, 1948.

For purity determination, the international rules for seed testing (International Seed Testing Association, 1959) prescribed a sample size of 25 g. For unit-weight determination, a sample size of 50 g was prescribed. For all *Abies* species, germination tests require a pre-test chilling period of 21 days at 3–5C. Recommended testing media were paper, sterile sand, or sand-loam mixtures. Other standardized rules require alternating temperatures with 16 hours at 20C and 8 hours at 30C, normal daylight, and a first count at 7 days with the final count at 28 days after beginning the test period.

Zon (1914) gave the germination capacity of balsam fir seed as 20–30 per cent, while the Forest Tree Seed Manual (U.S. Forest Service, 1948) cited a range of 1 to 74 per cent, with an average of 22 per cent. The 20-day period germination energy, however, was given by the same source as 40–70 per cent. Heit and Eliason (1940) found germination capacity to range from 27 to 55 per cent. These differences reflect great variability in the seed lots or in the test method and point up the need for standardized testing rules such as prescribed by the International Seed Testing Association.

The germination capacity of balsam fir seed stored in unsealed containers at room temperature was reduced from 34 per cent in the fall to 4 per cent by the following spring. By contrast, seed stored in sealed containers at 38–39F was said to retain viability for as long as five years (Toumey and Stevens, 1928). The storage of balsam fir seed in sealed jars for five years at 2–4C without loss of viability was reported by Heit and Eliason (1940).

Balsam fir seed production occurs at intervals of 2–4 years (Zon, 1914, U.S. Forest Service, 1948). A detailed study of balsam fir seed production in the Green River area of New Brunswick was carried out by Morris (1951), who noted a regular two-year periodicity in flowering followed by abundant seed production over a period of nearly 30 years. The same two-year interval was noted for balsam fir in the Boreal Forest Region of Ontario by MacLean (1960). Annual reports of seed production for 10 Lake States conifers and 20 hardwoods have been made since 1950 (Rudolf, 1950–63). These are summarized in Table 32 for balsam fir, white spruce, and black spruce. A crop of less than 15 per cent is considered a failure, 16 to 35 per cent is medium, and 61 to 90 per cent is a good crop. A bumper crop is considered to be over 90 per cent of a full seed crop. It appears from the table that the three species do not differ markedly in periodicity of seed production. More definite conclusions would require additional annual observations and statistical analysis.

There have also been reports of balsam fir seed crops in New England (Hart, 1958, 1959; McConkey, 1960, 1961). In general, good seed years occurred in 1958 and 1960 while in 1957 and 1959 there was a crop failure.

MacLean (1955) estimated that in 1954, a good seed year, there were 5–18 pounds of seed per acre in balsam fir stands on the Black River Experimental Forest in Ontario (Section B-8, after Halliday, 1937, and Rowe, 1959). A later report noted 1956 and 1958 as good balsam fir seed years in the same area. During these years the seed crop was 13–16 pounds per acre on fresh sites and 10–12 pounds per acre on wet sites. There were 73 trees on fresh sites and 68 on wet sites per acre (MacLean, 1959).

Ghent (1958b) investigated the production of balsam fir seed in connection with general studies on the effects of spruce budworm epidemics in the Black Sturgeon Lake area north of Lake Superior. Of interest is the report of a mixed stand (aspen-fir-spruce-birch) in which balsam fir accounted for 27 per cent of the basal area. Ages ranged upwards to 70 years. The balsam fir, during a producing period of 37 years, had yielded a total of 7.9 million seeds per acre or 5 seeds per square foot annually. This production resulted in the establishment of 5000 seedlings per acre, about one seedling from 1500 seeds. During the first half of the production period one seedling was established from each 600 seeds. The established seedlings showed a contagious pattern of distribution which yielded rather different density figures when measured with various sizes of sample plots (Ghent, 1963).

SEEDBEDS AND SEEDLING SURVIVAL

Variability in the number of seedlings which become established after a seed crop is very great. Establishment depends on size and dispersal of the seed crop, which can be affected directly by many factors. Germination and survival depend on combinations of many meteorological, edaphic, and biotic factors and on action by man.

Assuming favorable temperature, moisture is the most critical factor affecting germination. The seeds contain stored food for the first phase of seedling develop-

TABLE 32. ANNUAL SEED CROPS AS AN ESTIMATED PERCENTAGE OF A FULL SEED CROP IN THE LAKE STATES

Area	1949	1950	1951	1952	1953	1954	1955	1956	1957	1958	1959	1960	1961	1962
BALSAM FIR														
Northern Minnesota	7	75–95	7	75	7–25	7–75	25–75	7	25–50	7–50	7–25	50–95	7–50	75
Northeastern Wisconsin	7	95	7	50–75	75	7	75	7	25	75–95	75	75	7	75
Central Upper Michigan	7	50	25	25	7	50	50	75	7	95	7	95	50	75
Lower Peninsula, Michigan		95	25	25							7	95	7	25
WHITE SPRUCE														
Northern Minnesota	7–25	7–75	7	50–75	7–25	50–75	50–95	20–75	7–75	7–25	25–50	50–95	7–25	75
Northeastern Wisconsin	7	95	7	25–50	50	75	75	50	7	25–50	25	95	7	50
Central Upper Michigan	7	75	25	25	50	50	25	95	7	50	7	95	7	50
Lower Peninsula, Michigan	7	75	25	25	7–50	25	25	25	75	7	25	50	7	25
BLACK SPRUCE														
Northern Minnesota	50	50	25–50	50	50–75	50–75	75	50–75	7–50	50–75	50–75	50–95	7–25	75
Northeastern Wisconsin	7	95	25	75	50	50	75	50	7	50–75	75	75	7	50
Central Upper Michigan	7	75	25	25	50	50	7	75	25	75	7	75	7	75
Lower Peninsula, Michigan		75	25	25							7	50	7	25

Source: Rudolf, 1950–1962, 1963.

182

ment. Light requirements at the beginning are very low, reaching only 10 per cent of full sunlight (Place, 1955).

Weather largely governs the early success of reproduction. In New Brunswick rain in June and July is of special importance. The influence of day-to-day weather is greatest on very wet or very dry sites, and is usually much greater in the open than under vegetation. Seedlings germinating after mid-July seldom survive the following winter (Place, 1955). Weather conditions are partly responsible for temporary development of balsam fir outside the regionally suited habitats such as the islands of balsam fir in sugar maple–basswood stands in Itasca Park, Minnesota (Buell, 1956). The variability of balsam fir reproduction in marginal conditions in central Minnesota is demonstrated in Plate 16.

On hardwood slopes in the New York Adirondacks, Zon (1914) found 700–1000 balsam fir seedlings per acre and occasionally as many as 50,000. In stands of pure balsam fir there were as many as 300,000 per acre. In central Minnesota the number of seedlings 1–2 years old may reach 120,000 per acre; 3–5 years old, 10,000; 6–10 years old, 4000; and older than 10 years, 800 per acre (see Plate 16).

Forest litter has frequently been evaluated as a seed-storage medium. Bowman (1944) suggested that under a seed-producing stand the forest floor and slash contain enough seed to reseed the area after clearcutting. Nickerson (1958) did not find regeneration from balsam fir seed stored in the duff on a burned area in Newfoundland. Olmstead and Curtis (1947) reported a considerable amount of balsam fir seed, 44 in one and 378 in another 16-square-foot plot in addition to seed of other species stored in the duff under a forest stand in Maine. The conifer seed, however, did not germinate. No conifer seedlings were obtained in a greenhouse test of 270 surface soil samples collected from upland forests in central Minnesota in mid-July of 1961. These observations may indicate that it is not safe to rely on seed storage in litter as a factor in balsam fir reproduction.

Forest litter may play different roles depending on various conditions. Hardwood litter, especially in combination with poorly decomposed fibrous mor, is a poor seedbed, in which most spruce seedlings die within a few weeks after germination. Mortality of balsam fir seedlings has been somewhat less. Roots of the fir in partially decomposed hardwood leaves develop a characteristic and persistent sleigh-runner form. Seedlings grow slowly and are often mutilated by forest wildlife before they reach sapling size (MacLean, 1959).

Red spruce litter consists of hard needles which decompose slowly and develop an acid humus. Balsam fir needles are flat-lying, decompose more easily, and produce a humus of lower acidity. As a result the slowly developing red spruce seedlings persist better under balsam fir, but balsam fir reproduction predominates in red spruce stands containing a very few scattered balsam fir (Murphy, 1917).

Almost any kind of fresh litter constitutes a poor seedbed. Two inches of undecomposed litter tends to be limiting for the spruces, and more than three inches is limiting for balsam fir (Place, 1952a, 1955). Table 33 shows the effect of seedbed on germination, survival, and growth of balsam fir seedlings as compared with red and white spruce. The seedbed investigated had a *Calliergonella* (*Pleurozium*) *schreberi* cover and was shaded. In October, when the seedlings were removed, only 56 per cent of the balsam fir seedlings showed root growth as compared with

100 per cent for the spruce seedlings. Drought was the cause of nearly all seedling deaths, most of the mortality occurring between June 20 and July 12.

In uncut stands reproduction is most frequently found on rotten logs and stumps (Hosie, 1947). Spruce and fir reproduction may be found on logs showing only slight signs of decay without the occurrence of any previous successional stages (McCullough, 1948).

In the White Mountains of New Hampshire a good seed crop occurred prior to cutting in 1925. After the cutting a count was made of the number of seedlings established on untreated soil and on thoroughly exposed soil (Westveld, 1931). Results are given in Table 34. In this as in other comparisons of balsam fir and spruce reproduction, the greater abundance of balsam fir, at least during the early stages in the life of the stand, is clearly shown.

Cutting in a seed year is frequently advised but does not guarantee established reproduction. Drying of the exposed litter and humus such as reported by MacLean (1955) from the Black River Experimental Forest in Ontario (Section B-8) can cause mortality.

Mineral soil may have disadvantages under certain conditions. Light sands in open areas will usually be too dry except during wet years. Frost heaving can be serious on heavy mineral soils. In narrow, disked furrows, accumulating leaf litter may eliminate all reproduction. An exposed A_2 horizon may be responsible for poor growth of seedlings on mineral soil (Place, 1955). After scarification, exposed mineral soil may remain loose for the first summer and dry out rapidly (MacLean, 1959).

Aspen, spruce, and fir germinate better on exposed mineral soil than in litter and humus. Because of aspen competition, however, spruce and balsam fir may be more easily established in duff provided a light canopy of the parent stand remains (Bowman, 1944).

Microtopography is of considerable importance. In New Brunswick, Dixon and Place (1952) investigated spruce and fir reproduction in a stand clearcut eight years previously, and in an uncut stand. The regeneration survived better on the middle slopes of the hummocks which were from one-half to three feet above the water table. Under the canopy in the uncut stand, survival ranged from 3 to 17 seedlings per square foot whereas in the clearcut area, reproduction averaged only 1–3 seedlings per square foot. Scarification may also create unfavorable microtopography which may cause accumulation of water on moist sites or leaf litter in other places (MacLean, 1959).

The effect of canopy continues outside the stand boundaries. The northern and eastern exposures may be more favorable for establishment as compared with western exposures, as was observed in a spruce-fir stand developed on a former hardwood site after a short period of agricultural use at Dummer, New Hampshire (Baldwin, 1933).

MacLean (1959) investigated the effect of the width of clearcutting strips on reproduction of spruce and fir in the Black River Experimental Forest, Ontario (Section B-8). The direction of the cutting area was from northwest to southeast. The greatest abundance of spruce and fir seedlings was at the southwest boundary, reading 9000 per acre in the first 66 feet. Near the northwest boundary, seedlings numbered only 3000 per acre. The experimental clearcut width was 660 feet. The number of seedlings decreased sharply 200 feet from the southwest border and 130 feet from the northeast border of the clearcut strip. In the central part of the clearcut area there were 200 to 800 spruce and fir seedlings per acre. The area included both fresh and moist soils. It was cut in 1953, scarified in 1954, and the subsequent reproduction was counted in 1957 (see also Table 36).

LeBarron (1945) compared the establishment of different species of conifer seedlings five years after artificial seeding under upland forest conditions in northeastern Minnesota following clearcutting and partial cutting. In the partial cutting, 40 per cent of the tree canopy was left. The seedbeds investigated included mineral soil, duff, and duff covered with slash. The results are shown in Table 35. The advantage of exposed mineral soil in seedling establishment is obvious. No difference between the cutting methods was observed. This was explained by the local practice of "clearcutting" in which considerable amounts of unmerchantable standing timber and shrubs are left, thus not actually achieving complete stand opening.

However, MacLean (1959) obtained very different results in the Black River experimental area. The stands were mixed-wood, with balsam fir the predominant softwood, followed by white spruce, black spruce, and white-cedar. Paper birch was the main hardwood species. Average basal area was 83 square feet per acre. Soils were of great variability. Stand treatments were combined with ground scari-

TABLE 33. GERMINATION, SURVIVAL, AND GROWTH OF FIRST-YEAR SEEDLINGS OF BALSAM FIR AND SPRUCE ON DIFFERENT SEEDBEDS AT ACADIA FOREST EXPERIMENT STATION, NEW BRUNSWICK

Thickness of Horizons (In.)				Germination Percentage		Survival Percentage		Mean Dry Weight (Mg) of Seedlings	
L&F	H	Total	Texture of A_2	Spruce	Fir	Spruce	Fir	Spruce	Fir
¾–1	0–½	1–1½	Sandy loam ...54		36	44	50	12.4	22.7
1–2	¼–1	2–3	Sandy loam ...33		31	6	39	10.4	19.4
2–3	½–1½	2¾–3½	Sandy to clay loam ...46		28	15	93	8.1	18.6
2–3	1½–2½	3½–4½	Sandy loam ...40		33	12	15	8.6	14.9

Source: after Place, 1955.

TABLE 34. RELATIONSHIP OF SEEDBED CONDITION TO THE NUMBER OF SEEDLINGS SURVIVING PER ACRE FOR RED SPRUCE AND BALSAM FIR IN THE WHITE MOUNTAIN NATIONAL FOREST, NEW HAMPSHIRE

	Red Spruce		Balsam Fir	
Date of Examination	Exposed Seedbed	Untreated Seedbed	Exposed Seedbed	Untreated Seedbed
June 11, 1925	2,700	0	25,300	3,700
October 2, 1925	4,700	1,700	22,500	4,700
June 1, 1926	1,400	500	8,000	1,500

Source: Westveld, 1931.

fication. Cutting was done in the late summer and fall of 1953. Logs were skidded by horses to the strip roads which were bulldozed. Slash averaged 1–1.5 feet in depth and was widely distributed over the cutover strips. Where all species were cut about one-half of the ground was covered, but where softwoods alone were removed the coverage was about one-third. Poisoning of hardwoods was done with 2,4,5-T applied to frill girdles in August 1954. Scarification was done in July 1954 by a root-rake mounted on a D-6 Caterpillar tractor. On the fresh transects the prepared surface was 2 to 4 inches below the general level of forest floor. On moister sites humus was sometimes only partially removed. Spraying with an ester of 2,4,5-T in an oil carrier from a pressure pack can was done from July 28 to August 13, 1955. The herbicide, having 76.8 ounces of active acid per imperial gallon and a concentration of one quart to 40 gallons of oil, was applied mostly as a foliage spray. Results are shown in Table 36.

Table 36 indicates that scarification caused considerable increase in number of subsequent reproduction, and that there was a great variability depending on the treatment of the main stand. Advance reproduction was little affected by either the treatment of the main stand or by spraying of shrubs during the observation period. There was some reduction of height growth of balsam fir the year after cutting, but it was compensated by accelerated growth later. It was found particularly important to delay cutting until spruce and fir reproduction was fully established after scarification under the parent stand. The effects on the number of reproduction due to scarification were highly significant, but spraying effects were not. Subsequent reproduction of balsam fir was more abundant on fresh than on moist sites, while spruce did not show a clear indication. Advance reproduction of both fir and spruce was more abundant on moist sites.

The *Calliergon* (*Pleurozium*) and *Hylocomium* ground covers are fairly common on wet sites in the Black River experimental area. They are somewhat raised above the general level of the forest floor, and leaf litter does not accumulate on them. These mosses formed better seedbeds than forest litter. Even better seedbeds for spruce and fir are provided by *Hypnum* mosses which are smaller and more compact. *Polytrichum* mosses also provide good seedbeds (MacLean, 1959).

An attempt to follow the early development of balsam fir reproduction by investigating ground-cover influences on germination, first-year survival, and seedling growth was reported by Place (1955) from the Acadia Experimental Forest, New Brunswick (Table 37). The influence of different ground covers is complicated by the frequently mixed composition of mosses and by the occurrence of dwarf shrubs, bracken fern, and other vegetation. Thus no definite conclusions as to the effectiveness of specific ground covers can be made.

Bowman (1944) presented data from Michigan on the association of spruce-fir reproduction with varying species and densities of ground cover. The 73 per cent stocking of spruce-fir reproduction at 0–50 per cent ground-cover density decreased gradually to 24 per cent stocking at 90–100 per cent ground-cover density. Species differences were also noted. *Carex* sp. and *Ledum groenlandicum* ground covers were found with 16 per cent stocking. Comparable figures for other ground covers were *Sphagnum* sp., 66 per cent; *Cornus canadensis*, 72 per cent; *Pteridium aquilinum*, 83 per cent; and for *Alnus rugosa*, 37 per cent.

TABLE 35. PERCENTAGE OF SEEDS RESULTING IN ESTABLISHED SEEDLINGS UNDER DIFFERENT TREATMENTS IN AN UPLAND FOREST OF NORTHEASTERN MINNESOTA

Seedbed	Balsam Fir		Black Spruce		White Spruce		All Species	
	Clear-cut	Partial-cut	Clear-cut	Partial-cut	Clear-cut	Partial-cut	Clear-cut	Partial-cut
Mineral Soil	25.8	37.2	29.5	18.0	26.8	42.5	27.4	32.6
Duff	2.5	1.2	0.2	0.0	0.2	0.0	1.0	0.4
Covered duff	1.5	1.0	0.2	0.0	0.5	0.0	0.7	0.3

Source: LeBarron, 1945.

TABLE 36. NUMBER OF ADVANCE AND SUBSEQUENT REPRODUCTION OF SPRUCE AND FIR PER ACRE DEPENDING ON TREATMENT AND SITE IN THE BLACK RIVER EXPERIMENTAL FOREST, ONTARIO (SECTION B-8)[*]

Treatment of Stand and Soil	Spruce				Balsam Fir			
	Advance		Subsequent		Advance		Subsequent	
	Fresh	Moist	Fresh	Moist	Fresh	Moist	Fresh	Moist
CLEARCUTTING								
Scarified			990	210			2,440	1,940
Sprayed	125	350	110	0	1,000	1,725	80	50
Scarified and sprayed			500	420			1,720	900
Not treated	175	625	90	20	1,450	2,875	400	140
CLEARCUTTING AND HARDWOOD POISONING								
Scarified			760	1,510			34,840	26,930
Sprayed	75	500	260	500	1,025	2,600	15,380	24,480
Scarified and sprayed			1,500	2,320			55,250	21,700
Not treated	50	350	510	410	2,025	3,225	41,090	32,570
SOFTWOOD CUTTING								
Scarified			150	550			3,520	2,860
Sprayed	50	325	10	100	1,200	2,950	350	860
Scarified and sprayed			280	500			4,640	4,810
Not treated	125	275	60	40	1,525	3,125	2,240	1,450
NO CUTTING								
Scarified			1,250	1,600			38,240	8,510
Sprayed	175	425	540	100	1,100	3,925	29,610	6,190
Scarified and sprayed			1,750	2,790			43,900	23,990
Not treated	100	400	260	60	2,025	3,025	24,940	4,880

[*] Cutting was performed in summer and fall, 1953; hardwoods were poisoned in August, 1954; ground was scarified in July, 1954; shrubs were sprayed in July–August, 1955; and reproduction counts were made in August, 1957.

Source: after MacLean, 1959.

TABLE 37. GERMINATION, SURVIVAL, AND GROWTH OF FIRST-YEAR SEEDLINGS OF BALSAM FIR AND RED SPRUCE UNDER THE INFLUENCE OF DIFFERENT GROUND COVERS AT ACADIA FOREST EXPERIMENT STATION, NEW BRUNSWICK

Seedbed	Germination Percentage		Survival Percentage		Total Weight (Mg)		Length of Taproot (In.)		Total Root Length (In.)	
	Spruce	Fir	Spruce	Fir	Spruce	Fir	Spruce	Fir	Spruce	Fir
			UNDER SHELTER							
Sphagnum sp. ..	78	74	91	92	17.3	20.1	2.9	3.7	6.1	8.3
Polytrichum commune	85	90	90	94	13.8	18.9	2.6	3.1	6.1	7.4
Dicranum sp.	90	83	92	98	12.3	18.1	2.9	3.3	5.9	7.0
Calliergonella sp.	59	61	74	83	12.8	17.8	2.6	3.0	5.7	5.8
Litter	52	51	75	69	12.1	18.0	2.2	2.7	4.8	5.4
			OPEN							
Vaccinium sp. ..	23	46	81	89	9.0	15.0	2.4	2.8	4.4	5.4
Polytrichum juniperinum ..	13	69	50	61	26.3	23.9	3.3	3.5	8.9	7.7
Grass	11	62	38	82	10.0	13.0	2.6	3.1	4.2	4.6
Cladonia sp.	12	19	76	37	17.0	12.3	2.2	2.8	5.2	3.8

Source: after Place, 1955.

Bracken fern was found to impair the reproduction of balsam fir on wet sites, although no difference was detected on fresh sites (Place, 1952b). *Aspidium* ferns hindered reproduction in the Adirondacks, but the losses were not important since this fern occurred only in isolated patches (Kittredge and Belyea, 1923). Raspberry (*Rubus* sp.) may become beneficial to spruce and fir reproduction in New England since it hinders hardwood reproduction more than conifers, yet provides enough competition to prevent overstocking among the latter (Westveld, 1931). After investigating partially cut balsam fir stands on the Epaule River watershed in Quebec (Section B-1a, after Rowe, 1959), Hatcher (1960) agreed that dense raspberry does not materially affect reproduction. However, Baskerville et al. (1960) considered it detrimental for conifer reproduction of smaller size.

Various shrub species may play an important role in reproduction of the parent stand. Shrubs are particularly troublesome in prairie border areas, especially on the medium and more fertile forest soils. Cheyney (1946) called attention to the ability of balsam fir to penetrate hazel shrub. Hsiung (1951) recorded the dry-weight of 100 leaves of beaked hazel (*Corylus cornuta*) from various stands at Cloquet, Minnesota. The poorest vigor of this species was found under a spruce-fir canopy. That balsam fir can compete successfully with several shrub species has been noted by several authors (e.g., Cheyney, 1946). There is a possibility that understory balsam fir may help to secure the reproduction of overstory pine under proper management. To a limited extent, pine reproduction occurs naturally when a rather dense balsam fir understory becomes decadent and is blown down by wind and where a scattered pine overstory or seed trees are left standing.

In the main part of the range of balsam fir the most important interfering shrub species is apparently mountain maple (*Acer spicatum*) (Kittredge and Belyea, 1923; Westveld, 1931). Studies on mountain maple were published by Krefting (1953), Krefting et al. (1955, 1956), Vincent (1955, 1956), and others. The natural spread of mountain maple has been related to activities of the spruce budworm (Ghent et al., 1957), aspen mortality (Ghent, 1958a), cutting practices (Vincent, 1953, 1954), birch dieback, and other causes. In the northeast, striped maple (*Acer pensylvanicum*) also seriously affects balsam fir reproduction (Westveld, 1931).

Reproduction Methods

Patterns in the succession of tree generations can be observed in nature. Broadly speaking, regeneration may be in advance of or subsequent to the harvest cutting of the parent stand. Reproduction cuts may be devised to either take greater advantage of advance or to encourage subsequent regeneration. Usually harvest or reproduction cuts are considered as methods of natural reproduction by contrast with artificial seeding and planting. In general, balsam fir reproduces well as compared with other conifers. An abundance of balsam fir reproduction is often considered an undesirable competition for spruce and pine reproduction, and techniques are developed to control the balsam fir reproduction in such cases. This is one of the reasons why so little has been done to develop methods of artificial reproduction or even to find methods for encouraging natural reproduction of this species.

ADVANCE AND SUBSEQUENT REPRODUCTION

The problems of reproduction are intimately involved with forest succession, which was discussed in Chapter 2. Under natural conditions advance reproduction is a prerequisite for the maintenance of climax stands, subsequent reproduction being characteristic of pioneer stands.

In the optimum area of balsam fir, termed the "Black Forest of Gaspé" by Webb (1957), advance reproduction appears in stands reaching 40–50 years of age. In the Epaule River watershed area of Quebec, the process of stand deterioration and regeneration begins when trees reach 60 years of age and continues for 20–25 years (Hatcher, 1961). Advance reproduction is a general feature over a great part of the range of balsam fir from Newfoundland (Candy, 1951) to the New England states (Westveld, 1931; Recknagel et al., 1933), Michigan (Zasada, 1952), and Minnesota (Roe, 1953; Olson and Bay, 1955). It is scarce in Manitoba (Pike, 1955), Saskatchewan, and Alberta at the borders of the species' range (Kabzems, 1952; Rowe, 1955, 1961).

Natural catastrophes, often effective in inducing reproduction in unmanaged forests, also play a large role in extensively managed forests. Reproduction after fires, windfall, disease, and insect attacks has a specific pattern. Reproduction of balsam fir following natural destructive phenomena has not been studied intensively. Some information on the effects of fire will be discussed in a later section in connection with the regional problems of reproduction. Little systematic study has been made of reproduction after wind damage to the parent stands. It has been observed that on moist clay and clay loams, spruce and fir develop wide-spreading roots. Under such conditions trees are more frequently uprooted than broken when subjected to wind stress. Although spruce and fir seedlings will germinate on the exposed mineral soil and humus, survival will be better on decomposed humus, or especially, where some mixing of humus and mineral soils has occurred. On dry and fresh loams and sandy loams conifers root much deeper, and the trunks are usually broken when the trees are windthrown. If they do uproot, the resulting depressions are deep and irregular and provide a poor seedbed (MacLean, 1960).

On very shallow sites uprooted trees usually expose bedrock. Subsequently mosses, particularly *Polytrichum*, become established in the crevices and spruce and fir seedlings follow. As observed in the Black River experimental area, Ontario, shrubs may still provide strong competition in conditions created through uprooting of trees and on seedbeds provided by decaying trunks (MacLean, 1959).

Natural reproduction on areas following spruce budworm devastation was studied by Prebble (1949). He concluded that balsam fir seedlings predominated, but that there was little subsequent reproduction. The advance reproduction made rapid growth after defoliation of the overstory. Invasion by red maple, mountain maple, raspberry, and elderberry was intensive. In another study made after budworm attack, reproduction under one foot in height was little affected, while the effect on taller reproduction depended on the character of general defoliation (Vincent, 1956). Following budworm defoliation either balsam fir or white spruce may prevail, depending upon local conditions. Neither species appears to possess any inherent characteristics that would give it a particular advantage over the other in dominating such areas (Ghent et al., 1957). Further examples of naturally occurring advance and subsequent reproduction will be discussed together with other regional problems of reproduction later in this chapter.

REPRODUCTION CUTS

Man's activities can be directed to favor either advance or subsequent reproduction or a combination of both. Most commonly, reproduction cuts in virgin spruce-fir stands have depended primarily on the presence of naturally established advance reproduction. However, the second-growth forests are frequently denser than the preceding virgin stands. Application of a shorter rotation will require development of methods relying more heavily on subsequent reproduction, especially considering that the classical shelterwood and selection methods are not well suited for mechanization of the harvest operations. Difficulties resulting from adverse climatic and biotic conditions should be anticipated.

Commercial clearcutting with removal of all merchantable wood is successful if advance reproduction is present. In planning for the management of forests it is obviously important to have a knowledge not only of the reproduction of cutover areas but also of the condition of advance reproduction in stands to be harvested. Where advance reproduction is either small or lacking, commercial clearcutting is followed by rapid development of shrubs and hardwoods. On some sites in the Green River watershed area, mountain maple and other aggressive species can overtop conifer reproduction and suppress it for up to 40 years (Basker-ville, 1960).

A commercial clearcut is usually not a true clearcut. It is more like a shelter-wood where the last step of stand removal is left for nature to complete. The cull trees often left only interfere with reproduction, although under some conditions they may serve as protection or as seed trees. Rarely do they develop into future crop trees.

Bell (1962) investigated reproduction conditions after commercial clearcutting in northern Ontario and concluded from observations which covered a period of 20 years that nearly all the conifer reproduction for the next crop was present at the time of cutting. After commercial clearcutting of spruce and fir, it takes more than 20 years until the seedbed again becomes favorable for additional conifer seeding (MacLean, 1960).

Regulated clearcutting in strips was introduced in Europe about 1850 to rehabilitate the forest depleted by selective logging. It has also been successfully used during recent decades in North America, and as Moss (1962) suggests, this method may largely supersede the selective cutting systems in Canadian spruce forests. This type of cutting should have greater advantages in the Boreal Forest Region where temperatures are more limiting than in other areas (Holt, 1950).

A strip clearcutting method for reproducing spruce and fir in mature old-field stands was suggested by Woolsey as early as 1903 (Hawley and Hawes, 1912). The width of clearcut strips in balsam fir should not exceed twice the height of bordering trees if advance reproduction is inadequate. In judging the adequacy of advance reproduction, allowances should be made for losses due to logging, slash cover, and changes in environment (Zon, 1914).

It is recommended that cutting strips in a north-south direction should not exceed widths of 150 feet. In order to withstand hardwood competition, advance reproduction in spruce-fir stands should be ½ to 1 foot in height; in spruce-fir-hardwoods stands, about 2 feet in height; and in hardwood-spruce-fir stands, 4 to 5 feet in height. In hardwood-spruce-fir stands with little advance coniferous reproduction, the only way to secure conifers is through planting and release by elimination of hardwoods (Westveld, 1953).

Table 38 shows the marginal seeding of spruce and balsam fir in different direc-tions in burned and unburned areas in northern Michigan (Bowman, 1944). The prevailing wind caused the further distribution of seed in northern and eastern directions as compared with southern and western directions.

To secure subsequent reproduction from the sides, the cutting should be done in patches not exceeding 30 acres in second-growth balsam fir stands as experi-enced at Matane, Quebec (Ray, 1957).

From preliminary results of the Black River experiments in Ontario, MacLean (1959) concluded that a reasonable width of a clearcut strip should be 200 feet. Alternating strips 130 feet wide and extended in a northwest-southeast direction resulted in heavy losses from wind as has been experienced elsewhere.

Leaving scattered seed trees in spruce-fir stands is considered risky because of blow-down losses (Hawley and Hawes, 1912; Stott et al., 1942; Westveld, 1953). Seed-tree clumps located on ridges and knolls have been advocated (Hawley and

Hawes, 1912). From 6 to 10 windfirm, full-crowned, vigorous, short-boled white spruce seed trees per acre could be left in spruce-fir stands, but heavy slash and shrub invasion may delay the reproduction for 10 to 50 years (Stott et al., 1942).

Although reports noting the effects of leaving seed trees are rather abundant, the results are highly variable (Holt, 1950; Baskerville et al., 1960; and others). These reports deal primarily with spruce trees.

Most balsam fir seed is released early in the fall, but according to data by Roe (1946) from the Superior National Forest in Minnesota, about 40 per cent of the seed remained on the trees on November 15 and a few seed were still present by the end of the following May. This observation agrees with those made by Bonner (1941) on the drifting of balsam fir seed over the snowcrust in late winter in northern Ontario.

TABLE 38. SPRUCE-FIR REPRODUCTION IN PERCENTAGE OF FULL STOCKING
ADJACENT TO SEED-TREES IN NORTHERN MICHIGAN
10 TO 20 YEARS AFTER CUTTING

Direction from Seed Trees	Distance from Seed Trees (Ft.)									
	66	132	198	264	330	396	462	528	594	660
Unburned area										
South or west 90	70	90	70	40	30	50	40	30	30	
North or east 75	80	80	85	77	55	65	80	80	100	
Burned area										
South or west 57	58	48	42	36	20	5	0	0	0	
North or east 83	57	48	40	25	55	45	60	50	40	

Source: after Bowman, 1944.

Diameter limit cuts were developed as an improvement over commercial clearcutting. Under the diameter limit system, some of the merchantable trees are left for future cutting. Depending on the relative amount of residual stand, diameter limit cuts may resemble selection, shelterwood, or commercial clearcut methods. It has been generally believed, although not clearly demonstrated, that discriminate marking of diameter limits for different species might significantly affect the quantity and composition of reproduction. The usual aims are to increase spruce in the composition and to provide adequate stocking of balsam fir, yet to avoid stagnation due to overcrowding. A successful application of diameter limit cutting was carried out for 23 years at the Harpoon Experimental Forest in Newfoundland (Anonymous, 1948). An undisturbed stand contained 40,000 balsam fir seedlings per acre. After 23 years there were 5400 seedlings per acre on clearcut areas and 4400 seedlings per acre on areas cut to a 4-inch diameter limit. The number of seedlings decreased as the diameter limit was raised until only 800 balsam fir seedlings per acre were found on the 9-inch diameter limit cutting. Unfortunately, more detailed ecological information was not provided.

In northeastern spruce-fir types, the initial shelterwood cuttings should remove from 33 to 50 per cent of the volume, with the remainder being cut 10–15 years later (Westveld, 1953). In European practice the reproduction period in Norway spruce stands lasts 15–20 years and the stand is removed gradually in three or four cuttings.

The classical selection system, developed in Switzerland and rarely applied elsewhere, works well with tolerant species on fertile soils in areas of high precipitation where highly intensive management is feasible. Crude selection removes the largest and best trees at long intervals, and any resulting reproduction is incidental. These rough techniques are not expected to give the benefits of more refined selection methods (Hawley and Smith, 1954).

The greatest efforts at compromise in the direction of uneven-aged management in spruce-fir stands have been made in the northeastern United States. In his recommendations for this region, Westveld (1953) described both spruce and balsam fir as well adapted to the selection method. In the red spruce types the selection method definitely favors spruce at the expense of balsam fir. This is not true in the black spruce and white spruce types. Marking techniques can favor spruce reproduction and provide for the utilization of balsam fir before it becomes decadent.

Westveld also pointed out that the selection system discourages encroachment by hardwoods and shrubs and protects against windfall. Slash accumulation is negligible. Diameter limits can be established for the selection system but cannot substitute for marking. The Green River experiments have showed that it is possible to start conversion to selection forests in mature and overmature stands, but it can be done with less loss by starting with younger age classes (Baskerville, 1960).

Eichel (1957) suggested that balsam fir stands could be managed under Eberhard's shelter-wedge system, as used in mixed fir stands in Europe. This method combines gradual opening of the stand with clearcutting on numerous small areas. The system calls for a mobile landing unit in place of the concentrated landing and requires a dense network of trails and roads.

The success of different cutting systems may depend on several conditions. Under optimum conditions for reproduction success can frequently be achieved by any reasonable method. On the basis of his 30 years of observation, Waldron (1959) saw no differences in reproduction under diameter limit, seed tree, and clearcut systems in mixed-wood stands in Saskatchewan. The lack of differences due to treatments was attributed to the generally good conditions for regeneration of balsam fir and white spruce which prevailed at the beginning of the study. Different cutting methods produce the greatest diversity of results when conditions are less favorable. Under severe conditions only elaborate reclamation methods may secure reproduction.

EFFECT OF LOGGING ON REPRODUCTION

Logging may produce difficult problems, especially when dependence is based on advance reproduction. Economic conditions have discouraged elaborate and painstaking care in seeing that reproduction becomes established over the vast, extensively managed spruce-fir stands in Canada and in the United States with the result that reproduction is left largely to chance (Koroleff, 1954; Westveld, 1931).

The amount of logging damage to advance reproduction and the residual stand depends on the cutting system, total and merchantable stand volume, logging techniques, the season, care taken in logging, and other factors. Reports are many, and the figures shown differ widely. Some of the data show both the effects of logging damage and the slash cover. The total loss of advance reproduction due to logging damage and slash accumulation was estimated to be 84 per cent in Michigan (Bowman, 1944) and 89 per cent in Koochiching County, Minnesota (Jones, 1956).

According to estimates made by Koroleff (1954), logging damage causes 60 per cent loss of all advance reproduction in eastern Canada and up to 75 per cent of the older reproduction in commercial clearcutting operations. In Michigan, Westveld (1933) found that 60–75 per cent of advance reproduction was lost because of logging damage on clearcutting. With care the damage could be kept as low as 15–20 per cent (Bowman, 1944). Logging damage depends on the intensity of the cut as shown by data from the Pike Bay Experimental Forest in Minnesota. Loss of advance balsam fir reproduction under clearcutting was 66 per cent, while under light and moderate cuttings logging damage caused only 36 per cent loss. The growth of seedlings is faster after heavier cuts and the proportion of seedlings in different size classes changes with time — points which should be considered in damage appraisal (Roe, 1953, 1957).

Logging damage following the use of different diameter limit cuts in spruce-fir forests of the Adirondacks ranged from 42 to 53 per cent (Recknagel et al., 1933; Recknagel and Westveld, 1942). In a shelterwood operation in New Brunswick, 22–23 per cent of the advance reproduction was lost due to logging damage (Vincent, 1952). Shelterwood cuts are later repeated and some seedlings die following exposure. Stocking percentage is affected much less than the number of seedlings. In some cases logging roads may cover rather extensive areas. Westveld (1933) reported this as 39 per cent in an operation in Michigan. Although the roads are narrow, their number and location with respect to areas covered by heavy slash often result in uneven distribution of the remaining reproduction.

The slash-covered area, according to different sources, varied from 5 to 48 per cent of the total (Kittredge and Belyea, 1923; Westveld, 1931; Recknagel et al., 1933; Bowman, 1944). It takes 14–30 years for the coniferous slash to decompose (Kittredge and Belyea, 1923). Hardwood slash is largely decomposed within seven or eight years. In spruce-fir slash, fire danger is reduced to normal by about the twelfth year, but decomposition has not yet resulted in favorable conditions for the establishment of new seedlings (Westveld, 1931).

Slash disposal is most important to reproduction and fire protection needs. When subsequent rather than advance reproduction is the object, prescribed burning may be a means for slash disposal and may become a possibility for part of the spruce-fir stands in the future. Slash disposal for fire protection can be done only in specific situations. For control of insects and diseases no urgent need of slash disposal has yet been proved.

McArthur (1963) described stand conditions 20 years after partially mechanized logging in a uniformly stocked balsam fir stand in the Little Pabos River watershed, southeastern Gaspé, Quebec (Section B-2). The elevation was about

1000 feet above sea level. Soil was a shallow (8–14 in. to the parent material) but well-defined podzol with 60 per cent of silt and clay in the topsoil. Bulldozers were used to establish a network of hauling roads. Twenty years later there was a dense balsam fir stand with dominant trees 11–12 feet high developed from the undisturbed advance reproduction. The 12-foot-wide hauling roads had no reproduction in either wheel tracks or deeply scraped shoulders. The center of the road where only a part of the humus was scraped away was covered by very dense spruce reproduction, 4–5 feet high, with almost no balsam fir present, while upon the bulldozed debris beside the road a dense paper birch stand, 10–12 feet high, had developed. With better knowledge of the effects of mechanized logging, local cutting methods could be developed to work out the best compromise between the short- and long-range interests in the forest management of spruce-fir areas.

PROMOTION OF NATURAL REPRODUCTION

Slash disposal and protection against logging damage may assist in promoting natural reproduction by reducing negative effects. Positive silvicultural measures implementing reproduction include drainage, temporary agricultural use, soil scarification, and the use of herbicides.

Depending on the season, mechanical logging may automatically produce some soil scarification. Information on the influence of such logging on subsequent reproduction of balsam fir is still extremely limited. Plowing in irregular strips (Robbins, 1930) and disking (LeBarron, 1945) have been recommended as possible means of securing balsam fir and spruce reproduction. A disking experiment in a patchy balsam fir–paper birch stand in eastern Wisconsin was undertaken following a bumper seed crop of balsam fir. A heavy 6-foot-wide disk attached to a crawler tractor was used to disk about 60 per cent of the area. Disked strips were about 10 feet wide (Stoeckeler, 1952). Table 39 shows the results one year after disking.

In the second year the amount of total reproduction was reduced to 150,000 per acre, but balsam fir seedlings increased in number to 10,000 per acre (Stoeckeler, 1955). Balsam fir continued to gain through the third year when counts gave

TABLE 39. EFFECT OF DISKING ON REPRODUCTION IN
AN OPEN BALSAM FIR–PAPER BIRCH STAND IN
WISCONSIN, ONE YEAR AFTER DISKING

Species	Percentage of Milacre Quadrats Stocked in Disked Strips	Number of Seedlings Per Acre in Disked Strips	Number of Seedlings per Acre in Total Area
Paper birch	82.5	288,600	173,160
Balsam fir	57.5	14,100	8,460
Red maple	32.5	3,900	2,340
Aspen	12.5	1,200	720
Birch	2.5	100	60
All species	95.0	307,900	184,740

Source: Stoeckeler, 1952.

11,000 seedlings per acre, although total count for all seedlings had by then dropped to 100,000 per acre. Balsam fir stocking was 50 per cent after one year, 70 per cent after two years, and 65 per cent after three years. An average of only 2 balsam fir seedlings per milacre of unexposed mineral soil could be found, while 19 such seedlings were found on the 100 per cent exposed mineral soil. At 12 feet from the seed source there were only 3.5 seedlings per milacre. This declined slowly with increasing distance so that at 66 feet there were only 2.1 seedlings. The best time for disking was reported to be in August or September.

Scarification was considered a successful measure if it exposed mineral or organic soil suitable as a seedbed for germination in aspen-birch-spruce-fir forests in Ontario (Sections B-4, B-8, B-9). Sands, sandy loams, loams, and well-decomposed humus make good seedbeds, but silt loams, clay loams, silts, clays, and poorly decomposed humus do not. Grasses and sedges are the most aggressive pioneer plants to reinvade scarified areas. They are rather intolerant and to keep them out, scarification should first be made under a tree canopy a few years in advance of cutting. Such measures will also reduce the amount of aspen and birch reproduction. By cutting balsam fir before scarification and leaving the spruce, the latter species may be increased in the composition of reproduction. The irregular surface created by scarification also keeps the ground free from accumulation of hardwood litter, which frequently smothers newly-germinated conifer seedlings (MacLean, 1959, 1960; see Table 36).

In some situations disking can increase competition from aspen suckers. Scarification can also be used to reduce the amount of advance reproduction of balsam fir and to expose mineral soil for subsequent reproduction of pines, as was done by Horton (1962) in western Quebec (Section L-4b).

SEEDING AND PLANTING

In nurseries balsam fir is grown to a minor extent for Christmas tree plantings and for ornamental use. Very little stock is produced for reforestation purposes. Seed is stratified and sown in the spring, or more commonly is fall-sown in mulched seedbeds. A well-drained sandy loam-to-loam forms the best seedbed, and the seed may be sown broadcast or in drills at a rate to produce 60 to 80 seedlings per square foot. Seed should be covered with nursery soil to a depth of ⅛ to ⅜ inch (U.S. Forest Service, 1948).

Toumey and Neethling (1923) tested balsam fir germination in shaded seedbeds, open seedbeds, and mulched seedbeds. A hardwood leaf mulch, 2.5 inches thick was provided. On each seedbed, 1810 seeds were sown in early May. The results are given in Table 40.

At the time of the appearance of the shoots the roots had penetrated to a depth of 1.0 inch. By the fifth day they had reached 1.3 inches, but after 30 days the roots had not penetrated below 1.7 inches. Norway spruce and tamarack seedlings made somewhat better root development. Pine seedlings rooted much more deeply. In the nursery balsam fir develops slowly, and size of stock satisfactory for planting takes three to four years. Most present nursery practice is to grow seedlings as 2–1 or 2–2 stock (U.S. Forest Service, 1948), although Schenck (1912) advocated the use of plants five years old.

Temporary cold storage of balsam fir plants is most effective at temperatures slightly above freezing. Quick freezing and low temperatures (0–10F) may result in complete loss of plants. Balsam fir stores more successfully in the spring than in fall, but does not give as good survival as red pine, white pine, or spruce (Baldwin and Pleasanton, 1952).

European success with *Abies alba* prompted some interest in *Abies balsamea* as a species for planting in the early years of forestry in the United States. Apparently the first balsam fir plantation in the United States was established in Pennsylvania in 1903 (Illick, 1928). In 25 years it had reached an average height of 25 feet. No further report is available. In Canada, balsam fir was first planted about 1900 in a small experimental plot near Ottawa (Kittredge, 1929).

TABLE 40. NUMBER OF GERMINATED BALSAM FIR SEED AT DIFFERENT TIMES AFTER SEEDING AND PERCENTAGE OF TOTAL GERMINATION AND SURVIVAL AFTER ONE GROWING SEASON IN SOUTHERN NEW ENGLAND

Seedbed	Days after Seeding										Germination	Survival
Condition	23	27	31	41	44	51	58	83	96	116	Percentage	Percentage
Shaded			8	12	17	30	73	327	339		18.7	99.2
Open				8	8	12	14	14	14	17	0.9	23.6
Mulched	1	1	2	22	49	71	79				4.4	1.3

Source: after Toumey and Neethling, 1923.

Balsam fir is planted for Christmas trees in Ohio (Paton et al., 1944) and in West Virginia (Tryon et al., 1951), as well as in states within its natural range. It failed in shelterbelt and town plantings in the prairie region of Minnesota (Deters and Schmitz, 1936), Nebraska (Pool, 1939), and North Dakota (George, 1953). The few plantations in the Lake States have been moderately successful (Kittredge and Gevorkiantz, 1929), particularly because of the ability of balsam fir to compete successfully with hazel (Lake States Forest Experiment Station, 1953). Balsam fir was not, however, among the 15 principal species advocated by the U.S. Forest Service for plantations in the Lake States (Rudolf, 1946, 1950a). Plantation experience at Badoura, Minnesota, in the middle 1930's indicated that while survival was generally good, balsam fir in open fields may be subject to winter injury for several years. Balsam fir has also survived and grown well in an experimental Christmas tree plantation at the Rosemount Experiment Station of the University of Minnesota, on the Mayo Institute grounds at Rochester, Minnesota, and in a plantation of mixed exotic and native species at the Cloquet Forest (Schantz-Hansen and Hall, 1952). In the latter plantation balsam fir survived and grew better than *Abies concolor, A. grandis,* and *A. lasiocarpa.*

Although balsam fir has been planted largely for Christmas trees or experimental purposes, the new planting guide prepared by the Northeastern Forest Soils Conference (1961) listed a variety of potential reforestation sites for the species in the spruce-fir area of the northeastern states.

A series of direct seeding and planting experiments with balsam fir was undertaken in northern Wisconsin. Early results were reported by the Lake States Forest

Experiment Station (1950), and a more elaborate analysis was made by Stoeckeler and Skilling (1959). One series of experiments was conducted under an open aspen stand near Alvin, Wisconsin. The entire area was disked and seeded using about 27 seeds per spot. Some spots were left uncovered, others were covered by hand- or foot-scrape, and brush was dragged across the remainder. Results are given in Table 41.

Covering the seed by dragging brush over the spots proved to be a poor practice. After 11 years there were 755 seedlings per acre under shade as compared with 40 seedlings in the open. Thirty per cent of the shaded quadrats were stocked compared with 4 per cent in the open. Height growth averaged 1.8 feet in the open and 1.1 feet in the shade.

TABLE 41. RESULTS OF DIRECT SEEDING OF BALSAM
FIR ON A DISKED SITE WITH VARIOUS METHODS OF
SEED COVERING, NORTHEASTERN WISCONSIN

Seed Covering Method	Number of Seedlings per Acre		
	After 1 Year	After 2 Years	After 11 Years
Foot-scrape	2180	1600	940
Brush-drag	1610	990	230
No covering	2480	1350	570

Source: Stoeckeler and Skilling, 1959.

Other studies were initiated in 1949 on northern hardwood sites on the Argonne Experimental Forest in Wisconsin. Contrary to their conclusion on the aspen site study, Stoeckeler and Skilling (1959) considered the results of direct seeding on northern hardwood sites very poor or complete failures. Results of direct seeding of balsam fir on a swamp margin supporting an open stand of black spruce, white spruce, hemlock, red pine, and white pine were considered satisfactory. The soil was a sandy loam and scalping provided the ground preparation. After 10 years, 70 per cent of the fall-seeded spots and 40 per cent of the spring-seeded spots were stocked where screening protection had been continued until the second fall. Failure resulted where screening was not used.

Stoeckeler and Skilling (1959) also reported comparisons of planted balsam fir and white spruce on an understocked aspen site and on a northern hardwood site. Planting stock was 2–2 balsam fir and 2–3 white spruce. On the aspen site survival of balsam fir after 11 years was 58 per cent in both shade and open areas, whereas survival of white spruce was 78 per cent in the shade and 62 per cent in the open. Height in feet for balsam fir was 2.2 in shade, 5.0 in the open; for white spruce, 4.6 in shade, 4.8 in the open. On the northern hardwood site the survival of balsam fir after 11 years was 65 per cent with an average height of 4.5 feet. Survival of white spruce was 84 per cent with an average height of 5.7 feet. Balsam fir suffered heavily from browsing by the white-tailed deer.

From the foregoing studies it appears that planting was more satisfactory than seeding on open ground. On shaded areas direct seeding was less expensive. White spruce is more desirable and has other advantages over balsam fir on most planting sites.

Regional Problems

Regional forest classification systems furnish the basic units within which research can be evaluated and practical experience exchanged. The extensive reproduction study by Candy (1951) made use of the forest geographical classification of Canada by Halliday (1937). Many other authors have used the geographic or synecological classification systems of Clements, Braun, Heimburger, Hills, Westveld, Curtis, Rowe, Linteau, and others. Some investigators have made use of the forest cover type system developed by the Society of American Foresters. Although most recent studies give satisfactory local geographical and ecological descriptions of the localities where experiments are carried out, the value of some research is greatly reduced because of neglect in establishing a satisfactory ecological framework.

BOREAL FOREST REGION

The reproduction study by Candy (1951) clarified important problems of forest reproduction over much of Canada east of the Rocky Mountains. The data were collected under and grouped by cover types, e.g., softwood, mixed-wood, and hardwood; and by subtypes of the original stands. Further grouping was made according to stand treatment, including logging, fire, or both. Time elapsed since disturbance, was registered in five-year intervals. Conifer reproduction was given by species, and hardwoods were listed as tolerant and intolerant. Reproduction was further noted as advance growth or as subsequent reproduction. The survey was conducted primarily on a stocked milacre quadrat basis. For each quadrat the largest specimen of advance growth and subsequent reproduction for each species present were tallied. An actual count was made of the number of trees on every twentieth quadrat. The entire survey was organized according to Halliday's (1937) forest geographical sections of Canada. For each condition class 2000 quadrats were examined within each geographical section. Since the interest here is centered on balsam fir, the analyses presented have been limited to those which are significant to that species.

Table 42 presents Candy's data on reproduction in the Boreal Region in terms of the percentage of milacre quadrats stocked five years or more after logging. These data were derived from several thousand plots.

In cutover softwood stands the percentage of milacre quadrats stocked with balsam fir was highest in Section B-1 west (Quebec). In that section stocking of all species was 89 per cent and of balsam fir was 70 per cent. These percentages decreased westward toward Section B-18 (Saskatchewan) where balsam fir occurred on only 1 per cent of the quadrats, and only 45 per cent of the quadrats were stocked by some tree species (see Plate 15 and Table 42). This indicates that difficulty in obtaining suitable reproduction increases to the westward in Canada.

In cutover mixed-wood stands, the quadrat stocking of balsam fir ranged from 75 per cent in the east to 1 per cent in the west. The total quadrat stocking for any species was 91 per cent in the east and 32 per cent in the west. A possible difference in the average age of cutovers examined in the east and in the west could account for some of the differences in stocking. Balsam fir stocking in cutover hardwood stands seems to have been of some importance in certain western sections. Candy's data also illustrate the variation in coniferous reproduction in cutover softwood stands from 5000 trees per acre in Section B-1 (eastern Quebec) to 500 per acre in Section B-19a (north-central Alberta).

When fire has occurred in standing softwood timber, the resulting coniferous reproduction lies within a range of 1000–2000 trees per acre. Toward the western end of the range of balsam fir, fire in uncut softwood stands results in as much reproduction as does logging, and under certain conditions even more.

Balsam fir reproduction is more seriously limited by fire than is coniferous reproduction generally. The species is seldom found on more than 5 per cent of burned-over quadrats.

Logging followed by fire results in the least coniferous reproduction. In addi-

TABLE 42. PERCENTAGE MILACRE QUADRAT STOCKING OF ADVANCE AND
SUBSEQUENT REPRODUCTION FOLLOWING LOGGING
IN THE BOREAL REGION

Section of Province	Number of Milacre Quadrats Counted	White Spruce	Balsam Fir	Any Conifer	Any Hardwood	Any Species
ORIGINAL SOFTWOOD STANDS						
B-1 East, Quebec	5,080	38	69	88	9	89
B-1 West, Quebec	14,340	37	70	86	28	89
B-7 Noranda, Quebec	8,980	52	34	77	1	78
B-7 Clova, Quebec	28,020	50	20	65	5	68
B-4, Ontario	17,320	72	24	82	2	82
B-14, Manitoba	3,120	24	1	29	11	37
B-18, Manitoba	200	35	12	47		
B-18, Saskatchewan	3,540	27	1	30	23	45
B-18, Alberta	220	6	10	16	6	21
B-19, Alberta	3,780	13	2	20	12	30
ORIGINAL MIXED-WOOD STANDS						
B-1 East, Quebec	9,880	30	75	88	20	91
B-1 West, Quebec	3,240	16	65	78	26	83
B-7 Noranda, Quebec	3,920	32	75	85	9	86
B-7 Clova, Quebec	11,440	15	53	63	17	71
B-4, Ontario	9,980	22	68	75	19	81
B-14, Manitoba	2,580	16	8	22	37	49
B-18, Manitoba	3,760	11	12	24		
B-18, Saskatchewan	8,160	20	1	21	40	50
B-18, Alberta	20,220	10	8	18	19	34
B-19, Alberta	17,760	10	1	13	23	32
ORIGINAL HARDWOOD STANDS						
B-1 West, Quebec	380	6	62	64	17	72
B-14, Manitoba	540	1	0	1	73	73
B-18, Saskatchewan	820	7	0	7	32	36

Source: after Candy, 1951.

tion to mortality of advance reproduction and seed trees, the seedbed is damaged by the burning of the highly water-absorbent, partially decayed debris (such as windfalls) on which balsam fir and spruce reproduce readily. According to Hosie (1953), burned cutovers are frequently less than 50 per cent reproduced, and may not provide adequate stocking by the end of the rotation.

Coniferous reproduction in mixed-wood stands follows roughly the same trends as in softwood stands after disturbance. In hardwood stands occasional coniferous reproduction follows cutting, and no reliable trends could be observed in the reports.

TABLE 43. REPRODUCTION OF SPRUCE AND BALSAM FIR UNDER
VARIOUS CONDITIONS IN NEWFOUNDLAND

Section	Cover Type	Forest Type	Number of Milacre Quadrats Counted	Milacre Stocking Percentage	Spruce	Number of Stems per Acre	
						Balsam Fir	Spruce and Fir
		REPRODUCTION BEFORE DISTURBANCE					
A-8	Softwood	Cornus	2080	80	360	2226	2586
A-9	Softwood	Cornus	1130	42	274	1150	1424
A-8	Softwood	Sphagnum	980	48	918	348	1266
A-9	Softwood	Sphagnum	890	42	1955	1371	3326
A-8	Mixed-wood	Cornus	190	89	158	2474	2632
A-9	Mixed-wood	Cornus	450	57	156	1088	1143
		REPRODUCTION FOLLOWING LOGGING					
A-8	Softwood	Cornus	1680	62	755	2840	3695
A-9	Softwood	Cornus	3250	56	600	2437	3037
A-8	Softwood	Sphagnum	190	59	1526	474	2000
A-9	Softwood	Sphagnum	540	59	2056	1000	3056
A-8	Mixed-wood	Cornus	570	54	533	3860	4193
A-9	Mixed-wood	Cornus	1380	55	600	2000	2600
		REPRODUCTION FOLLOWING FIRE					
A-8	Softwood	Not specified	290	68	100	35	135
A-9	Softwood	Not specified	2160	41	907	324	1231
A-8	Mixed-wood	Not specified	60	80	167	167	334
A-9	Mixed-wood	Not specified	1390	47	1015	863	1878

Source: W. M. Robertson, as cited by Candy, 1951.

Robertson's data on reproduction in two forest sections in Newfoundland were included by Candy (1951) in his report. Information was presented on conditions present in several ecological forest types both prior and subsequent to disturbance (see Table 43).

In the Cornus type, balsam fir was found to predominate both before and after logging. Spruce (black spruce) was the principal species present in the Sphagnum type. Logging appeared to have decreased the reproduction in the Sphagnum type and to have increased it in the Cornus type. Fire resulted in a drastic reduction in the proportion of balsam fir reproduction. Conditions for reproduction in Newfoundland were considered to be generally good; however, it was felt that the relatively high proportion of balsam fir to spruce was undesirable.

Advance reproduction studies form a part of broader investigations on stand development which is the subject of Chapter 7. Such studies are vital for developing management techniques and for planning practical work in forest enterprises. A comprehensive study is currently under investigation in the Green River area (Section B-2). A five-year progress report was presented by Vincent (1955). The investigations were carried on over a block of 4023 acres classified into six cover types, i.e., mature softwood, mature softwood over immature softwood, mature mixed-wood, mature mixed-wood over immature softwood, cutover with immature softwood, and cutover with no immature softwood. Most of the spruce-fir

TABLE 44. STAND VOLUME AND REPRODUCTION IN 1945 AND 1950 IN THE GREEN RIVER EXPERIMENTAL AREA, NEW BRUNSWICK

| | Volume (Ft^3)* per Acre | | Number of Reproduction per Acre | | | | | |
| | | | Balsam Fir | | Spruce | | Hardwoods | |
Stand Description	1945	1950	1945	1950	1945	1950	1945	1950
Mature softwood	1933	2097	2250	2193	407	239	36	382
Mature softwood over immature softwood ..	1187	1486	2323	2085	192	197	338	500
Mature mixed-wood ...	1419	1469	1582	1700	136	173	127	382
Cutover with no immature softwood	168	235	1977	1573	298	196	88	303

*Including trees 3.6 in. dbh and over. Basis: about 530 plots, 0.1 acre in size.
Source: after Vincent, 1955.

overstory was 80 years old, with the understory 20–40 years old. The area was affected by birch dieback resulting in openings that are being invaded by mountain maple and hardwood reproduction. The changes in reproduction over a five-year period together with associated changes in stand volume are shown in Table 44. More detailed changes in volume are presented in Table 54 in Chapter 7 where change in stand structure is the main subject under consideration.

Stability in the number of softwood reproduction during the five-year period is largely explained by regular ingrowth. Reproduction was adequate on about 80 per cent of the plots in 1950.

A different picture emerges from the studies of Pike (1955) on the reproduction and stand development in an undisturbed white spruce-balsam fir stand at Duck Mountain, Manitoba (Section B-18). There the critical stage of stand decadence is not followed by thrifty reproduction. This is due to the more adverse climatic conditions. Data from 25 years' observation, in Table 45, show that seedlings present in 1921 all died. Balsam fir seedlings were 4–10 years old, suppressed, and browsed by deer in 1946. Additional data relating primarily to stand development are presented in Chapter 7.

Linteau's (1955) forest type classification for Section B-1 (Quebec) of the Boreal Region provides not only fundamental ecological information (see Chapter 2) but also information on reproduction conditions, with suggestions for management regarding the area.

A symposium on spruce-fir regeneration in Canada was organized in 1940 (Long, 1940). Many reports were prepared which dealt with both specific reproduction problems and general management considerations. Generally, in the eastern part of the region reproduction is satisfactory even under poor management practices, and balsam fir reproduces better than the spruces. Under more adverse conditions balsam fir loses its advantages over spruce. Openings in spruce flats and in swamps are restocked by black spruce. Removal of hardwoods in logging operations favors spruce more than balsam fir. In the western continental climate balsam fir finally drops out entirely while spruce persists.

TABLE 45. STAND BASAL AREA AND NUMBER OF SEEDLINGS PER ACRE IN AN UNDISTURBED SPRUCE-FIR STAND IN MANITOBA

Species	Basal Area (Ft²/Acre)		Number of Seedlings per Acre	
	1921	1946	1921	1946
White spruce	126	107	111	0
Balsam fir	25	11	592	550
Hardwoods	5	3		
All species	156	121	703	550

Basis: 1 plot, 0.6 acre in size.
Source: Pike, 1955.

GREAT LAKES–ST. LAWRENCE FOREST REGION

This region, represented in both Canada and the United States, displays a shorter but still well-expressed reproduction gradient roughly paralleling that in the Boreal Region. The broad features of balsam fir reproduction in the Canadian part of the region are shown in Plate 15.

According to Candy (1951), in the eastern part of Canada balsam fir quadrat stocking averages about 70 per cent in former softwood and mixed-wood stands, and less than 30 per cent in hardwood stands as shown in Table 46. Toward the western prairie the proportional and absolute stocking of balsam fir decreases considerably. There are about 4000 coniferous seedlings per acre in the eastern part of the region as compared with only 400–500 in the western portions.

In the east, fire in standing timber results in coniferous reproduction of about 2000 trees per acre, and in cutovers about 1000. Toward the prairies, the reproduction following fire, both in standing timber and on cutovers, becomes more comparable to reproduction after logging. Near the prairie there is a greater likelihood of situations where fire would be advantageous in obtaining coniferous reproduction. Balsam fir reproduction following logging and fire decreased in Section L-6 to 2–3 per cent of stocked quadrats. In Section L-9 in softwood types it was 4 per cent and in mixed-wood types 9 per cent.

Some of the earlier attempts to develop silvicultural methods on the basis of ecological forest types were made at the Lake Edward Forest in Quebec (Section

L-4). These early studies were begun in 1936. The types are discussed in Chapter 2. Ray (1941, 1956) presented data comparing the reproduction when a single cutting method was applied to various forest types (Table 47). The area was first logged for white pine and large spruce in 1890. In 1910, spruce was cut to a 10-inch diameter limit and balsam fir to a 7-inch limit. Hardwoods were not cut. Data were taken in 1925, 1936, and 1946. The number of balsam fir seedlings increased more than the number of spruce seedlings between 1936 and 1946 in the Cornus type. The volume of spruce also increased more than the volume of balsam fir during this period. In this type there is little danger of invasion by tolerant hardwoods or

TABLE 46. PERCENTAGE MILACRE QUADRAT STOCKING OF ADVANCE AND SUBSEQUENT REPRODUCTION FOLLOWING LOGGING IN THE GREAT LAKES– ST. LAWRENCE REGION IN CANADA

Section of Province	Number of Quadrats	White Spruce	Balsam Fir	Any Conifer	Tolerant Hardwoods	Intolerant Hardwoods	Any Species
ORIGINAL SOFTWOOD STANDS							
L-6, New Brunswick	1,420	21	63	67	32	10	78
L-7, Quebec	460	18	94	95	0	15	97
L-9, Ontario	10,560	9	33	48	7	19	63
L-12, Manitoba	3,520	4	0	48			
ORIGINAL MIXED-WOOD STANDS							
L-6, New Brunswick	1,940	17	56	60	45	10	79
L-7, Quebec	2,280	6	77	78	14	15	83
L-9, Ontario	21,140	10	39	50	17	23	69
ORIGINAL HARDWOOD STANDS							
L-6, New Brunswick	600	3	27	27	88	0	95
L-9, Ontario	160	1	24	24	92	2	97

Source: after Candy, 1951.

shrubs; therefore, Ray suggested a cut of spruce (mostly red spruce) to a 10-inch or 12-inch diameter limit plus all merchantable balsam fir. Residual trees from the diameter limit cut would provide some shade thus preventing the intolerant paper birch and pin cherry from dominating the larger openings. Since the soil is shallow, the selection system of cutting would create a high risk from windfall. In the Oxalis-Cornus type the number of balsam fir seedlings far exceeded the number of spruce seedlings. In volume, balsam fir was still greater in 1946 but the margin of difference over spruce was narrowing. Strong competition is a feature in this type. Spruce, which is preferred, should be allowed to grow to much greater size, with a rotation age twice as long as for balsam fir. In the Viburnum-Oxalis type, softwood seedlings were not so abundant; however, balsam fir seedlings had the highest representation. Spruce exceeded balsam fir by volume. In this type, only red spruce can successfully compete with yellow birch, maple, and beech. Ray suggested scarification of the ground to encourage coniferous repro-

TABLE 47. NUMBER AND VOLUME OF TREES IN CUTOVER STANDS ON THE LAKE EDWARD EXPERIMENTAL FOREST, QUEBEC*

Species and Size Class	Cornus Type						Oxalis-Cornus Type						Viburnum-Oxalis Type					
	Number of Trees			Volume (Ft³)			Number of Trees			Volume (Ft³)			Number of Trees			Volume (Ft³)		
	1925	1936	1946	1925	1936	1946	1925	1936	1946	1925	1936	1946	1925	1936	1946	1925	1936	1946
Spruce																		
Seedlings	419	352	272				247	344	321				132	350	245			
Saplings	106	318	364	84	91	108	55	148	155	48	44	45	38	103	136	23	22	31
Larger		194	263	414	791	1171		89	105	273	506	685		38	44	264	324	429
Balsam fir																		
Seedlings	534	342	2260				370	1275	2276				103	849	510			
Saplings	89	412	618	82	92	135	96	288	258	64	70	65	44	50	46	17	12	12
Larger		114	156	267	378	523		142	159	408	703	857		43	40	204	290	272
Other softwoods																		
Seedlings	83	107	113				67	67	102				10	2	2			
Saplings	17	73	81	13	20	21	11	35	33	10	9	9	2	2	0	1	1	0
Larger		34	43	94	134	165		15	18	87	122	125		2	1	14	23	5
Total softwoods																		
Seedlings	1036	801	2645				684	1686	2699				245	1201	757			
Saplings	212	803	1063	179	203	264	162	471	446	122	123	119	84	155	182	41	35	43
Larger		342	462	775	1303	1859		246	282	768	1331	1667		83	85	482	637	706
Yellow birch																		
Seedlings	10	0	0				39	8	62				46	63	158			
Saplings	5	9	6	4	4	1	38	23	17	14	11	9	68	21	18	18	9	8
Larger		14	8	33	106	88		43	42	660	746	782		18	65	1548	1606	1742
All hardwoods																		
Seedlings	166	66	47				129	56	302				113	818	1806			
Saplings	31	113	96	35	29	26	66	86	70	30	29	26	112	94	117	30	27	26
Larger		52	51	168	265	264		84	89	841	973	1015		109	120	2050	2099	2280
Total species																		
Seedlings	1202	867	2692				813	1742	3001				358	2019	2563			
Saplings	243	916	1159	214	232	290	228	557	516	152	152	145	196	249	299	71	62	69
Larger		394	513	943	1568	2123		330	371	1609	2304	2682		192	205	2532	2736	2986

*For summary see Table 67, Chapter 7. Saplings 1–3 in. dbh.
Basis: 343 plots, 0.2 acre in size.
Source: Ray, 1941, 1956.

duction. Balsam fir is unable to withstand prolonged suppression, but red spruce recovers well, attaining a size of 25 inches dbh at an age of 300 years.

Zasada (1952) studied reproduction on cutover swamplands of the Upper Peninsula of Michigan. He found that as a result of the aggressiveness of balsam fir, conifers maintained their position better under partial cutting than under commercial clearcutting. The stocking of spruce and fir was 76 per cent by milacre quadrats in areas where fires had been kept out after logging in northern Michigan (Bowman, 1944). To reach an adequate stocking, 8–10 years were needed.

In northeastern Minnesota balsam fir seedlings can be found over the entire range of physiographic conditions. In coniferous stands balsam fir reproduction ranges from none to 1100 per acre, in mixed-wood stands from 90 to 600, and in hardwood stands from 10 to 250 per acre (Schantz-Hansen, 1923, 1934). In north-central Minnesota 6–12 years after commercial clearcutting in a spruce-fir-birch cover type, 58 per cent of milacre quadrats were stocked with balsam fir, 9 per cent with white spruce, and total stocking was 85 per cent. There were 2000 balsam fir seedlings as compared with a total reproduction of 4700 seedlings per acre (Olson and Bay, 1955). The better-drained soils were mostly poorly stocked. Layering of balsam fir was insignificant. About 50 per cent of balsam fir and white spruce reproduction had become established before logging.

ACADIAN-APPALACHIAN FOREST REGION

Candy's data (Table 48) indicate that in general the Acadian Region possesses good reproduction conditions. The number of conifers is greater than in the Great Lakes–St. Lawrence Region to the west. However, the proportion of balsam fir is less than in the eastern parts of the Boreal Region and the Great Lakes–St. Lawrence Region. An important factor here is probably the superior tolerance of red spruce. Advance reproduction plays an important role. In number the red spruce comprises 50 per cent of the reproduction, and because of its greater size is better able to compete with hardwoods.

The relationships between balsam fir and pine reproduction are of less importance in this region than in the Great Lakes–St. Lawrence Region. The relationship between balsam fir and spruce is similar to that in the eastern part of the Boreal Region, although red spruce is better able to compete with balsam fir than is white spruce. Older stands having a considerable volume of hardwoods may have a substantial quantity of spruce and balsam fir advance reproduction; but, after logging, subsequent reproduction is primarily hardwoods. In some areas reproduction of balsam fir following logging and fire was found to be as low as 2 per cent of all milacre quadrats (Candy, 1951). After insect attacks conifer stocking was as high as 96 per cent in softwood and 76 per cent in mixed-wood types. Balsam fir stocking was 87 per cent and 46 per cent respectively (Section A-2).

Recommendations for the reproduction of four permanent and two temporary forest types in the spruce region of New England were developed by Hawley and Hawes (1912). To avoid wind damage, they suggested that spruce swamps, spruce slopes, and spruce flats on shallow soils should be clearcut in strips, or that carefully selected seed-tree clumps should be left. In stands of mixed hardwoods, spruce, and fir which occur on deeper soils the recommendation was for leaving spruce seed trees with one or two hardwoods by each to serve as protection against windthrow. For spruce flats on deep soils, the selection method based on a diameter limit guide of 10–14 inches for spruce and less for balsam fir, and a cutting cycle of 10–30 years was recommended.

A detailed analysis of the relationships of reproduction to parent stands, and of the changing structure of the reproduction with time for periods up to 40 years after logging, was presented by Westveld (1931) for several forest types in the northeastern states. In the spruce-flat type, logging made no great difference in the relative proportions of conifers and hardwoods that were reproduced; however, balsam fir increased at the expense of spruce. In the spruce-hardwood type, both hardwoods and balsam fir increased at the expense of spruce. In the spruce-slope type, hardwoods increased strongly, overtopping the abundant spruce and fir reproduction. As the young stand develops the coniferous component gains at the expense of the hardwoods, with balsam fir becoming more prominent than in the parent stand.

Data on composition of parent stands and reproduction by forest types in northwestern Maine have been presented by Oosting and Reed (1944). Balsam fir basal area attained 54 per cent of the total in the spruce-flat type, 14 per cent in

TABLE 48. PERCENTAGE MILACRE QUADRAT STOCKING OF ADVANCE AND
SUBSEQUENT REPRODUCTION FOLLOWING LOGGING IN THE
ACADIAN REGION OF CANADA

Section of Province	Number of Quadrats	Red Spruce and White Spruce	Balsam Fir	Any Conifer	Hardwoods Tolerant	Intolerant	Any Species
ORIGINAL SOFTWOOD STANDS
A-7, Nova Scotia	1,360	29	55	70	6	19	79
A-5, Nova Scotia	9,740	35	40	68	41	7	85
A-4, Nova Scotia	2,920	19	54	61	25	14	74
A-3, New Brunswick .	1,740	17	11	37	16	41	70
A-2, New Brunswick ..	9,200	58	35	80	4	7	81
A-4, New Brunswick ..	3,200	23	75	80	17	18	89
ORIGINAL MIXED-WOOD STANDS
A-7, Nova Scotia	1,720	17	68	74	11	18	81
A-5, Nova Scotia	15,060	27	29	59	50	11	84
A-4, Nova Scotia	1,820	11	43	48	56	17	86
A-3, New Brunswick ..	7,580	7	18	26	37	61	83
A-2, New Brunswick ..	4,120	35	46	70	19	30	86
A-4, New Brunswick ..	1,820	16	61	66	33	16	85
ORIGINAL HARDWOOD STANDS
A-7, Nova Scotia	180	16	62	69	28	8	80
A-5, Nova Scotia	400	18	6	35	58	32	89
A-4, Nova Scotia	1,820	1	13	13	97	0	98
A-3, New Brunswick ..	1,160	2	6	8	36	78	90
A-2, New Brunswick ..	100	20	54	62	3	64	78
A-4, New Brunswick ..	240	3	24	27	84	4	94

Source: after Candy, 1951.

the spruce-hardwood type, 10 per cent in the old field spruce type, and 4 per cent in the spruce-slope type. Basal area of red spruce was 16, 53, 89, and 90 per cent of total basal area respectively for these types. Density of balsam fir advance reproduction was greatest, totaling 50 seedlings per 16 m² plot in the old field spruce type, followed by the spruce-hardwood type with 47 seedlings, the spruce-flat type with 22 seedlings, and the spruce-slope type with only 2 seedlings. Density of red spruce seedlings was considerably lower, totaling only 6 seedlings per plot in the spruce-hardwoods and old field spruce types, 4 seedlings in the spruce-slope type, and only 0.3 seedlings per plot in the spruce-flat type. These results illustrate that the optima for reproduction numbers and volume production for a species do not necessarily coincide.

Recknagel et al. (1933) compared advance reproduction in the Viburnum-Oxalis and Oxalis-Cornus forest types in the Adirondacks. In the latter type there were 6400 balsam fir and 1700 red spruce seedlings per acre as compared with only 2000 balsam fir and 800 red spruce in the former type. Softwood advance reproduction was more numerous in the Oxalis-Cornus type. These results are similar to the data presented by Ray (1956) in Quebec (see Table 47). In this type undesirable hardwoods often occurred in clumps; in the Viburnum-Oxalis type, hardwoods were more scattered. Thus interference by hardwood reproduction was more serious to the development of spruce and fir in the latter type.

Conclusions

1. Balsam fir has great seeding ability, and its seedlings have the advantages of early establishment. Although in specific cases it has displayed pioneer species' characteristics, it reproduces generally as advance growth. Since it has great shade tolerance, advance reproduction is initiated relatively early. Abundant in the humid eastern areas, to the west the species becomes relatively scarce and more dependent on soil moisture conditions.

2. The amount of early reproduction fluctuates depending on weather conditions, and the maximum age reached is determined by the density and composition of parent stands for different climatic and edaphic conditions. The optimum environment for balsam fir reproduction and for productivity of the parent stands do not necessarily agree.

3. Attempts have been made to regulate the composition and amount of reproduction in order to favor spruce and pines rather than balsam fir, and to avoid future stagnation from overcrowding of balsam fir when it is important in management. For this purpose discriminatory marking or the establishment of varying diameter limits for the different species has been used, with inconclusive results. It has also been suggested that a dense understory of balsam fir may prevent the development of shrubs and dense herbaceous vegetation. Subsequent removal may then expose a seedbed for reproduction of the desired spruce or pine overstory.

4. Competing shrub vegetation can delay balsam fir reproduction for 30–50 years. Logging followed by fire may prevent balsam fir from regaining its former sites for an entire rotation.

5. Optimistic reports without background information have to be treated with

caution because, under favorable natural conditions, even crude treatments have resulted in satisfactory reproduction. The differences between treatments are more distinct under adverse conditions.

6. Reproduction cuts include commercial and regulated clearcutting, diameter limit and shelterwood cuts, and various selective logging methods. Differences between the systems rest upon creation of different microsite conditions and on differing patterns of seed sources and competing vegetation. Most studies have been carried out by determining the end results of different treatments, with some consideration being given to macroclimatic, edaphic, and biotic conditions. Microsite studies are rare.

7. Germination and seedling establishment have frequently been studied in relation to such complexes as litter, mineral soil, and open or closed crown cover. Little has been done to evaluate the moisture, nutrient, heat, and light factors involved.

8. Research in spruce-fir silviculture in the past has been largely based on conditions found in mature and overmature virgin forests, where low yields combined with abundant advance reproduction of balsam fir occur over vast areas. Second-growth forests with greater density but little advance reproduction are rapidly replacing virgin forests. Because of budworm attacks and development of rots, the rotation age has to be kept low. The classical shelterwood and selection systems are not compatible with mechanization of logging. This means that research on subsequent reproduction should be developed rapidly.

9. Intensive studies toward development of regulated clearcutting methods depending on subsequent rather than advance reproduction for different conditions are necessary. Direction of cut, width, and pattern of cutting sequence in space and time, and specific promotion measures should all be studied. Opinions have been expressed that equipment should be developed to serve both logging and reproduction purposes. New approaches to slash disposal may become necessary. Besides mechanical means, chemical and biological and combined possibilities could be considered. Difficulties can be anticipated, e.g., the possibility of frost damage, which under present conditions have been only casually reported.

10. Periodic surveys based on regional forest classification systems are of great practical and theoretical importance. Reproduction status before and after treatments should be determined.

11. The patterns of reproduction dynamics under different management systems should be studied with reference to regional forest classification systems on both permanent and temporary plots. The dynamics of population should be related to investigations of seasonal and long-range environmental changes and to the patterns of spatial distribution of forest stands.

7

STAND DEVELOPMENT

STAND development begins with the establishment of reproduction. It involves the development of stand structure, vertically and horizontally, by age, species, size, and crown class. The process is closely aligned with succession, especially secondary succession, with the growth of individual trees in stands, and with the growth of stands as a whole. Growth of trees and stands in terms of wood productivity is commonly expressed in the form of volume and yield tables. These and related problems will be discussed in Chapter 8. Viewed from a broad time scale, stand development is not only related to one growth and cut cycle, but continues in successive generations of stands, some of which may be similar or quite different from those preceding, and may even include, temporarily, cover types other than forest vegetation. Most research on stand development is static, in that it relates only to a fixed moment in the life of the stand. Some studies span short periods of stand history, but very few attempts have been made to analyze in detail the whole developmental cycle. Spatial distribution of forest stands and distribution of trees and other plants within the stands are of particular significance in understanding the complexities of forest interrelationships.

Fundamental studies on relationships between organisms and environment very often cannot be directly translated into management terms. Consolidation of individual studies is necessary, and for this purpose classification systems, models, and sets of models are used. By such means the new knowledge gained from fundamental investigations can be applied to different categories of forest management units.

Studies of stand development have produced a number of useful models in the past. The rapidly developing methodology of ecosystem research, with emphasis on the flow of matter and energy in space and time, can be expected to yield valuable applications in the future. New technical aids have also become available. Among them, the angle-count technique by Bitterlich has made the determination of stand basal area as simple as height and age measurement. Though problems in application still remain, it introduces a new principle of sampling and in the future may greatly facilitate both research and practical application in stand development and related problems.

Virgin Conditions and Natural Disturbances

The virgin forest as an object of research has attracted many botanists and foresters. Early ideas that such forests consisted of uneven-aged, mixed stands in a state of equilibrium were modified over several decades as understanding of the roles played by such natural disturbances as fire, windthrow, and insect and disease outbreaks in virgin forests developed. In addition, differences in species reproduction, characteristics, and longevity contribute to the cyclic development that characterizes virgin forests (Graham, 1941; Bonner, 1941; Kabzems, 1952; Kagis, 1954; Plochmann, 1956; Rowe, 1961).

TABLE 49. DEVELOPMENT OF NUMBER OF TREES AND TOTAL VOLUME 1–INCH DBH AND OVER PER ACRE IN VIRGIN MIXED-WOOD STANDS IN THE GOULAIS RIVER WATERSHED, ONTARIO

Species	Number of Trees				Volume (Ft³)			
	1920	1927	1946	1956	1920	1927	1946	1956
Spruce	29	32	33	29	177	224	234	226
Balsam fir	102	159	140	108	246	170	150	147
White-cedar...........	77	105	120	119	652	806	729	664
White pine	3	7	6	8	208	268	277	326
Softwoods	211	303	299	264	1283	1468	1390	1363
Paper birch	16	31	21	17	147	128	111	98
Yellow birch	42	50	58	55	1009	983	1223	1346
Maple	82	160	341	348	594	564	640	749
Hardwoods	140	241	420	420	1750	1675	1974	2193
Minor species *		9	7	15		13	24	30
All species	351	553	726	699	3033	3156	3388	3586

* Includes balsam poplar, American elm, black ash, red oak, and hornbeam; not tallied in 1920.

Basis: 55 plots, 0.1 acre in size.

Sources: MacLean, 1949; Jarvis, 1960.

UNDISTURBED VIRGIN CONDITIONS

"Undisturbed" virgin condition is a relative concept. Developments in two virgin stands involving balsam fir with no drastic disturbance for several decades have been reported from Canada. The first example is a virgin forest on the Goulais watershed in Ontario (Section L-10 after Halliday, 1937, and Rowe, 1959; see Plate 2) in an area of relatively high precipitation. The forest was described by Mulloy (1931), Sisam (1939), MacLean (1949), and Jarvis (1960). A summary of forest conditions for the period 1920–1956 is presented in Table 49.

For the 36-year period spruce remained constant by volume at between 6 and 7 per cent. Balsam fir decreased from 8 to 4 per cent and other softwoods (white-cedar and white pine) fluctuated slightly but started and ended the period at 28 per cent. Hardwoods also fluctuated but increased slightly over-all from 54 to 62 per cent by volume. The absolute volume of yellow birch, maple, and white pine also increased whereas balsam fir and paper birch decreased. During the period 1920–1956, total volume increased about 18 per cent or 550 cubic feet per acre. During the last decade, annual increment was 20 cubic feet per acre in the virgin mixed-wood stand, all of it in the hardwoods.

Somewhat similar information for the virgin forest at Duck Mountain, Manitoba, was given by Pike (1955) and is summarized in Tables 50 and 51. This forest represents an area characterized by a relatively small amount of precipitation (Section B-18).

In 1921 the age of white spruce varied from 30 to 140 years; its maximum diameter was 21 inches and its maximum height 85 feet. Balsam fir was 30–65 years old, having a maximum diameter of 12 inches at breast height and a maximum height of 60 feet. The spruce was sound but limby; by contrast the balsam

TABLE 50. COMPARISON BETWEEN MEASUREMENTS IN AN UNDISTURBED WHITE
SPRUCE–BALSAM FIR–HARDWOOD STAND IN 1921 AND 1946 AT
DUCK MOUNTAIN, MANITOBA

Diameter Class (In.)	White Spruce		Balsam Fir		Hardwoods		Totals	
	1921	1946	1921	1946	1921	1946	1921	1946
NUMBER OF TREES PER ACRE								
1–3	23	1	227	18	3	0	253	19
4–9	145	50	85	68	22	4	252	122
10–21	89	85	6	0	0	1	95	86
1–21	257	136	318	86	25	5	600	227
BASAL AREA (Ft^2) PER ACRE								
1–21	126	107	25	11	5	3	156	121
VOLUME (Ft^3) PER ACRE								
1–21	3,626	3,321	464	202	106	62	4,196	3,585
VOLUME (fbm) PER ACRE								
10–21	13,283	15,750	537	0	0	0	13,820	15,750

Basis: 1 plot, 0.6 acre in size.
Source: Pike, 1955.

TABLE 51. INCREMENT AND MORTALITY FROM 1921 TO 1946 IN AN UNDISTURBED
WHITE SPRUCE–BALSAM FIR–HARDWOOD STAND AT
DUCK MOUNTAIN, MANITOBA

Volume Changes	Periodic Increment [*] (Ft^3/Acre)		Periodic Increment [†] (fbm/Acre)	
	White Spruce	Balsam Fir	White Spruce	Balsam Fir
Gross increment	1101	244	6534	176
Mortality	1406	506	4067	713
Net increment	−305	−262	2467	−537

[*] Trees 1.0 in. dbh and larger.
[†] Trees 10.0 in. dbh and larger.
Basis: 1 plot, 0.6 acre in size.
Source: Pike, 1955.

fir was suppressed, and affected by heart rot. In 1946 the stand was disintegrating, and only the volume of white spruce trees 10 inches dbh or larger increased during the intervening 25 years. Even the reproduction did not make progress (see Table 45).

Plate 16 illustrates some stages in the development of balsam fir in central Minnesota as related to edaphic coordinates. Some background information is provided in Plates 11, 12, and 13. The origin of these communities is diversified, but they have been without appreciable disturbance for about 50 years. The edaphic range of the reproduction older than ten years agrees closely with the range of balsam fir possessing measurable basal area. The maxima of largest stand basal area, greatest heights, and lowest mortality agree, but the largest number of seedlings below ten years of age fall outside the maximum growth conditions. The edaphic range of balsam fir is in the area of minimal shrub cover (Plate 12). It barely reaches the edges of three shrub complexes: hazel at the dry side, alder at the wet side and mountain maple at moist-rich conditions. Balsam fir basal area also shows some indication of a bimodal distribution, which resembles the pattern of black spruce further north.

Plate 16 also presents weighted site indexes for the balsam fir in central Minnesota. These site indexes, computed proportionally to basal area for all diameter classes combine the joint effects of edaphic and stand conditions. Without the effect of competition, the normal site indexes would have maxima at higher effective nutrient levels and more mesic moisture conditions with some variations depending on species. The optimum niche for balsam fir under competition lies in a narrow band at the medium range of the moisture spectrum. It does not reach the highest nutrient levels which are occupied by sugar maple, basswood, and associated hardwoods. The optimum for balsam fir lies between the peaks for aspen and balsam poplar. Some basal area combinations of balsam fir and aspen may become very productive on certain sites. White spruce seldom coexists with highly productive aspen in the investigated area. The contact zone between black ash, white-cedar, and balsam fir is the place of greatest mortality for the latter species.

NATURAL DISTURBANCES

The examples discussed in this section originate in an area generally not under intensive management, and the disturbances described are a reasonable expression of natural forces unaltered by man. The influence of fire on balsam fir has been partially considered in preceding chapters, but will be discussed here in somewhat different terms.

The development of the aspen-birch-spruce-fir cover type has been frequently investigated both in Canada and in the United States. This type originates mainly after fire or cutting and has many variations due to stand ecological conditions and history. In 1950 there were 7.6 million acres of aspen cover type in Minnesota of which 570,000 would be converted to spruce-fir during the current rotation. In Wisconsin the predicted conversions are 310,000 of 5.9 million acres, and in Michigan, 550,000 of 6 million acres now in aspen. The percentage conversion for the Lake States as a whole was about 7 per cent (Heinselman, 1954).

In the Ontario Clay Belt (Section B-4), the spruce-fir-hardwood cover type reproduces predominantly to black spruce and hardwoods after fire (Bonner, 1941). In the first 30–40 years few balsam fir invade, but they increase steadily in later years to form 3 per cent of stand basal area at 100 years. The large buildup of balsam fir begins after 130–150 years when hardwoods die out and the spruce stands begin to open. After a 5-inch diameter limit cutting it was predicted that balsam fir would dominate the next generation.

A rapid invasion of balsam fir after fire in 1920 was described by MacLean (1949) and Jarvis (1960) in the Goulais watershed area (Section L-10) of Ontario. Development of the stands after fire is shown in Table 52.

There was a net annual increment of 30.2 cubic feet per acre from 1927 to 1946, and 31.4 cubic feet per acre from 1946 to 1956. Spruce increment increased from 2.3 cubic feet during the first period to 5.7 cubic feet in the second, while corresponding figures for balsam fir were 2.1 and 4.1 cubic feet. Over a 29-year period beginning seven years after fire, the proportion of spruce by volume increased from 4 to 10 per cent while balsam fir increased from 3 to 8 per cent during the same period.

MacLean (1960) investigated aspen-birch-spruce-fir stands in Ontario (Sections B-4, B-8, B-9) which had not been disturbed for over 200 years. Spruce was the major softwood component in the stands, though balsam fir seedlings were more common than spruce. Fir seedlings usually first appear when stands are 50 years old, but most of them invade 20 years later when more rotted wood is available. Spruce will still be common in the new stand, but aspen and birch will exist only as scattered specimens. In unfavorable conditions stands are likely to become more and more open, and shrubs may become the dominant vegetation for a period of time. As described by Ghent (1958) from the Lake Nipigon area of Ontario (Section B-10) overmature aspen stands, 125–140 years old, provide small openings suited to seedlings of mountain maple. Such seedlings compete with aspen, spruce, and fir. After 30–50 years or longer, spruce and fir eventually penetrate the cover of aging mountain maple, resulting in an increase of conifers in the new stand.

Table 53 incorporates major growth characteristics of the aspen-birch-spruce-fir cover type originating after fire on the best sites of the Ontario boreal forest (MacLean, 1960).

On dry sites only 80 per cent, and on shallow till over bedrock about 50 per cent, of the yields in Table 53 can be expected. Balsam fir comprises about 4 per cent of the total merchantable volume at a stand age of 70 years, and about 15 per cent at 130 years. Spruce and hardwoods remain reasonably even-aged, but balsam fir is younger and more apt to be many-aged. Advance growth consisting mostly of balsam fir reaches sapling size at a main stand age approximating 100 years.

Stand dynamics following fire in the mixed-wood cover type in Saskatchewan was studied by Kabzems (1952). The role of balsam fir is minor, reaching a peak of its importance at an average stand age of 60–80 years. Later, white spruce and especially black spruce become the important species at the expense of the declining hardwoods and balsam fir. In different climatic regions and under varying

TABLE 52. DEVELOPMENT OF NUMBER OF TREES 1-INCH DBH AND OVER AND TOTAL VOLUME PER ACRE FOLLOWING A 1920 FIRE IN THE GOULAIS RIVER WATERSHED, ONTARIO

Species	Number of Trees			Volume (Ft³)		
	1927	1946	1956	1927	1946	1956
Spruce	9	37	67	6	50	107
Balsam fir	11	58	83	5	45	86
White-cedar	5	34	39	13	27	71
Softwoods	25	129	189	24	122	264
Paper birch	91	252	147	43	215	260
Yellow birch	34	56	37	24	34	50
Maple	59	269	224	27	140	193
Poplar (aspen)	126	180	98	22	220	274
Other hardwoods ..	+	2	6	18	1	5
Hardwoods	310	759	512	134	610	782
All species ...	335	888	701	158	732	1046

Basis: 75 plots, 0.1 acre in size in 1927 and 1946; 61 plots in 1956.
Source: Jarvis, 1960.

TABLE 53. YIELD TABLE FOR ASPEN-BIRCH-SPRUCE-FIR COVER TYPE OF FIRE ORIGIN ON FRESH TO VERY MOIST PONDED AND LACUSTRINE DEPOSITS AND TILLS IN ONTARIO (SECTIONS B-4, B-8, B-9)

	Age (Years)										
	30	40	50	60	70	80	90	100	110	120	130
TREES 1–3 INCHES dbh											
Basal area (ft²/acre)											
Softwood	20	18	16	11	4	2	2	2	2	3	4
Hardwood	17	12	6	2	1	0.7	0.4	0.3	0.2	0.1	0.1
Number of trees/acre											
Softwood	1150	870	620	370	150	100	90	100	130	170	230
Hardwood	800	440	160	60	20	20	10	10	5	2	1
TREES 4 INCHES dbh AND OVER											
Basal area (ft²/acre)											
Softwood	1	22	40	56	66	72	74	72	68	62	57
Hardwood	27	60	78	86	89	89	86	81	77	73	68
Number of trees/acre											
Softwood	20	190	290	330	350	340	310	270	230	190	160
Hardwood	220	290	320	320	290	260	220	170	130	100	80
Average dbh (in.)											
Softwood	4.3	4.7	5.1	5.5	5.9	6.3	6.7	7.2	7.5	7.7	7.9
Hardwood	5.1	5.9	6.5	7.1	7.5	8.0	8.8	9.7	10.6	11.4	11.9
Merchantable volume/acre (cords*)											
Spruce	0.5	2.2	5.5	9.0	12.0	13.7	14.5	13.6	12.6	12.2	12.1
Balsam fir	0	0.2	0.6	1.1	1.7	2.7	4.2	5.1	6.1	6.5	7.0
Other softwoods	0.1	0.2	0.3	0.4	0.8	0.9	0.5	0.5	0.5	0.5	0.5
Hardwoods	5.3	13.4	21.3	26.9	28.2	28.2	27.5	27.7	27.7	27.6	27.4
All species	5.9	16.0	27.7	37.4	42.7	45.5	46.7	46.9	46.9	46.8	47.0

* Computed as 85 ft³ per cord.
Basis: 172 yield plots, 0.2 acre in size.
Source: MacLean, 1960.

edaphic conditions, stand development has been shown to follow distinctive patterns.

The effect of budworm outbreaks on stand development has been frequently although not very systematically investigated. Widely divergent conclusions have been reached. Budworm outbreaks have been credited on the one hand with creating balsam fir thickets and, on the other hand, with the creation of two-storied stands depending on conditions (see Chapter 5). They have been found to lead to periods of growth reduction as well as to several decades of accelerated growth. For example, a period of accelerated growth following spruce budworm attack in the Green River watershed of New Brunswick (Section B-2) was de-scribed by Vincent (1955). Furthermore, these attacks may have been responsible for both the wholesale elimination of balsam fir and the formation of pure balsam fir stands. Stand history and the development of epidemics play a great role. As Ghent et al. (1957) pointed out, this problem should be investigated on the basis of regional forest conditions.

Development of the stands described by Vincent (1955) was also influenced by birch dieback. The development of reproduction in this area is shown in Table 44. Changes in softwood volume over a five-year period (1945–1950) are shown in Table 54. In these circumstances balsam fir in budworm-affected stands composed of mature softwoods over immature softwoods had a net annual increase of 66 cubic feet per acre reduced in softwood total to 60 cubic feet per acre by a small decrease in spruce volume. Increment greatly exceeded that of other stands. In-growth played a relatively unimportant role in this type, since only 18 per cent of the increment occurred in the 4–6 inch dbh classes as compared with 69 per cent in the same dbh class in the cutover type. Mortality losses and losses by wind-fall were excessive.

A somewhat similar situation was described by Candy (1942) in the Upper Lièvre Valley of Quebec (Section B-7). Changes in number of trees, volume, and annual increment by species over a 10-year period are presented in Table 55.

The relatively large increment was explained as an aftereffect of disturbance. Net annual increment constituted approximately 50 per cent of the gross increment. Merchantable net increment was 36.0 cubic feet per acre including 11.0 cubic feet of spruce and 20.8 cubic feet of balsam fir.

In disturbed stands in the Upper Lièvre Valley, balsam fir added 92 per cent to its initial volume in the 10-year period prior to budworm infestation. Following the attack which took place in the 1940's, balsam fir lost 51 per cent of its volume. For all species the net annual growth for the first period was 30 cubic feet per acre and the net annual loss for the second period was 65 cubic feet per acre. In the spruce-fir type, the spruce was 160 years old and the balsam fir 40–60 years old. As a consequence of these attacks it was predicted that the newly developed second story would shortly reach another period of rapid growth (McCormack, 1953). These examples demonstrate variations in growth caused primarily by bud-worm attacks, although other interferences and age of stands also affected growth.

Additional details concerning changes in stand structure and the growth of individual trees have been presented by McLintock (1951, 1955). Mortality of the overtopped trees was higher during the budworm attack. After the attack, growth

TABLE 54. DEVELOPMENT OF BALSAM FIR AND SPRUCE VOLUME IN CUBIC FEET PER ACRE FROM 1945 TO 1950 IN THE GREEN RIVER WATERSHED, NEW BRUNSWICK

| | Merchantable Volume * | | | | Net Annual Increment of Softwoods | Percentage of Gross Increment Lost | |
| | Balsam Fir | | Spruce | | | | |
Cover Type	1945	1950	1945	1950		Mortality	Windfall
Mature softwood	1578	1673	355	424	33	42	16
Mature softwood over immature softwood ...	924	1253	263	233	60	15	18
Mature mixed-wood	1254	1332	165	137	10	35	43
Mature mixed-wood over immature softwood ...	1114	1240	9	13	26	9	44
Cutover with immature softwood	176	299	38	44	26	14	24
Cutover with no immature softwood	148	209	20	26	13	15	35

*Includes trees 3.6 in. dbh and over.
Basis: 641 plots, 0.1 acre in size.
Source: Vincent, 1955.

TABLE 55. NUMBER OF TREES, VOLUME, AND ANNUAL INCREMENT IN CUBIC FEET PER ACRE FROM 1930 TO 1940 BY SPECIES IN A COMPOSITE REPRESENTATIVE STAND IN THE UPPER LIÈVRE VALLEY, QUEBEC*

| | Number of Trees | | Volume (Ft³) | | Net Annual Increment (Ft³) |
Species	1930	1940	1930	1940	(1930–1940 Average)
Spruce	225	252	526	648	12.2
Balsam fir	661	872	441	750	30.9
Jack pine	48	22	331	202	−12.9
Other softwoods	5	4	46	51	0.5
Aspen	3	3	34	52	1.8
Paper birch	68	63	410	512	10.2
Yellow birch	13	17	140	187	4.7
Maple	5	4	8	7	−0.1
Other hardwoods	0	1	1	2	0.1
All species	1028	1238	1937	2411	47.4

* Includes trees 1 in. dbh and over.
Basis: 292 plots, 0.1 acre in size.
Source: Candy, 1942.

reduction up to 75 per cent of normal was evident for six to seven years in the surviving trees. Following defoliation, there was a three-year lag before diameter growth was affected. The same lag was also observed after crown recovery was complete.

Other reports on the effects of budworm upon changes in stand structure are discussed in Chapter 5. In general, budworm effects in spruce-fir stands following mild attacks resemble a thinning from above, with discrimination against balsam fir. In cases of heavy attacks the effect resembles a shelterwood cut or even clearcutting with only the dead trunks left standing. Balsam woolly aphid also introduces great changes in stand structure, about which the available information is mostly descriptive.

Insects may also attack tree species competing with balsam fir as shown by Duncan and Hodson (1958). In northern Minnesota they found that the forest tent caterpillar (*Malacosoma disstria* Hbn.) reduced the four-year growth of aspen by 2–6 cords per acre and the radial growth of aspen by 67 per cent. At the same time the radial growth of balsam fir increased by about 20 per cent, thereby improving the competitive advantage of balsam fir with respect to aspen. Neither aspen nor shrub mortality followed defoliation by this insect, and the aspen overstory tended to reclose, thus eliminating the temporary effect of understory balsam fir release.

The effects of diseases, particularly butt and heart rots, upon balsam fir are discussed in Chapter 4. Most studies have been made on the basis of tree classification rather than stand classification, with very little attention paid to site conditions. The effect of wildlife on stand development has been briefly discussed in Chapter 3. No detailed data were available to permit more thorough analysis.

Cleaning

Cleaning, also called weeding, includes those measures intended to favor established reproduction of seedlings or saplings over undesirable woody species of approximately the same age and size. Several shrub species and both tolerant and intolerant hardwood tree species are the principal competitors of balsam fir. Whether or not cleaning is done at all depends upon economic considerations and silvicultural objectives.

COMPETITION IN SEEDLING AND SAPLING STAGES

Regional ecological conditions and important competing species have been described in Chapter 2. Information here will emphasize early stages in the development of balsam fir.

The humid eastern part of the Boreal Region is favorable to the reproduction of balsam fir. Tolerant hardwoods do not penetrate this region. Quaking aspen, paper birch, balsam poplar, and sometimes black ash are present, spreading widely after disturbances. Balsam fir competes with white spruce on mesic-rich sites, with black spruce on wet sites, and with jack pine on dry sites. The main shrub competitors are *Corylus cornuta, Prunus pensylvanica, P. virginiana, Alnus rugosa, A. crispa, Viburnum cassinoides, V. edule, Cornus stolonifera,* and *Ribes* spp. In the southern part of the region mountain maple (*Acer spicatum*) and other intruders from the Great Lakes–St. Lawrence and the Acadian-Appalachian regions are important.

In the Great Lakes–St. Lawrence Region the tolerant hardwoods replace balsam fir on rich-mesic soils. On lowland soils balsam fir competes with black spruce and white-cedar. On moderately rich and moderately dry uplands it occurs as an understory beneath red pine and white pine. After disturbances it reproduces under intolerant hardwoods, competing with shrub species in the reproduction stage. Balsam fir also invades jack pine cover types on former red pine and white pine sites. In the eastern part of the Great Lakes–St. Lawrence Region, balsam fir competes with the more tolerant hemlock. Shrub competition is most severe in the prairie border areas.

In the Lake Edward Forest in Quebec (Section L-4) the Cornus type contains few shrubs, mainly *Acer spicatum, Nemopanthus mucronata,* and *Amelanchier bartramiana;* of the Oxalis-Cornus type the most prominent shrubs are *Viburnum lantanoides* (*V. alnifolium*), *Acer spicatum,* and *Corylus cornuta;* the Viburnum type has a dense development of *Viburnum lantanoides, Acer spicatum,* and *Lonicera canadensis. Sorbus americana* is present in all four types. Of the hardwood species, red maple, striped maple, and yellow birch are potential competitors of balsam fir in all types, paper birch may compete in the Cornus and Oxalis-Cornus types, and sugar maple is likely in the Viburnum-Oxalis and Viburnum types (Heimburger, 1941).

In the Lake States the following shrub species can be considered as constant and exclusive indicators of spruce-fir types: *Ribes glandulosum, R. triste,* and *Rhamnus alnifolia,* while *Lonicera hirsuta* and *Cornus stolonifera* can be considered as constant (Roe, 1935).

Forest types with the greatest abundance of balsam fir such as Clintonia-Lycopodium (on loam podzols) and Vaccinium-Cornus (on sandy podzols) have the least development of shrubs in Wisconsin (Wilde et al., 1949). The driest types having balsam fir are invaded by hazel and pin cherry. Species of *Ribes* penetrate the richer, more moist types. The situation closely resembles that in Minnesota discussed previously (Plates 5 and 12).

In the Acadian-Appalachian Region with its more humid character, balsam fir competes more effectively with tolerant hardwoods. Red maple is nearly universal. Sugar maple widely overlaps the site range of balsam fir. By reason of their tolerance, balsam fir, hemlock, and red spruce may be used to advantage in silvicultural control of shrubs. Logging and fires, however, render difficult the competitive position of conifers in relation to shrubs and to both tolerant and intolerant hardwoods. Heimburger (1934) presented the shrub problem in the New York Adirondacks briefly as follows: In the subalpine forest series balsam fir is present in all three types. No shrubs were noted in H-Co T (see Table 7 for type designations) but in O-HT and Asp-OT both mountain ash and paper birch are present. In the western series balsam fir is associated with mountain ash, red maple, striped maple, and *Amelanchier canadensis.* In forest types with *Viburnum alnifolium* (Vi-OT and Vi T) balsam fir is limited. In the eastern series balsam fir is associated mainly with *Diervilla lonicera* and *Viburnum alnifolium.*

To evaluate competition adequately in the seedling and sapling stages, comprehensive information on the height growth of balsam fir and competing species under different climatic, edaphic, and biotic conditions is necessary. Only a few approximate figures on height growth of balsam fir have been presented by different authors.

Zon (1914) investigated balsam fir height growth under average shade conditions and full light in Franklin County, New York (Table 56). After trees were five years old, their growth in the shade was less than half the growth of trees in full light.

Westveld (1931) noted that in the northeastern United States, balsam fir seedlings in the open may reach breast height in 15 years, while under a dense canopy 40 years may be required. Observations in the northern Clay Belt of Ontario revealed that open-grown balsam fir seedlings reached a height of about eight inches in the first five years (Bonner, 1941).

Similar results have also been reported from the Boreal Forest Region (Section B-1) in Quebec (Hatcher, 1960). For balsam fir, about 5 years were required to reach a 1-foot height and 7.5 years to reach the 3-foot level after cutting. Little difference was found between the first and second site class after cutting at this early age; however, before cutting 10 years were required to reach the 1-foot level in site class I, whereas 12.5 years were needed to attain the same height in site class II. Corresponding figures for a 3-foot height were 14 and 18 years.

TABLE 56. AVERAGE HEIGHT OF BALSAM FIR UNDER MEDIUM SHADE AND IN FULL LIGHT IN FRANKLIN COUNTY, NEW YORK

Age (Years)	Height (Ft)		Age (Years)	Height (Ft)	
	Shade	Light		Shade	Light
1	0.1	0.1	10	1.6	3.6
2	0.2	0.2	11	1.9	4.3
3	0.3	0.4	12	2.1	5.1
4	0.5	0.7	13	2.4	5.8
5	0.6	1.0	14	2.6	6.6
6	0.7	1.5	15	2.8	7.4
7	0.9	1.9	16	3.1	8.2
8	1.1	2.5	17	3.3	8.9
9	1.3	3.0	18	3.5	9.7

Basis: 324 trees in shade and 104 trees in full light.
Source: Zon, 1914.

MECHANICAL CONTROL OF COMPETING VEGETATION

Mechanical control of competing vegetation is the oldest control method and is still widely used. More recent recommendations are, however, largely concerned with chemical controls. With increasing concern over improper use of widespread chemical application a return to mechanical methods of control may again become popular in the future.

Westveld (1930b, 1931, 1953), working in the northeastern United States, has suggested that in spruce-fir types having intolerant hardwoods, early cleaning is not required. Under other conditions one treatment is sufficient. In spruce–fir–tolerant hardwood types one cleaning eight years after logging may be sufficient. In hardwood-spruce-fir types, the conditions are more difficult, and two to three cleanings at intervals of four or five years may be necessary. Relative height of both crop species and weed species should be considered. If advance reproduction on sites which reproduce easily averages one foot in height, the reproduction of spruce and fir will often gain dominance without weeding. On sites more difficult to reproduce, heights of desirable reproduction should average about twice that of undesirable plants. If conifers are adding six inches or more of height growth annually, there is no particular need for cleaning.

The most effective cleaning usually can be accomplished 5–8 years after girdling and cutting especially in the spruce-fir-hardwood type (Westveld, 1930b). Delay in cleaning increases costs and produces poorer results. The next choice would be cutover areas 8–15 years old in which hand tools, especially axes, can be employed. Cleaning under many conditions becomes prohibitive about 15 years after cutting. In order to favor 1800 spruce trees per acre, as many as 5000 hardwood stems may have to be removed. If spruce is present, balsam fir may not need to be given particular care, since economically, spruce is usually the more important species.

Balsam fir and red spruce seedlings 2–5 feet tall responded best to release by cleaning (Westveld, 1931). Species differences were noted, however, in that balsam fir was somewhat more responsive than red spruce, although the latter was more persistent under shade.

Under Michigan conditions, Westveld (1933) suggested early cleanings beginning at 5–10 years of age with one repetition 2–3 years later.

In the Viburnum type at the Lake Edward Forest in Quebec (Section L-4) the overtopping hardwoods were cut either at ground level or at three feet above the ground (Mulloy, 1941a). The number of conifers increased two to three times their number in uncut areas. Spruce benefitted more than balsam fir. No difference was noted between the plots where hardwoods were cut at different levels, but cutting at the three-foot level was cheaper.

Of all shrub species, mountain maple (*Acer spicatum*) is the most serious competitor of balsam fir. This is true not only because of similar site requirements but also because of the reproductive ability, and the somewhat greater shade tolerance of mountain maple as compared with balsam fir. Vincent (1953, 1954) investigated the effect of mountain maple on the height growth of spruce and balsam fir seedlings in a mixed spruce-fir-birch stand on rich loam soils in the Green River area of New Brunswick (Section B-2). Mountain maple developed nearly 100 per cent crown cover over the area 10 years after the removal of softwoods by logging. Such cover may remain closed for about 25 years, thus delaying softwood regeneration for 30 to 50 years. Mechanical or chemical treatment of mountain maple is difficult. Spruce and balsam fir were released by clearing a three-foot circle around each seedling on an area logged eight years earlier. Balsam fir shorter than six feet doubled its height growth as compared with its previous four-year average growth. Height growth of controls was only one-third of the released trees. Axe treatment required 44 man hours per acre to effect release.

Further investigation of the cleaning experiments begun in 1949 in the Green River area was reported by Baskerville (1959, 1961a). By 1953, shrubs had again reclosed above the softwoods, and height growth had tapered off although at a consistently higher level than for the control softwoods. In 1958 growth on treated plots was 4.5 times as great as the controls. The response of spruce after treatment was less than that of balsam fir, but the former continued to increase even after shrubs closed in so that by 1958 the growth rate of the two species was nearly equal. The author considered release work under these conditions as too expensive for economic justification, but suggested that satisfactory results probably could be achieved by cutting back the tops of competing shrubs.

CHEMICAL CONTROL OF COMPETING VEGETATION

Major developments in the synthesis and application of phytocides since World War II have made possible an entirely new array of silvicultural techniques. The discovery of 2,4-dichlorophenoxyacetic acid (2,4-D) as having selective phytocidal properties and its later synthesis in salt, amine, and ester formulations paved the way for its widespread use as a weed-killer which could be used safely for many agricultural crops. Early experience with 2,4-D applied to woody plants revealed that the conifers in general were much more tolerant than the hardwood species to its effects. This fortuitous selectivity in favor of the economically more valuable conifers stimulated extensive research on the ways by which such phytocides could be used in the manipulation of forest vegetation.

Unfortunately, little of this research has involved balsam fir directly. A few studies report on the relative phytocidal tolerance of the species. Stoeckeler (1948b) examined telephone line rights-of-way in Wisconsin which had been sprayed with 2,4,5-T, ammate (ammonium sulfamate), and 2,4-D, and reported balsam fir as being in the "hard to kill" class for all three materials used. In Minnesota, Leinfelder and Hansen (1954) surveyed numerous roadside areas which had been sprayed with 2,4-D and 2,4,5-T, using a great range of concentrations. Of the native conifers, balsam fir, black spruce, and white pine were found to be distinctly more resistant to injury. Balsam fir was uninjured by concentrations up to four pounds of acid equivalent per 100 gallons of diluent (4 lb ahg). A drench of the foliage using 12 pounds ahg induced curling and defoliation of new growth of balsam fir and killing of the leaders of other conifers. Since concentration of 2 pounds ahg is commonly adequate to kill most hardwood trees and shrubs, this tolerance of balsam fir provides a wide safety margin for any broadcast spraying designed to release the balsam fir from hardwood competition. Jankowski (1955) also rated balsam fir in the "high resistance" class in terms of aerial applications of 2,4-D and 2,4,5-T.

None of these reports refers specifically to the phytocidal tolerance of seedlings. Studies of weed control in forest nurseries have established (Stoeckeler, 1949) that balsam fir seedlings are intermediate between jack and red pines (more resistant) and white and Colorado spruces (less resistant) to Stoddard Solvent. Rates of 35–50 gallons per acre are recommended for weed control in first-year seedbeds up to July 1 and 50–75 gallons per acre after that and on older seedlings. In England (Woodford and Evans, ed. 1963) *Abies nobilis* and *A. grandis* were also rated in the most resistant class, but a rate of 25 gallons per acre was felt safe. On second-year seedbeds of these firs 2 pounds of simazine in 60–100 gallons of water per acre was recommended. Blanchard et al. (1957) in Minnesota rated balsam fir as more tolerant than most other conifers to amino-triazole applied 45 days after emergence in seedbeds. There was less than 5 per cent injury to this species from rates as high as 3 pounds per acre.

A considerable amount of phytocidal research has centered on or relates in part to some of the major competitors of balsam fir. While undoubtedly the species is more tolerant than most of its associates, nevertheless release from shrub competition is one of the serious silvicultural problems on large acreages of second-growth seedling stands. Three of the most important shrub species involved are mountain maple (*Acer spicatum*), beaked hazel (*Corylus cornuta*), and speckled alder (*Alnus rugosa*).

Speckled alder, a wet-site associate of balsam fir, has been evaluated for ease of control by phytocides. Melander chmn. (1948, 1949), in a pioneer effort to summarize what was known about woody plant reactions to phytocides at that early date, listed it as hypersensitive to sensitive in most tests to either 2,4-D or 2,4,5-T. Early summer foliage sprays gave somewhat better control than did later treatments. Zehngraff (1948) treated alder with 2,4-D (sodium salt) and ammate. He found it to be somewhat more resistant than hazel to the 2,4-D. Midsummer foliage spray applications gave less resprouting than earlier or later trials. Stoeckeler (1948a) rated alder "easy to kill" by either foliage sprays of 2,4,5-T at 5000 ppm or ammate applied to freshly cut stumps. Of eight different kinds of chemical treatments tested he rated 2,4,5-T foliage spray and ammate foliage spray as cheapest and most effective; cutting the alder and treating the stumps with ammate gave a 100 per cent kill of all stems with very little resprouting. A number of studies (Day, 1950; Stoeckeler and Heinselman, 1950; Day and Dils, 1951; and MacConnell and Bond, 1961) found 2,4,5-T to be effective in killing alder and somewhat better than 2,4-D for this purpose.

Rudolf and Watt (1956), summarizing research on chemical control of woody plants, list good control of mountain maple by basal sprays of 2 or 4 per cent solutions of 2,4,5-T in oil. Combined esters of 2,4-D and 2,4,5-T were similarly effective when applied to cut stumps, a measure involving considerable labor and probably impractical because of its cost.

Various applications of foliage sprays were tested at Basswood Lake, Minnesota (Hansen and Ahlgren, 1950b), but none gave a satisfactory kill of mountain maple. The best treatment (36 per cent kill) came with the use of 2,4,5-T at 3000 ppm, applied early in the season.

In a small-scale experiment in northern Minnesota, mountain maple clumps were killed and resprouting was kept at a low level by basal spraying in late fall with a 2–4 per cent solution of 2,4,5-T in fuel oil (Roe, 1953). Fuel oil alone or a 1 per cent solution of the herbicide also gave good top kill when applied in the fall, but the effect was partially offset by abundant vigorous sprouts the following spring. Chemical treatment or cutting of mountain maple for the purpose of encouraging coniferous reproduction must be done very carefully because modified techniques are used for the opposite purpose, i.e., to rejuvenate old mountain maple clumps for deer (Krefting, 1953; Krefting et al., 1955, 1956).

In the Upper Peninsula of Michigan, Day (1954b) compared 2,4,5-T; mixtures of 2,4-D and 2,4,5-T; and silvex (2-(2,4,5-trichlorophenoxy) propionic acid). These materials were used as foliage sprays on mountain, red, and sugar maples in mid-July. Silvex was reported most effective, but it left some green tissue.

In an experiment in the Green River area (Baskerville et al., 1960) a fog generator on the ground was used to drift atomized 2,4,5-T down-slope through the mountain maple crowns during the temperature inversion period in the evening. Mountain maple was killed satisfactorily. In a second study five 20-acre blocks were sprayed from the air, four with 2,4,5-T, and one with silvex at a rate of one pound of acid equivalent per acre. This resulted in a 45 per cent decrease in average height of the shrub stand and an 80 per cent increase in number of stems. Sutton (1958) in summarizing experiments in the chemical control of mountain maple pointed out the importance of investigating many factors which may influence the successful treatment of herbicide-tolerant species. The problem of regrowth is less difficult with mountain maple than with aspen. Cut-stump treatments are, however, capable of killing sapling maples (Rudolf and Watt, 1956).

Because of its tremendous regenerative potential, especially by means of sprouts from underground stems, the beaked hazel presents a most serious competitive problem to balsam fir within its area of overlapping ranges. Jensen (1948) conducted an early test of 2,4-D as an aid to preplanting site preparation. A light concentration (1000 ppm) was sprayed on the area to be planted in 1946 and again in 1947. By September of 1948, 95 per cent of the old stems were dead and regrowth was only 6–8 inches tall. The area was then clear for scalping and planting. Zehngraff and Bargen (1949) compared 2,4-D and ammate for hazel brush control in Minnesota. They reported 2,4-D to be superior to ammate in preventing stem regrowth. Best control was obtained from applications during July 1 to August 15.

From tests at the Quetico-Superior Wilderness Research Center in northern Minnesota, Hansen and Ahlgren (1950a) concluded that 2,4-D and 2,4,5-T were of about equal effectiveness on beaked hazel, and that June and July sprays gave 90–100 per cent control while in August the treatments were only 31–61 per cent effective. Studies by Roe (1950) confirmed the similar effectiveness of 2,4-D and 2,4,5-T and further demonstrated that 2400 ppm was as effective as 4800 ppm concentration. All treatments gave 90–100 per cent kill. Later reports (Roe, 1951, 1952b) indicated that resprouting was least from the midsummer treatments. Roe rated beaked hazel as "difficult to eradicate."

In an experiment at Itasca Park in Minnesota, Hansen (1955) found that while the general cover of hazel was checked by spraying with 2,4-D, *Rubus strigosus* and *Pteridium aquilinum* increased in abundance. According to Westveld (1931) and Bowman (1944) these species are not very detrimental to the establishment of balsam fir. Nonetheless, in some situations vegetation developing after chemical spraying could be more serious than that eliminated.

In Manitoba, Playfair (1956a,b) tested 2,3,6-T applied as both foliage and dormant sprays. While aspen, ash, willow and many other species died, hazel showed very heavy regrowth.

Klug and Hansen (1960) reported the results of comparing the resistance of beaked and American hazels to different concentrations of 2,4-D applied as dormant, basal sprays. Beaked hazel was more easily killed than American hazel. Concentrations of 5 pounds ahg gave excellent control (90 per cent), making unnecessary the higher concentrations usually recommended.

Thinning

Thinning refers to the reduction of competition in stands in late sapling and pole stages before they have reached maturity. The primary purpose of thinning is to increase harvest return. Since sawtimber production is seldom the goal of balsam fir management, thinning techniques are normally intended to improve the production of pulpwood.

Many of the earlier thinning studies had shortcomings because of failure to recognize fundamental forest site and stand characteristics. Other weaknesses stemmed from failure to recognize long-term dynamic processes and to distinguish between biological considerations and purely economic aspects of the problem.

Mechanical treatment predominates in thinning operations but there are specific situations where chemical treatments have been used.

Equal spacing between individual trees has usually been considered the ideal. Group patterns, however, are gaining increasing attention in more recent studies. Such patterns emerge from conditions favoring reproduction, from species competition, and from micro-environmental effects. Grouping patterns could be adapted to management goals for protection, natural pruning, growth, and reproduction purposes. At present very little is known about their significance, but methods of study are under development.

Background information for the regulation of stand structure has been discussed especially in Chapter 2. Information on growth and yield, treated in Chapter 8, has been obtained primarily for inventory purposes and has limited use in silviculture.

STAND STRUCTURE

Balsam fir in its climatic optimum may reach 90 feet in height. Under other conditions it may attain only 40 feet, while associated species, especially pines, become much taller. Climatic, edaphic, and biotic factors, and stand history are decisive in determining the position occupied by the species in a stand.

In different parts of the world workers have noted that after successful establishment of reproduction, especially in the Boreal Forest and on poor soils, differentiation among individuals is slow. This results in growth stagnation of entire stands until heavy mortality sets in. Similar phenomena have often been observed in pure stands on old fields which have been naturally reseeded or artificially reproduced, and on other sites where there is great uniformity in the environment and little genetic variability.

Kittredge and Gevorkiantz (1929) investigated the height growth of several species associated with aspen in Minnesota and Wisconsin on 50–60 foot site indexes for aspen. Height growth of balsam fir and other species was measured under categories labeled "crowded" and "not crowded," depending on the degree of aspen suppression. Results are shown in Table 57.

Aspen held a considerable lead in height growth up to 40 years of age, but by 60 years balsam fir in not crowded conditions had nearly attained the height of aspen and had surpassed all other species. The difference in height of balsam fir under the two degrees of suppression was 18 feet. However, the age of aspen and that of balsam fir and other species are not necessarily coincident, thus producing considerable variation in stand structure. On the average, balsam fir attained the height of 4½ feet in not crowded conditions in 16 years, while under crowded conditions 24 years were necessary.

In northeastern United States most spruce-fir pulpwood originates from stands containing 25–75 per cent hardwoods (Westveld, 1937). Great differences exist between spruce and balsam fir growing under intolerant hardwoods (aspen, birch, pin cherry, etc.) as compared with such tolerant hardwoods as sugar maple and beech.

Bowman (1944) believed that spruce and fir in northern Michigan could hardly be considered overtopped unless the overstory density exceeded 50 per cent. He considered it sufficient to recognize only two degrees of overtopping, i.e., 50–75 per cent and 75–100 per cent, in overstory density. His yield tables (see Chapter 8, Tables 103 and 104) indicate that a 50–75 per cent hardwood overstory had beneficial effects on spruce-fir growth when compared with the growth of spruce and fir in pure uneven-aged coniferous stands. On the other hand, a 75–100 per cent hardwood overstory decreased the growth of spruce-fir understories relative to that of pure uneven-aged spruce-fir stands.

TABLE 57. HEIGHT IN FEET OF BALSAM FIR AND OTHER SPECIES IN STANDS DOMINATED BY ASPEN IN WISCONSIN AND MINNESOTA *

Age (Years)	Aspen Site Index		Balsam Fir		White Pine		White Spruce		Red Pine	
	50	60	NC*	C*	NC*	C*	NC*	C*	NC*	C*
10 ...	14	15	4	1	4	2	3	2	5	3
20 ...	26	29	11	4	13	6	11	5	17	9
30 ...	36	40	25	10	28	15	25	12	31	20
40 ...	44	49	43	22	39	25	37	22	41	28
50 ...	52	58	54	33	46	32	46	30	49	34
60 ...	58	65	60	42	51	36	52	35	54	38

*NC = not crowded, refers to trees in understocked aspen stands or in small openings. C = crowded, refers to trees in comparatively dense stands or in groups of larger aspen.
Basis: 158 balsam fir, 103 white pine, 33 white spruce, 37 red pine.
Source: Kittredge and Gevorkiantz, 1929.

From the information presented by several authors and from general silvicultural knowledge it can be concluded that a dense balsam fir understory may successfully prevent dense shrub and herbaceous development. Its subsequent removal may provide an adequate seedbed for the pine or spruce overstory. The understory may also serve to facilitate natural pruning in the overstory, thus increasing the sawlog qualities of conifers and improving the veneer logs of hardwoods. In addition it will yield income from pulpwood.

THINNING IN ONE-STORIED STANDS

Balsam fir thinning information will be discussed in two arbitrary groups, i.e., one-storied and two-storied stands, but without making a sharp distinction between them. The material treated here includes not only studies in clearly defined single-storied stands but also experiments in stands where a scattered overstory was present which exercised little effect or which was ignored by the authors.

Some general recommendations on thinning in even-aged, second-growth spruce-fir were made by Westveld (1953). He advised removing 20–25 per cent of the basal area every 10 years. Basal area should seldom be reduced below 100 square feet per acre, a safe level being 100–150 square feet per acre. To reduce hazard from windthrow, no more than 35 per cent of the basal area should be removed at any one time.

In the Green River area of New Brunswick stagnation of balsam fir begins at an age of 50–55 years (Holt, 1949), although Thomson (1949a) reporting from the same area indicated that it begins considerably earlier. Extremely high density of balsam fir reproduction following sudden disturbances, such as insect infestation, blow-down, clearcutting, or fire, was said to cause such stagnation. Thomson measured sample plots in a 60-year-old stand on a well-drained northern slope in the Acadia Experimental Forest, where variations in density were related to increasing distance from seed source. There were 2500 trees per acre in the densest part but only 755 in the open part. Merchantable volume (trees 4.0 in. dbh or larger with 3.0 in. top diam.) reached 41 cords per acre in the open stand and 16.5 cords per acre in the dense stand.

To reduce the quantity of reproduction the diameter limit in partial cuts should be increased (Anonymous, 1948). Bulldozing and winching (Northeast Pulpwood Research Center, 1952) and chemical treatments (Hart, 1961) have been suggested as more economical than conventional methods of cutting.

More recent studies, also in the Green River area, have placed the stagnation problem in a somewhat different light. The outbreak of spruce budworms between 1910 and 1920 affected more than 60 per cent of the forest area of New Brunswick and resulted in the establishment of dense young stands of balsam fir, white spruce, and black spruce. In such stands comprising 47,000 acres, or about 20 per cent of the productive forest land in the area, there was usually an overstory of about 100 mature and overmature balsam fir, white spruce, black spruce, and paper birch trees per acre (Vincent, 1962). The overstory was considered to be sufficiently open to have no significant effect upon the growth of the understory except for a few nearby trees. Mean annual increment increased from 2 cubic feet per acre in 1922 to 42 cubic feet per acre in 1952 in stands having 3000 stems or less per acre. More heavily stocked stands yielded 60 cubic feet per acre in 1952. The periodic annual increment of total volume reached 132 cubic feet per acre in more open stands and 176 cubic feet per acre in denser stands at the end of the experiment. The soil in the investigated area was moderately deep, loamy, and stony till over permeable shales and sandstones, thus assuring good drainage conditions. No stagnation was observed and the investigation showed that even without thinning there would be a good crop of pulpwood in 30–40 years. An early removal of the mature and overmature overstory, which could yield about 10 cords per acre, was considered silviculturally and economically desirable.

Baskerville (1961b) studied the effect of crown release on the development of immature balsam fir in the same area. The stand was 40–50 years old, with a light overstory of black and white spruce that survived the budworm epidemic. Trees of 3, 4, and 5 inches in diameter had been released previously when at 1–3, 3–5, and 5–7 feet in height. Five years after release, basal area had increased 7 per cent in the first, 33 per cent in the second, and 47 per cent in the third release class as compared with untreated controls. Basal area increment was related to the initial diameter in inches at breast height (dbh), crown width in feet (CW), and a competition factor (CF).[*] The following equation was derived for predicting basal area increment in square feet per acre per year:

BA increment $= -0.0314 +0.0072$ dbh $+0.00283$ CW $+0.000077$ CF ± 0.001.

A report on a 19-year thinning experiment at the Petawawa Forest Experiment Station (Section L-4) was prepared by Daly (1950). The thinning study was established in a mixed stand of conifers and hardwoods 45 years of age in Cornus-Maianthemum type on moist sandy or loamy flats. Spruce and balsam fir made up 60–70 per cent, hardwoods 15–25 per cent, of the stand. Overtopping hardwoods were removed and conifers were thinned. After 19 years the 100 largest trees on the thinned plots showed 50 per cent greater diameter increment than comparable controls. No information regarding total or merchantable growth on a stand basis was reported.

Day (1945) reported a combined thinning experiment in a 30-year-old predominantly balsam fir stand at the Dunbar Experimental Forest in northern Michigan. In this operation 230 trees were removed, of which 210 were sold as Christmas trees, and the remainder, 3 cords, sold as pulpwood. The residual stand contained 1420 trees per acre 1.5 inches dbh and larger. In further reports on thinning from the experimental forest, Day (1954, 1961) related the results in a 30-year-old balsam fir stand thinned to spacings of 6 x 6, 8 x 8, and 10 x 10 feet to an unthinned control. Additional thinning was done in 1955 when the experiment was enlarged. The results are shown in Table 58. The original spacing, 8 x 8 feet, which was later enlarged to 9.2 x 9.2 feet, showed the greatest annual increment. Since this plot possessed the greatest initial volume in 1940, when the series of experiments was started, the results are not conclusive.

THINNING IN TWO-STORIED STANDS

Thinning in two-storied stands is more complicated than in one-storied stands. The thinning can be made in either of the two stories or in both. Furthermore, either the overstory or the understory may be completely killed by natural causes, girdled, or poisoned, leaving the other story intact. This is actually a specific type of silviculture, some phases of which could be considered either here or under partial cutting. Information on this complex situation is scarce. Mulloy (1941b), working in the Lake Edward Experimental Forest in Quebec (Section L-4), found no difference in height growth of balsam fir and red spruce growing under overstories of 21, 43, and 64 square feet in basal area per acre. Evidently the overstory was not heavy enough to cause suppression of height growth under conditions there.

In the same area Robertson (1942) reported on the effect of removal of the yellow birch overstory on stand composition in an Oxalis-Cornus type (Table 59). In 20 years following release the number of balsam fir increased 30 per cent whereas there was no change in the number of red spruce. The volume of balsam fir increased 70 per cent as compared with a 15 per cent increase for red spruce.

In the Acadia Experimental Forest of New Brunswick release of 50 scattered

[*] $CF = \dfrac{\text{mean distance to surrounding trees in ft}}{\text{mean basal area of surrounding trees in ft}^2}$

TABLE 58. RESULTS OF THINNING IN A SPRUCE-BALSAM FIR STAND ON THE DUNBAR EXPERIMENTAL FOREST IN NORTHERN MICHIGAN[*]

Plot Number	Spacing (Ft)	Before Treatment 1940	After Treatment 1940	1945	1950	Before Treatment 1955	After Treatment 1955	1960
			NUMBER OF TREES PER ACRE					
1	6 x 6	2910	1210	1190	1160	1150	790	770
2	10 x 10	1940	440	430	430	430	430	430
3	8 x 8	2320	680	660	650	640	510	500
4	Control	2210	2210	2040	1690	1420	1420	1140
			BASAL AREA (Ft²) PER ACRE					
1	6 x 6	103	54	94	126	157	128	155
2	10 x 10	107	38	67	93	122	122	145
3	8 x 8	121	61	97	133	164	141	162
4	Control	105	105	145	164	187	187	204
			VOLUME (Cords) PER ACRE					
1	6 x 6	0.36	0.36	4.26	10.47	19.42	17.76	24.44
2	10 x 10	2.28	1.32	6.70	13.66	22.72	22.72	29.69
3	8 x 8	4.68	3.84	11.07	20.52	30.51	26.56	33.30
4	Control	3.72	3.72	9.36	15.72	25.05	25.05	33.18

SUMMARY OF GROWTH, 1941–1960

Plot Number	Spacing (Ft)	Periodic Annual Growth (Cords per Acre) 1941–1945	1946–1950	1951–1955	1956–1960	Mean Annual Increment at 50 Years (Cords per Acre)
1	6 x 6	0.78	1.24	1.79	1.34	0.52
2	10 x 10	1.08	1.39	1.81	1.40	0.61
3	8 x 8	1.45	1.89	2.00	1.35	0.76
4	Control	1.13	1.27	1.87	1.62	0.66

[*] Includes all trees 1-in. dbh and larger. Plot size, 0.2 acre.
Source: after Day, 1961.

TABLE 59. EFFECT OF REMOVAL OF A YELLOW BIRCH OVERSTORY ON THE STAND COMPOSITION IN AN OXALIS-CORNUS TYPE, LAKE EDWARD, QUEBEC

Species	Under Undisturbed Yellow Birch Overstory No. Trees 1-in. dbh and Over	Volume (Ft³/Acre)	Yellow Birch Overstory Removed 20 Years Previously No. Trees 1-in. dbh and Over	Volume (Ft³/Acre)
Yellow birch .	65	1094	8	1
Spruce	192	82	192	94
Balsam fir	1668	596	1513	1013
Paper birch ..	25	150	70	21
Cherry	25	21	265	48
All species ..	1975	1943	2048	1177

Basis: 2 plots, presumably ¼ acre or larger.
Source: Robertson, 1942.

spruce and balsam fir trees 1–4 inches dbh per acre from overtopping hardwoods (mainly 20–30-year-old paper birch, red maple, and yellow birch) was reported by Thomson (1949b). Ten years later the volume of released trees had increased four times, that of the controls only three times. The greatest response was in the smallest diameter classes. The release work required one man-day per acre.

In 1935 at the Pike Bay Experimental Forest in Minnesota, balsam fir was released by the removal of 15 cords of overstory aspen and paper birch per acre. In 1949 from 4 to 14 cords of balsam fir per acre were available on the treated area. No merchantable volume of balsam fir was reported from the surrounding uncut aspen-birch stand. Sample tree measurements indicated that average tree volumes were 4.5 and 2.5 cubic feet on the treated and untreated areas respectively (Roe, 1952).

In the Adirondacks the effect of release from a white pine overstory on the growth response of young balsam fir was studied in conjunction with changes in stand structure (McCarthy, 1918). At the time of release the age of the understory ranged from 6 to 55 years, averaging 27 years. The diameter of the oldest age class averaged 3.8 inches at the stump, grading downwards for younger age classes, and averaging 2 inches for all age classes. Despite the initial differences, fifty years later when the stand was cut, balsam fir of all age classes had practically the same merchantable height (47–50 ft), total height (63–69 ft), crown length (31–34 ft) and dbh (9–10 in.). The study provides an excellent illustration of the structural development of a homogeneous stand from what was initially a very heterogeneous understory.

Thinning in an understory of spruce-fir may also have effects similar to a release cutting. A mixed stand of spruce, balsam fir, hemlock, white-cedar, and hardwoods growing beneath a white pine overstory on a flat, poorly drained, coarse-textured soil in Maine was thinned at 80 years of age. Balsam fir and spruce responded well, especially in areas where the white pine overstory was most open (Chapman, 1954). A similar response was noted in Michigan by Day (1942).

GIRDLING

A modified thinning measure, girdling differs from thinning in that the treated stems are left standing, thus creating no immediate slash problem and giving a measure of protection to the stand that slowly diminishes over time. Damage to the released stand is minimal. Girdling is a relatively old practice used primarily to eliminate unwanted hardwood weed trees.

Girdling has been practiced in the United States since about 1900 (Cary, 1928). Early projects were carried out by the U.S. Forest Service at Corbin Park, New Hampshire (Westveld, 1930a), and by private companies in the Adirondacks (Churchill, 1927). The girdling experiments at Corbin Park were begun in 1905 when the overstory, consisting largely of big tooth aspen, paper birch, and yellow birch, with some sugar maple and beech, was 60 years old and carried approximately 100 square feet of basal area per acre. Spruce-fir reproduction, mainly red spruce and balsam fir was 2–6 feet high with some saplings attaining diameters of 1–3 inches dbh. Two girdled plots had 70 and 54 square feet per acre of hardwood basal area remaining after the first treatment. The control plot had 97 square feet per acre. A second girdling was made in 1915. At this time 65 per cent of all hardwoods on the first plot and 90 per cent of those on the second plot were girdled. Growth of conifers in the heavily treated plot over a 30-year period (1905–1935) was 50 cubic feet per acre annually. On the partially released plot annual growth was 29 cubic feet per acre, and on the control plot it was only 3 cubic feet per acre. For the period 1925–1935, the annual increment per acre on the three plots was 98, 58, and 2 cubic feet respectively. It should be noted that release was effected at an early age in this experiment.

The first Canadian experiments at Grand' Mère, Quebec, in 1917, also generally showed positive results except where too many trees were girdled (Plice and Hedden, 1931). Girdling was done to kill hardwoods oppressing both planted and natural coniferous reproduction.

TABLE 60. EFFECT OF HARDWOOD GIRDLING INTENSITY ON CONIFER DIAMETER INCREMENT IN INCHES OVER THE FIRST SIX-YEAR PERIOD AT THE LAKE EDWARD FOREST IN QUEBEC

	Diameter Classes (Inches)									
	2		4		6		8		10	
Treatment	Spruce	Balsam Fir	Spruce	Balsam Fir	Spruce	Balsam Fir	Spruce	Balsam Fir	Spruce	Balsam Fir
Not girdled	0.10	0.44	0.57	0.73	0.31	1.50	0.83	1.65	....	1.00
40% girdled	0.39	0.67	0.87	1.50	0.97	1.16	0.72	1.71	....	1.40
100% girdled	0.59	0.97	1.09	1.72	1.54	2.24	1.04	2.13	....	1.86

Basis: Presumably 100 trees of each species.
Source: Plice and Hedden, 1931.

A test on the effect of intensity of girdling upon diameter growth of spruce and fir at the Lake Edward Experimental Forest (Section L-4a) was described by Plice and Hedden (1931). The results are presented in Table 60. The authors concluded that girdled trees required 2–5 years to be killed. Trees died approximately in the order of their tolerance, i.e., paper birch, yellow birch, red maple, sugar maple, and beech.

Clarke (1940) reported on girdling investigations at Fairfield, St. John County, and Schoales Dam, Kings County, New Brunswick. Experiments in each location covered 50 acres of treatment with 10 acres left as control. In ten years, conifer increment at Fairfield on the treated area averaged 115.6 cubic feet annually as compared with 87.6 cubic feet for the control. The treatment raised the percentage of annual growth from 4.4 to 5.2 per cent. At Schoales Dam the annual increment was 57.7 cubic feet for the treated area and 33.3 cubic feet for the control over the 10-year period. The growth rate for spruce rose from 2.9 to 4.7 per cent following girdling. Balsam fir on both areas was predominantly of a mature age and girdling could not even retain previous increment. Recknagel and Westveld (1942) also demonstrated that girdling cannot rejuvenate old mature trees. Their data with discussion are presented in Table 63 in the section on the effect of partial cuts.

These experiments have shown that girdling is most successful in terms of growth if applied to release young reproduction. Westveld (1953) noted that from the point of view of immediate returns, the most "economic" girdling releases trees that are already of merchantable size. Trees which do not suppress conifers and trees with deteriorating crowns should not be girdled. Girdling and poisoning should be done selectively, to eliminate trees of poor form and quality.

As in all silvicultural operations, detailed technical guidance should be developed upon an ecological forest classification basis. Heimburger (1934) has already pointed out that the mere existence of a hardwood cover over a coniferous understory does not furnish sufficient information for guiding girdling techniques. The same results cannot be expected from girdling hardwoods in the Viburnum-Oxalis and Oxalis-Cornus types. Westveld (1930a) has stated that in the northeastern United States the long-range effects of girdling may increase the coniferous component, particularly in the spruce-flat type and in the yellow birch–spruce subtype.

CHEMICAL THINNING

Where desirable, balsam fir can also be removed by chemical means. Atkins (1956) applied 2,4,5-T and 2,4-D mixed with 2,4,5-T in an oil carrier as a basal spray on the lower 12–18 inches of bark. A kill of 75 per cent was achieved on 7–10 inch dbh balsam fir with both spring and late summer treatments.

Hart (1961) investigated the use of soil sterilants in order to thin balsam fir thickets in the Penobscot Experimental Forest in Maine. The treated stand was 35 years old, 16 feet tall, and had 8000–9000 stems per acre, 90 per cent of which were 0.5–2.5 inches dbh. Such thickets occur on poorly drained soils with a hardpan beneath and the number of stems may reach 30,000 per acre.

Three formulations were tested: Telvar W, containing 80 per cent 3-(p-chlorophenyl)-1, 1-dimethylurea; arsenic trioxide; and Chlorea, containing 40 per cent sodium chlorate, 57 per cent sodium metaborate, and 1 per cent 3-(p-chlorophenyl)-1, 1-dimethylurea. Two different dosages of each were used: Telvar W at 40 and 80 pounds per acre, arsenic trioxide at 800 and 1200 pounds per acre, and Chlorea at 900 and 1300 pounds per acre. The goal was to create 7-foot-wide openings. Chemicals were applied in 7-foot circles during the first week of July. The first effects appeared in September and were clearly visible in November, continuing for three years. Telvar W killed 78 per cent of the trees in 7-foot circles. Results with Chlorea and arsenic trioxide showed kills of 28 and 31 per cent respectively. Telvar W killed to a maximum distance of 9 feet outside the circle of application, Chlorea to 4.5 feet, and arsenic trioxide to 3.6 feet. The chemical applications gave results that were considered a partial success, but further testing was needed before final recommendations could be made.

Clearcutting and Partial Cutting

With respect to balsam fir silviculture the regulated clearcutting system is an unexplored possibility. The widely-used commercial clearcutting ends in complete removal of the residual stands only in exceptional cases. Provision has seldom been made for adequate blocks of mother stands to serve as a seed source and to provide the necessary favorable microclimatic conditions for germination and establishment of reproduction. The clearcutting system is basically a reproduction cutting;

however, every type of cutting provides the pre-conditions for the development of the new stand. Clearcuttings of various types should be investigated from the point of view of stand development as well as reproduction.

The term "partial cuttings" has very broad applications. From commercial clearcutting to the selection system in its classical form, there is always a residual stand which cannot be considered reproduction. This residual stand can respond and become a valuable part of the new stand; it can survive but interfere with the reproduction; or it can be lost through mortality within a few years after exposure. Partial cuttings attempt to solve both growth and reproduction problems.

Discussion in this section will refer to the response to release and will also consider the reaction of stands to exposure. The chief problem in partial cuttings is the cutting cycle and its implications. Present information on partial cuttings mostly concerns cutting in conifers with the retention of hardwoods. Very little information is available on integrated cutting of conifers and hardwoods.

CLEARCUTTING

Such questions as cutting width, direction of cut, spatial and temporal cutting sequence, and even rotation cannot be discussed with specific reference to balsam fir at present. Development of rots and budworm epidemics call for a relatively short rotation of balsam fir (60–70 years), although there are great regional and local differences dependent upon edaphic conditions. MacLean (1960) suggested an 85-year rotation if utilization was for spruce and fir only, 65 years for hardwood utilization, and a 70-year rotation if utilization of all species in the aspen-birch-spruce-fir cover type of northern Ontario was contemplated.

Present experience is limited to commercial clearcutting, predominantly in virgin stands usually with abundant advance reproduction of balsam fir.

Hatcher (1960) investigated the development of balsam fir following commercial clearcutting in Laurentides Park, Quebec (Section B-1). The investigated area covered five square miles of which 70 per cent was undisturbed virgin forest, 25 per cent was of fire origin, and 5 per cent originated from blow-down. Balsam fir was 60–90 years old and the spruce about 100 (50–200) years. There were 1900–3000 cubic feet of spruce and fir and 20–600 cubic feet of paper birch per acre before 87–94 per cent of all spruce and fir 4 inches dbh and over was removed. Small quantities of paper birch were used for fuel for the adjacent logging camps. The forest recovered well and formed well-stocked sapling stands predominantly of balsam fir with some paper birch, white and black spruce, seven years after cutting. About 80–90 per cent of the balsam fir was less than two feet high at the time of cutting. Most of it had been established 20 years before logging although some was 50 years of age. None of the seedlings were established later than five years after cutting. The results showed that seedlings released after suppression soon attained a growth rate comparable to that of new seedlings. No relationships were found between the length of suppression period and the growth rate after release. Height growth above breast height was greatest in the Dryopteris-Oxalis forest type, followed by growth in the Hylocomium-Oxalis and Hypnum-Cornus types. (Types according to Linteau, 1955; see Table 5.)

In investigating cutover aspen-birch-spruce-fir cover types in Ontario (Sections

B-4, B-8, B-9), MacLean (1960) concluded that clearcutting will increase hardwoods, particularly aspen sprouts, which are then able to compete with shrubs. The new stands, however, will produce more hardwoods and perhaps more conifers than the crop that follows slowly disintegrating old-growth stands.

RELEASE AND EXPOSURE

Since partial cuttings cover a wide range of intensities, their effect on the residual stands is sometimes more aptly phrased "reaction to exposure" rather than "response to release." The same intensity of treatment may produce different effects under different climatic, edaphic, and biotic conditions. Under optimum growth conditions balsam fir can withstand rather drastic treatments. As a shade-tolerant tree it responds well to release, but is short-lived in the western parts of its range.

In the Green River area the average length of balsam fir suppression has been 30 years, but it may last 100 years. Morris (1948) found that length of suppression had no effect on later cone production and incidence of rots. It may be that suppression keeps the tree in a juvenile stage and produces wood which is more resistant to rots. These conclusions were drawn from observations in an optimum growth area of balsam fir.

In an experimental area in the Ontario Clay Belt (Section B-4) 43 per cent of the trees remaining after a 5-inch diameter limit cut were suppressed but were able to recover as long as suppression had not lasted for more than 50 years. Residual advance growth after such cuts generally ranged from 8 to 40 years of age (Bonner, 1941).

Balsam fir on good sites (site index 42.5) in northern Michigan 20 years after cutting had a total diameter growth of 2.5 inches as compared with 2.0 inches in the 20 years prior to the cut. White spruce reacted even more favorably, the diameter increment increasing from 1.9 to 2.9 inches (Bowman, 1944). On poor sites both balsam fir and black spruce had a diameter growth during the same period of 1.5 and 1.7 inches, respectively. Since Bowman (1944) did not present an age analysis of the trees, it is possible that at least part of them had not reached the period of accelerated growth before the cutting was made. The results, therefore, may not represent the effect of release alone.

Only recently have a few reports (Turner, 1952; Mott et al., 1957) appeared on the changes in radial growth in different sections of the stems. These were concerned with the response to defoliation by the spruce budworm. No general studies on stem growth of balsam fir are known. Conclusions based on diameter growth response at stump height and even at breast height may prove misleading since such measurements may not always be representative of the growth of the total stem. Changes in competition and exposure often result in adjustments in tree form which last for a considerable time.

The effect of partial cutting may last from a few years to several decades. Some effects can be traced throughout the following generation. Exposure to sun, drought, frost, and wind are of the greatest importance. Wind effects are particularly heavy. There are examples of partial cutting resulting in total loss of the residual stand in a short time (Baskerville et al., 1960). In partial cuts, logging damage is done not only to the advance reproduction but also to the residual stand. Trimble (1942), working in New Hampshire, found that damage was kept relatively low by careful work and by using horse-skidding. About 3.3 per cent of residual balsam fir were lost and 9.2 per cent irreparably damaged. Of residual spruce about 1.1 per cent were lost and 5.2 per cent heavily damaged. Balsam fir is more susceptible than many other species because its bark is more easily peeled by abrasion in logging.

CUTTING CYCLE

The cutting cycle is a key element in systems of partial cutting. Under an intensive selection, cutting cycles do not exceed 6–10 years. In extensive forest management, cutting cycles are determined more by economic conditions than by natural potentialities. The minimum number of cords per acre for an economically feasible operation is often decisive in determining the cutting cycle. In 1956 on large holdings in Canada the minimum volume per acre ranged from 7 to 10 cords. Climatic, edaphic, and biotic conditions, degree of admixture of nonmerchantable hardwoods, and stand-age structure are other factors affecting the cycle. Partial cutting is a method of trial and error based largely on local experience.

When Graves (1899) planned partial cuts in spruce-fir stands in Whitney Park, New York, some experience from private companies was already available. He estimated that if all spruce were cut to a 6-inch stump diameter another cut would be possible in 75 years, at 8 inches the cycle was predicted to be 50 years. Special "yield tables" were prepared based on diameter growth, specifying dbh limits of 10, 12, and 14 inches for the cut. According to Meagher and Recknagel (1935), the cut was finally set at 12 inches dbh, and 36 years later another cut was available, just as predicted by Graves. It should be noted that the major species in this area have considerable longevity. Graves found red spruce at 9 inches dbh to average 163 years, and 27-inch red spruce to average 310 years. Balsam fir 10.5 inches dbh was 134 years old.

Another example of cutting cycle and softwood diameter limits, for cutting only softwoods, was presented by Recknagel (1933) for a forest at Newcomb in the Adirondacks. Diameter limits of 10 inches for spruce and 6 inches for balsam fir were recommended. According to the computations, an equally large cut could be made 25 years later.

Robertson (1951) determined cutting cycles from a hypothetical model of a selection forest with a maximum tree age of 100 years. At this age there were 19.5 cords of merchantable volume per acre in diameter classes of 4 inches and above, and 15.2 cords in classes of 6 inches and above. It was assumed that following clearcutting, the remaining 1–3 inch seedlings and saplings would be destroyed and that reproduction from seed would start the stand anew. Cutting to a 6-inch diameter limit, the 4–5 inch saplings together with ingrowth should make two cuttings possible on a 90-year rotation, each cut providing about 15 cords. Using selective marking three cuts of 15 cords each would be possible in a period of 90 years. With clearcutting, the annual profit would be $6.06 per acre; with diameter limit cutting, $8.94 per acre; and with selective cutting, $12.03 per acre, based on 1951 costs and returns.

In a balsam fir–white spruce stand in Newfoundland, Robertson (1949) compared clearcutting and diameter limit cutting. Using the 5-inch diameter limit, 15 cords per acre were removed leaving 2.2 cords; under commercial clearcutting, 16.5 cords were removed with 0.7 cords remaining. The advance growth averaged 35 years in age. At age 65 following clearcutting the volume reached 21.1 cords per acre. After the 5-inch diameter limit cut it was computed to be 29.0 cords per acre. The figures indicated a smooth rise of the volume curve from the very beginning without losses due to exposure in early years.

In the northeastern United States, under a 60-year cutting cycle, the average cut per acre is 6.0 cords as compared with a mortality of 6.2 cords. If a 30-year cutting cycle could be applied, 3.6 cords per acre could be saved; if a 20-year cutting cycle, 4.2 cords (Westveld, 1946).

Aside from those examples cited above, many authors have advocated shortening the cutting cycle. At present the cutting cycle in the spruce-fir area in Canada is 40–50 years because of abundant reproduction in the virgin stands. This utilizes a commercial clearcut having a 4-inch top diameter.

SELECTIVE CUTTING OF CONIFERS

About 1900, foresters in eastern North America became aware that removal of merchantable conifers and retaining hardwoods in mixed stands would soon lead to a decrease in softwood production. At that time the first girdling studies, previously discussed, were initiated. Graves (1899) devised a partial cutting system which would maintain better softwood production under the residual hardwoods.

Early cuttings (1890–1892 and 1917–1920) of softwoods on a partial-cut basis in softwood, mixed-wood, and hardwood cover types were made in New Hampshire. Details of the cutting regulations were not specified and the data are based on measurements made in 1941. Results are shown in Table 61. Fifty years after the oldest cut, the spruce-fir absolute volumes in both softwood and mixed-wood types were not much less than the volume in the virgin forest. In the hardwood cover type they had decreased appreciably. Especially in the softwood cover type the percentage of balsam fir had greatly increased at the expense of spruce (Recknagel and Abel, 1942).

Recknagel (1936, 1942) investigated an area partially cut in 1906 at Newcomb, New York, measuring the volume, mortality, and growth after 25, 30, and 35 years. Original volume was not given. Wind damage caused great loss of balsam fir. From the results, presented in Table 62, Recknagel concluded that the cutting cycle should not exceed 25–30 years.

Recknagel (1933) and Recknagel and Westveld (1942) described another example of a partial cutting, of 1895, near Newcomb, New York. An experiment for comparing different cutting methods and hardwood girdling was established in 1930. The stand consisted of spruce (mainly red spruce), balsam fir, beech, yellow birch, and sugar maple. Measurements were made after the cut in 1931 and again in 1940. Results are shown in Table 63. After the 5-inch diameter limit cut, only 4.7 per cent of the original spruce-balsam fir volume remained; after the 6- and 8-inch limit cuts, 17.1 per cent remained; and after the selection cut, 44.0 per cent remained. Hardwood girdling was done selectively, reducing the basal area of live hardwoods from 83 to 48 square feet per acre, no softwoods being cut. The original paper presents a stand structure analysis by diameter classes of 1–4, 5–7, 8–10, and 11 inches and larger. Noticeable losses of merchantable volume occurred due to exposure during the first ten years after treatment. Hardwood girdling failed to stimulate the growth of conifers.

Development of mixed-wood stands after partial cutting was studied by Candy (1938) at the north shore of Lac Tremblant, Quebec. Spruce and balsam fir were cut in 1901 and 1919 for pulpwood, leaving hardwoods intact. Development of this partially cut stand is presented in Table 64. The trend toward hardwoods is clear. The losses in softwood volume were accompanied by a considerable reduction in the number of conifer saplings and seedlings.

TABLE 61. EFFECT OF PAST CUTTING OF SOFTWOODS ON THE PRESENT VOLUME OF SPRUCE-FIR AND THE RELATIVE AMOUNT OF SPRUCE AND BALSAM FIR IN NEW HAMPSHIRE

Cover Type and Stand Origin	Volume (Cords) per Acre in 1941	Percentage of Softwoods in 1941	
		Spruce	Balsam Fir
Softwoods			
Virgin	22.49	53.6	46.4
Cut 1890–92	18.70	28.3	71.7
Cut 1917–20	8.33	14.0	86.0
Mixed-woods			
Virgin	11.98	50.0	50.0
Cut 1890–92	10.32	36.9	63.1
Cut 1917–20	5.50	28.2	71.8
Hardwoods			
Virgin	2.05	38.5	61.5
Cut 1890–92	0.65	50.7	49.3
Cut 1917–20	0.52	53.8	46.2

Basis: About 3000 plots, 0.2 acre in size.
Source: Recknagel and Abel, 1942.

TABLE 62. VOLUME AND MORTALITY OF SPRUCE-FIR STANDS CUT IN 1906 TO A 10-INCH DIAMETER LIMIT AFTER 25, 30, AND 35 YEARS AT NEWCOMB, NEW YORK [*]

Type and Species	Volume (Ft³) per Acre			Mortality (Ft³) per Acre	
	1931	1936	1941	1931–36	1936–41
Softwood-flat					
Balsam fir	442.5	474.5	469.8	76.3	16.5
Spruce	351.7	373.7	301.5	14.3	8.9
Hardwood Type					
Balsam fir	214.3	203.4	207.9	75.8	3.2
Spruce	252.8	428.0	318.9	5.9	13.3

[*] Includes trees 5-in. dbh and over.
Basis: 20 plots, 0.25 acre in size.
Sources: Recknagel, 1936, 1942.

TABLE 63. EFFECT OF DIFFERENT CUTTING METHODS ON THE SUBSEQUENT GROWTH OF SPRUCE-FIR UNDERSTORY IN SPRUCE-HARDWOOD STANDS IN THE ADIRONDACKS

| | Unmerchantable (1–4 in. dbh) | | | | | | Merchantable (5-in. dbh and over) | | |
| | Number per Acre | | | Cubic Feet per Acre | | | Cords [*] per Acre | | |
Year	Spruce	Balsam Fir	Total	Spruce	Balsam Fir	Total	Spruce	Balsam Fir	Total
5-inch limit cut									
1930	123	249	372	26	46	72	5.70	4.52	10.22
1931	81	131	212	20	31	51	0.30	0.18	0.48
1940	128	199	327	28	48	76	0.18	0.18	0.36
6- and 8-inch limit cut									
1930	96	331	427	22	43	65	5.22	4.34	9.56
1931	57	236	293	14	37	51	1.13	0.50	1.63
1940	100	205	305	24	55	79	0.66	0.66	1.32
Selection cut									
1930	113	184	297	34	23	57	5.22	2.14	7.36
1931	79	124	203	22	20	42	2.68	0.56	3.24
1940	107	190	297	27	35	62	2.47	0.55	3.02
Hardwood girdling									
1930	105	87	192	25	16	41	6.94	2.69	9.63
1940	185	103	288	38	30	68	7.08	1.96	9.04
Control									
1930	86	58	144	23	14	37	6.44	1.93	8.37
1940	178	25	203	58	10	68	6.73	1.63	8.36

[*] 1 cord = 95.6 ft³ wood.
Basis: 5 plots for unmerchantable trees, 1.4 acres in size; 5 plots, 28.0 acres in size.
Source: Recknagel and Westveld, 1942.

238

Partial cuttings in softwood and mixed-wood stands were made on experimental areas in the Goulais River watershed (MacLean, 1949; Jarvis, 1960) about 1910. Spruce and balsam fir were cut to a 6-inch diameter level and white pine was cut for sawtimber. The volume of spruce and fir left after cutting was somewhat larger than the volume measured in 1920 (Table 65).

Cutting and subsequent losses from exposure reduced the percentage of spruce from 32 to 12 per cent in 1920, after which it slowly recovered to attain 21 per cent of total volume in 1956. The percentage of balsam fir was little affected, being 28 per cent before cutting and 29 per cent in 1956. Budworm attack from 1928 to 1946 had some effect on growth. The total net annual periodic increment was 52 cubic feet per acre in the first period, 61 in the second, and 39 in the third. Development in number of trees and volume after cutting in mixed-wood stands is shown in Table 66.

TABLE 64. DEVELOPMENT OF THE RESIDUAL STAND AFTER SOFTWOOD LOGGING IN 1901 AND 1919 AT LAC TREMBLANT, LABELLE COUNTY, QUEBEC

Year	Main Stand				Number of Saplings		Number of Seedlings		Net Increment per Acre, 1919–1929 (Ft³)		
	Spruce		Balsam Fir			Balsam		Balsam		Balsam	Hard-
	No.	dbh	No.	dbh	Spruce	Fir	Spruce	Fir	Spruce	Fir	wood
1919 45	45	7.6	26	5.2	32	142	249	624			
1924 30	30	6.7	26	5.6	47	179	342	412			
1929 35	35	6.7	25	5.8	30	97	291	262	−2.2	−1.3	+46.5

Basis: Plot number and size unknown.
Source: Candy, 1938.

TABLE 65. DEVELOPMENT OF NUMBER OF TREES AND TOTAL VOLUME IN PARTIALLY CUT (1910) SOFTWOOD STANDS IN THE GOULAIS RIVER WATERSHED, ONTARIO

Species	Number of Trees per Acre				Volume (Ft³) per Acre*			
	1920	1927	1946	1956	1920	1927	1946	1956
Spruce	74	99	125	121	144	213	366	489
Balsam fir	195	381	444	403	329	514	630	702
White-cedar	48	101	124	141	256	344	408	522
White pine	4	4	6	5	130	72	125	155
Softwoods	321	585	699	670	859	1143	1529	1868
Paper birch	26	58	62	56	202	230	228	233
Yellow birch	7	10	11	10	93	110	143	165
Maple	13	41	61	68	20	36	59	83
Hardwoods	46	109	134	134	315	376	430	481
Minor species † ..		10	11	37		23	29	33
All species ..	367	704	844	841	1174	1542	1988	2382

* The cubic-foot volume per acre before cutting in 1910 was 551 for spruce, 477 for balsam fir, and 1700 for all species; the cubic-foot volume per acre after cutting in 1910 was 164 for spruce, 332 for balsam fir, and 1200 for all species.

† Includes balsam poplar, tamarack, and jack pine; not tabulated in 1920.

Basis: 96 plots, 0.1 acre in size.

Sources: MacLean, 1949; Jarvis, 1960.

Spruce volume, originally 13 per cent of the total, decreased to 5 per cent after the partial cut in 1910 and then slowly increased, making up 8 per cent of the total in 1956. Balsam fir volume, originally 17 per cent, was reduced to 15 per cent after the cut and maintained approximately this volume through 1956. Total annual net periodic increment per acre for all species was 24 cubic feet for the period 1921–1927, 29 for 1928–1946, and 47 for 1946–1956. The possibility for another partial cut of spruce and fir in the mixed-wood type was considered remote, but in

TABLE 66. DEVELOPMENT OF NUMBER OF TREES AND TOTAL VOLUME IN PARTIALLY CUT (1910) MIXED-WOOD STANDS IN THE GOULAIS RIVER WATERSHED, ONTARIO

Species	Number of Trees per Acre				Volume (Ft³) per Acre *			
	1920	1927	1946	1956	1920	1927	1946	1956
Spruce	31	45	54	52	95	120	190	239
Balsam fir	113	205	234	194	258	266	329	388
White-cedar	61	78	89	94	518	575	486	533
White pine	2	3	3	2	71	31	46	63
Softwoods	207	331	380	342	942	992	1051	1223
Paper birch	24	49	43	52	214	235	266	309
Yellow birch	37	54	55	54	774	802	912	1056
Maple	53	128	188	220	317	366	421	516
Hardwoods ...	114	231	286	326	1305	1403	1599	1881
Minor species † ..		8	5	23		23	25	44
All species ..	321	570	671	691	2247	2418	2675	3148

* The cubic foot volume per acre before cutting in 1910 was 338 for spruce, 421 for balsam fir, and 2530 for all species; the cubic foot volume per acre after cutting in 1910 was 101 for spruce, 337 for balsam fir, and 2200 for all species.

† Includes balsam poplar, American elm, and black ash; not tabulated in 1920.

Basis: 168 plots, 0.1 acre in size.

Source: MacLean, 1949; Jarvis, 1960.

the softwood type total volume by 1956 already had exceeded the volume of the virgin stand.

One of the most exhaustive reports on stand development was prepared by Ray (1941, 1956) from the Lake Edward Experimental Forest. Some of the information covers 60 years, with a period of 30 years being covered very intensively. The information is based upon ecological forest types, and an extensive analysis of stand structure is presented. The findings were previously discussed from the reproduction standpoint in Chapter 6 (see Table 47).

The investigated area was cut for sawlogs in 1890 and for pulpwood in 1910, with only conifers being cut. Spruce was cut to a 10-inch diameter limit and balsam fir to a 7-inch limit. Age of balsam fir ranged from 55 to 80 years, whereas the spruce was older at the time of cutting. The pulpwood cut in the Cornus type amounted to 8 cords per acre, in the Oxalis-Cornus type to 10 cords, in the Viburnum-Oxalis type to 4.5 cords, and in the Viburnum type to 3 cords. Development of the partially cut stands from 1915 to 1946 is presented in Table 67.

The effect of exposure was noted for the first 10–15 years after cutting. During

TABLE 67. DEVELOPMENT OF TOTAL VOLUME IN CUBIC FEET PER ACRE AFTER LOGGING IN 1890 AND 1910 AT LAKE EDWARD EXPERIMENTAL FOREST, QUEBEC *

Species	Cornus Type				Oxalis-Cornus Type				Viburnum-Oxalis Type			
	1915	1925	1936	1946	1915	1925	1936	1946	1915	1925	1936	1946
Spruce	492	498	882	1279	342	321	550	730	305	287	346	460
Balsam fir	413	349	470	658	540	472	773	922	296	221	302	284
Other softwoods	120	107	154	186	138	97	131	134	21	15	24	5
Softwoods	1025	954	1506	2123	1020	890	1454	1786	622	523	672	749
Yellow birch	50	37	110	89	620	674	757	791	1464	1566	1615	1750
Other hardwoods	160	166	184	201	198	196	246	250	436	514	511	556
Hardwoods	210	203	294	290	818	870	1003	1041	1900	2080	2126	2306
All species	1235	1157	1800	2413	1838	1760	2457	2827	2522	2603	2798	3055

* For structural analysis see Table 47, Chapter 6.
Basis: 343 plots, 0.2 acre in size.
Source: Ray, 1941, 1956.

the 30-year period after cutting, spruce improved its relative position in all types and balsam fir remained at about the same level. A new cutting was scheduled for 1950, with a 14-inch diameter limit for spruce and a 7-inch limit for balsam fir. Total cut between 1890 and 1950 amounted to 11.7 cords per acre in the Cornus type, 18.0 cords in the Oxalis-Cornus type, and 8.6 cords in the Viburnum-Oxalis type. No cut was planned in the typical hardwood Viburnum type (for type descriptions see Chapter 2 and Fig. 7). Ray (1956) concluded that cutting of softwoods alone offered no serious threat to their recovery in the Cornus and Oxalis-Cornus types. Red spruce persisted quite well also in the Viburnum-Oxalis type, but an integrated softwood-hardwood management program now offers even better possibilities for these stands.

A series of partial cuttings was carried out in the Epaule River watershed area of Quebec (Section B-1b). The forest consisted of about 75 per cent balsam fir and had been cut prior to 1925. Composition of the experimental stands in 1953 is shown in Table 68 (Hatcher, 1961).

The volume of spruce and paper birch increased heavily with increasing age of the stands, while the volume of balsam fir increased little. The total basal area was unchanged at 157 square feet per acre in both age classes in the Dryopteris-Oxalis type (types after Linteau, 1955; see Chapter 2). The average stand diameter increased 0.6 to 0.8 inches. In the Hylocomium-Oxalis type, basal area changed from 137 to 152 square feet per acre with increasing age. The gains in the longer rotation under these conditions depend on changes in species composition, on larger diameter classes, and on increased height of the stems.

Cutting operations in 1953 removed 50 per cent of the conifer volume or about

TABLE 68. TOTAL VOLUME IN CUBIC FEET PER ACRE BY SPECIES, AGE CLASS, AND DIAMETER GROUP IN TWO FOREST TYPES IN THE EPAULE RIVER WATERSHED AREA, QUEBEC

Diameter Group (Inches)	Even-Age 50 Years			Even-Age 70 Years			Uneven-Aged		
	Spruce	Balsam Fir	Paper Birch	Spruce	Balsam Fir	Paper Birch	Spruce	Balsam Fir	Paper Birch
DRYOPTERIS-OXALIS TYPE									
1-3	7	180	22	2	125	16	6	220	20
4-9	94	2417	104	108	2444	271	93	1728	179
10+	39	466	340	160	513	437	109	596	469
Total	140	3063	466	270	3082	724	208	2544	668
Per cent ...	4	83	13	7	75	18	6	74	20
HYLOCOMIUM-OXALIS TYPE									
1-3	1	166	35	2	97	21	10	225	61
4-9	202	2018	95	230	2407	296	216	1641	267
10+	47	309	82	137	201	177	83	510	83
Total	250	2493	212	369	2705	494	289	2376	411
Per cent ...	9	84	7	10	76	14	9	78	13

Basis: 87 plots, 0.1 acre in size.
Source: Hatcher, 1961.

TABLE 69. VOLUME, INCREMENT, AND MORTALITY IN CUBIC FEET PER ACRE FOLLOWING PARTIAL CUTTING IN TWO FOREST TYPES IN THE EPAULE RIVER WATERSHED AREA, QUEBEC

Volume Category	Even-Age 50 Years			Even-Age 70 Years			Uneven-Aged		
	Spruce	Balsam Fir	Paper Birch	Spruce	Balsam Fir	Paper Birch	Spruce	Balsam Fir	Paper Birch
DRYOPTERIS-OXALIS TYPE									
Volume									
After cut .	97	1253	427	120	1258	652	96	880	562
5 years									
later	106	1053	322	114	853	536	96	1086	601
Net incre-									
ment	9	−200	−105	−6	−405	−116	0	206	39
Mortality ..	30	449	120	23	614	150	17	119	43
Gross in-									
crement .	39	249	15	17	209	34	17	325	82
HYLOCOMIUM-OXALIS TYPE									
Volume									
After cut .	184	954	129	137	1260	436	161	820	339
5 years									
later	265	988	149	122	842	428	190	786	350
Net in-									
crement ..	81	34	20	−15	−418	−8	29	−34	11
Mortality ..	0	228	6	22	573	48	14	302	48
Gross in-									
crement ..	81	262	26	7	155	40	43	268	59

Basis: 87 plots, 0.1 acre in size.
Source: Hatcher, 1961.

15 cords per acre. Reaction to the treatment by age classes and types is shown in Table 69. Balsam fir had greater mortality than spruce and paper birch. Mortality was higher in 70-year-old stands than in 50-year-old and uneven-aged stands. The effect of forest type was inconclusive.

The main conclusion was that partial cutting hastens the process of deterioration and reproduction of mature stands. Most losses occurred in the first three years following cutting. The variation among individual plots indicated that the maximum limit for partial cuttings should be set at 40 per cent of conifer volume if they are to be retained under these conditions (Hatcher, 1961). This is higher than the maximum limit for partial cuttings suggested by McLintock (1954), who indicated that no more than 25 per cent of the basal area of trees 4 inches dbh and larger should be removed in such a cut.

CUT OF CONIFERS AND HARDWOODS

In contrast to the number of references on softwood cuttings, only two were found dealing with integrated cutting of hardwoods and softwoods in mixed stands.

Day (1942) compared partial cutting of three different intensities as applied to a mixed stand of balsam fir, white spruce, aspen, and paper birch in Upper Michigan. There was also an admixture of 6 per cent white pine. The experiment covered a 10-year period and tested cuts which removed 60, 25, and 14 per cent of the merchantable volume. The heaviest cut resulted in losses by windfall and serious reproductive competition by aspen suckers. The lightest cut showed no growth response. It was concluded that in stands with 20 cords of merchantable volume per acre, a proper partial cut should be 20–40 per cent by volume. If less volume is present, the cut should be below 20 per cent. The cut should also be less when aspen is not present or if the aspen is girdled. Thrifty white pine and white spruce should be preserved for future cuts.

Another example covering seven years of experience was presented by Hart (1956) from the Penobscot Experimental Forest, Maine. Cutting was done in 1945 on a 20-acre tract which was burned over in 1885 and which before the cut had 21 cords of merchantable volume per acre. The main species were balsam fir, aspen, red spruce, and white spruce. As guides in marking, the following minimum diameter limits were used: spruce, 9.5 inches; balsam fir, 7.5 inches; aspen, 6.5 inches. Ten cords of spruce and fir and 3.8 cords of aspen were removed. In 1953 the volume had increased to 11.8 cords. Net annual growth over the period was 55.1 cubic feet per acre; for balsam fir alone, 12.8. Annual mortality per acre for all species was 15.4 cubic feet; for balsam fir it was 7.27. To reduce mortality a cutting cycle of less than 20 years was advocated. Originally the experiment was planned to test the effect of spruce budworm attack (see page 169).

Integrated management has improved the possibilities for developing methods of partial cuts to fit the economic requirements for greater output from unit areas. Problems of mechanizing such cuts still remain. It appears that for management of extensive areas, special attention should be given to developing better regulated clearcutting methods or shelterwood variations.

Conclusions

1. Regional forest classification systems are gradually becoming the basis for silviculture over the wide range of balsam fir. The optimum conditions for reproduction and for high productivity are not necessarily identical within a region, but may change from one region to another.

2. A valuable source of information on natural factors, the virgin forest also poses a major problem of practical forest management throughout much of the range of balsam fir. It cannot be visualized as an equilibrium of uneven-aged stands which leads to accumulation of maximum yields. Because of differences in reproductive habits, longevity, and natural catastrophes, even-aged, uneven-aged, pure, and mixed stands occur in various stages of development.

3. The effects of budworm epidemics and fire in virgin and extensively managed stands have been studied to some extent, but the effects of wind, diseases, changes in water level, and other natural disturbances have been reported only casually.

4. After fire and clearcutting, balsam fir recovers lost territory over wide areas, frequently expanding after disturbances when the more economically valuable species — particularly spruces and pines — have been removed.

5. Chemical treatment of competing shrubs and hardwoods has been developed greatly in recent years, but the approach needs more ecological consideration. Little can be gained following successful treatment of one shrubby species if an-

other species in the same area is rejuvenated or if the eliminated shrub vegetation is merely replaced by a dense herbaceous vegetation.

6. In the past, few studies on stand development have given the necessary attention to stand structure. Age structure frequently has been assumed to be identical to diameter distribution. The relationships between stand overstory and understory frequently have been overlooked in planning thinning studies. Little consideration has been given to the theoretical and practical aspects of spacing and grouping of trees within the stands. The role of spatial distribution in stand protection, growth, and development of timber quality has not been properly investigated.

7. The role of balsam fir in forest stands may be either primary or secondary. As an understory species, balsam fir aids natural pruning of both overstory conifers and hardwoods. A dense balsam fir understory prevents shrub development and preserves an adequate seedbed for regeneration of more desired species.

8. Treatments used for release may cause positive responses in growth or negative reactions because of exposure. These two aspects of release need to be investigated simultaneously. Response to release or reaction to exposure may appear in very different ways. A single measurement such as diameter growth at stump height or at any other single point may, under certain conditions, have little correlation with the stem growth of the entire tree or with the response of the whole stand.

9. Some attention has been paid to stand stagnation following overcrowding. Little information is available, however, on the extent of stagnation, its duration, its causes, and on preventive measures which may reduce its effect. It can be expected that in addition to preventive silvicultural measures such as modification of the reproduction cut, chemical treatment in noncommercial thinnings and mechanization of thinning operations will be used increasingly.

10. Although partial cuttings have been studied for many years, new avenues of research have been opened by changing economic conditions particularly as demonstrated in integrated softwood and hardwood management, by the need for mechanization, by increasing appreciation of the importance of basic forest classification, and by thorough analysis of stand structure. Alternatives to partial cutting such as improved clearcutting techniques and the shelterwood system coupled with intensified thinning will also need to be considered.

8

GROWTH AND YIELD

The general ecological background of balsam fir growth and yield was presented in Chapters 2 and 3, and some phases of stand structure were reviewed in Chapter 7. The present chapter is concerned largely with the basic mensurational aspects of growth and yield. Included within its scope are studies of the form and volume of individual trees; the composition, structure, and productivity of forest stands; area and age-distribution of spruce-fir stands; and growth and yield of balsam fir. Special attention is given to balsam fir volume and yield tables.

Most tree-form data for balsam fir are rather old, although volume-table investigations and composite volume-table studies have been conducted more recently. A number of valuable yield studies were completed in the period 1940–1949. Recent investigations have placed emphasis on individual production elements of forest stands and their interrelationships. Detailed physiological studies, forest classification schemes, and ecological forest mapping investigations now in progress will eventually form the basis of improved forest production and management practices.

A few attempts have been made to evaluate forest production by stands and forest types within administrative units. Published reports, however, frequently combine data from several species or lump together stands of very different ecological character. For balsam fir, then, only very general summaries can be given, and few scientific or practical conclusions can be drawn from the available data.

Tree Form, Tree Volume, and Stand Volume

Stem form and volume of individual trees are determined both by physiological processes of growth inherent in the species, and by modifications resulting from varied environmental conditions. Typical physiologically defined growth patterns of forest trees occur under the most stabilized growing conditions such as are found in fully stocked, even-aged stands on homogeneous sites. Secondary stabilized growing conditions may be found in the growth patterns developed by open-grown trees. Both types of patterns vary by forest type and region. Volume-table construction was little influenced by biological considerations in the past. Lately

246

it has become a joint problem in which the biological aspects of growth are coordinated with mathematical and statistical procedures.

Tree volume tables inadvertently incorporate some information about stem form which is revealed by comparing actual tree volumes with the volumes of simple geometrical forms. The most common of these are cylinders constructed upon dbh and height. The ratio of tree volume to this type of cylinder gives the dbh-form factor. Zon (1914) prepared special tables for balsam fir dbh-form factors in Maine and New York. For trees 7–16 inches dbh the breast-height-form factors ranged from 0.553 to 0.412 in Maine and from 0.527 to 0.435 in New York. The trees with smaller diameters had the largest form factors, as expected. Regional differences are perhaps due to the better growing conditions for balsam fir in Maine than in New York as Zon (1914) has repeatedly emphasized. Breast-height-form factors are now conceded to be poor indexes of stem form since dbh varies with butt swell and other conditions which make tree-to-tree variability at breast height much greater than in the upper portion of the tree.

Behre (1927) measured the "absolute form factor," a ratio of tree volume above breast height to a cylinder volume of equal height with dbh as diameter. The absolute form factors ranged from 0.595 to 0.776 for balsam fir in different areas. Though less objectionable than breast-height-form factors because butt swell is excluded, they still poorly represented the true form of the tree.

The form quotient, $q_{.5}$, was originally defined by Schiffel at the end of the nineteenth century as the ratio between diameter at one-half the total height to dbh. Since then it has been used in many different ways, e.g., to exclude or include bark, and to allow various corrections for butt swell and other conditions. Clark (1903) determined that the form quotient for balsam fir in the Adirondacks was 67.6 per cent (apparently including bark), the breast-height-form factor being 45.7 per cent. The difference between these two ratios, known as Kuntze's constant, was 21.9 per cent.

Behre (1927) introduced the "normal form quotient," a modification of the form quotient corrected for bark and butt swell. Normal form quotients for balsam fir in the northeastern United States ranged from 59.5 to 77.6, averaging 70.7. In Canada the range was 61.6 to 75.5. The Canada Forest Service (1930, 1948) form-class volume tables are based on form quotients computed for diameters inside bark but without correction for butt swell. Volume tables based on form quotients of 60, 65, and 70 for balsam fir were prepared (see Table 76). Gevorkiantz and Olsen (1955) computed the original Schiffel's form quotient by measuring diameters with bark and without using butt-swell corrections. The average form quotient for 17 tree species in the Lake States was 68.

Taper tables express tree form more completely. In North America they have been primarily used for the preparation of board-foot volume tables. Taper tables list diameters, in absolute or relative units, at fixed levels or at relative heights above dbh or the ground. The oldest taper tables for balsam fir were prepared by Zon (1914) for several forest types in Maine and New York. Those for New York are reproduced in Table 70. The table shows that balsam fir growing in the swamp

TABLE 70. TAPER OF BALSAM FIR AS A PERCENTAGE OF DIAMETER INSIDE BARK BREAST HIGH IN HARDWOOD SLOPE AND FLAT TYPES (HF), AND SWAMP TYPES (S), NEW YORK *

dbh (In.)	Height above Ground (Ft)													
	10		20		30		40		50		60		70	
	HF	S	HF	S	HF	S	HF	S	HF	S	HF	S	HF	S
20-FOOT TREES														
2	77.8	72.2												
4	78.4	70.3												
30-FOOT TREES														
2	84.2	84.2	47.4	52.6										
4	89.5	86.5	60.5	56.8										
6	91.2	87.5	64.9	58.9										
40-FOOT TREES														
2	94.7	89.5	73.7	68.4	47.4	42.1								
4	92.1	91.9	73.7	73.0	50.0	43.2								
6	93.0	91.1	75.4	73.2	50.9	46.4								
8	92.1	92.0	76.3	74.7	52.6	46.7								
10	91.6		76.8		53.7									
50-FOOT TREES														
6	93.1	93.0	81.0	77.2	63.8	59.6	37.9	31.6						
8	93.4	93.3	81.6	78.7	64.5	58.7	36.8	33.3						
10	92.6	91.5	80.0	77.7	63.2	57.4	36.8	33.0						
12	92.1	91.2	78.9	77.0	63.2	56.6	36.8	31.9						
60-FOOT TREES														
6	94.8	94.7	86.2	82.5	72.4	66.7	53.4	47.4	29.3	24.6				
8	93.5	93.4	83.1	81.6	70.1	65.8	51.9	46.1	27.3	23.7				
10	93.7	93.6	83.2	80.9	69.5	66.0	50.5	45.7	26.3	23.4				
12	93.0	92.0	81.6	80.5	69.3	65.5	50.0	46.0	26.3	23.9				
14	91.7		81.2		68.4		48.9		25.6					
16	91.4		81.5		68.2		49.0		25.8					
70-FOOT TREES														
8	94.8	94.7	87.0	82.9	75.3	69.7	61.0	52.6	42.9	35.5	23.4	17.1		
10	94.7	93.7	85.3	83.2	74.7	70.5	60.0	54.7	42.1	37.9	22.1	18.9		
12	93.9	93.0	84.2	82.5	73.7	71.1	58.8	56.1	40.4	38.6	21.1	20.2		
14	92.5		82.7		72.9		58.6		39.8		21.1			
16	92.1		82.9		73.0		57.9		40.1		20.4			
80-FOOT TREES														
10	94.7		87.4		78.9		66.3		51.6		34.7		16.8	
12	93.9		85.1		76.3		64.9		50.9		34.2		17.5	
14	93.3		83.6		75.4		63.4		50.0		34.3		17.9	
16	92.8		83.0		74.5		63.4		49.7		35.3		18.3	

* Basis: presumably 1109 trees in hardwood slope and flat types, 341 trees in swamp type. Source: after Zon, 1914.

type has a more rapid taper than trees growing in the hardwood slope and flat types.

Canada Forest Service (1930, 1948) form-class tables, which were developed from the taper studies of Wright (1923, 1927), are shorter because the values are listed according to percentages of the total tree height and not by absolute heights as presented by Zon (1914). The higher form classes have the least taper, as expected (Table 71). Relatively, taper differences are greater between the two highest form classes than between any other adjacent form class.

Gevorkiantz and Olsen (1955) prepared a composite taper table for average conditions in Lake States species (Table 72). When the diameter inside bark at half the height is 63 per cent of dbh, the diameter outside bark is 68 per cent. This composite taper table underestimates the volume for balsam fir by 3 per cent.

TABLE 71. TAPER IN PERCENTAGE OF DIAMETER BREAST HIGH
INSIDE BARK AT TEN-PER-CENT INTERVALS ABOVE THE
GROUND FOR BALSAM FIR IN CANADA, BY FORM CLASS *

Percentage of Total Height	Form Classes			
	60	65	70	75
Breast height	100.0	100.0	100.0	100.0
10	93.3	94.0	95.0	95.8
20	86.9	88.2	89.8	91.6
30	79.4	82.0	84.0	87.3
40	70.5	74.6	77.5	81.8
50	60.0	65.0	70.0	75.0
60	49.3	54.4	59.0	65.1
70	38.0	42.2	46.6	53.0
80	26.1	29.5	33.0	38.7
90	13.4	15.5	17.9	22.0

* For definition of form class, see text.
Basis: 368 trees in Manitoba, Quebec, and New Brunswick.
Source: Canada Forest Service, 1948.

TABLE 72. TAPER IN PERCENTAGE OF DIAMETER INSIDE BARK BREAST HIGH
AT TEN-PER-CENT INTERVALS ABOVE THE GROUND FOR BALSAM
FIR IN THE LAKE STATES

Percentage of Total Height	Total Height of Trees (Ft)								
	20	30	40	50	60	70	80	90	100
Ground line	100	101	102	104	107	109	110	110	110
10	99	96	94	91	89	88	88	87	86
20	94	90	85	81	80	80	81	81	81
30	87	82	76	75	75	75	76	76	76
40	78	73	69	69	69	69	70	70	70
50	66	63	63	63	63	63	63	63	63
60	54	52	55	56	56	56	56	56	56
70	39	40	44	47	48	48	48	48	48
80	26	26	31	36	38	38	38	38	38
90	14	14	18	22	24	24	24	24	24

Basis: presumably 47 stands with 17 species represented.
Source: after Gevorkiantz and Olsen, 1955.

Stump-dbh taper tables have been prepared to estimate dbh classes of the trees that were logged. Canadian form-class tables include such data, an example of which is reproduced in Table 73.

Although Zon (1914) attempted to relate stem form of balsam fir to different climatic and edaphic conditions, later investigators largely ignored these influ-

TABLE 73. RELATIONSHIP BETWEEN STUMP DIAMETER OUT-SIDE BARK AND DIAMETER BREAST HIGH OUTSIDE BARK OF BALSAM FIR IN CANADA *

Stump Diameter outside Bark (In.)	Diameter Breast High outside Bark (In.)
6	5.3
8	7.2
10	9.1
12	10.9
14	12.6
16	14.4
18	16.2

* Stump height 1.5 ft.
Basis: 308 trees.
Source: after Canada Forest Service, 1948.

ences. In the following section stand conditions such as structure, composition, density, and age which influence stem form will be considered in relation to stand volume.

STEM VOLUME

In general the tendency is to prepare average volume tables for large administrative areas or regions. For the vast balsam fir range in Canada, the Canada Forest Service (1930, 1948) prepared composite form-class tables, assuming that three form classes were adequate for most of the variations of climate, soils, and stand conditions that might be encountered. In the United States volumes have been averaged primarily by regions, e.g., the Lake States and the northeastern United States. Tables have also been prepared for individual states. With respect to measurements volume tables have been prepared for total stem volume and merchantable volume, the latter often in terms of specified stump-heights and minimum top-diameters. Volumes may be given in cubic feet, board feet, or cords. Tree volume tables that include branch volume are rare. Stump and top volumes as percentages of total volume have been reported by Morawski et al. (1958), but information on the amount of foliage is lacking.

The volume of individual stems is so influenced by stand conditions that the influence of site becomes secondary. Site plays a more decisive role in total yield, and its mensurational aspect will be discussed in more detail in conjunction with yield tables.

Total-stem volume tables. Zon (1914) presented total-stem volume tables in cubic feet for balsam fir in several northeastern states. Volume for New York trees is shown in Table 74. Average volumes of balsam fir of the same diameter and height were greater in Maine than in New York, but the differences may not have been statistically significant. Local cubic-foot volume tables for balsam fir in central Maine were published by Funking and Young (1955). A volume table by Brown and Gevorkiantz (1934) for total peeled balsam fir in Wisconsin included the stump and top volume (Table 75). Brown and Kaufert (1936) also prepared a volume table for balsam fir in Minnesota and Wisconsin.

The form-class volume tables used in Canada were developed to eliminate the need for local volume tables, which are difficult to compare and of variable quality. To determine the average form quotient, Canadian foresters recommended that the diameters inside bark of 50 trees in the stand be measured at breast-height and at one-half the total height. Canadian tables issued in 1930 were amended and corrected by Mulloy and Kearney (1941) and revised in 1948.

TABLE 74. TOTAL CUBIC-FOOT VOLUME OF BALSAM
FIR IN NEW YORK [*]

dbh (In.)	Total Height (Ft)						
	20	30	40	50	60	70	80
4 ...	0.96	1.43	1.91				
6 ...		3.15	4.19	5.23	6.24	7.24	
8 ...			7.25	9.01	10.76	12.51	14.24
10 ...				13.59	16.23	18.86	21.47
12 ...					22.38	26.06	29.72
14 ...					29.12	33.98	38.74
16 ...					36.53	42.52	48.59

[*] Volume with bark, includes stem and stump.
Basis: presumably 947 trees.
Source: after Zon, 1914.

TABLE 75. TOTAL PEELED CUBIC-FOOT VOLUME OF
BALSAM FIR IN WISCONSIN [*]

dbh (In.)	Height of Trees (Ft)							
	10	20	30	40	50	60	70	80
1	0.03	0.07	0.11					
2	0.12	0.22	0.32					
3		0.48	0.69	0.91				
4		0.85	1.21	1.56				
5		1.27	1.82	2.37	2.92			
6		1.81	2.59	3.37	4.15			
7			3.56	4.52	5.50	6.48		
8			4.50	5.80	7.10	8.40		
9				7.20	8.80	10.5	12.1	
10					10.8	12.7	14.7	16.6
11					12.9	15.3	17.7	20.1
12					15.1	17.8	20.5	23.3
13					17.5	20.6	23.8	26.9

[*] Basis: 60 trees.
Source: after Brown and Gevorkiantz, 1934.

In Table 76 the total volume of balsam fir, including stump and top, is given for form-class 65. The volumes for the latter are about 5 per cent less than the volumes for form-class 70.

Total volume in cubic feet can be more accurately related to stand and site conditions than merchantable volume, especially when the latter is expressed in cords or board feet. In 1914 Zon attempted to assess the effects of stable site factors upon taper and tree volume (Table 70), but he was almost alone in this. Prielipp (1956) investigated the influence of soil moisture conditions on the volume of balsam fir in northern Michigan. Of the three soil moisture regimes examined, intermediate or transition sites had the tallest stems and upland sites the shortest stems, for similar diameter classes. Volume tables showed that wet sites produced the greatest individual tree volumes, especially for dbh classes above 10 inches.

Merchantable volume tables. A great number of merchantable volume tables based on different top diameters, and expressed in cubic feet, board feet (fbm), or cords, is available for balsam fir. Brown and Gevorkiantz (1934) prepared a table for peeled merchantable volume of balsam fir in second-growth stands in Wisconsin (Table 77). Merchantable volume ranged from 60 to 90 per cent of total stem volume, the amount depending on dbh. Brown and Kaufert (1936) prepared a merchantable volume table for balsam fir in Minnesota and Wisconsin.

Eight merchantable cubic-foot volume tables for balsam fir, form classes 65 and 70, were prepared by the Canada Forest Service (1930, 1948). The volume tables were constructed for 1.0- and 1.5-foot stump heights, and with 3–4 inch or

TABLE 76. TOTAL CUBIC-FOOT VOLUME OF BALSAM FIR,
FORM CLASS 65, IN CANADA *

dbh (In.)	Total Height of Trees (Ft)									
	10	20	30	40	50	60	70	80	90	100
1	0.04	0.06	0.08							
2	0.15	0.23	0.31	0.39						
3	0.33	0.52	0.70	0.89						
4	0.59	0.92	1.25	1.59	1.92					
5		1.44	1.95	2.48	2.98					
6			2.80	3.57	4.30	5.02				
7			3.81	4.85	5.86	6.85	7.94			
8			4.99	6.34	7.68	8.96	10.3	11.7		
9			6.31	8.04	9.70	11.4	13.1	14.8		
10				9.95	12.0	14.0	16.0	18.3		
11				12.1	14.5	17.0	19.5	22.2	24.6	
12				14.3	17.2	20.3	23.2	26.4	29.2	
13					20.3	23.7	27.3	30.9	34.3	
14					23.5	27.6	31.6	35.9	39.8	
15					27.0	31.6	36.2	41.1	45.7	
16					30.7	35.9	41.2	46.8	52.0	57.8
17					34.7	40.6	46.6	52.8	58.8	65.1
18					38.8	45.4	52.2	59.3	66.0	73.2
19						50.6	58.2	66.1	73.6	81.6
20						56.1	64.5	73.2	81.5	90.5

* Basis: 368 trees in Manitoba, Quebec, and New Brunswick.
Source: Canada Forest Service, 1948.

varying top diameters. Merchantable volume averaged about 95 per cent of total volume.

Board-foot volume tables based on the Scribner Decimal C, Dimick, Bangor, and Maine log rules were prepared for balsam fir by Zon (1914). The Scribner Decimal C volume table was reprinted by Brown and Gevorkiantz (1934). Form-class tables used in Canada (Canada Forest Service, 1930, 1948) also included board-foot volume tables. Duerr, chmn. (1953) prepared board-foot volume tables, using dbh and total height, for balsam fir and other species in the Adirondack and Catskill areas of New York. These volume tables were based on the International ¼-inch log rule for 16.3-foot logs and half-logs, with a 1-foot stump, and a top diameter of 6 inches inside bark. Error estimates at the 5 per cent

TABLE 77. MERCHANTABLE PEELED CUBIC-FOOT VOLUME
OF BALSAM FIR IN WISCONSIN *

dbh	Total Height of Trees (Ft)											
(In.)	20	25	30	35	40	45	50	55	60	65	70	75
4	0.50	0.60	0.70	0.80	0.90	1.00						
5		1.13	1.33	1.53	1.73	1.93	2.13					
6		1.80	2.12	2.44	2.76	3.08	3.40	3.72				
7			3.10	3.52	3.93	4.36	4.76	5.18	5.60	6.02		
8				4.63	5.20	5.76	6.32	6.88	7.46	8.01		
9					6.48	7.20	7.92	8.66	9.40	10.1		
10						8.91	9.83	10.7	11.6	12.5	13.4	
11							11.9	13.0	14.1	15.2	16.3	
12							14.0	15.3	16.6	17.9	19.1	20.4
13							16.3	17.7	19.2	20.6	22.1	23.5

* Stump height 1 ft; top diam. 3 in. inside bark.
Basis: 58 trees.
Source: Brown and Gevorkiantz, 1934.

level were computed to be ±35 per cent for individual trees and ±3.5 per cent for a 100-tree sample.

Canadian form-class tables (Canada Forest Service, 1930, 1948) include 20 different board-foot volume tables for balsam fir. For pulpwood operations, tables for form classes 65 and 70 are given for each log rule. For sawlog operations, tables were prepared for form classes 60, 65, and 70. The tables were based on the Doyle, New Brunswick, Quebec, and Scribner log rules.

A standard cord in the United States and Canada is generally accepted as a pile of wood 4 x 4 x 8 feet, i.e., 128 cubic feet of wood, bark, and air space, or wood and air space if peeled. Zon (1914) recommended the use of conversion factors to change standard cords of balsam fir to a cubic-foot measure, as follows: standard cords of 4-foot billets, 4–7 inches in diameter, 0.72; 8–12 inch billets, 0.74; "average-size" pulpwood, 0.74.

The Lake States Forest Experiment Station (1935) published a table showing that the solid content of stacked cordwood bolts including bark ranges from 60 to 100 cubic feet per standard cord. To exclude bark, the table values should be reduced by 12 cubic feet.

Total and merchantable tree volumes of balsam fir in cords were given by Zon (1914). Brown and Gevorkiantz (1934) and Brown and Kaufert (1936) prepared volume tables for unpeeled merchantable volume in cords for balsam fir in Wisconsin and Minnesota.

Merchantable volume tables are highly diversified in being prepared not only for very different stand conditions but to serve different types of utilization practices. Therefore only one table is presented here to serve as an illustration (see Table 77).

Composite-species volume tables. The principle underlying a composite or average volume table for several species in an area is not new. Composite volume tables originated when a table prepared for one species was applied to another, a common practice since the beginnings of forest mensuration several centuries ago. In 1918 Cary presented a cubic-foot volume table for spruce, based on 2500 trees in Maine, New Hampshire, and New York, with the suggestion that the same table, if reduced by 8 per cent, could be used for balsam fir. Composite pulpwood tables were published by the Lake States Forest Experiment Station (1943 — these were prepared by Gevorkiantz) and by Gevorkiantz (1945).

Spurr (1952) prepared composite total and merchantable volume tables for four different spruces and balsam fir applicable in both the Lake States and the northeastern United States. The composite total volume tables need to be reduced by 7 per cent for balsam fir, 9 per cent for Norway spruce, 12 per cent for white spruce, and 13 per cent for black spruce. No correction was needed for red spruce. The composite merchantable volume table needs to be reduced by 6 per cent to make it applicable to balsam fir.

Gevorkiantz and Olsen (1955) prepared composite volume tables based on the average volumes of 17 Lake States species measured in 47 stands and including several thousands of trees. The basic formula for peeled cubic foot volume was given as $V = 0.42\ BH$, where V is total peeled volume in cubic feet, B is basal area outside bark at breast height in square feet, and H is total height of tree in feet. The composite tables can be used without correction if an error allowance of ± 6 per cent of stand volume is tolerated. The standard deviation for individual tree cubic-foot volume was estimated at ± 10 per cent. The composite tables can be adapted to differing form quotients ($q_{.5}$), bark volume, and species taper, where these vary from the standards that are built into the tables. For example, the table assumes a bark volume of 14 per cent. For balsam fir, this value is too high. Balsam fir bark volume constitutes 12 per cent of the total cubic-foot volume on the average in the Lake States. To obtain the actual balsam fir bark volume for more precise volume corrections, it is necessary to multiply the bark percentage at dbh by 2.0. After volume corrections are made for bark thickness and differences in form quotient, the residual correction is for species taper. Thus, in order to obtain balsam fir volume which is corrected for species taper, the composite table values must be increased by 3 per cent.

Gevorkiantz and Olsen (1955) also prepared separate volume tables based on dbh and total height, and on dbh and number of 8-foot bolts to 3- and 4-inch top diameters inside bark. They also compiled two board-foot volume tables, one each for the International ¼-inch and Scribner Decimal C log rules.

Special methods for entire stands. The most common practice followed to obtain stand volume is to add the volumes obtained for the individual trees in the stand. Tree heights can be determined for each tree separately, or more commonly, height curves are constructed for whole stands by dbh-classes. Instead of constructing special height curves for each stand, standard height curves can be used. For the last decades mensuration techniques (tariff methods) have been developed which make use of the advantages of standardization in the construction of volume tables. Although the actual term "tariff" is seldom used in American forestry literature, the principles behind such methods have been applied to a number of forest stands, including balsam fir, in New York. Workers at the New York State College of Forestry at Syracuse have been particularly active in the construction of tariff tables. These tables can be used for stand volume determination from ground surveys (Duerr, chmn., 1953), stand volume estimation from aerial photographic surveys (Ferree, 1953), and growth-rate determinations for several growth zones in New York (Ferree and Hagar, 1956).

In the New York study, tariffs (the actual term used was "site," defined in a broad manner) were first determined from dominant and codominant trees, 19 inches dbh or larger, by estimating the number of 16-foot logs per tree up to a 4-inch top diameter inside bark. Tariff I ("site 1") trees had at least 5 logs, Tariff II, 3–4½ logs, and Tariff III, less than 3 logs per tree (Duerr, chmn., 1953). Originally, a swamp site was included (see Table 78). Its later exclusion actually improved the conceptual homogeneity of the method by eliminating the mechanical mix-up of mensurational and ecological procedures. As the data indicate, the swamp site had an intermediate position between Tariff Classes II and III. The same tariffs can be applied over a wide range of sites within a region, provided the mensurational definitions (in this case, number of 16-foot logs per tree) are the

TABLE 78. PEELED MERCHANTABLE VOLUME IN CUBIC FEET OF BALSAM FIR AND RED SPRUCE BY TARIFF AND SITE CLASSES IN THE ADIRONDACKS AND CATSKILLS, NEW YORK *

dbh (In.)	Tariff Classes			Swamp Site
	I	II	III	
6	2.0	1.9	1.9	2.1
8	6.1	5.2	3.9	5.2
10	12.5	11.2	8.0	9.4
12	21.5	19.6	14.7	15.5
14	33.5	29.6	22.9	23.4
16	47.2	41.6	32.0	33.4
18	64.4	53.6	42.0	
20	84.6	67.3	53.2	
22	109.2	81.3		
24	138.2	95.7		
26	174.0	112.2		
28	215.6			

* Tariffs express the relationship between dbh and height, originally all classes were called "sites"; stump height 1 ft, top diam. 3.6 in. inside bark.
Basis: presumably 97 plots, 0.2 acre in size.
Source: after Duerr, chmn., 1953.

same. A tree volume table for red spruce and balsam fir in New York, using the tariff approach, is given in Table 78.

Ferree and Hagar (1956) prepared tariffs by using total height rather than number of logs as the basis for volume estimation. A two-tariff table showing height and diameter relationships for red spruce and balsam fir in New York is given in Table 79. The tables show that red spruce and balsam fir do not agree in the two tariffs. In Tariff I, balsam fir is taller than red spruce for given diameters. In Tariff II, the reverse is true. The error of estimate at the 5 per cent level was computed to be ±25 per cent for single trees and ±2.5 per cent for a 100-tree sample.

TABLE 79. MINIMUM TOTAL HEIGHT IN FEET BY DIAMETER-BREAST-HIGH CLASSES OF DOMINANT AND CODOMINANT BALSAM FIR AND RED SPRUCE FOR DETERMINATION OF TARIFF ("SITE") CLASSES IN NEW YORK *

dbh (In.)	Tariff I		Tariff II	
	Balsam Fir	Red Spruce	Balsam Fir	Red Spruce
6...............	20	19	16	16
8...............	35	32	27	27
10..............	49	43	34	36
12..............	59	53	40	45
14..............	69	62	45	49
16..............	70	68	48	52
18..............	71	71	50	54
20..............		74		55

* Tariffs express the relationship between dbh and height, originally called "sites"; stands below the requirements of Tariff II belong to Tariff III.
Basis: 97 plots, 0.2 acre in size.
Source: after Ferree and Hagar, 1956.

Local stand and stock tables for balsam fir have been prepared by different workers concurrently with tables for other species. Stand tables showing the distribution of trees per unit area by diameter classes have been prepared for balsam fir in central New Brunswick by Baskerville (1960). Combined stand and stock tables giving number of trees and volume per acre by diameter classes have been prepared by Ray (1956) for several site types in the Lake Edward Experimental Forest, Quebec, and by Jarvis (1960) for the Goulais watershed in Ontario. These tables, prepared for uncut, cutover, and burned softwood and mixed-wood types, show trees and volumes per acre for balsam fir, spruce, paper birch, yellow birch, maple, and other species.

Formulas are also available for determining the volume of entire stands. Bedell and Berry (1955) developed a stand-volume formula, directly applicable to balsam fir: $V = 0.45 H - 3.1$, where V is merchantable volume in cubic feet per square foot of stand basal area, and H is average stand height in feet. Merchantable volume was defined as all trees 4 inches dbh and over, above a 1-foot stump, and up to a 3-inch top diameter inside bark.

USE OF AERIAL PHOTOGRAPHS

Species identification is one of the major problems encountered in the use of aerial photographs. The species commonly associated with balsam fir include spruces, pines, hemlock, northern white-cedar, tamarack, and various hardwoods. Aerial photographic descriptions of balsam fir in comparison with other species have been given by Spurr (1948) and Sayn-Wittgenstein (1960, 1961).

According to these sources, the identification of balsam fir on aerial photographs is not too difficult. Hardwoods can, of course, be easily separated on winter photographs. Hardwoods can also be distinguished on summer photographs taken with infrared film. Softwoods reflect less infrared radiation and therefore have darker tones than hardwoods. On summer panchromatic films identification is more difficult, but softwoods are frequently darker with a finer and more regular texture, and with more pointed crowns than hardwoods. Fall photographs taken with panchromatic film easily distinguish between softwoods and hardwoods, since the colored leaves of the latter register a light tone that makes them quite conspicuous.

Distinguishing among conifers presents more of a problem. Balsam fir has a very dense, narrow, cone-shaped, symmetrical crown capped by a compact, rigid, spire-shaped tip. On infrared photographs red spruce appears in a darker tone than balsam fir. The long conical crown also serves to identify red spruce. The tone of black spruce is darker than balsam fir, but white spruce usually appears in a lighter tone. Balsam fir branches appear less prominent than those of white spruce. The tone of jack pine is much like balsam fir, but may be distinguished by its irregular crown, its thinner foliage, and its association with certain recognizable sites. The tone of red pine is also similar to balsam fir, although the crowns of the former species are large and spreading. White-pine crowns are more star-shaped, and the tone is lighter than balsam fir. White-cedar, though similar in tone to balsam fir, may be distinguished by its more rounded crowns. On infrared photographs hemlock and tamarack appear in a lighter tone than balsam fir, and may be nearly as light as hardwoods.

On large-scale color photographs the lighter cones of white spruce can be easily distinguished from the purplish-green cones of balsam fir, and the small purple (immature) or brown (mature) cones of black spruce.

Stem and crown diameter relationships of balsam fir, taken in Minnesota and Michigan, show that stem dbh in inches averages 2–3 inches less than crown diameter in feet. Since balsam fir is commonly an understory tree and not always visible in aerial photographs, this rough relationship can only be used where crowns can be measured. In volume tables prepared by aerial photogrammetric methods, crown diameters substitute for dbh. Heights can be measured from aerial photographs, however, by appropriate scales based on stereoscopic displacement (Spurr, 1948).

The Northeastern Forest Experiment Station (1947) prepared a table showing peeled merchantable cubic-foot volume for red spruce, balsam fir, and eastern hemlock from measurements of 1646 trees in stands in the White Mountain National Forest, New Hampshire (Table 80). The standard error of estimate was given as ±32 per cent for individual trees. As computed by Spurr (1948), correlation indexes of 0.715 for crown diameter and cubic-foot volume, 0.706 for total height and volume, and 0.870 for tree volume and both height and crown diameter were obtained for the data in Table 80.

An aerial-photo volume table for spruce and balsam fir was prepared by Spurr (1952) for the eastern United States, the Lake States, and Canada. The computed standard error of estimate was ±51 per cent.

Ferree (1953) devised a method for estimating timber volume from aerial photographs based upon cover types, tariffs, and stand-size classes for different photographic scales. Adjustments for understory trees were also included.

TABLE 80. PEELED MERCHANTABLE VOLUME IN CUBIC FEET
FOR RED SPRUCE, BALSAM FIR, AND EASTERN HEMLOCK IN
NEW HAMPSHIRE *

Visible Crown Diameter (Ft)	Total Height (Ft)				
	30	40	50	60	70
4	1.1	2.7	4.3		
6	2.6	4.2	6.2	9.5	
8	3.6	5.5	8.2	12.4	
10	4.3	7.3	11.1	15.7	22.8
12	4.5	9.4	14.6	19.3	25.6
14		12.5	17.7	22.7	29.2
16			21.3	26.4	34.4
18			24.2	29.6	41.8
20			26.3	32.4	47.1
22				35.1	52.0

*Stump height 1 ft, top diam. 4.0 in. inside bark.
Basis: 1646 trees in White Mountain National Forest.
Source: Northeastern Forest Experiment Station, 1947.

UNUSED AND DEFECTIVE MATERIAL

Schenck (1939) noted that in virgin balsam fir stands, the volume of standing dead and down timber was equivalent to the volume of living trees. In extensively managed forests of central Minnesota the total volume of standing dead and down balsam fir varied 50–150 per cent of the live volume on wet sites and 10–50 per cent on mesic sites (Bakuzis and Hansen, 1961). In Koochiching County, northern Minnesota, natural mortality was estimated at 49 per cent of total growth in the spruce-fir-hardwood forest type. Thinning experiments revealed that up to 50 per cent of this volume could be harvested (Hubbard, 1956). Bickford et al. (1961) presented mortality data by species groups for several stand-condition classes in spruce-fir forests of New England (see Table 121). Mortality data with respect to stand development are discussed in Chapter 7. The mensurational aspects are reviewed below in connection with yield tables.

In addition to mortality losses, production losses accumulate from unused bark, stumps, branches, foliage, and damage by tree diseases, insects, and climatic agencies. Although systematic appraisals of production losses have not been made for balsam fir, a scattered amount of information is available. Some of these losses are discussed at greater length in preceding chapters.

In Ontario the volume in cubic feet of the top and stump as a percentage of total stem volume inside bark averaged 70 per cent for trees 20–40 years old, but decreased to about 25 per cent for 150-year-old trees. As shown in Table 26, cull due to decay increased with age, reaching a maximum of 30 per cent for the oldest trees (Morawski et al., 1958).

The cull percentage for balsam fir and red spruce averaged 12 per cent of total cubic volume at the highest elevation in the Adirondacks of New York. At lower elevations corresponding cull percentages were 9 per cent on medium sites and 8 per cent or less on the best sites, depending on the size class of trees (Ferree and Hagar, 1956).

Zon (1914) found bark volume of balsam fir in Maine (Table 81) to average

TABLE 81. BARK VOLUME IN CUBIC FEET FOR
BALSAM FIR IN MAINE *

dbh	Total Height (Ft)				
(In.)	40	50	60	70	80
7....................	0.6	0.7	0.9	1.0	
8....................	0.7	0.9	1.1	1.3	1.5
9....................	0.9	1.1	1.4	1.6	1.9
10...................		1.3	1.6	2.0	2.3
11...................		1.6	1.9	2.3	2.7
12...................			2.2	2.7	3.1
13...................			2.5	3.1	3.6
14...................			2.8	3.4	4.0
15...................			3.2	3.8	4.5
16...................			3.5	4.2	5.0

* Basis: presumably 330 trees.
Source: after Zon, 1914.

10.6 per cent of the total volume of the stem. Gevorkiantz and Olsen (1951) determined that bark volume of balsam fir in the Lake States averaged 12 per cent of the total cubic-foot volume. The difference between these volumes may be attributed to inherent differences in the trees from the two regions, or to varying sampling techniques.

Double bark thickness of balsam fir at dbh was computed in Canadian form-class tables as reproduced in Table 82. Bark thickness is strongly influenced by diameter. Measurements of single bark thickness at different heights above the ground for 381 trees from the Clay Belt of Ontario (Bonner, 1941) are given in Table 83.

Gevorkiantz and Olsen (1950) studied defects in merchantable balsam fir in the Lake States (see Table 84). Stems having more than 60 per cent cull were considered unmerchantable and were not included in the tally. The primary variable associated with increased cull was age of the trees.

The amount of unusable material in balsam fir, exclusive of roots and foliage, was computed by Zon (1914) from stands in Maine and New York to average nearly 30 per cent of the total volume. This figure included 8.4 per cent for the stump and top less than 4 inches in diameter, 10.6 per cent for bark, and 11.2 per cent for cull in the usable portion of the bole.

TABLE 82. DOUBLE BARK THICKNESS IN INCHES AT BREAST HEIGHT FOR BALSAM FIR IN CANADA [*]

dbh (In.)	Double Bark Thickness	dbh (In.)	Double Bark Thickness
3	0.1	12	0.7
4	0.2	13	0.7
5	0.2	14	0.8
6	0.3	15	0.9
7	0.4	16	0.9
8	0.4	17	1.0
9	0.5	18	1.0
10	0.6	19	1.1
11	0.6	20	1.2

[*] Basis: 368 trees in Manitoba, Quebec, and New Brunswick.
Source: Canada Forest Service, 1930.

TABLE 83. SINGLE BARK THICKNESS IN INCHES FOR BALSAM FIR IN THE CLAY BELT, ONTARIO [*]

dbh (In.)	Height above Ground (Ft)						
	1	10	20	30	40	50	60
4	0.17	0.10					
5	0.20	0.12	0.11				
6	0.22	0.15	0.13	0.11			
7	0.25	0.17	0.15	0.13	0.11		
8	0.27	0.20	0.18	0.16	0.13		
9	0.30	0.22	0.21	0.18	0.16		
10	0.32	0.25	0.24	0.21	0.19		
11	0.35	0.28	0.26	0.25	0.22	0.18	
12	0.38	0.31	0.29	0.28	0.25	0.21	
13	0.40	0.34	0.33	0.31	0.28	0.25	0.19
14	0.43	0.39	0.37	0.35	0.32	0.28	0.22

[*] Basis: 381 trees.
Source: after Bonner, 1941.

TABLE 84. AVERAGE PERCENTAGE DEFECT IN MERCHANTABLE UPLAND BALSAM FIR STANDS IN THE LAKE STATES [*]

Stand Age at dbh (Years)	Average Defect (%)	Stand Age at dbh (Years)	Average Defect (%)
10	0.5	60	8.0
20	1.4	70	10.3
30	2.6	80	12.8
40	4.1	90	15.4
50	6.0	100	18.0

[*] Defect is all culled material not usable for pulpwood.
Basis: number of plots unknown.
Source: after Gevorkiantz and Olsen, 1950.

Doyle (1952), working in the Gaspé Peninsula, New Brunswick, and Nova Scotia, found that 26.6 per cent of all merchantable volume (excluding bark) of balsam fir was left unused in the forest because of defects or inconveniences. This figure included 2.1 per cent for high stumps, 5.1 per cent for long butts, 1.8 per cent for defective trees, and 17.6 per cent for tops. Closer utilization and better cutting practices could perhaps have reduced this high degree of loss.

Yield Tables

Attempts have been made to prepare yield tables for pure, fully stocked balsam fir stands, spruce-fir stands, and stands of very diversified conditions. Existing yield tables have been mainly prepared from temporary plots and forest surveys. In the future, data from permanent plots and repeated surveys will give rise to better yield tables for balsam fir and other species. Both normal and special-use yield tables need to be developed for various site conditions.

MEASUREMENT OF SITE

Yield depends not only on site potential, but on age, past treatment of the stand, stand composition, and stand structure. Since species react differently to the same conditions, results obtained with one species are often not directly applicable to another. In practice, because of the need for simplification, this is not always observed. Duerr, chmn. (1953) combined merchantable yield data for balsam fir and red spruce. Linteau (1955) did not differentiate between balsam fir and spruce within forest types. Bowman (1944) likewise did not differentiate balsam fir from spruce in his site measurement studies. He also suggested that aspen might serve as a site indicator for other species. Westveld (1941, 1953) also combined balsam fir and spruce in his yield tables.

Separate yield tables for balsam fir, red spruce, and white spruce were prepared by Meyer (1929). Separate site-index curves are available for balsam fir, white spruce, and black spruce in the Lake States (Gevorkiantz, 1956; 1957a,b). Differences between red spruce and balsam fir were also recognized by Ferree and Hagar (1956).

Within regions, differences in site have been studied by a number of authors using different methods. Ferree and Hagar (1956) gave quantitative measures of site for comparing several growth zones in the spruce-fir types of the Adirondacks in New York.

Differences in the growth of balsam fir arising from edaphic variations within a geographical region have been recognized by several workers, but systematic studies are lacking. In 1962 a major investigation by Linteau, of edaphic effects on the growth of several species in the northeastern part of the Boreal Forest Region, was in progress.

In spite of their deficiencies, site indexes based on height-age relationships are still widely used for growth and yield studies. Yield tables based on site index have been prepared for balsam fir by Meyer (1929), and for spruce-fir by Bowman (1944). Because reference ages used by different workers do not agree, comparisons are rather difficult to make. Additional complications arise when the reference

age is to stump height or to breast height. For example, Meyer (1929) used 65 years
at stump, Bowman (1944) and Linteau (1955) used 50 years at stump, and Gevorkiantz (1956) used 50 years at breast height as reference ages for site-index
curves. In general a reference age of 65 years at a one-foot stump height corresponds to 50 years at breast height (see Figure 11).

Linteau (1955) assumed a common site index for balsam fir and spruces. He
found that forest types for Section B-1 of the Boreal Region in Canada (after the
classification by Halliday, 1937) could be assigned to one of four 10-foot site-
index classes at 50 years. About 88 per cent of the stands sampled were found in
the yield class to which the type was assigned. Linteau also presented detailed
type descriptions, listing for each the physiographic and soil properties, shrub
species, and ground vegetation which could be used to identify types under modified forest conditions (see Table 5).

The difficulties of age determination in balsam fir have made site-index curves
based on stump height rather questionable. To exclude part of the variability
caused by suppression of advance reproduction or by butt rot, the age of balsam
fir for site-index purposes is determined at breast height.

The shortcomings of height measurements in site-index determinations were
discussed by Vincent (1961), who based his investigations on measurement of
trees in New Brunswick. Vincent noted that 77 per cent of all balsam fir leaders
were repeatedly damaged by spruce budworm attacks. Winter drying and late
frosts may also affect height growth. Suppression of advance reproduction may
last for 30 years, and in exceptional cases for 100 years. For example, in a stand
having 5300 stems per acre, the average height of dominants was 32 feet, but in a
stand having 1000 stems per acre, the height was 43 feet. The stands were of the
same age and grew on similar sites. Vincent suggested substituting a growth-
intercept method for total height in site-index determination. The five-year period
of height growth, beginning at breast height was specifically recommended.

In 1950 Gevorkiantz and Olsen suggested that a volume index be used in place
of total height. The volume index is the product of the average basal area and the
average height of codominant and dominant trees. It is directly applicable when
stand "competition index" is normal, i.e., when the latter value is 0.8.

The competition index is $c = d/p$, where c is the index, d is total stand basal
area in percentage of a "normal" basal area, and p is basal area of balsam fir in
percentage (exclusive of suppressed trees) of total basal area. The stand-volume
index is adjusted to the competition index by a variable factor which is 1.26 for
$c = 0.2$, 1.00 for $c = 0.8$, and 0.75 for $c = 1.6$, a total range of ± 25 per cent.
Table 85 reproduces the site-class table, based on volume indexes, for upland
balsam fir in the Lake States (Gevorkiantz and Olsen, 1950). Table 86 shows correction factors for volume index. Table 87 shows the assumed standard or normal
basal area for stand density determinations.

For balsam fir in the Lake States several sources indicate that, on the average,
10 years are needed to attain breast height on good sites and 15 years on poor sites.
However, these figures should be used with caution.

Simple guides for the approximate evaluation of site index, based on height at
50 years under modified natural stand conditions, were presented by Bowman

TABLE 85. VOLUME INDEXES OF AVERAGE DOMINANT AND CODOMINANT UP-LAND BALSAM FIR FOR SITE CLASS DETERMINATION IN THE LAKE STATES *

Stand Age at dbh (Years)	Site Class				
	Poor	Fair	Medium	Good	Excellent
10	Less than 0.2	0.2	0.3	0.4	0.5 or more
20	" " 1.6	1.6–2.5	2.6–3.2	3.3–4.4	4.5 " "
30	" " 3.5	3.5–5.6	5.7–7.4	7.5–9.9	10.0 " "
40	" " 6.3	6.3–10.0	10.1–13.2	13.3–17.6	17.7 " "
50	" " 10.0	10.0–15.9	16.0–20.9	21.0–27.9	28.0 " "
60	" " 14.2	14.2–22.7	22.8–29.9	30.0–39.9	40.0 " "
70	" " 16.5	16.5–26.5	26.6–34.8	34.9–46.5	46.6 " "
80	" " 18.0	18.0–28.9	29.0–38.0	38.1–50.7	50.8 " "

* Volume index is the product of basal area of the average tree in square feet and total height in feet.
Basis: number of plots unknown.
Source: after Gevorkiantz and Olsen, 1950.

TABLE 86. CORRECTION FACTORS FOR THE FIRST ESTIMATES OF VOLUME INDEXES OF AVERAGE DOMINANT AND CODOMI-NANT UPLAND BALSAM FIR IN THE LAKE STATES *

Competition Index	Correction Factor	Competition Index	Correction Factor
0.2	1.26	1.0	0.91
0.4	1.17	1.2	0.84
0.6	1.09	1.4	0.79
0.8	1.00	1.6	0.75

* For volume index and competition index definitions see text.
Basis: number of trees or plots unknown.
Source: after Gevorkiantz and Olsen, 1950.

TABLE 87. ASSUMED NORMAL BASAL AREA IN SQUARE FEET PER ACRE FOR UPLAND BALSAM FIR STANDS IN THE LAKE STATES *

Average Stand dbh (In.)	Stand Basal Area (Ft²/Acre)	Average Stand dbh (In.)	Stand Basal Area (Ft²/Acre)
2	84	7	169
3	117	8	174
4	137	9	177
5	151	10	179
6	161	12	180

* Basis: number of plots unknown.
Source: after Gevorkiantz and Olsen, 1950.

(1944) for northern Michigan. For the major soil conditions, these are outlined in the accompanying tabulation.

Balsam Fir *Site Index*	*Soil Conditions*
40–50.	Loams, clays, clay loams, and silty soils if well drained, especially on limestone areas
35–40.	Sandy loams and sands
35　.	Mucks
30–35.	Peats of average thickness and drainage
15–25.	Very wet peats

Where tolerant hardwoods are admixed with white spruce and balsam fir, very high site indexes, often 50–60, occur. Aspen or other poplars are usually mixed with white or black spruce and balsam fir on well-drained soils where site indexes of 40 are found. Swamp sites, when balsam fir is mixed with black spruce or white-cedar, indicate a site index of 35. With respect to ground cover, predominance of bracken fern, dwarf dogwood, upland blue grass, and upland blueberry indicate site indexes of 40 or more. If the ground cover is mainly sphagnum, Labrador tea, sedgegrass, lowland blueberry, and alder, the site index rarely exceeds 35. Annual ring measurements made on old stumps or residual trees, and height growth of young reproduction have also been considered as indicators of site.

In his relative stand-density yield tables, Westveld (1941, 1953) divided the spruce-fir stands of the northeastern United States into two broad site groups: (1) softwoods predominating and (2) softwoods as secondary species. The spruce-swamp type, upper spruce slope, old field spruce, balsam fir flat, red spruce–balsam fir–paper birch, and red spruce–yellow birch types were included in the first group. Types in which red spruce and balsam fir were variously mixed with tolerant hardwoods, especially yellow birch, sugar maple, and beech on lower slopes and better drained soils were included in the second group.

NORMAL YIELD TABLES

Normal yield tables give yields and other data for fully stocked stands growing in undisturbed conditions or under specific types of management. The tables are actually quantitative summaries of the growth conditions in the forest stands to which they apply. They can be used to make comparisons of growth between different species growing under various climatic, edaphic, and biotic conditions, including changes introduced by man. With the aid of correction factors, normal yield tables can also be applied to understocked and overstocked stands. The data given in balsam-fir yield tables together with additional information from other sources are reviewed in the following sections.

Height growth. The maximum height reached at maturity by balsam fir under optimum growing conditions in the eastern part of its range is about 90 feet. Yield tables for the northeastern states (Meyer, 1929) show site indexes of 40–70 feet at an age of 65 years. In the Lake States site-index range is 30–70 feet at 50 years (Gevorkiantz, 1956). The site-index curves are roughly comparable, since the former are based on age at a 1-foot stump, and the latter on age at breast height. An analysis of the two site-index curves shows that height growth culminates in the Northeast at an earlier age than in the Lake States (see Figure 11). According to Bowman (1944), however, the trends of the site-index curves for Michigan follow closely those of Meyer (1929) for the Northeast. Other data also suggest that the heights at maturity shown for balsam fir in the Lake States may be too high. Some of the differences noted may be more apparent than real, their discrepancies arising from sampling errors or curve-fitting techniques.

Zon (1914) compared the height growth of balsam fir on three forest types in New York. The best height growth was made by balsam fir in the hardwood-slope type. Mean annual height growth culminated at 45 years in the spruce-flat type and at 40 years in the hardwood-slope and swamp types. Annual periodic growth culminated at 20–25 years in the spruce-flat type, 25–30 years in the swamp type, and 30–35 years in the hardwood-slope type. The apparent differences in growth culmination ages within a geographic region may result from edaphic conditions, species competition, stand history, and sampling errors.

The effect of aspen competition on height growth of balsam fir and other conifers in the Lake States was shown by Kittredge (1931). Under a light aspen overstory the height growth of balsam fir was only slightly less than that of aspen. A heavy aspen overstory reduced the balsam fir height growth equivalent to a 20-foot difference in site index. The most rapid height growth of balsam fir occurred at 30–40 years of age (Table 57). At 60 years balsam fir in competition with aspen was taller than white pine, red pine, or white spruce.

Plate 16 shows the combined effect of site and competition on the average weighted heights for different species in comparison with balsam fir in Minnesota. Under the limited conditions of central Minnesota, balsam fir is capable of reaching a site index of 60 feet. Even on such favorable sites balsam fir must compete strongly with balsam poplar and aspen, the latter species often overtopping balsam fir. The height-growth curves of balsam fir and several spruces adjusted to a common site-index reference age of 65 years at a 1-foot stump are compared by regions in Figure 11. The basic data for the northeastern states are reproduced in Tables 88, 94, and 96. Data for Michigan (Bowman, 1944) are in Tables 89, 95, and 97. Height curves for the Lake States are redrawn from Gevorkiantz (1956; 1957a,b).

The site-index curves of balsam fir vary only slightly from those of white spruce and red spruce in the northeastern states as shown in Figure 11 and Table 88.

Heimburger (1934) and Loucks (1956) showed that on similar sites, balsam fir has a higher site index than red spruce or white spruce. Bowman (1944) considered the pattern of height growth of balsam fir, white spruce, and black spruce to be nearly the same. His combined data for the three species are given in Table 89.

Lake States data (Gevorkiantz, 1956; 1957a,b) show that balsam fir height growth is more rapid than white spruce up to about 50 years, after which white spruce develops faster. In the Port Arthur district of Ontario, Burgar (1961) found that balsam fir height growth follows that of white spruce up to about 70 years, although diameter growth of balsam fir was less for this period.

In some of the earlier studies height growth was believed to be a good index of volume production. Although this may be true if all conditions are equal, most

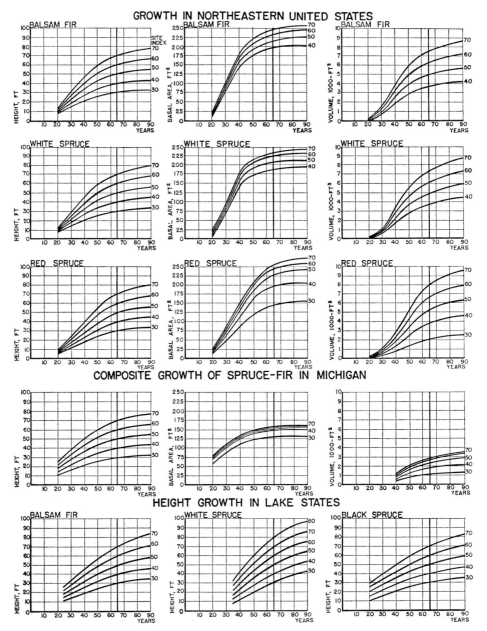

Figure 11. Height, basal area, and volume growth of balsam fir and spruces in the northeastern United States, composite growth of spruce-fir in Michigan, and height growth in the Lake States, by site classes based on a common reference age of 65 years at the stump in fully stocked, even-aged stands. (After Meyer, 1929; Bowman, 1944; and Gevorkiantz, 1956, 1957a,b.)

266

TABLE 88. TOTAL HEIGHT IN FEET OF AVERAGE DOMINANT AND CODOMINANT TREES OF BALSAM FIR, WHITE SPRUCE, AND RED SPRUCE IN FULLY STOCKED AND EVEN-AGED STANDS IN THE NORTHEASTERN UNITED STATES BY SITE INDEXES *

Total Age (Years)	Site Index of Balsam Fir					Site Index of White Spruce					Site Index of Red Spruce				
	30	40	50	60	70	30	40	50	60	70	30	40	50	60	70
20	8	9	10	12	13	8	9	10	11	12	6	7	8	9	10
30	14	18	22	26	30	14	18	21	25	28	12	14	17	20	23
40	21	27	34	40	47	20	26	32	38	45	18	23	28	34	39
50	26	34	43	51	60	26	34	42	50	58	24	31	39	47	54
60	29	38	48	58	67	29	38	48	58	67	28	38	47	57	66
70	31	41	52	62	72	31	41	52	62	72	31	42	52	62	73
80	32	43	54	65	76	32	44	54	66	76	33	44	55	66	77
90	33	44	56	67	78	34	45	56	68	80	34	45	56	68	80

* Site index is the height in feet at 65 years measured at the ground.
Basis: 201 plots, including 159 plots of pure red spruce, 28 plots of red spruce and balsam fir, 12 plots of red spruce and white spruce, 2 plots of red spruce, white spruce, and balsam fir; plot size includes 100 trees at the minimum.
Source: after Meyer, 1929.

TABLE 89. HEIGHT IN FEET OF AVERAGE DOMINANT SPRUCE AND BALSAM FIR IN FULLY STOCKED AND EVEN-AGED STANDS IN NORTHERN MICHIGAN BY SITE INDEXES *

Total Age (Years)	Site Index				
	20	30	40	50	60
20	8	12	16	20	24
30	12	18	24	30	36
40	16	24	33	41	48
50	20	30	40	50	60
60	22	33	44	55	66
70	23	35	46	58	70
80	24	36	48	60	73
90	25	37	50	62	75

*Site index is height in feet at 50 years at the ground.
Basis: 209 plots, 0.2 acre in size.
Source: after Bowman, 1944.

of the time they are not. Therefore recommendations based on this assumption should be treated with caution.

Number of trees, diameter, and basal area per acre. Diameter, number of trees, and basal area per unit area are closely related to each other. Reineke (1933) established a relationship among them with the formula: $\log N = -1.605 \times \log D + k$, where N is the number of trees per acre, D is average stand dbh, and k is a constant for species. Later investigations showed that the coefficient was not a constant, but varied with site and age. This formula has been used rather frequently in Canada in a number of studies on balsam fir. Some of them will be reviewed in the discussion of special-use tables.

Wilson (1946) related stand height to number of trees per unit area, but this relationship has not yet been applied to balsam fir. The relationships between number of trees per unit area and average diameter or average height have not produced an objective numerical measure of stand density, although they are valuable aids in yield-table construction.

Table 90 shows for the northeastern United States the average number of balsam fir per acre in even-aged stands when compared with white and red spruce. The number of white spruce stems per acre decreases more rapidly with age than does balsam fir. Reduction in numbers of red spruce stems is less than that for either of the other species.

Table 91 shows the change with age in the number of balsam fir and spruce stems per acre in Michigan. Although the data of Tables 90 and 91 are not exactly comparable, they suggest a slower reduction of spruce and fir in Michigan than in the Northeast.

Table 92 shows the average diameters of balsam fir, white spruce, and red spruce in the northeastern United States. The average stand diameters of the three species differ only slightly, but for older trees, white spruce is increasing in size faster than either balsam fir or red spruce.

Table 93 shows the average diameters by age classes for spruce-fir stands in Michigan. A comparison with Table 91 shows that except for very young stands, the diameters of spruce and fir are smaller in Michigan than in the Northeast.

Relationships of stand basal area to age for balsam fir and spruce are shown in Figure 11. The original data are presented in Table 94 for the northeastern states and in Table 95 for Michigan.

In the Northeast the stand basal area of balsam fir is less than that for red spruce up to age 30, exceeds red spruce in the 30–60 year age period, and declines again after 60 years. In comparison with white spruce, balsam fir basal area is greater in all stands up to age 90. In Michigan the differences between species are not distinguished. This is reflected in Table 95, which lumps together white spruce, black spruce, and balsam fir. Additional studies are needed to trace out this relationship in the Lake States. When the two regions are compared, spruce-fir stands in the Northeast show up to 60 per cent more basal area than those in the Lake States.

The introduction of Bitterlich's angle-count technique has popularized stand basal area measurements. About 50 years ago Ostwald suggested that different elements of forest stands, biological and economic in nature, including volume, increment, piece-product ratios, relative market values of products, and others could be related to a unit of stand basal area. This idea has been recently revived by several workers. MacLean and Bedell (1955) used unit basal area as the reference measure in a growth and yield study in the Clay Belt of Ontario. Spurr (1952) found that stand basal area per acre is more closely related to growth than other stand-density measures. Bowman (1944) used basal area as the basic measurement for the construction of yield tables for uneven-aged stands and for balsam fir understories.

In the northeastern states the periodic stand basal area growth of balsam fir, white spruce, and red spruce culminates at 30–40 years (Meyer, 1929). In Michi-

TABLE 90. NUMBER OF TREES PER ACRE OF BALSAM FIR, WHITE SPRUCE, AND RED SPRUCE IN FULLY STOCKED AND EVEN-AGED STANDS IN THE NORTHEASTERN UNITED STATES, BY SITE INDEX [*]

Total Age (Years)	Site Index of Balsam Fir				Site Index of White Spruce				Site Index of Red Spruce				
	40	50	60	70	40	50	60	70	30	40	50	60	70
20	7280	4980	3540	2510	7530	5880	4470	3370	9945	5285	3690	2935	2495
30	5060	3460	2460	1745	4840	3780	2880	2170	7670	4075	2845	2265	1925
40	3300	2260	1605	1135	2980	2360	1800	1320	5745	3055	2130	1695	1440
50	2180	1490	1060	755	1820	1420	1090	790	4240	2255	1575	1250	1065
60	1600	1095	780	555	1210	945	720	540	3115	1655	1155	920	782
70	1340	920	652	465	965	750	570	420	2290	1215	849	676	575
80	1225	835	595	420	845	655	500	365	1940	1030	719	572	486
90	1160	790	560	400	785	615	470	345	1825	969	676	538	458

[*] Site index is height in feet at 65 years at the ground; minimum dbh class 1 inch.
Basis: 201 plots (see Table 88).
Source: after Meyer, 1929.

TABLE 91. NUMBER OF TREES PER ACRE IN FULLY STOCKED AND EVEN-AGED SPRUCE-FIR STANDS IN NORTHERN MICHIGAN, BY SITE INDEX [*]

Total Age (Years)	Minimum dbh Class — 1 Inch Site Index					Minimum dbh Class — 4 Inches Site Index				
	20	30	40	50	60	20	30	40	50	60
20		5950	4575	3575	2675					
30	4900	3450	2650	2050	1550			70	100	200
40	3175	2250	1750	1350	1025		465	560	720	670
50	2575	1800	1425	1100	825	60	670	755	715	550
60	2400	1700	1325	1015	775	200	760	830	665	460
70	2300	1650	1270	990	750	330	810	850	620	380
80	2275	1625	1260	975	740	445	840	850	580	320
90	2250	1610	1250	965	740	540	865	845	540	270

[*] Site index is height in feet at 50 years at the ground.
Basis: 209 plots, 0.2 acre in size.
Source: after Bowman, 1944.

TABLE 92. AVERAGE DIAMETER IN INCHES AT BREAST HEIGHT FOR BALSAM FIR, WHITE SPRUCE, AND RED SPRUCE IN FULLY STOCKED AND EVEN-AGED STANDS IN THE NORTHEASTERN UNITED STATES, BY SITE INDEX [*]

Total Age (Years)	Site Index of Balsam Fir				Site Index of White Spruce				Site Index of Red Spruce				
	40	50	60	70	40	50	60	70	30	40	50	60	70
20	0.6	0.8	1.0	1.2	0.6	0.8	1.0	1.1	0.5	0.8	1.0	1.2	1.4
30	1.7	2.1	2.6	3.2	1.6	2.0	2.3	2.8	1.0	1.7	2.2	2.5	2.8
40	2.8	3.6	4.4	5.4	2.9	3.5	4.2	5.0	1.6	2.6	3.4	3.9	4.4
50	3.8	4.9	6.0	7.3	4.2	5.0	5.9	7.1	2.2	3.6	4.7	5.4	6.0
60	4.7	5.9	7.3	8.9	5.3	6.3	7.5	8.9	2.8	4.6	5.9	6.8	7.6
70	5.2	6.6	8.1	9.9	6.0	7.2	8.6	10.2	3.4	5.5	7.0	8.2	9.1
80	5.5	7.0	8.6	10.5	6.4	7.7	9.2	11.0	3.8	6.1	7.8	9.1	10.1
90	5.7	7.3	9.0	10.9	6.7	8.0	9.5	11.4	3.9	6.3	8.1	9.4	10.5

[*] Site index is height in feet at 65 years at the ground. Minimum dbh class 1 inch.
Basis: 201 plots (see Table 88).
Source: after Meyer, 1929.

TABLE 93. AVERAGE DIAMETER IN INCHES AT BREAST HEIGHT FOR FULLY STOCKED AND EVEN-AGED SPRUCE–BALSAM FIR STANDS IN NORTHERN MICHIGAN, BY SITE INDEX *

Total Age (Years)	Minimum dbh — 1 Inch Site Index					Minimum dbh — 4 Inches Site Index				
	20	30	40	50	60	20	30	40	50	60
20	0.7	0.8	0.9	1.2	1.4					4.2
30	1.7	2.1	2.5	2.8	3.3			4.2	4.6	5.1
40	2.3	2.9	3.5	4.1	4.7		4.2	4.7	5.3	5.8
50	2.7	3.4	4.1	4.8	5.5		4.6	5.2	5.8	6.4
60	2.8	3.6	4.3	5.1	5.8	4.1	4.8	5.4	6.1	6.7
70	3.0	3.7	4.5	5.3	6.0	4.2	4.9	5.6	6.2	6.9
80	3.1	3.8	4.6	5.4	6.2	4.3	5.0	5.7	6.3	7.0
90	3.1	3.9	4.7	5.5	6.2	4.3	5.0	5.7	6.4	7.0

* Site index is height in feet at 50 years at the ground.
Basis: 209 plots, 0.2 acre in size.
Source: after Bowman, 1944.

TABLE 94. STAND BASAL AREA IN SQUARE FEET PER ACRE FOR BALSAM FIR, WHITE SPRUCE, AND RED SPRUCE IN FULLY STOCKED AND EVEN-AGED STANDS IN THE NORTHEASTERN UNITED STATES, BY SITE INDEX *

Total Age (Years)	Site Index of Balsam Fir				Site Index of White Spruce				Site Index of Red Spruce			
	40	50	60	70	40	50	60	70	40	50	60	70
20	16	18	20	21	17	19	22	24	19	22	24	25
30	78	87	93	99	71	79	86	90	64	73	78	82
40	143	158	170	180	140	158	172	180	115	132	142	149
50	175	194	209	221	172	192	209	219	162	186	200	210
60	189	210	226	238	183	204	222	232	189	217	233	245
70	196	217	234	248	187	210	228	238	200	230	247	260
80	201	223	240	253	189	212	230	241	207	238	256	269
90	205	227	245	259	191	213	232	243	211	243	261	275

* Site index is height in feet at 65 years at the ground. Minimum dbh class 1 inch.
Basis: 201 plots (see Table 88).
Source: after Meyer, 1929.

gan the basal area growth of fir and spruce culminates at 30 years on poor sites and 20 years on the best sites (Bowman, 1944). Little difference in stand basal area occurs for spruce and fir of site indexes 30–60 in Michigan. A comparison of the two areas reveals that stand basal areas in Michigan exceed those of the Northeast for age classes up to 30 years, but after that period, the reverse situation occurs. The rapid early growth of balsam fir in the Michigan area needs further study. The fact that balsam fir has on the whole a higher basal area in the more humid eastern states agrees with observations that have been compiled for other species inhabiting similar climatic regions.

More recently attempts have been made to investigate the interrelationships among the different constituents of balsam fir growth. McLintock (1948) measured increment cores taken at breast height of 188 balsam fir, classifying the trees according to crown ratio, vigor, and crown characteristics. Each individual class was assigned from 1 to 3 points, allowing the total rating to vary from 3 to 9. The validity of McLintock's classification system was tested by Gagnon (1963) by weekly measurements of radial growth of 38 balsam fir trees during one growing season at Valcartier, Quebec. The results indicated no significant correlation with the total height of the trees ($r = +0.03$), with crown class ($r = +0.09$), or with live crown ratio ($r = +0.14$), but a significant relationship existed with vigor class ($r = +0.33$), and a highly significant relationship with the point rating of McLintock's tree classes ($r = +0.58$).

TABLE 95. STAND BASAL AREA IN SQUARE FEET PER ACRE FOR FULLY STOCKED AND EVEN-AGED SPRUCE–BALSAM FIR STANDS IN NORTHERN MICHIGAN, BY SITE INDEX *

Total Age (Years)	Minimum dbh Class — 1 Inch Site Index					Minimum dbh Class — 4 Inches Site Index				
	20	30	40	50	60	20	30	40	50	60
20	46	67	71	73	75					
30	69	100	109	110	111		2	3	6	8
40	82	122	130	132	134	15	59	71	75	77
50	91	133	142	144	145	24	94	116	120	123
60	96	139	149	151	152	27	106	127	134	137
70	98	143	152	155	157	28	110	133	140	143
80	99	145	154	157	158	28	113	136	144	146
90	100	146	155	158	159	29	115	139	147	149

*Site index is height in feet at 50 years at the ground.
Basis: 209 plots, 0.2 acre in size.
Source: after Bowman, 1944.

Baskerville (1961) developed a formula for basal area growth using dbh, crown width, and a competition factor (see Chapter 7). In addition, he found that dbh and crown surface are highly significantly correlated ($r^2 = 0.769$). Other correlations were also highly significant, although the coefficients were lower: dbh and crown volume ($r^2 = 0.517$); dbh and crown width ($r^2 = 0.380$); and tree basal area and crown width ($r^2 = 0.605$). The study by Baskerville (1961) was based on 120 trees measured at Green River watershed in New Brunswick.

Vezina (1963) measured 53 open-grown trees in Quebec and found a very strong correlation ($r = +0.973$ or $r^2 = 0.947$) between crown width and dbh. Encouraged by these results and following the lead of other authors, Vezina attempted to use this relationship for determination of a "competition factor" with an extended application to even-aged stands of various ages and site indexes. Preliminary results indicated that there are some difficulties in applying the competition factor in younger stands, that a similar factor can be computed for evenly spaced plantations or stands with natural regeneration in clumps, and that the factor was dependent on site index and stand age.

Volume growth. Volume growth varies considerably in different geographical regions. The yields of site-index 60 in Michigan are about half the corresponding yields in the northeastern states (compare Tables 96 and 97). Although some of the regional differences may be due to varying yield-table construction techniques, most of the yields shown actually reflect natural conditions.

Table 96 gives the yield for balsam fir and two eastern spruces under varying conditions in the northeastern states. From an analysis of the data shown in the table, it is possible to compute that for balsam fir, net periodic annual growth culminates at 40–50 years, and net mean annual growth at 50–60 years. White spruce shows a similar growth pattern, but the slower-developing red spruce shows growth maxima 10 years later.

Tables 97, 98, and 99 show similar growth and yield data for Michigan. Net periodic annual growth for balsam fir culminates at or before age 40. Net mean annual growth reaches a maximum at 60 years.

TABLE 96. TOTAL VOLUME IN CUBIC FEET PER ACRE FOR FULLY STOCKED AND EVEN-AGED BALSAM FIR, WHITE SPRUCE, AND RED SPRUCE STANDS IN THE NORTHEASTERN UNITED STATES, BY SITE INDEX [*]

Total Age (Years)	Site Index of Balsam Fir				Site Index of White Spruce				Site Index of Red Spruce			
	40	50	60	70	40	50	60	70	40	50	60	70
20	80	110	135	165	90	106	143	170	80	120	160	200
30	720	960	1210	1455	540	715	890	1060	500	680	860	1060
40	1940	2600	3270	3940	1750	2310	2860	3430	1350	1860	2350	2850
50	2910	3920	4920	5910	2840	3750	4640	5550	2550	3510	4430	5350
60	3500	4700	5900	7100	3510	4630	5740	6860	3490	4780	6030	7240
70	3820	5140	6450	7760	3920	5180	6420	7680	4050	5590	7020	8430
80	4080	5480	6870	8270	4210	5560	6880	8240	4400	6060	7610	9150
90	4290	5760	7230	8700	4450	5880	7280	8700	4600	6340	7970	9580

[*] Site index is height in feet at 65 years at the ground; minimum dbh class 1 inch.
Basis: 201 plots (see Table 88).
Source: after Meyer, 1929.

Neither the tables of Meyer (1929) and Bowman (1944) nor the special-use tables of Gevorkiantz and Olsen (1950) show that the culmination-time of growth varies with site. The methods of construction are such that the rate of change for the various site-index curves are equalized so that growth maxima are shown to occur simultaneously. In a sense this presents a false picture, since — as Zon (1914) and many other workers have pointed out — site does have an effect on the culmination-time of growth in height, diameter, and volume, not only for balsam fir but for many other species as well.

SPECIAL-USE YIELD TABLES

Special-use tables are compiled to predict future development of stands under widely different conditions. The latter include variations in age structure, stand composition, and density for different geographic regions. Special-use yield tables are based primarily on accumulated growing-stock data, and attempt to project the future development for a limited period, commonly 10 years.

Age, species composition, and stand density. In balsam fir studies age has been treated in a variety of ways. Mulloy (1944, 1947) and others have used the stand-

TABLE 97. TOTAL VOLUME IN CUBIC FEET OF PEELED WOOD PER ACRE FOR FULLY STOCKED AND EVEN-AGED SPRUCE–BALSAM FIR STANDS IN NORTHERN MICHIGAN, BY SITE INDEX *

Total Age (Years)	Site Index				
	20	30	40	50	60
40	120	500	780	920	1020
50	240	1010	1580	1860	2040
60	300	1290	2000	2400	2580
70	360	1460	2270	2700	2920
80	400	1590	2490	2930	3200
90	410	1700	2650	3130	3400

* Site index is height in feet at 50 years at the ground; minimum dbh class 1 inch.
Basis: 209 plots, 0.2 acre in size.
Source: after Bowman, 1944.

TABLE 98. TEN-YEAR PERIODIC ANNUAL GROWTH IN CUBIC FEET OF PEELED WOOD PER ACRE IN FULLY STOCKED AND EVEN-AGED SPRUCE–BALSAM FIR STANDS IN NORTHERN MICHIGAN, BY SITE INDEX *

Total Age (Years)	Site Index				
	20	30	40	50	60
40	12.5	47.7	72.0	95.0	105.5
50	9.0	38.5	60.5	70.0	74.3
60	5.0	22.0	33.0	38.8	41.2
70	3.1	15.8	24.2	27.8	30.0
80	2.8	11.9	19.0	21.1	22.0
90	2.0	8.9	14.2	16.2	17.6

* Site index is height in feet at 50 years at the ground; minimum dbh class 1 inch.
Basis: 209 plots, 0.2 acre in size.
Source: after Bowman, 1944.

TABLE 99. TOTAL MEAN ANNUAL GROWTH IN CUBIC FEET PER ACRE FOR FULLY STOCKED AND EVEN-AGED SPRUCE–BALSAM FIR STANDS IN NORTHERN MICHIGAN, BY SITE INDEX *

Total Age (Years)	Site Index				
	20	30	40	50	60
40	3.0	12.5	19.8	23.0	26.0
50	4.5	20.0	31.5	36.8	40.5
60	5.0	21.5	34.0	39.5	43.0
70	4.9	20.9	32.7	38.2	41.8
80	4.7	19.8	31.1	36.5	39.9
90	4.1	18.9	29.5	34.5	37.8

* Site index is height in feet at 50 years at the ground; minimum dbh class 1 inch.
Basis: 209 plots, 0.2 acre in size.
Source: after Bowman, 1944.

density, yield-table method, classifying their data by total age. Gevorkiantz and Olsen (1950) determined age at breast height in order to exclude the early suppression period and the commonly occurring butt rot. Westveld (1941) used time elapsed since cutting, later (1953) modified to "effective age." The latter term refers to time since cutting adjusted for residual volume or losses in volume since cutting that result from unfavorable interferences with normal growth. Effective age can be computed by interpolation from density yield tables or from alinement charts (see Westveld, 1953). Bowman (1944) ignored age in dealing with uneven-aged understories and cutover spruce-fir mixtures, classifying his data by basal area per acre. Ferree and Hagar (1956) ignored age by substituting present stand volume and "ingrowth index," the latter term referring to the number of trees in the 1-inch diameter class just below the minimum limits of poletimber and sawtimber sizes.

Methods used for mixed stands vary considerably. In their growth study Gevorkiantz and Olsen (1950) limited their investigations to balsam fir. Since the projections were only for 10–20 years, they assumed no species composition changes during that period. Ferree and Hagar (1956) classified their data by cover types, and for the 10-year projection period no change in species composition was anticipated. In the data of Bowman (1944) and Westveld (1941, 1953), hardwoods mixed with spruce and fir were shown to be relatively unimportant, hence recognition of only two broad species composition groups was necessary. Mulloy (1944, 1947) and later authors of stand-density yield tables recognized the great changes that hardwood admixtures undergo during a full rotation. In general, the range of percentage of hardwoods decreases from a maximum amplitude (0–100 per cent) at 20 years of stand age to a very narrow amplitude (8–15 per cent) at 120 years. Yield tables useful for prediction methods in such stands must account for changes in species composition with age. MacLean (1960) compiled a yield table for characteristic stands of the aspen-birch-spruce-fir type on the best sites in Ontario (Table 53). On drier sites yields are 80 per cent of table values, but stands growing on shallow till over bedrock yield only 50 per cent of table values. Balsam fir constitutes 4 per cent of the total merchantable volume at 70 years, and 15 per cent at 130 years. In overmature mixed stands, however, much balsam fir volume occurs on trees considerably younger than the spruce and hardwood components.

Variation in stand density is accounted for in several ways. Perhaps the simplest is to use stand basal area directly (see Bowman, 1944). Closely related to this method is the use of accumulated volume (Ferree and Hagar, 1956). Gevorkiantz and Olsen (1950) introduced a more complicated method, considering stand density as a ratio of actual stand basal area to an assumed normal basal area, and expressed as a percentage. Normal basal area depends only on the average dbh of all species present in the stand.

Westveld (1941, 1953) considered the total number of trees corresponding to average diameters in average stands (Table 100). His original unabridged table listed the average number of trees per acre for average diameters measured to tenths of inches. The range of stand densities found on predominantly softwood sites was 10–250 per cent of the average. For secondary softwood sites, the corre-

TABLE 100. AVERAGE NUMBER OF TREES PER ACRE FOR DETER-
MINING RELATIVE STAND-DENSITY INDEX IN THE SPRUCE–FIR
REGION OF THE NORTHEASTERN UNITED STATES [*]

dbh of Average Stand (In.)	Average Number of Trees	dbh of Average Stand (In.)	Average Number of Trees
1	1660	6	543
2	1345	7	433
3	1075	8	352
4	850	9	283
5	685	10	225

[*] To determine the density of any given stand, divide the total number of trees per acre by the total number of trees indicated for the same average diameter in the table.

Basis: 365 plots, 0.125 acre in size.

Source: Westveld, 1953.

TABLE 101. CHANGES IN DENSITY PERCENTAGE OVER A TEN-YEAR PERIOD FOR SPRUCE–BALSAM FIR STANDS OF DIFFERENT DENSITIES AND EFFECTIVE AGES IN THE NORTHEASTERN UNITED STATES [*]

Present Stand Density (Per Cent)	Effective Age of Stand at Beginning of Period (Years)						
	0	10	20	40	60	80	100
DOMINANT SOFTWOOD SITES							
20	56	41	28	7	2	0	−2
50	68	51	33	7	2	0	−2
90	68	51	30	7	2	0	−2
130	60	42	24	7	2	0	−2
SECONDARY SOFTWOOD SITES							
20	28	23	20	20	19	19	18
50	27	22	16	14	13	13	12
90	27	22	9	5	4	4	3
130	24	20	2	−4	−5	−5	−6

[*] For definition of density see Table 100, and for effective age see text; for density percentage at the end of ten-year period add amount of change to density at the beginning of the period.

Basis: 365 plots, 0.125 acre in size.

Source: Westveld, 1953.

sponding range was 10–150 per cent. When the reference values of Gevorkiantz and Olsen (1950) and Westveld (1953) are converted to a common base, Westveld's 2-inch diameter class has only 35 per cent as many trees as shown in the Lake States data, but his 8- and 10-inch diameter classes have 70 per cent as many trees.

Westveld's (1941) first tables were improved in his later work by introducing corrections for varying stand densities (Table 101). To obtain future density for periods longer than 10 years the computation procedure should be repeated by 10-year intervals.

Development of a density index independent of site and age has been the goal of several investigators. Mulloy (1944, 1947) and Bedell and MacLean (1952) have applied the "stand-density index" of Reineke (1933) to stands containing balsam fir. The Reineke index considers the basal area of one 10-inch tree (0.545 ft²) as the density unit. However, the absolute measure of the stand density unit cannot be fixed in units of square feet per acre. If the average tree diameter is less than 10 inches, actual basal area in square feet per acre is not used; instead, Reineke's formula is used to compute the equivalent number of 10-inch trees, or if plotted, to read them from the graph. Reineke's formula was derived from data taken from fully stocked stands. Subsequent investigations have shown that average stocking can be used to compute the basic reference curve. These computations of stand density are complicated. Originally the density

TABLE 102. YEARLY PERCENTAGE INCREASE OF DENSITY IN UNIFORM AND PATCHY EVEN-AGED SPRUCE–BALSAM FIR STANDS IN NORTHERN MICHIGAN, BY SITE INDEX *

Site Index	Uniform Stand	Patchy Stand
20	0.3	0.15
30	0.6	0.30
40	0.9	0.45
50	1.2	0.60
60	1.5	0.75

* Site index is height in feet at 50 years at the ground; to compute future stand density, the percentages shown in the table should be multiplied by the number of years in the prediction period and the amount added to the present stand density percentage.
Basis: number of plots unknown.
Source: after Bowman, 1944.

index was assumed to be applicable to many species and to be independent of age and site. Although the stand-density formula is inadequate for such purposes, it does serve more limited uses. Mulloy (1947) concluded that under certain conditions all density indexes would merge at 120 years. To construct stand-density yield tables, a basic or average yield curve is necessary. In addition, a series of permanent sample plots must be established to determine the changes with age of the density index.

The trend of stand density with age was also investigated by Bowman (1944) in Michigan (Table 102) for several site indexes in even-aged stands.

Examples of special-use yield tables. A number of special-use yield tables in abridged form are included in this section in order to give some idea of their usefulness and versatility. Bowman (1944) based his uneven-aged-stand yield table on present basal area and site index, using these to predict yields by 10-year intervals for 50 years in the future. Data in Table 103 are for the 10- and 20-year predictions. Using similar principles, Bowman (1944) also prepared tables for overtopped spruce-fir stands, projecting yields ahead for 50 years. Table 104 includes the data for 10- and 20-year predictions.

Comparison of Tables 103 and 104 indicates that an overstory with a density of 75–100 per cent suppresses the growth of the spruce-fir understory, whereas an

TABLE 103. FUTURE MERCHANTABLE VOLUME IN CUBIC FEET OF PEELED WOOD PER ACRE FOR ALL DEGREES OF STOCKING IN UNEVEN-AGED SPRUCE–BALSAM FIR STANDS IN NORTHERN MICHIGAN, BY SITE INDEX *

Present Stand Basal Area per Acre (Ft²)	10 Years after Measurement Site Index					20 Years after Measurement Site Index				
	20	30	40	50	60	20	30	40	50	60
20	45	110	245	330	370	65	180	510	690	750
40	160	290	535	670	780	200	405	980	1280	1410
60	325	530	900	1150	1280	380	740	1580	2020	2240
80		870	1400	1730	1840		1170	2220	2805	3060
100		1285	1990	2415	2700		1610	2610	3130	3410
120		1640	2405	2845	3185		1800	2775	3255	3570
140			2655	3080	3420					

* Site index is height in feet at 50 years at the ground; stump height 1 ft; top diam. 3 in. inside bark.
Basis: 92 plots, 0.2 acre in size.
Source: after Bowman, 1944.

TABLE 104. FUTURE MERCHANTABLE VOLUME IN CUBIC FEET OF PEELED WOOD PER ACRE FOR OVERTOPPED SPRUCE–BALSAM FIR IN MIXED HARDWOOD–CONIFER STANDS IN NORTHERN MICHIGAN, BY SITE INDEX *

Present Stand Basal Area per Acre (Ft²)	10 Years after Measurement Overstory 75–100% Site Index			10 Years after Measurement Overstory 50–75% Site Index			20 Years after Measurement Overstory 75–100% Site Index			20 Years after Measurement Overstory 50–75% Site Index		
	40	50	60	40	50	60	40	50	60	40	50	60
10	95	160	160	135	195	205	300	420	450	395	560	595
20	210	265	300	280	360	395	460	590	645	595	790	850
30	325	385	455	420	510	610	600	775	850	795	1010	1145
40	450	540	620	590	705	840	790	1010	1105	1040	1330	1500
50	580	705	805	750	920	1070	1000	1275	1400	1310	1680	1870
60	720	880	1020	960	1180	1345	1250	1595	1775	1660	2120	2350
70	885	1110	1260	1180	1460	1640	1570	2000	2160	2085	2640	2930

* Site index is height in feet at 50 years at the ground; stump height 1 ft; top diam. 3 in. inside bark; overstory density expressed as basal area in percentage of normal basal area.
Basis: 23 plots, 0.2 acre in size.
Source: after Bowman, 1944.

overstory density of 50–75 per cent has a stimulating effect. Possibly this effect is the result of the occupation of better sites by mixed stands. Comparable tables were prepared for cutover stands. Table 105 is an abridged form of one of Bowman's (1944) tables. A comparison of Tables 103 and 105 shows that cutover stands develop less volume than uneven-aged stands at higher basal areas. These prediction tables of Bowman seem not to have been widely used. No further application of his methods was found in later studies.

The yield tables and their accessories prepared by Westveld (1953) are presented in an abridged form (Tables 100, 101, 106, and 107). Westveld held that the earlier tables of Meyer (1929) were applicable only under special conditions, e.g., pure and even-aged stands growing on the best sites. To estimate future yield it is necessary to know the broad site group, effective age (time elapsed since cutting adjusted for leftover volume and subsequent losses), percentage of spruce-fir in stand composition, and stand density. Adjustments also need to be made for changes in density. The yield tables show merchantable volume of spruce and fir and indicate that dominant softwood sites produce more spruce-fir volume than secondary softwood sites. Accuracy was estimated at ±12 per cent. Additional tables were also prepared for volumes in cords, assuming 95 cubic feet of solid peeled wood in a 128-cubic-foot stacked cord.

TABLE 105. FUTURE MERCHANTABLE VOLUME IN CUBIC FEET OF PEELED WOOD PER ACRE FOR CUTOVER SPRUCE–BALSAM FIR STANDS IN NORTHERN MICHIGAN, BY SITE INDEX *

Present Stand Basal Area per Acre (Ft²)	10 Years after Cutting Site Index					20 Years after Cutting Site Index				
	20	30	40	50	60	20	30	40	50	60
10		20	190	300	330	45	140	500	725	820
20	35	130	330	440	470	125	300	710	930	1040
30	90	240	470	585	630	210	470	920	1160	1270
40	150	400	620	730	795	290	645	1140	1380	1520
50	195	485	770	880	965	340	830	1365	1610	1780
60	255	620	930	1050	1145	390	1030	1605	1850	2050
70		760	1095	1210	1330		1235	1850	2100	2320
80		900	1260	1390	1515		1410	2100	2365	2600
90			1390	1580	1720			2285	2630	2910
100					1915					3155

* Site index is height in feet at 50 years at the ground; stump height 1 ft; top diam. 3 in. inside bark.
Basis: 86 plots, 0.2 acre in size.
Source: Bowman, 1944.

According to Mulloy (1944, 1947), stand-density yield tables should be prepared for both cover types and site types. Mulloy (1944) presented an average yield curve for stand-density index (S.D.I.) 100, in the spruce-fir cover type, Timagami district, Ontario (Table 108). To construct these basic yield curves, stands of different densities are measured, although Mulloy suggested avoiding stands with irregular tree distributions and those subjected to major disturbances. The volumes are simply reduced to 100 S.D.I. Predictions cannot be made from this table alone, because neither the stand-density index nor species composition remains constant. Changes in stand characteristics over time should be determined by establishing a large number of permanent sample plots under a wide variety of conditions. Mulloy noted that the culmination of mean annual increment comes later for stands having a higher S.D.I. This is also true for stands carrying a higher percentage of hardwoods.

Application of competition-index yield tables (Gevorkiantz and Olsen, 1950) makes use of present volume and competition indexes and a knowledge of site to determine the future competition index (Table 109). Table 109 shows that for a competition index of 0.40, a stand 10 years old at breast height increases to

0.55 during the next 10 years. Continuing the step-wise analysis until the stand reaches 70 years of age, its competition index would be computed as 0.74. Starting with a competition index of 1.00 the changes would be insignificant during the same period of time. Beginning with an index of 1.60 at age 10, a stand will have an index of only 1.21 at age 70. These data show that over a span of 70 years gaps in understocked stands tend to fill in, and high mortality in initially overstocked

TABLE 106. FUTURE MERCHANTABLE VOLUME IN CORDS OF PEELED WOOD PER ACRE FOR SPRUCES AND BALSAM FIR ON DOMINANT SOFTWOOD SITES IN THE NORTHEASTERN UNITED STATES, BY STAND DENSITY AND COMPOSITION PERCENTAGES *

Present Stand Density (%)	Present Composition Percentage of Spruces and Balsam Fir									
	10	20	30	40	50	60	70	80	90	100
10 YEARS AFTER CUTTING										
10			0.02	0.3	0.5	0.8	1.2	1.5	1.9	2.3
50	0.04	0.2	0.4	0.7	1.1	1.4	1.8	2.2	2.6	3.0
90	0.3	0.6	0.9	1.3	1.6	2.0	2.4	2.9	3.3	3.8
130	0.8	1.2	1.5	1.9	2.3	2.7	3.2	3.6	4.1	4.5
20 YEARS AFTER CUTTING										
10	0.6	0.9	1.3	1.6	2.0	2.4	2.8	3.3	3.7	4.2
50	1.1	1.5	1.9	2.3	2.7	3.1	3.6	4.0	4.5	5.0
90	1.7	2.1	2.6	3.0	3.4	3.9	4.3	4.8	5.3	5.8
130	2.4	2.9	3.3	3.7	4.2	4.6	5.1	5.6	6.1	6.6
30 YEARS AFTER CUTTING										
10	2.1	2.5	3.0	3.4	3.8	4.3	4.7	5.2	5.7	6.2
50	2.8	3.2	3.7	4.1	4.6	5.1	5.5	6.0	6.5	7.0
90	3.6	4.0	4.5	4.9	5.4	5.9	6.4	6.9	7.4	7.9
130	4.3	4.8	5.3	5.7	6.2	6.7	7.2	7.7	8.3	8.8
40 YEARS AFTER CUTTING										
10	3.9	4.4	4.9	5.4	5.9	6.3	6.8	7.3	7.9	8.4
50	4.7	5.2	5.7	6.2	6.7	7.2	7.7	8.2	8.7	9.3
90	5.5	6.0	6.5	7.0	7.5	8.1	8.6	9.1	9.7	10.2
130	6.4	6.9	7.4	7.9	8.4	9.0	9.5	10.0	10.6	11.1
50 YEARS AFTER CUTTING										
10	6.0	6.5	7.0	7.5	8.0	8.5	9.1	9.6	10.2	10.7
50	6.8	7.3	7.8	8.4	8.9	9.4	10.0	10.5	11.1	11.7
90	7.7	8.2	8.7	9.3	9.8	10.4	10.9	11.5	12.0	12.6
130	8.6	9.1	9.7	10.2	10.7	11.3	11.9	12.4	13.0	13.6
60 YEARS AFTER CUTTING										
10	8.2	8.7	9.2	9.8	10.3	10.8	11.4	12.0	12.5	13.1
50	9.1	9.6	10.1	10.7	11.2	11.8	12.4	12.9	13.5	14.1
90	10.0	10.5	11.1	11.6	12.2	12.8	13.3	13.9	14.5	15.1
130	10.9	11.4	12.0	12.6	13.2	13.7	14.3	14.9	15.5	16.1

* For density percentage, see Table 100; composition percentage is determined by number of trees of spruces and balsam fir in relation to the total number of trees in the stand; minimum dbh class 6 in.; stump height 1 ft; top diam. 3 in. inside bark.
  Basis: 207 plots, 0.125 acre in size.
  Source: Westveld, 1953.

stands reduces density, with the result that variations in stocking become less marked with time. Although there is always a tendency toward normality in understocked and overstocked stands, normality is seldom attained because of the many accidental disturbances to forest stands over time.

TABLE 107. FUTURE MERCHANTABLE VOLUME IN CORDS OF PEELED WOOD PER ACRE FOR SPRUCES AND BALSAM FIR ON SECONDARY SOFTWOOD SITES IN THE NORTHEASTERN UNITED STATES, BY STAND DENSITY AND COMPOSITION PERCENTAGES *

Present Stand Density (%)	Present Composition Percentage of Spruces and Balsam Fir								
	10	20	30	40	50	60	70	80	90
10 YEARS AFTER CUTTING									
10	0.1	0.2	0.3	0.4	0.5	0.5	0.5	0.5	0.5
50	0.3	0.4	0.6	0.8	0.9	0.9	0.9	0.9	0.9
90	0.5	0.8	1.1	1.4	1.5	1.5	1.5	1.6	1.6
130	0.9	1.4	2.2	3.3	3.7	3.8	3.9	4.0	4.0
20 YEARS AFTER CUTTING									
10	0.2	0.4	0.6	0.8	0.8	0.8	0.9	0.9	0.9
50	0.5	0.8	1.1	1.3	1.4	1.5	1.5	1.5	1.5
90	0.9	1.4	2.1	3.0	3.5	3.6	3.7	3.7	3.8
130	1.6	3.2	5.0	5.7	5.9	6.0	6.0	6.1	6.1
30 YEARS AFTER CUTTING									
10	0.5	0.8	1.0	1.3	1.4	1.4	1.4	1.4	1.5
50	0.9	1.3	1.9	2.8	3.3	3.4	3.5	3.5	3.6
90	1.5	3.0	4.8	5.6	5.8	5.9	5.9	5.9	6.0
130	4.0	5.7	6.6	7.1	7.2	7.3	7.3	7.3	7.3
40 YEARS AFTER CUTTING									
10	0.9	1.3	1.9	2.6	3.1	3.2	3.3	3.3	3.3
50	1.5	2.9	4.7	5.5	5.7	5.8	5.8	5.9	5.9
90	3.8	5.6	6.5	7.0	7.2	7.2	7.3	7.3	7.3
130	6.1	7.1	7.8	8.1	8.3	8.3	8.3	8.3	8.3
50 YEARS AFTER CUTTING									
10	1.5	2.6	4.6	5.4	5.7	5.7	5.8	5.8	5.8
50	3.9	5.6	6.5	7.0	7.1	7.2	7.2	7.3	7.3
90	6.0	7.1	7.7	8.1	8.2	8.3	8.3	8.3	8.3
130	7.4	8.2	8.7	9.0	9.1	9.2	9.2	9.2	9.2

* For definitions of density percentage and composition percentage, see Tables 100 and 105; minimum dbh class 6 in.; stump height 1 ft; top diam. 3 in. inside bark.

Basis: 158 plots, 0.125 acre in size.

Source: Westveld, 1953.

Gevorkiantz and Olsen (1950) presented tables to predict the growth and yield of balsam fir 10 and 20 years in the future. The production of total gross volume increases with increasing competition index. However, high competition index indicates reduction in merchantable volume, particularly in stands on poorer sites or in younger age classes. The net growth becomes negative at 70 years in stands on poor and medium sites, while on the best sites the growth reduction occurs 10 years later. Accordingly, their data show that growth culminates on all sites at 20 years (breast-height age). The foregoing data for a 10-year prediction period are summarized in Tables 110 and 111.

The yield tables prepared by Ferree and Hagar (1956) are based on standard dbh-height curves (tariffs) compiled for cover types within regions. Growth rate is computed from accumulated volume and an ingrowth index for specific size classes. Because the average accumulated volumes are based on data from different combinations of sites, mixtures of species, stand density, and past history, they do not reflect only site differences. Tariffs were applied to different spruce-fir stands in the Adirondack Mountains of New York. Tariffs I and II are applicable to stands at medium elevations where the growing season is 105–134 days (Zone B). For growing seasons of 135–159 days (Zone C), at lower elevations, 9 per cent must be added to the cubic-foot volume of Tariff I and 2 per cent to Tariff II. Tariff III is applied to stands at higher elevations where the growing season is 90–100 days (Zone A).

TABLE 108. TOTAL VOLUME AND GROWTH IN CUBIC FEET PER ACRE FOR STANDS WITH 100 STAND-DENSITY INDEX UNITS ON AVERAGE SITES OF BALSAM FIR–SPRUCE COVER TYPES IN THE TIMAGAMI DISTRICT, ONTARIO *

Age (Years)	Total Volume	Current Annual Growth	Mean Annual Growth
10	90		9.0
20	250	16.0	12.5
30	415	16.5	13.8
40	556	14.1	13.9
50	668	11.2	13.4
60	764	9.6	12.7
70	840	7.6	12.0

* For definition of stand density, see text.
Basis: number of plots unknown.
Source: after Mulloy, 1944.

TABLE 109. COMPETITION INDEX OF BALSAM FIR TEN YEARS HENCE ON UPLAND SITES IN THE LAKE STATES *

Present Age at dbh (Years)	Present Competition Index						
	0.40	0.60	0.80	1.00	1.20	1.40	1.60
10	0.55	0.76	0.90	1.06	1.20	1.25	1.35
20	0.49	0.70	0.87	1.03	1.20	1.29	1.41
30	0.46	0.67	0.84	1.01	1.20	1.30	1.44
40	0.44	0.64	0.82	1.00	1.20	1.32	1.47
50	0.42	0.62	0.81	1.00	1.20	1.33	1.48
60	0.41	0.61	0.80	1.00	1.20	1.34	1.49
70	0.40	0.60	0.80	1.00	1.20	1.34	1.50

* For definition of competition index, see text.
Basis: number of plots unknown.
Source: Gevorkiantz and Olsen, 1950.

TABLE 110. GROSS VOLUME IN CUBIC FEET PER ACRE IN UPLAND STANDS WITH BALSAM FIR IN THE LAKE STATES [*]

Stand Age at dbh (Years)	Present Competition Index						
	0.4	0.6	0.8	1.0	1.2	1.4	1.6
POOR SITE							
20	440	570	710	830	920	960	980
40	1110	1420	1770	2070	2290	2390	2400
60	1460	1890	2360	2790	3090	3200	3230
80	1440	1860	2320	2720	3060	3170	3190
FAIR SITE							
20	520	680	830	980	1090	1130	1140
40	1250	1640	2010	2360	2620	2720	2750
60	1640	2140	2640	3120	3460	3560	3630
80	1660	2130	2650	3130	3490	3640	3650
MEDIUM SITE							
20	600	780	960	1140	1260	1300	1330
40	1400	1820	2270	2670	2960	3080	3110
60	1830	2380	2970	3470	3850	3980	4050
80	1880	2420	3050	3510	3940	4100	4150
GOOD SITE							
20	680	890	1100	1300	1430	1500	1520
40	1580	2060	2530	3010	3310	3430	3490
60	2050	2680	3280	3920	4320	4470	4510
80	2140	2780	3460	4070	4500	4720	4720
EXCELLENT SITE							
20	790	1030	1280	1480	1650	1710	1740
40	1820	2350	2900	3410	3780	3910	3980
60	2360	3080	3790	4470	4920	5110	5210
80	2490	3250	4030	4730	5210	5490	5490

[*] For mixed stands, reduce the values according to the percentage of balsam fir; volume is inside bark including stem and stump; for competition index, see text.
Basis: number of plots unknown.
Source: Gevorkiantz and Olsen, 1950.

Table 112 shows the range of growing-stock volume, net growth, and mortality for different stand-size classes. Cubic-foot volume is measured inside bark from a 1-foot stump, to a top diameter of 3.6 inches. Poletimber trees are 5.6–8.5 inches dbh, and sawtimber trees 8.6 inches dbh or larger. Table 113 shows gross accretion and total gross growth with respect to gross volume, i.e., volume before defect is deducted. Table 114 presents ingrowth by ingrowth-index classes. The ingrowth index is the total number of live trees per acre in the 1-inch diameter class just below the minimum dbh for poletimber. Accretion is the increase in timber volume due to growth of trees that were above minimum dbh at the beginning of the growth period. The total growth consists of accretion and ingrowth added.

The foregoing tables were taken from the studies of Ferree and Hagar (1956). These workers also presented growth and yield tables for board-foot volumes by the International ¼-inch log rule.

TABLE 111. PREDICTED PERIODIC GROWTH IN CUBIC FEET PER ACRE IN THE NEXT TEN YEARS IN UPLAND STANDS WITH BALSAM FIR IN THE LAKE STATES *

Stand Age at dbh (Years)	Present Competition Index						
	0.4	0.6	0.8	1.0	1.2	1.4	1.6
POOR SITE							
20	440	550	620	660	690	700	710
40	290	370	430	480	500	510	520
60	60	70	70	80	80	90	100
80	−150	−200	−230	−280	−320	−330	−330
FAIR SITE							
20	520	640	730	780	830	850	870
40	320	400	450	490	520	530	540
60	80	90	90	100	100	100	100
80	−90	−130	−170	−190	−210	−210	−220
MEDIUM SITE							
20	580	740	830	890	920	940	950
40	370	430	490	540	550	560	560
60	80	100	110	120	120	130	140
80	−50	−70	−90	−100	−110	−110	−110
GOOD SITE							
20	680	840	950	1010	1050	1070	1080
40	400	470	520	580	620	630	640
60	120	130	150	160	160	170	170
80	−10	−30	−50	−60	−60	−70	−70
EXCELLENT SITE							
20	760	950	1080	1140	1180	1200	1210
40	450	530	600	660	690	700	700
60	150	170	200	220	230	230	230
80	−10	−20	−30	−40	−40	−40	−40

* For competition index, see text; volume is inside bark including stem and stump.
Basis: number of plots unknown.
Source: Gevorkiantz and Olsen, 1950.

Regional Growth and Yield Data

The summarization of balsam fir area, volume, and growth data by regions presents a difficult problem. Balsam fir is widely distributed, occurs in a wide variety of mixtures with other species, and varies in abundance from place to place. Most of the statistical data is available for spruce-fir cover types, and because of the manner of presentation the contribution of balsam fir in such mixtures can be extracted only with difficulty. The geographic distribution of balsam fir is shown in Plates 1 and 2.

AREA DISTRIBUTION

According to the Canada Bureau of Statistics (1960), the total forest-land area of Canada is 1,092,888,000 acres of which 611,327,000 acres are productive forest. Softwoods comprise 62 per cent, mixed-woods 24 per cent, and hardwoods 13 per cent of the accessible forest area. Balsam fir may occur in any of these

TABLE 112. GROWING STOCK, NET GROWTH, AND MORTALITY IN CUBIC FEET
PER ACRE PER YEAR IN THE SPRUCE–FIR COVER TYPE,
ADIRONDACK MOUNTAINS, NEW YORK [*]

Stand Size Class	Tariff I			Tariff II			Tariff III		
	Growing Stock	Net Growth	Mortal-ity	Growing Stock	Net Growth	Mortal-ity	Growing Stock	Net Growth	Mortal-ity
Seedlings-saplings .	0–400	39	1	0–400	25	3	0–400	11	3
Light poles	400–800	61	4	400–800	38	8	400–800	14	9
Heavy poles	800–920	74	6	800–1000	45	11	800–1100	14	14
Light saw-timber ..	920–1350	81	8	1000–1400	51	14	1100–1440	13	20
Medium saw-timber ...	1350–2080	86	12	1400–2180	52	18	1440–2260	10	29
Heavy saw-timber ...	2080–3580	67	20	2180–3730	41	25	2260–3800	4	35
Very heavy sawtimber	3580+	34	33	3730+	26	31			

[*] Tariff classes express stand diameter and height relationships, originally called "sites" (see Table 79). Cubic-foot volume without bark, stump height 1 ft, top diameter 4 inches inside bark.

Basis: 97 plots, 0.2 acres each.

Source: after Ferree and Hagar, 1956.

TABLE 113. ANNUAL GROSS ACCRETION AND TOTAL GROSS GROWTH IN CUBIC
FEET PER ACRE IN RELATION TO GROSS VOLUME IN THE SPRUCE–BALSAM FIR
COVER TYPE, ADIRONDACK MOUNTAINS, NEW YORK [*]

Present Gross Volume	Tariff I		Tariff II		Tariff III	
	Gross Accretion	Total Gross Growth	Gross Accretion	Total Gross Growth	Gross Accretion	Total Gross Growth
200	16	40	14	29	6	14
400	30	55	24	41	14	21
600	42	67	33	50	18	27
800	52	77	41	57	21	31
1000	60	85	47	63	24	34
1200	66	91	53	68	26	36
1400	71	95	58	72	27	38
1600	75	97	62	74	28	39
1800	78	97	64	75	29	40
2000	80	96	66	75	30	40
2400	82	92	67	73	31	41
2800	81	88	66	70	32	42
3200	78	83	63	66	32	42
3600	74	78	59	61		
4000	69	73	53	56		
4400	64	68				
4800	58	63				
5200	52	58				

[*] Tariff classes express stand diameter and height relationships, originally called "sites" (see Table 79).

Basis: 97 plots, 0.2 acres each.

Source: after Ferree and Hagar, 1956.

TABLE 114. ANNUAL GROSS INGROWTH IN CUBIC FEET PER ACRE IN RELATION TO INGROWTH INDEX IN THE SPRUCE–BALSAM FIR COVER TYPE, ADIRONDACK MOUNTAINS, NEW YORK *

Ingrowth Index	Gross Ingrowth		
	Tariff I	Tariff II	Tariff III
10	2	2	
20	5	5	1
30	8	8	2
40	12	11	4
50	15	13	6
60	18	15	8
70	20	17	10
80	21	19	11
90	23	20	12
100	24	21	12
110	25	22	13
120	26	23	13

* The cubic-foot ingrowth index is the total number of live trees per acre in the 5-inch dbh class; tariff classes express stand diameter and height relationships, originally called "sites" (see Table 79 and footnote in Table 112).
Basis: 97 plots, 0.2 acres each.
Source: after Ferree and Hagar, 1956.

large groupings. Unfortunately no detailed data for balsam fir or spruce-fir types in Canada are available, although some idea of the relative abundance of balsam fir in the different parts of Canada can be obtained from Halliday and Brown (1943); their data are reproduced in part in Plate 2.

The total forest area of the United States according to the U.S. Forest Service (1958) is 664,194,000 acres, of which 486,609,000 acres are commercial forest land. The area in spruce-fir and spruce-fir-hardwood types in seven states having considerable balsam fir is given in Table 115. Maine, with the largest spruce-fir acreage, also has the largest percentage of total forest area in the spruce-fir types.

Timber Resource Review figures (U.S. Forest Service, 1958) show that spruce-fir types generally had good stocking. In New England 90 per cent of all spruce-fir types were considered well-stocked. Corresponding figures were 85 per cent for the middle Atlantic States, and 50 per cent for the Lake States.

GROWING STOCK

The available growing-stock data contain somewhat more specific information for balsam fir than the area-distribution data which frequently represent a combination of several species.

Table 116 shows the volume of accessible merchantable balsam fir by provinces in Canada. Balsam fir volume for inaccessible areas, not included in the table, amounts to about 920 million cubic feet.

Approximately 20 per cent of all accessible merchantable balsam fir volume is in dbh-size-classes of 10 inches or larger. The largest volumes of balsam fir are in Quebec, New Brunswick, Ontario, and Newfoundland. In the latter province

TABLE 115. DISTRIBUTION OF SPRUCE–BALSAM FIR COVER TYPE AREA IN SELECTED STATES HAVING IMPORTANT AMOUNTS OF BALSAM FIR[*]

| State and Year | Millions of Acres | | | | Percentage Spruce-Fir of: | |
	Seedlings [†] and Saplings	Pole-timber	Saw-timber	Total	Total Softwood Area	Total Forest Area
Maine, 1958[a]	688	3,446	2,927	7,061	70	41
New Hampshire, 1948[b]	78	271	224	573	29	12
Vermont, 1948[c]	119	214	238	571	55	18
New York, 1950[d]	130	240	221	591	30	5
Total North-east	1,015	4,171	3,610	8,796	46	24
Minnesota, 1953[e]	596	528	109	1,233	28	7
Wisconsin, 1956[f]	269	102	2	373	18	2
Michigan, 1955[g]	358	320	56	734	20	4
Total Lake States	1,223	950	167	2,340	23	4
Total Northeast and Lake States	2,238	5,121	3,777	11,136	38	12

[*] Areas by stand-size-classes in the northeastern states are recomputed to include only spruce-balsam fir stands.

[†] Understocked seedling and sapling areas also are included.

Sources: [a]Ferguson and Longwood, 1960; [b]Northeastern Forest Experiment Station, 1950; [c]McGuire and Wray, 1952; [d]Northeastern Forest Experiment Station, 1955; [e]Cunningham et al., 1958; [f]Stone and Thorne, 1961; [g]Findell et al., 1960.

TABLE 116. MERCHANTABLE-SIZE GROWING STOCK OF ACCESSIBLE BALSAM FIR IN CANADA, 1959

| Province | Material 4–9 in. at dbh (Thousand Cords)[a] | Material 10 in. dbh and Larger (Millions Ft³) | Total Merchantable Volume (Millions Ft³) | Percentage Balsam Fir in Total Volume of: | | |
				Spruce-Fir	Soft-woods	All Timber
Newfoundland Labrador	6,952	116	706	11	11	10
Island	39,977	725	4,123	61	61	58
Prince Edward Island	276	13	36	39	36	24
Nova Scotia	24,267	416	2,480	45	38	25
New Brunswick ...	43,812	1,427	5,151	50	43	31
Quebec	124,633	2,105	12,699	33	29	21
Ontario	41,495	892	4,419	14	9	5
Manitoba	3,357	47	333	8	6	4
Saskatchewan	1,641	48	187	5	3	1
Alberta[b]	7,645	446	1,096	7	4	2
Total Canada ...	294,055	6,235	31,230	12	7	5

[a]1 cord = 85 ft³ solid wood; [b]includes *Abies lasiocarpa* and *A. balsamea*.
Source: Canada Bureau of Statistics, 1960.

balsam fir is the most important species, comprising 58 per cent of the merchantable growing stock. For Canada as a whole, balsam fir constitutes 5 per cent of all accessible growing stock, 7 per cent of merchantable softwood volume, and 12 per cent of merchantable spruce-fir volume. Comparatively little balsam fir growing stock exists in the western provinces. All balsam fir volumes listed for British Columbia and a large proportion listed for Alberta belong to different species of *Abies*, of which *Abies lasiocarpa* is perhaps the most important.

Estimates of total volumes for balsam fir in the United States were prepared by Zon (1914) and Betts (1945). The estimated balsam fir growing stock in 1914 is given in Table 117.

Betts (1945) estimated balsam fir volume to be 6 billion board feet, of which 1 billion was in the Lake States and 5 billion in the Northeast. Table 118 shows both the merchantable balsam fir volume and the percentage that balsam fir comprises of the merchantable volume of the spruce-fir forest type, softwoods, and all species in the northeastern United States. About 83 per cent of balsam fir growing stock in the latter area was in Maine. The annual growth of balsam fir sawtimber in Maine was estimated by Ferguson and Longwood (1960) at 513.2 million fbm, and the total annual growth of all material was 319 million cubic feet. No growth data were available for other states in the Northeast.

TABLE 117. ESTIMATED BALSAM FIR GROWING STOCK IN THE UNITED STATES IN 1914*

State	Millions of Board Feet	State	Millions of Board Feet
Maine	3000	Michigan	200
New York	250	Vermont	110
New Hampshire	400	Minnesota	1000
Wisconsin	395	Total	5355

*The log rule not indicated.
Source: Zon, 1914.

TABLE 118. GROWING STOCK OF BALSAM FIR IN THE NORTHEAST*

	Maine 1958[a]	New Hampshire 1948[b]	Vermont 1948[c]	New York 1950[d]	Total
Poletimber (thousand cords)	33,105	1,425	762	1,460	36,752
Sawtimber (million fbm)† .	3,340	386	362	229	4,317
Total merchantable volume (million ft³)	3,249	291	189	173	3,902
Percentage of total merchantable volume:					
Spruce-fir	47	45	30	25	44
Softwoods	30	15	16	7	24
All species	20	8	5	2	11

*Volumes by stand-size-classes are recomputed to include only spruce-balsam fir stands.
†Board-foot volume by the International ¼-inch log rule.
Sources: [a]Ferguson and Longwood, 1960; [b]Northeastern Forest Experiment Station, 1950; [c]McGuire and Wray, 1952; [d]Northeastern Forest Experiment Station, 1955.

More detailed data were available for the Lake States (Table 119). The estimated annual mortality of balsam fir was 2.5 per cent of the growing stock in
Minnesota, 2.7 per cent in Wisconsin, and 1.2 per cent in Michigan. Balsam fir
mortality computed from U.S. Forest Service Resource reports in Minnesota, Michigan, and Wisconsin averaged 2.0 per cent per year. For all softwoods average
mortality was 1.9 per cent per year.

Balsam fir comprised 45 per cent of the spruce-fir growing stock and 61 per
cent of the cubic-foot-volume growth in the Lake States. Both allowable and
actual cut of balsam fir were consistent with the level of growing stock. The
greatest growth increase was found in Minnesota, where balsam fir comprising 38
per cent of spruce-fir growing stock contributed 58 per cent to the annual cubic-
foot growth (see also Plate 8).

In the seven states in which merchantable volume of balsam fir occurs, the
merchantable growing stock was estimated in 1955 as 4909 million cubic feet. This
is a larger figure than reported by either Zon (1914) or Betts (1945), although the
three estimates cannot be compared directly. Estimates of the present volume of
balsam fir indicate that it makes up about 1.4 per cent of all softwoods and about
1.0 per cent of the total growing stock in the United States. Although this is a
relatively small percentage it is well to keep in mind that 45 per cent of all growing
stock in the United States consists of many minor species (U.S. Forest Service,
1958). The total merchantable volume of balsam fir in both Canada and the
United States, as indicated in Tables 116 through 119, can be estimated at about
36 billion cubic feet. This can be considered a conservative estimate.

TABLE 119. BALSAM FIR GROWING STOCK, ANNUAL NET PERIODIC GROWTH,
ALLOWABLE AND ACTUAL CUT, AND AS PERCENTAGE OF ALL GROWING STOCK
IN THE LAKE STATES

	Minnesota 1953[a]	Wisconsin 1956[b]	Michigan 1955[c]	Totals and % of Stock
Poletimber (thousand cords)				
Growing stock	5,510	1,609	3,638	10,757
Annual growth	236.3	85.2	191.8	513.3
Allowable cut	136.0	37.6	103.0	276.6
Sawtimber (million fbm)[*]				
Growing stock	504	111	577	1,192
Annual growth	78.2	17.4	66.2	161.8
Allowable cut	30.2	4.3	22.9	57.4
Total merchantable volume (million ft³)				
Growing stock	441	151	415	1,007
Annual growth	37.7	10.4	29.6	77.7
Allowable cut	18.0	3.9	13.2	35.1
Actual cut	13.4	3.8	11.5	28.7
Percentage of balsam fir in growing stock of:				
Spruce-fir	38	65	55	45
Softwoods	16	11	16	21
All species	6	2	4	8

[*]Board-foot volume by the International ¼-inch log rule.
Sources: [a]Cunningham et al., 1958; [b]Stone and Thorne, 1961; [c]Findell et al., 1960.

YIELD AND GROWTH PER ACRE

As previously indicated, area data are available only for spruce-fir forest types. These latter frequently include cedar-tamarack-spruce types where balsam fir is scarce or lacking entirely. Separate yields per acre for balsam fir are therefore not available.

According to Zon (1914) the merchantable volume based on small sample plots in pure balsam fir stands, 70–100 years old in New York, was 992 cubic feet or 10.2 cords per acre in the swamp type, 690 cubic feet or 7.3 cords per acre in the mixed hardwood-slope type, and 1342 cubic feet or 14.4 cords per acre in spruce-fir flats. In Maine volumes were considerably higher, 2575 cubic feet or 25 cords per acre on spruce-fir flats and 1329 cubic feet or 12.6 cords per acre on hardwood slopes. However, volumes that high were not found over large areas.

TABLE 120. POTENTIAL YIELD PER ACRE OF SPRUCE–BALSAM FIR
STANDS IN NORTHERN MINNESOTA *

Total Age (Years)	Sawtimber Volume (fbm)	Additional Pulpwood (Cords)	Total Volume (Cords)
20		0.1	0.3
40	400	1.4	5.6
60	1400	4.0	14.6
80	3550	5.2	22.2

* International log rule, ¼ in. saw kerf.
Source: Allison and Cunningham, 1939.

The mean annual growth of balsam fir in Maine was 30.5 cubic feet per acre on the flat type, and 15.2 cubic feet per acre on the hardwood slope type. Comparable figures for New York were 14.9 and 9.6 cubic feet per acre. Swamp types in New York produced 11.5 cubic feet per acre. None of these figures refer to pure balsam fir stands (Zon, 1914).

In the 1930 Forest Survey, net mean annual growth of merchantable volume per acre in the spruce-fir types of the northeastern United States was 20 cubic feet (Dana and Greeley, 1930). Under extensive forest management, annual growth was predicted to increase to 24 cubic feet per acre, and under intensive management to 45–80 cubic feet per acre.

In the spruce-hardwood type in the Adirondacks, Donahue (1940) found that balsam fir comprised 63 cubic feet per acre out of a total for all species of 3402 cubic feet per acre. In the spruce-flat type, balsam fir accounted for 759 of 1734 cubic feet per acre; in the spruce-swamp type, for 744 of 1456 cubic feet per acre. All figures were for merchantable volume.

Allison and Cunningham (1939) presented potential yield data for spruce-fir cover types in northeastern Minnesota (Table 120).

Allison and Brown (1946) found that mean annual growth per acre (merchantable volume) of balsam fir at the Cloquet Experimental Forest, Minnesota, was 55 cubic feet in the spruce-fir-hardwood type and 46 cubic feet in the lowland spruce-fir type. Comparable figures for periodic annual growth were 40 and 21 cubic feet per acre, respectively. Hubbard (1956) found that periodic net annual growth in mature and overmature spruce-fir-hardwood type in Koochiching County, Minnesota, was 0.24 cords per acre of merchantable wood (including bark). Ingrowth and accretion (gross growth) added 0.47 cords per acre per year.

Bickford et al. (1961) investigated the average growth rates in the spruce-fir area of Maine and northern New Hampshire. The stands were classified as softwood (66–100 per cent conifers); mixed-woods (21–65 per cent conifers); and hardwoods (0–20 per cent conifers). Three height classes were recognized: less than 35 feet; 35–64 feet; and over 64 feet. Stand density was determined by percentage of crown closure in three classes: 71–100 per cent; 41–70 per cent; and 11–40 per cent. For practical purposes accretion, ingrowth, and mortality were determined for three species groups. The results are shown in Table 121. No specific data for balsam fir were given. Mortality was least in the spruce-fir-hemlock type. Ingrowth for this type was 48 per cent, differing only slightly from that of other species groups.

TABLE 121. AVERAGE GROWING STOCK AND NET ANNUAL PERIODIC INCREMENT IN CUBIC FEET PER ACRE FOR DIFFERENT STANDS IN THE SPRUCE-FIR ZONE OF MAINE AND NORTHERN NEW HAMPSHIRE*

Height (Ft)	Crown Density (%)	Initial Volume (Ft³)	Mortality of Gross Increment in Spruce-Fir-Hemlock (%)	Net Increment in Cubic Feet per Acre per Year			
				Spruce-Fir-Hemlock	All Conifers	All Hardwoods	All Species
Softwoods							
0–34	70+	849	6.4	48.4	52.7	0.5	53.2
35–64	70+	1858	14.8	58.6	61.1	0.7	61.8
65+	70+	1966	46.7	33.2	27.0	6.1	33.1
Softwoods							
0–34	11–70	430	7.4	42.4	46.6	−1.6	45.0
35–64	11–70	1559	19.3	51.3	53.1	−0.2	52.9
Mixed-woods							
0–34	70+	947	0.0	44.9	46.3	1.6	47.9
35–64	70+	1698	21.3	44.9	47.4	2.9	50.3
65+	70+	2496	81.7	4.0	4.7	−35.5	30.8
Mixed-woods							
0–34	11–70	273	2.4	24.6	24.6	12.0	36.6
35–64	11–70	1428	21.6	38.4	32.5	3.1	41.6
65+	11–70	1910	2.8	40.7	27.5	15.9	43.4
Hardwoods							
0–34	70+	247	0.0	6.7	8.7	20.6	29.3
35–64	70+	896	38.8	7.4	7.6	12.2	19.8
65+	70+	2132	72.0	3.6	3.6	10.3	13.9
Hardwoods							
0–34	11–70	492	0.0	0.0	0.0	24.9	24.9
35–64	11–70	903	17.6	13.1	13.2	9.0	22.2
65+	11–70	1308	3.6	10.6	10.9	−8.3	2.6

*Total stem volume without bark.
Basis: 865 plots of unknown size.
Source: Bickford et al., 1961.

According to Ferguson and Longwood (1960) about 75 per cent of all balsam fir growing stock in Maine 6 inches dbh or over, is in the 6–8 inch dbh class. These small diameters are associated with low and medium heights.

Growth data for Canada were compiled by Candy (1938) for broad cover types (softwoods, mixed-woods, hardwoods), age classes, and geographic areas. The averages of the 20-year periodic annual growth by type groups varied from 15 to 55 cubic feet per acre. The best growth in softwood types was found in Nova Scotia, New Brunswick, and part of Ontario. Mixed-woods showed the most rapid growth in Quebec, Alberta, and parts of Ontario. Hardwoods showed the best growth in Manitoba and Saskatchewan. The 20-year periodic annual growth culminated at age 40–60 years for all forest types in Nova Scotia, New Brunswick, Quebec, and Alberta, and at age 20–40 years in Ontario. In Manitoba and Saskatchewan periodic growth culminated at 40–60 years in the softwood types, and at 20–40 years in the mixed-wood and hardwood types. No doubt more detailed growth data will become available in the future from the many permanent sample plots that have been established in spruce-fir forest types.

Conclusions

1. Growth and yield data for balsam fir, though considerable, have been compiled mainly to serve local needs and by various methods that make comparison of results difficult. Standardization of methods and techniques is needed to make possible more general conclusions.

2. The application of volume tables relies predominantly on dbh and height measurements for construction of individual height curves, although a few tables use standard height curves (tariff methods). Tariff methods could be more widely exploited in studies of balsam fir and associated species.

3. In yield tables heights of dominant trees at a reference age have been most frequently adopted as a satisfactory index of site and rate of growth. Attempts have also been made to substitute diameter for age, but very often the substitution has proved to be inconsistent with sound biological concepts and has not served practical needs satisfactorily.

4. Considerable effort has been expended to develop numerical indexes for stand density independent of age and site. Although these specific goals have not been reached, the studies have contributed to the general knowledge of balsam fir. More recently, Bitterlich's angle-count technique for estimating stand basal area has focused attention on this important element of stand volume. Wider use of direct measurements of stand basal area for new and different purposes can be expected.

5. Some applications of mathematical and statistical methods to heterogeneous data from different populations frequently lead to models that not only contradict basic biological relationships but are also misleading in practice. A number of resulting misconceptions of this type have not yet been replaced.

6. The physiological background of tree growth and the matter-energy budget of forest ecosystems have seldom been considered in growth and yield studies of forest trees. Some of the investigations now in progress, however, indicate that these subjects will be thoroughly explored in the future.

7. The effect of variations in stand density, species composition, and age structure of balsam fir stands has seldom been studied. Although foresters are very much concerned about balsam fir dominating present spruce-fir sites, little is known about the relative growth of the competing species.

8. To date, the regional and local variations of growth and yield have not been adequately investigated. The recent development of regional ecological forest classifications should stimulate further study and enable more reliable comparisons of growth and yield to be made.

9

UTILIZATION

PROPERTIES of a tree species largely determine its specific utility. The utilization of balsam fir for pulp and paper is determined by its fiber quality, pulp yields, and pulp quality. The limited utilization of the species for lumber relates to its low density and small size. On the other hand, balsam fir makes a highly desirable Christmas tree because of its needle retention, color, shape, and fragrance.

The properties of balsam fir which affect the extent of its utilization for specific purposes are treated in this chapter. Mostly this is concerned with the properties and utilization of wood, since the latter is the primary product; a discussion of minor products—including bark properties and utilization, an increasingly important subject—is also presented.

Utilization in its broadest sense covers the procurement of forest products and their technology. No attempt has been made here to consider aspects of utilization beyond the manufacture of the primary goods, i.e., lumber, pulpwood, and some minor products.

Utilization also includes aesthetic enjoyment of the species in its natural setting, or in a detached environment such as a formal garden or around a home. The value of balsam fir for Christmas trees and in landscaping is considered in a later portion of this chapter. Indirect uses by wildlife, and other effects and functions of the species as an influence in the forest environment, have been considered from an ecological point of view in Chapter 3 and will not be further discussed here.

Wood Properties

The anatomical and morphological properties of the wood of balsam fir were considered in Chapter 1. Tree form and size were covered in Chapter 8. In this section the basic physical and mechanical properties of the wood of balsam fir that determine its use for lumber, and the structural and chemical properties that are most important to its use for pulp, will be discussed.

The physical properties of wood are closely related to the mechanical, and no hard or fast distinction can be made between them. Generally the physical properties refer to the distinctive properties of wood as a material, e.g., color, density, size and structure of its component parts; and the manner in which wood reacts to the application of external forces, i.e., to moisture, heat, sound, light, and electricity. The mechanical properties of wood are those that relate to the strength of its materials. As such they are a part of the subject of wood engineering. For further distinction between the physical and mechanical properties of wood reference can be made to Brown et al. (1952).

PHYSICAL PROPERTIES

The sapwood of balsam fir is usually described as creamy white in color with little odor and taste, and composed of bands of light-colored springwood and darker bands of summerwood (Betts, 1945; Sudworth, 1916; U.S. For. Prod. Lab., 1955). The heartwood is usually difficult to distinguish from the sapwood, but in some trees the former is gray to gray-brown in color. Distinctly colored heartwood is usually the result of fungus infection or chemical stain discoloration. High moisture content pockets or zones may have a blue-gray cast in freshly cut trees. The occurrence of such high moisture content zones or wetwood, which occur primarily in heartwood, are treated in Chapter 4.

The differentiation of balsam fir sapwood and heartwood through the application of various chemical stains has received some attention. A 45 per cent solution of perchloric acid was reported by Brown and Eades (see Hale, 1950) to clearly differentiate balsam fir heartwood and sapwood. Kutscha and Sachs (1962), working with a wide range of stains and indicators on seven western species of *Abies*, found several effective sapwood-heartwood differentiating chemicals including Triplex Soil Indicator, 40 per cent perchloric acid, Bromphenol Blue, Bromcresol Green, and Alizarine Red. Because of the resemblance of the wood of other true firs to balsam fir, some of these chemicals, in addition to perchloric acid, should also be effective in differentiating the sapwood and heartwood of balsam fir. The two kinds of wood also differ with respect to permeability of liquids, the heartwood normally being far more impervious (Wallin, 1951). Wallin also consistently isolated bacteria from heartwood and occasionally encountered bacteria in the sapwood. These areas are generally considered to be responsible for the erratic results obtained in preservative treatment of balsam fir.

As is true for most conifers, tracheids make up about 95 per cent of the wood of balsam fir. Tracheid-length information on balsam fir has been gathered by Zon (1914), Casey (1952), and the U.S. Forest Products Laboratory (1953), and is shown in comparison with several other conifers in Table 122.

The more recent tracheid-length determinations indicate little difference between balsam fir, spruces, and jack pine. Although the pulps that result when pulping by various processes may differ considerably, this does not relate to fiber (tracheid) length of the original wood but is due to other factors.

The U.S. Forest Products Laboratory (1955) reported the density (volume weight) of air-dry (15 per cent moisture content) balsam fir wood as 26.9 pounds per cubic foot. The corresponding densities for white spruce, black spruce, and red spruce at the same moisture content were 29.4, 28.8, and 28.4 pounds respectively. Densities of 24 pounds per cubic foot for air-dry balsam fir (12 per cent moisture) and 45 pounds per cubic foot for green (109 per cent moisture) were noted for balsam fir in Canada (Rochester, 1933). The moisture-free weight of pulpwood per cord for several species has been given as follows: balsam fir, 1790 pounds; spruce, 2040 pounds; jack pine, 2040 pounds; paper birch, 2890 pounds; and aspen, 1870 pounds (Schafer et al., 1955).

Wide differences in moisture content are the principal causes of density variations. Wallin (1951) reported moisture contents ranging from 120 to 230 per cent based on oven-dry weight. Heartwood in general had the lower moisture levels while sapwood had moisture levels mostly between 160 and 200 per cent.

TABLE 122. TRACHEID LENGTH OF BALSAM FIR AND FOUR OTHER SPECIES

| Species | Tracheid Length in Millimeters | | | | |
	Minimum[a]	Maximum[a]	Average[a]	Average[b]	Average[c]
Balsam fir	1.68	3.75	2.52	3.5	3.5
White spruce	2.52	4.70	3.56	3.7	3.5
Black spruce	2.14	3.74	2.60		3.5
Red spruce	1.89	4.16	3.23	3.4	3.7
Jack pine				2.9	3.5

Source: [a]Zon, 1914; [b]Casey, 1952; [c]U.S. Forest Products Laboratory, 1953.

The highest moisture content normally occurred in the wetwood areas. Gibbs (1935) found the average moisture content of balsam fir sapwood to be about 200 per cent, based on oven-dry weight. Moisture contents of 260–300 per cent were also encountered, usually in wetwood spots in the heartwood. Demarcation between wet and dry heartwood was usually very distinct.

The average moisture content of green balsam fir pulpwood (with sapwood and heartwood averaged) was reported to be 117 per cent based on oven-dry weight (U.S. For. Prod. Lab., 1955). Similar material from Canada averaged 109–123 per cent in moisture content (Rochester, 1933).

Clark and Gibbs (1957) noted that the wood of live balsam fir trees had the lowest moisture content in late summer. According to Truman (1959), seasonal variation in moisture content of balsam fir was 4.5 per cent as compared with 6.5 per cent for spruce. Moisture content increased from the base to the top in individual trees, there being as much as 13 per cent greater moisture in the tops.

The specific gravity (sp-gr) of balsam fir wood has been determined by several investigators using varying bases for determination. These are briefly summarized as shown in the accompanying tabulation.

Source	Sp-gr	Basis for Determination
Sargent (1884)	0.3819	not specified
Brown et al. (1949)	0.34	green volume
	0.36	air dry (12% moisture) volume
U.S. Forest Products Laboratory (1955)	0.34	green volume
	0.36	air dry (12% moisture) volume
	0.41	oven-dry volume
Hale and Prince (1940) .	0.31–0.34	oven-dry wood per cc of saturated wood
Rochester (1933)	0.33	green weight
	0.37	oven-dry weight

Geographic differences have also been investigated. Hale and Prince (1940) concluded from a rather extensive study in Canada that the densest wood occurred in the eastern provinces.

The effect of rate of growth on specific gravity of balsam fir, white spruce, and black spruce has also been studied. For balsam fir from a spruce-fir type, readings ranging from 0.309 to 0.339 (based on green volume) for samples having 9–19 growth rings per inch were obtained. The results were not consistent, however, and Hale and Prince (1936) concluded that there was little or no relationship between growth rate and specific gravity of balsam fir, possibly because of the little variation between the summerwood and springwood portions of the annual ring. These workers also reported that white spruce having 11 and 16 growth rings per inch had values of 0.333 and 0.345 respectively, and black spruce having 21 and 28 rings per inch, 0.407 and 0.431 respectively. These findings bear out the contention of many paper-mill technologists that the weight per cord of balsam fir is only slightly less than that of white spruce but considerably lower than that of black spruce.

The wood of balsam fir has about the same specific gravity, 0.41 based on oven-dry volume, as several other true firs, i.e., *Abies grandis*, *A. nobilis*, *A. magnifica*, *A. amabilis*, and *A. concolor*. Balsam fir wood is apparently somewhat heavier than that of *A. lasiocarpa*, and *A. arizonica* (Markwardt and Wilson, 1935).

In Ontario balsam fir was found to produce in 60 years the same wood volume produced by black spruce in 90 years (Godwin, 1960). Consequently, even though the dry weight per cubic foot of balsam fir was only 21 pounds as compared with 25 pounds for black spruce, the production of dry weight per year by balsam fir exceeded the production by black spruce by 25 per cent. Donahue (1940) made a similar comparison of wood production by balsam fir from different forest types. Although spruce-swamp type balsam fir produced less volume, because of its higher specific gravity it produced more wood substance annually than the faster growing spruce-flat balsam fir.

Compression wood is distributed rather widely in balsam fir (Kaufert, 1935; Hale and Prince, 1940). Most compression wood occurs in suppressed, slow growing, and leaning trees, and in stands that have been adversely affected by windstorms. Compression wood is commonly found in trees growing on steep slopes. Attacks by the balsam woolly aphid cause similar changes in wood structure (see Chapter 5). In lumber the presence of compression wood results in brittleness and splitting (Balch, 1952, 1956). Its effect on pulping properties is discussed in another section.

The persistence of balsam fir branches after dying is well recognized. These branches dry out and may remain attached for 15–20 years. They have been shown by several workers to be entrance courts for heartrot fungi (see Chapter 4). As affecting wood properties, the persistent branches give rise to lumber that is full of knots and consequently weak and brittle.

The sapwood and heartwood are completely lacking in natural durability when subjected to conditions favorable to decay. In this respect balsam fir ranks with the least durable woods such as cottonwood, basswood, and white fir (Betts, 1945). Use under conditions conducive to decay thus requires preservative treatment if serviceability beyond a few years is desired. The literature contains little information on the treatability of balsam fir. Hunt and Garratt (1953) include the true firs with those species possessing heartwoods that are very difficult to treat. This observation has been borne out in treating trials on balsam fir made at the University of Minnesota School of Forestry from 1945 to 1955 (unpublished results). In these tests treatment of dry posts with oil-borne preservatives was difficult and penetration was very erratic. Retentions normally were satisfactory but distribution of preservative was extremely variable. Treatment of freshly cut green posts with water-borne preservatives generally gave better results. Brush applications of oil- and water-repellent solutions of pentachlorophenol on peeled balsam fir cabin logs is commonly practiced. Although superficial, such treatment may greatly extend the life of cabin logs, particularly if joints and checks are well soaked and construction is such as to keep the logs dry. In ability to take and hold paint, balsam fir ranks with white fir and the spruces and is placed in the third of four wood groups (Betts, 1945).

The shrinking and swelling of balsam fir wood with loss or gain in moisture is similar to that of other species having like specific gravity. Shrinkage studies on balsam fir have been made by Koehler (1924), Hawley and Wise (1926), Rochester (1933), and the Canada Forest Products Laboratory (1956). The results of these studies compare rather closely with results reported by the U.S. Forest Products Laboratory (1955), which are shown in Table 123. Seasoning of balsam fir by kiln drying is reported to offer no particular problems if reasonable care is used in kiln operation (Betts, 1945; Brown et al., 1952).

TABLE 123. SHRINKAGE OF BALSAM FIR UPON DRYING

Degree of Drying	Shrinkage in % of Green Dimension		
	Radial	Tangential	Volumetric
Green to 20% moisture	1.0	2.3	3.7
Green to 6% moisture	2.3	5.5	9.0
Green to 0% moisture	2.9	6.9	11.2

Source: U.S. Forest Products Laboratory, 1955.

Risi and Arseneau (1957, 1958) treated balsam fir wood with acetic anhydride after it was allowed to swell in a pyridine bath. Dimethylformamide was used as a catalyst. The same authors made stabilization tests on balsam fir wood with phthalic anhydride, using dimethylformamide as a solvent. The latter gave a high degree of dimensional stability although problems arose because of leaching and side reactions.

MECHANICAL PROPERTIES

Early tests and reports on the strength of balsam fir wood were made by Sargent (1884) and Penhallow (1907). Rochester (1933) made extensive strength tests on balsam fir in Canada. The strength values for balsam fir included in the *Wood Handbook* (U.S. For. Prod. Lab., 1955) are taken from earlier tests reported by Markwardt and Wilson (1935). These values are based on a single collection of material from five trees in Wisconsin, however they do not differ greatly from the results presented for balsam fir from Canada (Rochester, 1933; Can. For. Prod. Lab., 1956). Representative strength data from two basic sources are shown in Table 124.

While balsam fir compares favorably with other true firs in specific gravity, it generally ranks below *A. grandis, A. nobilis,* and *A. concolor* in such important strength properties as modulus of rupture and modulus of elasticity (Markwardt and Wilson, 1935).

The wood of balsam fir has been characterized as soft, weak in bending and compressive strength, low in shock resistance, possessing good splitting resistance, but low in nail-holding capacity (Betts, 1945). Balsam fir is considered relatively easy to work with hand tools (U.S. For. Prod. Lab., 1955).

The U.S. Forest Products Laboratory (1955) has also provided basic stress figures for structural sizes. Reduced from laboratory figures in order to provide a margin of safety, these figures do not allow for the effect of knots or any other strength-reducing features. Reduction of basic stresses to take into account knots, cross-grain, and other defects affecting strength is expressed as "strength ratio"

TABLE 124. STRENGTH PROPERTIES OF GREEN AND AIR-DRY BALSAM FIR WOOD
IN THE UNITED STATES AND CANADA

Characteristic	United States		Canada	
	Green[*]	Air-Dry [†]	Green[*]	Air-Dry[†]
Static bending				
Fiber stress at proportional limit				
(p.s.i. = lb/in.²)	3000	5200	2800	4300
Modulus of rupture (p.s.i.)	4900	7600	5300	8200
Modulus of elasticity (1000 p.s.i.)	960	1230	1140	1380
Work to proportional limit (in.-lb/in.³)	0.52	1.23	0.42	0.83
Work to maximum load (in.-lb/in.³)	4.7	5.1	7.4	7.9
Impact bending — height of drop (in.) of 50 lb				
hammer, causing complete failure	16	20	17	20
Compression parallel to grain				
Fiber stress at proportional limit (p.s.i.) ...	2080	3970	1690	3330
Maximum crushing strength (p.s.i.)	2400	4530	2400	4890
Compression perpendicular to grain				
Fiber stress at proportional limit (p.s.i.)	210	380	240	460
Shear parallel to grain				
Maximum shearing strength (p.s.i.)	610	710	rad. 630	870
			tan. 700	970
Tension perpendicular to grain				
Maximum tensile strength (p.s.i.)	180	180	rad. 270	260
			tan. 320	360
Hardness				
Load required to embed a 0.444 in. ball to				
half of its diameter (lb)				
End	290	510	320	680
Side	290	400	rad. 270	360
			tan. 290	410

[*]Moisture content of green wood: U.S. = 117%; Canada = 109%.
† Moisture content of dry wood: U.S. = 12%; Canada = 12%.
Basis: 5 trees from Wisconsin; 26 trees for green tests and 21 trees for air-dry tests, from 4 localities in Quebec, Manitoba, and Saskatchewan.
Sources: U.S. Forest Products Laboratory (1955); Markwardt and Wilson (1935); Canada Forest Products Laboratory (1956).

in per cent. Working stresses are obtained by multiplying basic stresses by this ratio. The basic stresses for clear, green balsam fir lumber are presented in the accompanying tabulation.

Extreme fiber stress in bending or tension parallel to
grain, p.s.i. 1300
Maximum horizontal shear, p.s.i. 100
Compression perpendicular to grain, p.s.i. 110
Compression parallel to grain L/d * = 11 or less, p.s.i. 950
Modulus of elasticity in bending, 1000 p.s.i. 1000

*L = length; d = least dimension.
Source: U.S. Forest Products Laboratory, 1955.

The various Canadian species used for structural timbers are divided into four strength groups. Douglas fir and western hemlock are placed in the upper group. Balsam fir together with Pacific silver fir and grand fir, eastern hemlock, lodgepole pine, ponderosa pine, and all spruces are placed in the third group. The allowable unit stresses for select structural grades in the third group are as follows: bending, 1300 p.s.i.; shear, 90 p.s.i.; and modulus of elasticity, 1,100,000 p.s.i. (Can. For. Prod. Lab., 1956).

The Canada Forest Products Research Branch (1961) published span tables for 15 major species including balsam fir. These figures show that balsam fir compares favorably in most respects. As a service to architects, builders, building inspectors, and others, the tables show maximum spans for wood joists and rafters.

CHEMICAL COMPOSITION

There are available several analyses of balsam fir wood showing the major components and their constituents. Most of the chemical analyses have been made in connection with utilization by the pulp and paper industry. Analyses made by different workers vary in their sampling and analytical methods. In none of the chemical analyses reported is there information on variability in composition of the four major constituents, i.e., extractives, ash-forming minerals, lignin, and cellulose, that might be used as guides by silviculturists seeking to improve the quality of balsam fir.

The earliest determination of the chemical composition of balsam fir wood appears to be the work of Johnsen in 1917 as cited by Witham (1942). Lauer and Youtz (1933) presented a proximate analysis. Hajny and Ritter (1941) also furnished an analysis of this species. These early reports are primarily of historic interest and will not be detailed here.

The U.S. Forest Products Laboratory has provided the information shown in Table 125, in which balsam fir is compared with four commonly associated conifers and with quaking aspen. This information was published later by McGovern et al. (1947) and Casey (1952). The proximate analysis given in Table 126 for balsam fir wood from New Brunswick as compared with three other conifers and aspen are reported by Clermont and Schwartz (1951). In Table 126 values for lignin are corrected for ash; those for holocellulose for ash, lignin, and extractives; those for alpha cellulose for ash and lignin; and those for hemicellulose for ash. There is good agreement between these chemical analyses on the content of holocellulose, lignin, ash, and extractives. The analyses also show that the basic composition of balsam fir differs little from that of three eastern spruces. Chemical composition differs strikingly from that of aspen with respect to alpha-cellulose and lignin content. In comparison with aspen, balsam fir has about 10 per cent less holocellulose and 10 per cent more lignin.

Timell (1957) investigated the carbohydrate portion of balsam fir and other woods in terms of monosaccharide units. For determination of monosaccharide constituents the wood was converted to the corresponding mixture of sugars by

TABLE 125. CHEMICAL COMPOSITION IN PERCENTAGES OF SIX
WOODS IN THE UNITED STATES

Composition	Balsam Fir	Red Spruce	White Spruce	Black Spruce	Jack Pine	Quaking Aspen
Holocellulose						
Total	70	73	73		72	82
Alpha	44	48	49		49	51
Cross-and-Bevan cellulose						
Total	58	60	61	61	58	64
Alpha	42	43	44	44	41	48
Lignin	29	27	27	27	27	17
Total pentosans	11	12	11	11	13	23
Solubility in ether	1	1.5	1.5	1	2	1.0
1% NaOH	11		12	11	13	19
Hot water	4		3	3	4	3
Ash	0.5	0.2	0.3	0.3		0.3

Source: U.S. For. Prod. Lab., cited by McGovern et al., 1947, and Casey, 1952.

TABLE 126. PROXIMATE ANALYSIS OF WOOD OF SEVERAL
CANADIAN PULPWOOD SPECIES [*]

Composition	Balsam Fir	White Spruce	Black Spruce	Jack Pine	Aspen
Moisture content	6.62	6.69	7.67	6.64	5.14
Solubility in:					
Cold water	2.70	1.36	1.41	2.18	1.48
Hot water	3.59	2.22	2.46	3.69	2.75
Ethyl ether	1.80	2.12	1.03	4.30	1.89
1% NaOH	13.40	12.50	12.30	16.30	19.30
Acetyl	1.52	1.08	1.14	1.08	3.41
Methoxyl	5.47	5.07	5.07	4.97	5.47
Pentosans	6.97	8.00	7.56	10.13	17.20
Ash	0.40	0.22	0.21	0.19	0.38
Lignin	27.70	26.96	27.25	27.38	18.12
Holocellulose	70.00	72.00	71.70	68.00	80.30
Alpha cellulose	49.41	50.24	51.10	47.52	49.43
Hemicellulose	15.41	16.39	15.18	16.18	21.18
Uronic anhydride	3.84	4.48	3.67	3.67	4.97
In hemicelluloses:					
Pentosans	31.9	33.5	36.3	33.7	69.1
Uronic anhydride	16.1	16.3	16.7	15.0	16.5
Hexosans (by difference)	52.0	50.2	47.0	51.3	14.4

[*] In percentage of moisture-free unextracted wood.
Source: Clermont and Schwartz, 1951.

treatment with sulfuric acid, following the method of Saeman and co-workers. An attempt was made to estimate the "true" cellulose by correcting the conventional alpha-cellulose values for the presence of mannose, xylose, and noncellulosic glucose residues. Part of his data is shown in Table 127. Balsam fir is characterized by a relatively high proportion of xylan as compared with other conifers.

In a comparison of the chemical composition of sound and partially decayed, stored balsam fir pulpwood, the stored wood showed (a) much higher cold water, hot water, and NaOH solubility; (b) somewhat increased copper number; and (c) greatly increased beta cellulose (Kress et al., 1925). The observed chemical changes (Table 128) are typical of those occurring in wood during the early stages of decay before the major wood constituents are seriously affected.

TABLE 127. CHEMICAL COMPOSITION OF SEVEN NORTH AMERICAN PULPWOOD SPECIES [*]

Composition	Balsam Fir	White Spruce	White-Cedar	Jack Pine	Tamarack	Paper Birch	Quaking Aspen
Alpha cellulose ...	47.7	48.5	48.9	45.0	47.8	44.5	56.5
Alpha cellulose corrected[†]	44.8	44.8	45.4	41.6	43.9	41.0	53.3
Lignin	29.4	27.1	30.7	28.6	28.6	18.9	16.3
Acetyl	1.5	1.3	1.1	1.2	1.5	4.4	3.4
Ash	0.2	0.3	0.2	0.2	0.2	0.2	0.2
Uronic anhydride .	3.4	3.6	4.2	3.9	2.9	4.6	3.3
Galactan	1.0	1.2	1.5	1.4	2.3	0.6	0.8
Glucan	46.8	46.5	45.2	45.6	46.1	44.7	57.3
Mannan	12.4	11.6	8.3	10.6	13.1	1.5	2.3
Araban	0.5	1.6	1.3	1.4	1.0	0.5	0.4
Xylan	4.8	6.8	7.5	7.1	4.3	24.6	16.0

[*] In percentage of extractive-free wood.
[†] Corrected for nonglucan material.
Source: Timell, 1957.

TABLE 128. CHEMICAL COMPOSITION IN PERCENTAGE OF SOUND AND PARTIALLY DECAYED BALSAM FIR WOOD

Composition	Fresh (Sound)	Stored (Partially Decayed)
Cold water soluble	0.5	6.5
Hot water soluble	1.3	9.4
Ether soluble	1.0	0.7
1% NaOH soluble	10.1	22.8
7.14% NaOH soluble	18.0	30.6
Copper number	6.1	8.3
Lignin	31.5	30.5
Cellulose	50.5	52.8
Alpha cellulose	58.5	54.8
Beta cellulose	18.0	27.4
Gamma cellulose	23.5	17.8
Pentosans	10.2	9.1
Methylpentosan	2.4	2.6
Ash	0.4	1.2

Source: Kress et al., 1925.

Pulping Properties

Balsam fir is pulped by all pulping processes. The groundwood or mechanical process is widely used for the pulping of balsam fir, although some difficulty has been encountered due to pitch problems. Data from the U.S. Census Bureau (1918–1932) for the period 1916–1930 revealed that about one-half of the balsam fir used for pulp and paper was pulped by the sulfite process. Today balsam fir is also extensively pulped by the sulfate process and by a variety of semichemical processes.

GROUNDWOOD OR MECHANICAL PULPING

Because of its long fibers and light color, balsam fir is well suited to mechanical pulping (U.S. For. Prod. Lab., 1953). Balsam fir and spruce (white, black, and red) groundwood pulps are used with chemical pulps in the manufacture of a wide range of papers, especially newsprint, which is normally 75–85 per cent groundwood or mechanical pulp and 15–25 per cent bleached sulfite or sulfate (Betts, 1945). Balsam fir groundwood pulps are considered soft and of excellent color (Betts, 1945). The brightness of pulps produced from balsam fir and other species is shown in Table 129. In these tests only cottonwood pulps exceeded balsam fir in brightness. Perry (1960) indicated that where balsam fir was not available, the same brightness and softness could be obtained by using poplar or aspen.

TABLE 129. COLOR CHARACTERISTICS OF GROUNDWOOD PULPS

Species	Bausch and Lomb Opacimeter Reading (% Whiteness)	Brightness (%)
Balsam fir	85.0	58.6
Cottonwood	85.5	63.0
Spruce	81.5	56.4
Jack pine	75.0	42.4

Source: after Wynne-Roberts, 1937.

The groundwood pulp yield for balsam fir is 1500 pounds per standard cord of 128 cubic feet, as compared with 1800 pounds per cord for spruce (Zon, 1914). Similar data presented by Thickens (1916) gave the yield of balsam fir pulp per cunit (100 ft^3) of round wood as 1910 pounds, 490 pounds less than for an equal volume of spruce. Since pulp yields are correlated with wood density (Wynne-Roberts, 1937), the above comparisons were probably made with black spruce, which is generally somewhat denser than white or red spruce. Wynne-Roberts' pulp yield data are shown in Table 130.

Pitch problems encountered in the use of balsam fir groundwood pulps have frequently been mentioned (Zon, 1914, and others). More recent studies have shown that the pitchy materials that clog screens, felts, and coat cylinder faces are pectinous materials from the bark, probably deposited on the surface of the wood during hand or mechanical peeling. Pitch problems can be eliminated or minimized by using seasoned wood or by adding certain chemicals to the pulp suspensions (Schafer, 1960).

The grinding characteristics of balsam fir as compared with other species have been studied by several workers (Wynne-Roberts, 1937; Hyttinen and Schafer, 1958; and Schafer, 1960). In the tests by Wynne-Roberts (1937) balsam fir ranked second to spruce in power consumption and freeness of pulp (Table 131).

Schafer (1960) presented additional data developed at the U.S. Forest Products Laboratory on the grinding characteristics of balsam fir as compared with white and black spruce (Table 132). In these studies the spruces ranked above balsam fir in most characteristics although balsam fir generally ranked high. Though not as strong as those of spruce, balsam fir groundwood pulps apparently are stronger than those of southern pine. To obtain maximum strength properties, stones duller than those used for spruce are more desirable in grinding balsam fir (Hyttinen and Schafer, 1958).

Hyttinen and Schafer (1958) and Schafer (1960) presented data on the effect of various conditions of balsam fir wood on pulp quality. Stored pulpwood does not dry uniformly, and variations in its moisture content also cause variations in the quality of groundwood pulps. Wood of low moisture content, when ground at low pressure, produced pulp of good strength at moderate consumption of energy (66 hp-days) although the pulp had lower freeness (65 ml) than is desired for some purposes. Pulps of higher freeness were readily obtained by increasing the grinding pressure, which reduced energy consumption but also lowered the strength. Pulps made from remoistened wood were appreciably

TABLE 130. GROUNDWOOD PULP YIELDS OF
DIFFERENT SPECIES

Species	Density (Oven-Dry Weight of Green Volume)	Yield of Screened Pulp per Cunit (Pounds)
Balsam fir	0.34	2050
Spruce	0.38	2375
Jack pine	0.41	2500
Poplar	0.43	2575
Cottonwood	0.44	2700

Source: Wynne-Roberts, 1937.

TABLE 131. GRINDING CHARACTERISTICS OF
DIFFERENT WOODS *

Species	Freeness (cm³)	Production (Tons/10 Ft² per Day)	Power Consumption (Hp/Tons per Day)
Balsam fir	68	9.6	80
Spruce	115	12.0	70
Jack pine	58	8.0	105
Poplar	37	6.0	140
Cottonwood	22	3.2	215

* Bursting strength constant at 8 p.s.i. 32-pound basic weight.
Source: after Wynne-Roberts, 1937.

higher in freeness than dried stored wood, and were stronger when compared on the same basis, and longer fibered.

Storage of pulpwood reduces the brightness of groundwood pulp. This reduction is linearly related to the time that the wood had been in storage. A brightness of 70 G.E.-equivalent can be obtained by bleaching the pulp from fresh wood with about 1 per cent zinc hydrosulfite or sodium peroxide, whereas with an increasing wood storage age, from two to three times this amount of bleaching agent is required to reach the same brightness (Whitman, 1955, 1957).

Satisfactory groundwood pulps can be produced from small and knotty balsam fir wood from the tops of trees (Kincaid, 1945). Utilization of tops to a 2-inch diameter, as is sometimes done with black spruce, is impractical with balsam fir because of the abundance of sizable branches.

TABLE 132. TYPICAL GROUNDWOOD PULPING CHARACTERISTICS OF BALSAM FIR, BLACK SPRUCE, AND WHITE SPRUCE

Characteristic	Balsam Fir	White Spruce	Black Spruce
Density of wood based on moisture-free weight and green volume (lb/ft³)	21	24	25
Stone surface condition	Sharp	Medium	Sharp
Pressure of wood on stone (p.s.i.)	20	25	20
Energy consumed per ton of moisture-free wood (hp-days)	46	50	52
Freeness (Schopper-Riegler) (ml)	497	460	345
Screen analysis (% retained on 24-mesh)	9.4	12.7	2.9
Screen analysis (% passing 150-mesh)	44.6	34.5	47.4
Burst factor	14.3	19.3	17.6
Tear factor	38	73	41
Tensile strength on air-dry basis (p.s.i.)	1614	1926	1706
Breaking length (m)	3050	3530	3650
% whiteness (Ives parts blue)	65		65
% brightness (G.E.-equivalent)		60	5

Source: after Schafer, 1960.

Even small amounts of decay developed in balsam fir pulpwood during storage cause serious reductions in yields and brightness of groundwood pulps (Glennis and Schwartz, 1952). Decay in pulpwood stored for more than one year could be a serious factor in the production of groundwood as well as chemical pulp (Shema, 1955). After one year, the annual loss of pulpwood owing to decay amounts to about 10 per cent of the volume when fresh. Dead wood, although satisfactory if not decayed, should be piled separately from green wood since the former, if mixed with green wood, often becomes overcooked in the pulping process, producing pulp of low yield and low grade (Stewart, 1943).

Wood deformation similar to compression wood (*Rotholz*) caused by insect attacks (see Chapter 5) on balsam fir produces short-fibered, brittle pulp with low burst and tear factors (Balch, 1956). The effects of the higher lignin content, darker color, and low fibril angle of compression wood are reduction in brightness and in strength properties. However, compression wood can be better utilized in groundwood pulping than in chemical pulping.

CHEMICAL PULPING

Because of its lack of resins and long fibers, balsam fir is well adapted to pulping by the sulfite process (U.S. For. Prod. Lab., 1953). The unbleached sulfite pulp produced from balsam fir is light in color and has excellent strength, rating next to spruce among northern conifers in this respect. Balsam fir sulfite pulps are used in many of the highest quality papers.

According to Wells and Rue (1927) yields of sulfite pulp, based on the weight of moisture-free wood, range from 45 to 50 per cent for balsam fir. Zon (1914) gave the yield of balsam fir sulfite pulp as 1000 pounds per cord as compared with 1200 for spruce. Kellogg (1923) gave the yield of balsam fir sulfite pulp as 970 pounds per 100 cubic feet of pulpwood. Yields of balsam fir as compared with

TABLE 133. SULFITE PULP YIELDS AND SOME CHARACTERISTICS
OF PULPS OF DIFFERENT SPECIES

Species	Yield of Screened Pulp		Paper Strength	
	% of Moisture-Free Wood	Pounds per Cord	Relative Bursting Strength	Relative Tearing Strength
Balsam fir	47	790	90	100
Black and white spruce ..	48	920	100	100
Jack pine	45	865	75	100
Paper birch	46	1250	75	90
Aspen	52	915	50	60

Source: McGovern et al., 1947.

some other species, and certain other characteristics of sulfite pulp as given by McGovern et al. (1947), are shown in Table 133. A recent U.S. Forest Products Laboratory experiment with a multistage sulfite pulping process indicated that balsam fir could be satisfactorily pulped (Sanyer et al., 1962).

Ostrowski (1943) and Wells and Rue (1927) also concluded from their studies on sulfite pulping that this process is excellently adapted to balsam fir and produces soft pulps of light color that are readily bleached and rank close to spruce sulfite pulps in strength. They point out that the sulfite process requires non-resinous, bark-free, clean, sound wood. Bark, dirt, resins, and decay seriously affect the pulps produced, more so than for any other pulping process. Irwin and Lauer (1961) specifically studied the effects of dirt and knots as factors in the pulping of balsam fir and found that serious reduction in pulp quality resulted when these were present.

Increasing quantities of balsam fir and other wood species are being pulped by the sulfate process (McGovern et al., 1947). The lower cost of pulping chemicals used in the sulfate process, combined with the production of strong pulps that are bleachable to a high degree of whiteness, in addition to the high recovery rate of chemicals, have been largely responsible for the rapid growth of sulfate pulping. Balsam fir pulps made by the sulfate process are suitable for a wide range of papers and other fiber products (Betts, 1945).

Sulfate pulp yields for balsam fir reported by Kellogg (1923) were 1010 pounds per 100 cubic feet (cunit) of solid wood. Wells and Rue (1927) gave the yield of pulp as 40–50 per cent based on the dry weight of wood.

More recent sulfate pulping trials on balsam fir have been reported by Bray et al. (1944) and Schafer et al. (1955). Pulp yields and quality of balsam fir sulfate pulps are compared with other northern pulpwood species in Tables 134 and 135.

TABLE 134. YIELD OF MOISTURE-FREE SCREENED SULFATE PULP

| | Yield of Pulp (Lb) per: | | Cords Required per Ton of Pulp | |
| | 100 Lb Moisture-Free | | | |
Species	Wood	Cord	Moisture-Free Pulp	Air-Dry Pulp
Balsam fir	48	785	2.55	2.29
White spruce ..	48	900	2.22	2.00
Black spruce ..	49	920	2.17	1.96
Jack pine	47	880	2.28	2.05
Aspen	56	960	2.08	1.87

Source: after Bray et al., 1944.

TABLE 135. YIELD OF MOISTURE-FREE SCREENED PULP PER 100 POUNDS OF MOISTURE-FREE WOOD, AND STRENGTH OF UNBLEACHED SULFATE PULP

| | Yield in Pounds | Relative Strength Values (Spruce = 100) | | | |
Species		Bursting Strength	Tearing Strength	Folding Endurance	Tensile Strength
Balsam fir	50	96	91	106	105
Spruce	50	100	100	100	100
Jack pine	51	90	92	94	96
Paper birch ...	50	77	62	41	86
Quaking aspen .	54	65	65	30	87

Source: after Schafer et al., 1955.

Recently the U.S. Forest Products Laboratory (Laundrie, 1961) investigated the possibility of further developing the sulfate pulping of balsam fir. Experiments were made to produce sulfate pulps with permanganate numbers of 16.5 to 20.9 for bleachable grades and 26.8 for brown paper grades. The pulps of lower permanganate numbers were bleached to a brightness of about 89 per cent by a five-stage Cl–ClO$_2$ bleaching process without significantly changing their strength. The brown paper grade was stronger in bursting, tensile strength, and folding endurance than a comparable pulp made from southern pine, but was lower in tear resistance in the freeness range from 300 to 500 ml (Canadian standard). By beating the balsam fir pulp lightly to a freeness of approximately 675 ml, the pulp developed strength properties about equal to those of the best southern pine kraft.

SEMICHEMICAL PULPING

A wide variety of semichemical processes have been developed for the production of pulps of greater strength than groundwood pulps and of lower cost and higher yield than chemical pulps. In one of these processes pulpwood is impregnated with pulping agents in order to soften it prior to grinding. In other processes chips are soaked at room temperature in sodium hydroxide and then defibrated by mechanical means.

One of the earliest developed and still most commonly used semichemical processes involves the cooking of chips with sodium sulfite buffered with sodium bicarbonate. This neutral sulfite process was used by McGovern and Keller (1948) in pulping balsam fir and several other species. These cooks were carried far beyond those of the usual semichemical process and resulted in complete pulping of the chips with consequent reduction in the yield of pulp produced. The results of these tests are given here not because they are typical for presently operated semichemical pulping processes but because they show comparative yields of balsam fir and other northern pulpwood species (Table 136). Yields in the range of 85–95 per cent are expected when chips are cooked only enough to soften them and pulping is completed by mechanical means.

TABLE 136. COMPARISON OF PULPS OF DIFFERENT SPECIES
OBTAINED WITH SODIUM SULFITE BUFFERED WITH
SODIUM BICARBONATE

Pulp Yield and Composition	Balsam Fir	Black Spruce	White Spruce	Quaking Aspen
Pulp yield (%)	44.2	41.6	43.8	46.0
Chemical composition (%)				
Lignin	1.1	1.3	1.5	2.1
Cross-and-Bevan cellulose .	94.8	95.4	96.0	94.2
Alpha cellulose	80.6	80.2	83.6	78.4
Total pentosans	8.2	10.0		15.2
Solubility in 1% NaOH		3.7		4.3

Source: after McGovern and Keller, 1948.

Chidester et al. (1960) made tests with neutral sulfite under more normal semichemical pulping conditions and obtained yields of 90–97 per cent for balsam fir. By reducing the pH of the cooking liquor from 10.4 to 7.6 the yield of pulp was increased from 90.3 to 94.8 per cent, and brightness from 47.6 to 52.5 G.E.-equivalent. The results of tests on papers made from these pulps indicated that for certain types of paper these high-yield semichemical pulps could be substituted for certain low-yield and more expensive chemical pulps.

Although balsam fir has also been pulped successfully by the soda process and the chemi-groundwood process, the literature contains little specific information on these processes as far as their application is concerned, and they will not be further dealt with here.

Lumber and Pulpwood Products

The specific wood properties and small size of the species are determining factors in the utilization of balsam fir. Small size and low density are the primary factors in limiting its greater use for lumber. Low density and strength, abundance of knots, and poor treating qualities limit its use for poles, posts, ties, piling, and mine timbers. Use for cabin logs in Canada and the United States has been quite extensive, and fairly successful where the logs are periodically treated by spray or brush with oil solutions of pentachlorophenol. Such use is normally determined more by the availability of balsam fir than by its possession of desirable characteristics, since it is not especially well suited to the purpose.

Only in its use for pulp and paper is the wood of balsam fir reasonably desirable. Good fiber length (3–4 mm) and fiber quality, combined with past, present, and potential future abundance at reasonable costs, are its chief advantages. Balsam fir compares poorly with other pulpwood species in wood density, the property which largely determines the yield of pulp per cord.

HARVEST AND PROCUREMENT

Most balsam fir harvested for logs and pulpwood is commercially clearcut. Although the silvicultural characteristics of the species, particularly its tolerance to shade, indicate its suitability for partial cutting, economic considerations have normally dictated clearcutting as the prevailing harvest practice. Mechanization of logging has been largely responsible for the lack of partial cutting as a more common harvest practice for balsam fir. Belotelkin et al. (1942) showed that, by logging with horses and removing 10–60 per cent of stand volume, logging costs were not appreciably greater than for clearcutting. He attributed this to the fact that the diameter of material removed was greater than for clearcutting, and logging costs per unit of product produced are less for large than for small wood. Information on balsam fir logging costs has also been reported by Holt (1950).

The marked decrease in logging costs with increase in diameter of logs or bolts has been clearly shown in time studies by Jensen (1940). Jensen also found that the same harvest operations for balsam fir required somewhat less time than for spruce. Some of Jensen's results are shown in Table 137.

The Canada Forest Products Laboratory (Doyle, 1957) has also carried out rather extensive studies of the effect of tree size on logging costs with results similar to those reported by Jensen (1940). On the average, less time was required

TABLE 137. PRODUCTION TIME IN MINUTES PER CORD IN HARVEST OPERATIONS FOR BALSAM FIR OF DIFFERENT DIAMETERS IN THE NORTHEASTERN UNITED STATES *

dbh (In.)	Felling	Limbing	Peeling	Bucking At Landing	Bucking In Woods
4	178	194	196	182	220
6	108	159	116	129	180
8	77	137	80	113	154
10	59	121	60	107	140
12	49	111	46	103	133

*The figures for felling, limbing, and peeling are based on time studies from 875 trees; the figures for bucking from 1388 trees. Data from work of one-man crews.

Source: Jensen, 1940.

to saw 1000 board feet of lumber from balsam fir than from spruce logs of the same size. Maximum efficiency for lumber sawing was reached for 12-foot balsam fir logs which had diameters of 11–12 inches at the small end.

Canadian research workers have been particularly active in studies aimed at reducing the moisture content of balsam fir pulpwood and sawlogs to improve their floatability. Because of the emphasis on water transport of pulpwood in Canada as compared with the United States, floatability is an acute problem in parts of that country. Peeling and piling for several months prior to "watering" pulpwood is common practice and is reasonably effective in reducing sinkage losses during the pulpwood drive.

Gibbs (1935) investigated the effect of tree-girdling on moisture loss. After girdling, the sapwood above the girdle lost most of its free water during two to three months in summer although the heartwood still showed wet spots. From the standpoint of improving floatability and decreasing sinkage losses, girdling was found to be only partially effective. The top logs of balsam fir are notoriously poor floaters. Since moisture content for balsam fir does not appear to vary seasonally in the same way as for aspen and other hardwoods, floatability of balsam fir pulpwood is related largely to sapwood-heartwood ratio and position in the tree. Butt logs or bolts and larger logs usually have a higher heartwood content, which is low in moisture content and relatively impervious, than sapwood. Top logs or bolts not only have greater amounts of water-permeable sapwood, but more access points for water, especially in branch scars. McIntosh (1949a, 1951) also noted the effect of girdling on moisture loss of balsam fir.

Weight reduction of wood can also be achieved by leaving felled trees with crowns intact for some period during the summer. According to McIntosh (1949a) balsam fir felled in July showed an average weight reduction of 6 pounds per cubic foot in 36 days, the loss of weight being somewhat reduced by the shade of surrounding trees. Trees felled in an open area in August lost 7 pounds per cubic foot in 17 days and 11 in two months. Peeling such trees 2–3 weeks after felling was more difficult than peeling freshly felled trees. Evidence of bark-insect attack, but no bluestain, was found in trees with limbs intact for two weeks after felling.

Chemical debarking to reduce the moisture content of wood and increase the ease of peeling has recently found much favor in pulpwood operations, especially in Canada. Arsenical compounds were found to facilitate removal of tree bark by Tanger in 1937. Extensive use of chemical debarking in Canada was proposed by White in 1943 (cited by Wilcox et al., 1956) but not until the studies of McIntosh (1948, 1949a,b, 1951) was this procedure placed on a sound footing. Chemical debarking is less costly, more rapid, and less laborious than sap-peeling, and is less wasteful than mechanical peeling (Wilcox et al., 1956).

Even where part or all of the bark is left on the bolt after the chemically treated tree is felled, drum or machine barking at the mill is faster than barking of green wood, and requires less recycling of the bolts. The lack of resins and sap at peeling time and the weight reduction are additional advantages. Investigations by McIntosh (1949a,b, 1951) showed that chemical debarking treatments for balsam fir, when applied in June and July, provided good reduction in weight by October. The weight per cubic foot of balsam fir treated with sodium arsenite in June was reduced to 39.9 pounds by September as compared with a weight of 51.7 pounds per cubic foot for untreated trees. Bark removal was also facilitated by the treatment. Balsam fir and jack pine responded equally well to sodium arsenite treatment and both responded better than black spruce and several hardwoods. Wilcox et al. (1956) reported on experiments extending over four years with chemical debarking of several species. Application of 40 per cent sodium arsenite in the middle of the sap-peeling season maintained peelability for several months, but since balsam fir bark breaks into small patches after chemical treatment, the total labor for hand-peeling chemically treated trees was greater than for sap-peeling. However, treated balsam fir peels readily in a drum barker. When the arsenic source was monoethanolamine arsenite-creosote, chemical peeling was easier than sap-peeling, the bark coming off readily when stripped with a peeling spud. Peeling of balsam fir treated with ammonium sulfamate was found more difficult than when soluble arsenic compounds were used (McIntosh and Hale, 1949). In a time study comparison it was found possible to girdle 22.3 trees per hour and to girdle and apply chemicals to 10.3 trees per hour (McIntosh, 1951). The girdling and chemical treatment of balsam fir was somewhat easier than for jack pine and much easier than for black spruce.

Balsam fir needles contained 4.0 ppm of arsenic two weeks after chemical treatment and 2.5 ppm after five weeks (Webb et al., 1956). The development of decay following treatment is a serious problem in chemically debarked balsam fir, but decay losses can be minimized by harvesting within 4–6 months after treatment (Wilcox et al., 1956). No adverse effect on pulp quality was found to result from chemical debarking. Bark beetles and woodborers do some damage to chemically killed trees if the latter are left for long periods before harvesting (Mott, 1954).

Attempts have been made to salvage balsam fir killed by fire, insects, diseases, and wind. Although wood-boring insects and decay fungi normally invade balsam fir trees soon after they succumb, small trees that dry out rapidly and fire-killed stands may deteriorate rather slowly. Skolko (1947) found that decay losses in some fire-killed stands were not serious up to five years after fire-killing. Stillwell (1958) observed that commercial salvage operations in fire-killed balsam fir were possible in Newfoundland and New Brunswick for a number of years after fires. However, the presence of charred wood on pulpwood from fire-killed balsam fir makes such wood undesirable for groundwood and sulfite pulping. Since most of the balsam fir pulping is done by these processes — rather than by the sulfate process, in which some charcoal can be tolerated — relatively little fire-killed balsam fir is pulped.

Some information on balsam fir stumpage, pulpwood, and sawlog prices is shown in Table 138. Because of its generally lower specific gravity and greater availability, balsam fir commands lower prices, as a raw material, than spruce.

To provide information for the design of harvesting equipment Keen (1963) prepared tables showing weight and volume relationships for balsam fir in Quebec and Ontario and providing information on slash and bark, seasoning weight losses, and the location of centers of gravity of the stems. Specifications are provided for different dbh and height classes. Tables are based on measurements of 264 balsam fir in 10 sets of samples. The data indicate that for a 6-inch dbh tree, merchantable wood weight comprises 50 per cent; bark, 11 per cent; and slash, 39 per cent of the total weight. For a 14-inch tree, merchantable wood weight is 62 per cent; bark, 14 per cent; and slash, 24 per cent. Moisture content was 29 to 69 per cent of green weight, crown diameters were 8 to 32 feet, while crown volume varied from 46 to 9048 cubic feet for all tree sizes. There was only a slight difference between the centers of gravity of balsam fir, black spruce, white spruce and jack

TABLE 138. PULPWOOD PRICES IN WISCONSIN, NOVEMBER 1962 *

	Stumpage per Cord (Standing Tree)	Delivered Mill Price		F.O.B. Car Price	
		Rough	Peeled	Rough	Peeled
Balsam fir	$4.00–8.00	$19.00–22.50	$24.00–27.50	$17.50–22.00	$ 27.00
Aspen	1.80–4.00	11.00–14.50	19.00–20.00	14.00	17.00–19.00
Birch, white ..	1.00–2.50	12.00–14.50	21.00		17.00–20.00
Hardwoods, mixed	1.00–2.00	12.00–16.00	20.50–23.00		20.00–21.00
Hemlock	4.00–6.45	19.00–19.50	24.00–25.00	17.00–18.50	23.00–25.00
Oak	1.00–2.00	15.00			
Oak, chemically treated		16.50			
Pine, jack and red	2.50–6.00	17.50–18.00	22.50–24.00	17.50–18.50	22.50
Pine, white ..		15.00	20.00–21.00	14.50	
Spruce	6.00–12.00	25.00–28.50	30.00–33.50	24.00–27.00	32.00

* Cord size = 4′ x 4′ x 100″.
Source: Peterson, 1962.

pine. For 5-inch dbh trees the center of gravity for balsam fir was located about 15 per cent from the butt, and for 18-inch trees, 33 per cent from the butt of the total tree length.

LUMBER

In his historical summarization of lumber production in the United States, Steer (1948) presented information on balsam fir for the period 1905–1944. By the latter date production had declined to such an extent that the species was no longer listed separately, but was included together with other species produced in small volume. Figure 12 and Table 139 show, for the period 1905–1944, the balsam fir lumber production totals and trends for the United States and for individual states in which the species is important. Horn (1950, 1952a, 1954, and 1957) reported the balsam fir lumber production for the Lake States for 1948–1954 (Table 140). These data show clearly that balsam fir in recent years has made up such a small part of national lumber production (ca. 0.03 per cent) and even of Lake States production (ca. 0.6 per cent) that it can be discounted, except possibly in a few local situations, as a lumber-producing species in the United States. In contrast to the situation in the United States, balsam fir lumber produc-

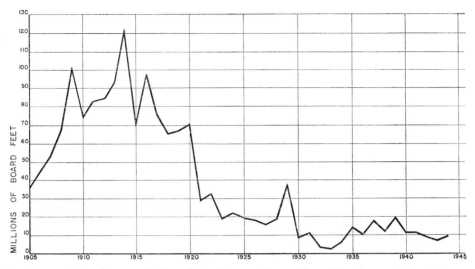

Figure 12. Production of balsam fir lumber in millions of board feet from 1905 to 1943 in the United States. (Redrawn from Betts, 1945. Courtesy of U.S. Forest Service.)

tion in Canada has remained fairly stable. Production figures for Canadian provinces for the period 1910–1955 have been reported by Picard (1924), Canada Bureau of Statistics (1906–1962), and Canada Forestry Branch (1956). Some of this information is summarized in Table 141. Currently, the heaviest cut of balsam fir for sawtimber comes from Quebec, where balsam fir constituted 8 per cent of all lumber produced in 1955. On a relative scale, however, balsam fir was most important in lumber production in Newfoundland (67 per cent in 1955), followed by Prince Edward Island (23 per cent), New Brunswick (8 per cent), and Nova Scotia (7 per cent).

The larger total and continued high production of balsam fir lumber in eastern Canada no doubt relate to its larger size, better quality, and greater abundance in that part of its range as compared with the situation in the United States.

Low density and resultant low strength limits the uses that can be made of balsam fir lumber. In the past, according to Koehler (1924) and Betts (1945), it was used extensively for boxes, crating, barrel headings, dimension stock, sheathing, and for similar purposes not demanding great structural strength. Together with northern white-cedar it has recently found increased use as knotty panel lumber. Its lack of taste, odor, and resin has made balsam fir a favorite wood for fish boxes in Canada and the Lake States. Balsam fir lumber in the United States is graded under the rules of the Northern Hemlock and Hardwood Association (1947) or the Northern Pine Manufacturers' Association (1939). The same requirements applying to northern conifers apply to balsam fir. In Canada the Canadian Lumbermen's Association (1955) grading rules, which also apply to the eastern spruces and jack pine, are followed in grading balsam fir lumber.

Balsam fir lumber production and utilization in the United States probably will continue to decline because of the greater demand for pulpwood. Production and utilization of the species for lumber in Canada will probably continue at somewhat near past levels because of its abundance, size, and quality. The volume of sawlog

TABLE 139. PRODUCTION OF BALSAM FIR LUMBER IN THOUSANDS OF BOARD
FEET IN SIX STATES, AND BALSAM FIR LUMBER COMPARED WITH TOTAL
LUMBER PRODUCTION IN THE UNITED STATES

Year	States						United States	
	Maine	New Hampshire	Vermont	Michigan	Minnesota	Wisconsin	Balsam Fir	Total Lumber
1905 ...	21,899	1,702	8,468	1,531			35,506	43,500,000
1910 ...	42,836	3,863	8,298	4,787	10,147	4,196	74,580	44,500,000
1915 ...	37,279	3,705	8,849	4,491	14,159	2,446	71,358	37,011,656
1920 ...	31,042	2,332	4,440	5,321	12,377	13,903	70,511	34,999,800
1925 ...	8,540	1,270	2,439	1,926	4,194	733	19,686	40,999,641
1930 ...	2,381	565	4,042	404	368	121	7,916	29,358,021
1935 ...	9,854	37	2,778	571	1,047	213	14,524	22,943,833
1940 ...	3,124	527	5,162	610	2,711	363	12,583	31,159,126
1944 ...	2,270	631	2,459	483	1,836	512	8,465	32,937,549

Source: Steer, 1948.

313

and lumber quality balsam fir available today and predicted for the future in Canada far exceeds the present rate of cutting (Canada Forestry Branch, 1956).

PULPWOOD

Were it not for its major present use as pulpwood, balsam fir would probably be numbered among the least used and lowest in value of timber species. Increased use for pulp and paper has developed in spite of early prejudices against the tree. In addition to its low density and the pitch problems encountered in pulping, prejudice against this species nonetheless persists because of its susceptibility to

TABLE 140. BALSAM FIR SAWLOG AND BOXBOLT TIMBER DRAIN IN THOUSANDS OF CUBIC FEET FROM 1948 TO 1954 IN THE LAKE STATES

State	1948	1950	1952	1954
SAWTIMBER				
Minnesota	570	620	422	374
Wisconsin	180	140	105	87
Michigan	530	370	89	103
Lake States	1280	1130	616	564
POLETIMBER				
Minnesota	480	320	76	184
Wisconsin	40	70	19	50
Michigan	110	190	16	96
Lake States	630	580	111	330
ALL TIMBER				
Minnesota	1050	940	498	558
Wisconsin	220	210	124	137
Michigan	640	560	105	199
Lake States	1910	1710	727	894

Source: after Horn, 1950, 1952a, 1954, 1957.

spruce budworm attack (Chapter 5), which makes it a somewhat hazardous raw material on which to base a permanent industry.

Utilization standards for pulpwood encourage greater use of the tree than do the standards for sawtimber. In sawtimber operations in the Lake States in 1944, only 30 per cent of the wood harvested appeared in the final product. The comparable figure for pulpwood was 78 per cent (Demmon, 1951). These differences are largely due to more complete utilization in the woods. Irrespective of other requirements, pulpwood is usually utilized to a 3.5-inch top diameter; sawlogs are taken only up to 9 or 10 inches (Watson, 1955).

No standardized grading rules have been developed for balsam fir pulpwood, although grade standards have been proposed. Three pulpwood grades proposed by Kellogg (1923) for several New York species included balsam fir. Kellogg based his specifications for pulpwood grading upon the minimum diameter for 4-foot bolts, maximum size of knots, percentage of decay, and proportion of balsam fir in spruce-fir mixtures. These rules were never found to be practicable. Most companies and individual mills have their own specifications, which are kept flexible to fit different local conditions.

TABLE 141. BALSAM FIR LUMBER PRODUCTION IN THOUSANDS OF BOARD FEET IN CANADA

Province	1910	1915	1920	1925	1930	1935	1940	1945	1950	1955
Newfoundland *									27,587	22,142
Prince Edward Island	1,127	2,340	1,771	231	1,030	792	942	2,106	2,604	2,241
Nova Scotia	4,938	7,091	10,982	2,577	3,972	4,126	8,521	11,320	17,906	25,787
New Brunswick	15,256	45,659	53,150	17,216	7,237	11,376	15,758	10,086	20,489	21,629
Quebec	87,292	170,794	47,116	20,266	35,711	58,230	91,295	57,876	69,466	87,066
Ontario	15,307	4,341	7,102	2,921	1,412	1,179	3,418	2,497	4,498	4,585
Manitoba			10	35		364	709	338	469	326
Saskatchewan						60	41		571	104
Alberta *		20	875	40	191	172	216		3,661	7,274
British Columbia *		3,276	11,384	14,937	31,246	28,090	4,525	9,927	17,421	52,687
Total volume (M fbm)	123,920	233,521	132,390	58,223	80,799	104,389	125,425	94,150	164,672	223,841
Percentage balsam fir of total lumber	2.8	6.2	3.1	1.5	2.0	3.5	2.7	2.1	2.5	2.8

* Newfoundland data from 1910 to 1945 not available; Alberta data include A. *balsamea* and A. *lasiocarpa*; British Columbia data include A. *lasiocarpa*.

Sources: after Picard, 1924; Canada Bureau of Statistics, 1906–1955, 1962; Canada Forestry Branch, 1956.

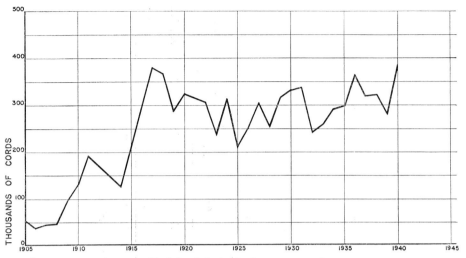

Figure 13. Production of balsam fir pulpwood in the United States, 1905 to 1940.
(Redrawn from Betts, 1945. Courtesy of U.S. Forest Service.)

Betts (1945) presented a graph (Figure 13) showing the production of balsam fir pulpwood in the United States from 1905 to 1945. After 1940 the national statistical reports on pulpwood production and consumption no longer differentiated between balsam fir and spruce. The increase in the consumption of balsam fir pulpwood in the period 1900–1940, with estimates for 1950 and 1960, is shown in Table 142. More recent information on utilization in the Lake States is shown in Table 143.

The total amount of balsam fir pulpwood produced in 1942, 257,000 cords, was the same in 1946. This amount comprised 14.7 per cent of the total pulpwood production in the Lake States. In the early 1940's the Lake States production of balsam fir was somewhat less than half the national total, the remainder being accounted for by New York and New England. If the same ratio between the Lake States and the Northeast still prevails, present production of balsam fir pulpwood in the United States would be about 700,000 cords, or about 1.7 per cent of the national total. This information reveals that although production and

TABLE 142. CONSUMPTION OF PULPWOOD OF ALL SPECIES AND
OF BALSAM FIR IN THE UNITED STATES FOR SELECTED YEARS

Year	All Species Cords	Balsam Fir Cords	Balsam Fir % of All Species
1905	3,192,002	56,744	1.8
1910	4,094,306	132,362	3.2
1920	6,114,072	328,882	5.4
1930	7,195,524	379,483	5.3
1940	13,742,958	472,186	3.4
1950	23,627,000	600,000 (est.)	2.5
1960	40,485,000	700,000 (est.)	1.7

Source: after U.S. Bureau of Census (1906–1962); and other sources.

consumption have increased, this increase has been much less than that of other species, e.g., the southern pines, which now make up over 50 per cent of the pulpwood produced and consumed by the paper industry in the United States.

Horn (1962) has shown that even in the Lake States, where production and consumption of pulpwood have not increased as rapidly as they have nationally, the percentage of balsam fir decreased between 1950 and 1960 (Table 143). This decrease has been due largely to the greatly increased use of aspen, which now makes up about 46 per cent of the pulpwood produced in the Lake States. Although there is probably a similar trend in the northeastern states, recent data on this subject are lacking.

Canadian statistical yearbooks have presented information on pulpwood production for several decades. The early reports, from 1909 to 1922, gave separate information for balsam fir. For the years 1923–1925, information on balsam fir pulpwood was given as rough percentages. Later the various species contributing to pulpwood production were merely listed according to their importance, and finally even this information was discontinued, and the amount of balsam fir and spruce consumed by Canadian mills was given as a single figure (see Table

TABLE 143. PRODUCTION OF BALSAM FIR PULPWOOD IN
THOUSANDS OF CORDS IN THE LAKE STATES

State	1946	1950	1955	1960
Minnesota	93	126	120	145
Wisconsin	47	69	48	50
Michigan	117	123	122	135
Total	257	318	290	330
Percentage balsam fir of total pulpwood ..	11.4	16.9	12.1	9.9

Source: after Horn, 1952b, 1956, 1962.

TABLE 144. PRODUCTION AND CONSUMPTION OF PULPWOOD IN CANADA *

Year	Total Production (Cords)	Percentage Exported	Consumed in Canada (Cords)	
			Balsam fir †	Spruce and Fir
1910	1,541,028	61.2	120,475	590,705
1915	2,355,550	40.3	307,219	1,305,375
1920	4,024,826	31.0	687,519	2,560,543
1925	5,092,461	28.0	748,619	3,391,342
1930	5,997,183	22.3		4,415,255
1935	6,095,016	18.2		4,558,931
1940	8,716,541	17.8		6,388,998
1945	10,973,096	15.3		6,470,478
1950	13,424,358	13.3		9,569,848
1955	16,087,951	11.7		10,911,661
1960	16,612,000	6.9		12,000,000 ‡

* Newfoundland included in 1950 and subsequently.
† Data not available for after 1925.
‡ Estimated.
Sources: Canada Bureau of Statistics, 1911–1955; Shaw, 1962; and others.

144). The total production of pulpwood exceeded 16 million cords in 1955 and 16.5 million cords in 1960 (Can. Bur. Statistics, 1962; Shaw, 1962). A few individual reports for selected years refer to balsam fir. Balsam fir accounted for 20 per cent of Canadian pulpwood exported from 1917 to 1922 (Picard, 1924). In 1935 Canada produced 2.4 million cords of balsam fir, about 40 per cent of the pulpwood produced in that year (Mulloy, 1937).

Minor Products

Some balsam fir wood not suited for lumber or pulpwood is used for fuel and small amounts may be used locally for fence posts. According to Panshin et al. (1950), the wood is also suitable for cross ties if treated. Aries (1947) suggested that waste material could be used for manufacturing wood flour. Balsam fir wool can be made from paper-mill screenings and sawmill waste (Allison and Brown, 1946). These latter uses are minor; in total they account for little utilization of the species. The miscellaneous products of balsam fir that have some value are bark, Canada balsam, and needle or essential oils, each of which are discussed separately.

BARK PROPERTIES AND USES

The fibrous components and chemical composition of balsam fir bark have received relatively little attention from research workers in Canada and the United States. Since bark makes up about 12 per cent of the volume of balsam fir trees (Gevorkiantz and Olsen, 1951), it is of considerable importance and should receive much more study.

TABLE 145. AMOUNT OF OVEN-DRY BARK IN POUNDS PER 100 CUBIC FEET OF SOLID PEELED WOOD FOR SIX PULPWOOD SPECIES IN CANADA

Amount	Balsam Fir	Black Spruce	White Spruce	Red Spruce	Jack Pine	Poplar
Average	395	353	331	361	242	521
Maximum ...	433	412	371	417	274	585
Minimum	357	294	291	305	210	457

Source: Millikin, 1955.

Millikin (1955) determined the weight of bark per 100 cubic feet of solid peeled wood of balsam fir and associated pulpwood species (Table 145). He also found that the weight of bark per unit of peeled pulpwood was higher for balsam fir originating in the Rainy River district of Ontario than for that from the north shore of Lake Superior or the St. Maurice valley in Quebec.

The heating value of oven-dry balsam fir bark was found to be 9100 BTU [*] per pound as compared to 8610 BTU for black spruce, 8530 for white spruce, and 8630 BTU per pound of oven-dry red spruce bark. Only the bark of birch showed higher heating values, 10,310 BTU per pound for paper birch and 9200 BTU for yellow birch

[*]British thermal units; 1 BTU = 0.252 Cal.

(Millikin, 1955). The heating value of balsam fir bark was determined by Rue and Gleason (1924) as 8970 BTU. Ash content was found to be 5.2 per cent of oven-dry weight.

Information on the chemical composition of balsam fir bark and its pulp yields when cooked with NaOH are presented in Table 146 (Clermont and Schwartz, 1948). The high extractive content, high ash content, and low fiber content as indicated by pulp yields are typical for coniferous tree barks generally.

Clermont and Schwartz (1948) also determined the composition of reducing and fermentable sugars in balsam fir, black spruce, and white spruce barks (Table 147). Species differences were not striking.

Moisture-free bark was found to contain a total of 31.5 per cent extractives

TABLE 146. COMPOSITION IN PERCENTAGES OF MOISTURE-FREE BARK OF BALSAM FIR AND ASSOCIATED SPECIES IN CANADA

Composition and Pulp Yield	Balsam Fir	Black Spruce	White Spruce
Cold water soluble	22.6	25.9	33.9
Hot water soluble	26.7	32.3	41.8
Ether soluble	23.6	25.5	22.9
Ash	2.0	2.4	3.1
Successive extractions with alcohol- benzene, alcohol and hot water	36.2	38.7	47.7
Lignin (insoluble in 72% sulfuric acid) ..	32.3	27.4	27.3
Ash in lignin	0.2	0.5	1.0
1% sodium hydroxide soluble	35.3	39.2	44.0
Pentosans	12.5	17.4	18.3
Pulp yield (5% NaOH, 1 hour at 170C) ..	39.5	33.7	30.3
Lignin in pulp	15.1	8.9	5.7
Pentosans in pulp	5.1	6.9	5.3
Pulp yield (10% NaOH, 4 hours at 170C)	10.3	16.6	17.2
Lignin in pulp	19.1	18.4	19.0
Pentosans in pulp	1.3	5.3	2.3

Source: Clermont and Schwartz, 1948.

TABLE 147. PERCENTAGES OF REDUCING AND FERMENTABLE SUGARS IN THE BARK OF BALSAM FIR AND ASSOCIATED SPECIES IN CANADA

Sugars Present	Balsam Fir	Black Spruce	White Spruce
Reducing sugars from extracted bark ..	61.3	61.6	57.5
Fermentable sugars from extracted bark	47.3	43.4	39.7
Fermentable sugars on unextracted bark basis	30.1	26.6	20.8
Reducing sugars in pulp from extracted bark	91.5	99.1	99.6
Fermentable sugars in pulp from extract- ed bark	85.0	91.0	92.5
Fermentable sugars in pulp on unextract- ed bark basis	21.4	18.8	14.6

Source: Clermont and Schwartz, 1948.

distributed as follows: 0.8 per cent of steam distillable oils; 14.1 per cent of ethyl-ether-solubles (resenes, phytosterols, resins, and fatty acids); 2.3 per cent of ethyl alcohol-soluble tannins, phlobaphenes, and alcohol-soluble sugars; and 14.3 per cent of water-soluble salts, mucilage (or pectin), and carbohydrates. The extractive-free bark (68.5 per cent) contained 36.7 per cent Cross-and-Bevan cellulose and 24.8 per cent of a fraction isolated as lignin (Hay and Lewis, 1940).

The difference in results reported in Table 146 and by Hay and Lewis (1940) could be due to such factors as age of trees, season of year when trees were cut, sampling procedures, and other variables not controlled in the analyses.

Additional information on the chemistry of balsam fir bark was provided recently (U.S. For. Prod. Lab., 1961). Material soluble in 1 per cent NaOH comprised 49.4 per cent of the weight of oven-dry unextracted bark. The amount of reducing sugars, when hydrolyzed with 72 per cent sulfuric acid was 46.4 per cent of the unextracted bark, 45.3 per cent of extractive-free bark, and 32.9 per cent of 1 per cent NaOH-extracted bark. The composition of reducing sugars from extractive-free bark was as follows: glucose, 64 per cent; galactose, 5 per cent; mannose, 12 per cent; arabinose, 9 per cent; and xylose, 7 per cent.

The tannin content of balsam fir bark was reported to be relatively low, only 3.4–4.1 per cent as compared with 13.0 per cent for red spruce and 24.4 per cent for Sitka spruce (Rogers, 1952). The tannins are difficult to extract with hot water, even after removal of resinous materials with ethyl ether, probably because of the presence of and interference with extraction by pectic and mucilaginous materials (Graham and Rose, 1938).

The pectinous materials are considered to be the cause of the so-called pitch problems encountered in sulfite pulping of balsam fir (Hay and Lewis, 1940). Bark pectins may contaminate the surface of balsam fir pulpwood during hand peeling in the natural peeling season. Some bark fragments containing stone cells may also remain on the surface. Such stone cells have protoplasts containing tannins and mucilaginous materials. Stone cells are thick-walled and are penetrated only with difficulty by pulping liquors. The pectic, mucilaginous, and stone-cell-containing materials cause the gelatinous specks in paper, formerly attributed to pitch. The fact that balsam fir is less resinous than the spruces, and pitch problems are not encountered with the latter species, lends credence to this explanation for the pitch problems associated with paper production from what is essentially a nonresinous wood.

The pectinous and mucilaginous constituents of balsam fir bark make it difficult to handle this material in chippers, grinders, hammer mills, and disc refiners. However, with a "protessor" (screw-type extruder) it is possible to remove enough of these objectionable materials to make the bark processable in conventional equipment (Bender, 1959).

The principal present use of balsam fir bark is for fuel in some paper mills where it has accumulated as a waste product. Fiber content is low (see Table 146), and extractives do not appear to be sufficiently abundant or valuable to offer much encouragement for future utilization. If left in the forest the bark contributes to the humus and organic content of the forest floor.

CANADA BALSAM

Detailed information on Canada balsam is presented by Marriott and Greaves (1947), who include a considerable list of references in their paper. This work is often quoted in textbooks and other references on the subject.

The oleoresin of balsam fir is rich in terpenes and does not belong to the true balsams. Oleoresin develops in canals formed by the separation of cells in the bark and collects in small cavities under the periderm. These resin-filled cavities appear as prominent blisters on the smooth, thin bark of young trees and branches.

The resin is obtained by puncturing and draining the blisters with a hollow metal tube about 10 mm in diameter, through which the balsam drains into a can. Another way of collecting resins is by means of a sharpened tube or simply a spout projecting from the rim of a can into which the resin can flow. The best and cleanest collection method is through use of a steel-pointed glass syringe fitted with a rubber bulb. Large trees may yield a pound of resin although the average is nearer 8 ounces. Blisters higher in the trees contain less oleoresin, but it is of higher quality. About 150 blisters are worked, on the average, to yield a pint (0.473 l) of resin. One man working alone can collect about half a gallon (1.892 l) of resin per day.

Steam distillation yields about 24 per cent volatile oil and leaves a residue of hard yellow resin. An alkaloid soluble in water is also present. The volatile oils are composed largely of laevo-pinene and various terpenes including, possibly, limonene. The resinous substance consists largely of four acid resins, i.e., canadinic, canadolic, and alpha- and beta-canadinolic acids, with the last two predominating, and canadoresene, an indifferent resene.

The oleoresin is a transparent liquid, pale greenish-yellow in color, and slightly fluorescent. A thin colorless film is present. When exposed to the air the resin gradually loses the volatile oil, becomes thicker and more yellowish and darker in color. Ultimately it dries to a transparent, resinous mass. The resin is completely soluble in ether, chloroform, carbon tetrachloride, benzol, and turpentine; and partially soluble in absolute alcohol and methyl hydrate. When mixed with 20 per cent MgO_2 that has been previously moistened with water, the oleoresin solidifies. Its solubility in alcohol and its behavior with magnesia distinguishes Canada balsam from other coniferous resins. Another distinction is the large refractive index of Canada balsam (Marriott and Greaves, 1947). A grain of starch laid in Canada balsam remains clearly visible, whereas in other balsam resins, starch grains similarly placed become indistinct or invisible. Some specific properties of Canada balsam are listed in Table 148.

The oleoresin of *Abies lasiocarpa*, but not that of *A. grandis*, may serve as a substitute for Canada balsam (Bickford et al., 1934). Canada balsam has long been used as a medium for the permanent mounting of microscopic specimens (Marriott and Greaves, 1947), many persons regarding it as unsurpassed in general usefulness for this purpose. About 150 years ago Canada balsam was introduced into optical work, where even today it is widely used as a cement for glass.

Canada balsam export from Canada in 1947 was valued at $50,000, prices ranging from $24.50 to $27.00 per gallon. Balsam was one of the first American contributions to *Materia medica*, and its use as a simple drug is very old. Canada balsam was used by the Algonquin Indians as a poultice on burns (Marie-Victorin, 1919). Although still collected in the United States locally, it has become a negligible item, most needs in the United States being met through imports from Canada. The recent wide use of substitute synthetic resins for mounting media and optical work will probably serve to hold the production of Canada balsam rather low.

NEEDLE OILS

The quantity of balsam fir needles and their mineral element content have been investigated by Chandler (1944). The fragrance of the needles is due to

TABLE 148. PROPERTIES OF CANADA BALSAM

Constants	Marriott and Greaves (1947)	Bickford et al. (1934)	Parry (1918)
Specific gravity at 15.5C	0.987–0.994	0.900–0.995	0.985–0.995
Optical rotation, 100 mm tube	$+1° - +4°$	$+1° - +4°$	$+1° - +4°$
Refraction index at 20C	1.518–1.521	1.520	1.518–1.521
Acid value	84–87	80–87	84–87
Ester value	5–10	4–10	5–10
Saponification value	87.5–105	84–97	
Bromine number	65.9		
Percentage of volatile oil		16–25 (ave. 24)	15–25
Adhesive strength (tensile strength in pounds per square inch)		340	

Sources: as listed.

their essential oils (Jenness and Caulfield, 1941). The needle oil content of several coniferous species in Quebec was found to vary as follows: balsam fir, 1.0–1.4 per cent; white-cedar, 0.6–1.0 per cent; white pine, 0.6–1.0 per cent; black spruce, 0.5–0.7 per cent; and hemlock, 0.4–0.6 per cent (Risi and Brûlé, 1945a,b, 1946). The essential oil content of balsam fir needles was estimated to be 0.59 per cent (Shaw, 1953). Marshall (1955) estimated that for each cord of balsam fir cut, 500 pounds of needles are discarded. From these about 5 pounds (1 per cent) of essential oils could be extracted if there were markets and economic uses for them.

Miscellaneous Uses and Values

In this section the properties which make balsam fir a widely used Christmas tree are reviewed together with a discussion of its ornamental and recreational values.

CHRISTMAS TREES

Firs together with spruce have been traditional Christmas trees of northern people in both Europe and North America. The inherent properties of balsam fir make it admirably suited to Christmas tree use. The boughs and needles are pleasantly aromatic, and the needles are dark green and well retained even after several weeks indoors. The tree itself has a pleasing shape, bearing a symmetrical narrow crown which terminates in a spire-like tip. Nearly all balsam fir Christmas trees originate in naturally grown stands, and they can often be purchased at prices lower than those which generally prevail for plantation-grown trees. Because of its soft flexible needles and branches, balsam fir handles well and can be packed tightly, enabling the hauling of more trees per truck or car load than is possible with the bulkier, stiff-branched pines. It retains its color well into the fall and in this respect is superior to many races of both Scotch pine and red pine. Balsam fir can also be grown on a stump-culture basis in which a branch or adventitious shoot is used to regenerate a new tree below the point of removal.

The species, however, is not ideally suited to culture as a plantation Christmas tree (Lawrence, 1957). Its relatively slow early growth results in a Christmas tree rotation of 10–15 years, about double that required for northern pines. Its site requirements, particularly south of its normal range where the Christmas tree industry is largely located, are more restrictive than those for pines. Wilde (1953) recommended that Wisconsin growers plant balsam fir on soil having at least 25 per cent silt and clay and a minimum of 3 per cent organic matter. The water table should be within 3–5 feet of the surface, or there should be a compacted B-horizon with a minimum air permeability of 70 mm.

Shearing recommendations for balsam fir and spruces grown for Christmas trees were given by Trenk (1960). Shearing should be done in the dormant season. Each leader on a lateral should be cut back to the axil of the laterals at its base, since this produces new growth from the laterals which will tend to cover up gaps in the foliage pattern. Lack of cultural attention, especially shearing, is considered an important reason for the species' loss of Christmas tree markets to the pines in recent years (Hansen, 1963).

Christmas trees dry out very rapidly indoors if their cut ends are not placed in water. Freshly cut balsam fir at 115 per cent moisture content declined to 20 per cent in 16 days at indoor temperatures of 68–78F. On the other hand, when the butt ends were placed in water, moisture content remained above 100 per cent for the entire 20-day test period (Wagner, 1961). Flammability was also tested at various moisture contents and with several chemical solutions. Complete resistance to ignition was found on trees with moisture contents above 100 per cent, and effective resistance at moisture contents of 60–100 per cent. Balsam fir became flammable at 50 per cent and extremely flammable at 20 per cent. Butt soaks in solutions of calcium chloride, ammonium sulfate, ammonium sulfamate, and ammonium phosphate gave no added protection to that of water used alone, and sometimes resulted in such unfavorable side effects as discoloring of foliage and early needle drop (Wagner, 1961; U.S. For. Prod. Lab., 1941).

Balsam fir was included with other species in cold storage tests conducted to determine the feasibility of early fall cutting if followed by proper storage conditions (McGuire et al., 1962). It was concluded that trees stored at 22F retained needle moisture as well as freshly cut trees. However, trees cut earlier than November 1 and stored at 22F lost needle moisture rapidly after removal from storage even though there was no moisture loss during storage. This result was attributed to the fact that the trees cut in September and October had not yet hardened off and were injured by the cold storage conditions. Storage at 32F resulted in considerable moisture loss. Use of polyethylene film, either perforated or continuous, prevented moisture loss; trees thus stored were nearly equal in keeping quality to freshly cut trees.

Numerous publications of a popular nature dealing with the general problems of Christmas tree culture and marketing, most of them published by state agricultural extension groups, have appeared in recent years. Few of them stress balsam fir culture. Cope (1949) wrote the first of such bulletins, relating it to his considerable experience in New York. Tryon et al. (1951) in West Virginia and Simonds (1953) in Pennsylvania reported on Christmas tree problems in their respective states.

The popularity of balsam fir in the Christmas tree market has been shown by local species distribution studies in Ohio (Mitchell and Quigley, 1960), Michigan (James and Bell, 1954; James, 1958, 1959), and Minnesota (Duncan et al., 1960; Hansen, 1963), and by a harvest study in Wisconsin (Horn, 1959). These reports are briefly summarized in Table 149.

TABLE 149. CHRISTMAS TREE SPECIES DISTRIBUTIONS FOR SELECTED STATES

Species	Ohio *	Michigan †	Wisconsin ‡	Minnesota § City Lot	Suburban Lot
Balsam fir	50.8	35	54.3	59	43
White spruce			12.0	6	8
Black spruce	3.7	23	7.2	6	1
Norway spruce				10	8
White pine	3.3		1.7	3	3
Scotch pine	29.6	33		8	23
Red pine	7.3	6	24.8	8	14
Douglas fir	5.3	1			
Others		2			
Total	100.0	100	100.0	100	100

* In percentage of trees sold in Canton, Columbus, and Washington Court House, 1956 (Mitchell and Quigley, 1960).

† In percentage of trees sold in statewide survey, 1956 (James, 1958).

‡ In percentage of number of trees cut for sale, 1957 (Horn, 1959).

§ Consumer preference, based on 286 interviews in a city lot and 74 in a suburban lot (Duncan et al., 1960).

In 1957 Canada exported 12 million Christmas trees valued in excess of 6 million dollars. Balsam fir comprised 72 per cent of the export. The total domestic consumption in Canada was 4 million trees (Babcock and Nicolaiff, 1958).

While balsam fir continues to be one of the most popular of all Christmas tree species in the United States, during recent years it has lost its position of preeminence while the pines have relatively gained. In 1948 over 6 million balsam fir were sold in the United States (Sowder, 1949), ranking first among all species and representing 30 per cent of the total market. In 1955 it slipped to second place, next to Douglas fir, for 24 per cent of the market (Sowder, 1956). By 1960 Douglas fir was in first place, Scotch pine second, and balsam fir — accounting for 16 per cent of trees sold — was third (Sowder, 1962). A report for Minnesota (Hansen, 1963), while indicating a similar general trend, showed balsam fir as still leading in sales in 1962 after dropping from 48 per cent in 1960 to 40 per cent in 1962. Indications were cited that there would be continued loss of sales to Scotch and red pines.

Although no systematic analysis of the reasons for the decline in use of balsam fir for this purpose has been published, consumer preferences appear to be running more and more to cultured plantation trees, particularly Scotch pine. The availability of balsam fir from natural stands has discouraged plantation production. In tracing the trends of species planted for Christmas trees from 1947 to 1957 by 76 growers in Minnesota, Sullivan (1959) did not report any balsam fir plantations. Of 700,000 balsam fir Christmas trees harvested in Wisconsin in 1957, only 3000 came from plantations (Horn, 1959). The number of balsam fir grown in plantations in Michigan amounted to only a fraction of one per cent of all plantation Christmas trees (James, 1959). The existing market suggests that some new approach, either in natural stands or plantations, should be made to secure future supplies of balsam fir Christmas trees.

Federal grades of Christmas tree quality are prescribed for application to balsam fir along with other species (U.S. Agric. Marketing Serv., 1957).

ORNAMENTAL AND RECREATIONAL USE

The landscape architect visualizes trees in a more or less abstract manner. Their beauty is related to form, to patterns of light and shade, to color, and to their arrangement in the pictorial setting (Bracken, 1932). Harmony in space, in habit, in flower and fruit, and in color, size, and form of branches is important in tree composition (Kuphaldt, 1927).

Balsam fir has not generally been recommended for its beauty in cultivation. Fernow (1910) felt that it lacked the ornamental beauty of other species of firs, and would admit the use of balsam only where others failed because of excessive soil moisture. Bailey (1933) noted that although attractive in its native northern environment, it ultimately loses its attractiveness in cultivation. For shorter term effects, however, balsam fir may be a useful ornamental. Oliver (1957) considered it desirable as a temporary tree among other species. It remains attractive for the first 20 years. The species is seldom listed in commercial nursery catalogues and it is even more difficult to find the special horticultural forms described by earlier authors (see Chapter 1).

In keeping with a universal inclination to prefer exotics for landscaping, balsam fir has been more widely planted for this purpose in Europe than in North America. In Europe it can be found in ornamental gardens and arboreta from Portugal to Finland and from the British Isles to Moscow. There are also reports of its being planted in Australia (Spaulding, 1956). The species has not found much use in cities on the West Coast although *Abies concolor* and many other western conifers have been used in the natural range of balsam fir (Waterman et al., 1949).

Balsam fir has not been recommended for planting in the South or in the Great Plains (Taylor, 1952). Furthermore, the tree should not be planted in large cities where smoke may be injurious. In New England it should be planted north of Connecticut and then only on moist or wet sites (Coffin, 1940).

Though not adapted to foundation plantings because it quickly becomes too large and must be replaced, this species is often chosen for its decorative value as an outdoor Christmas tree. Firs are suitable for individual specimen plantings only if established at some distance from buildings, or from other plantings, to allow them to develop unhampered in full normal habit (Kumlien, 1946).

The importance of keeping the trees in vigorous condition in order to maintain the natural symmetrical form was stressed by Wyman (1951). He especially warned against pruning any species of fir to reduce its size and growth.

Coffin (1940) considered firs suitable for planting in formal avenues, set at closer intervals than most hardwoods. Firs can also be used in group plantings along informal drives. Their use in background plantings or as heavy screens is often suggested (Bailey, 1933; Snyder et al., 1948; Kumlien, 1946). Firs form good backgrounds for lower plantings of shrubs and flowers. They should be used only with great care in windbreaks because of their relatively high moisture requirement.

Dwarf forms of balsam fir are suitable for rock gardens (Kuphaldt, 1927). These include such varieties (cultivars) as 'Compacta,' 'Prostrata,' 'Nana,' and the naturally occurring high altitude 'forma *hudsonica.*'

The use of balsam fir on the typical small residential lots of large cities and suburbs is rather limited. It is more appropriately used, where site conditions permit, on country estates where natural woodlands and mass plantings can be integrated into a pleasing, well-balanced landscape.

Since balsam fir is a major element in the composition of the Boreal Forest, it plays a part in any extensive recreational use made of northern lands. For this purpose it has both desirable and undesirable characteristics. The major attractive qualities are its gracefulness, pleasing symmetry, and fragrance as a young tree. These values are recognized by the visitor not so much in monotonous pure stands but when placed in contrast with other tree species, especially birch, aspen, and pine, and with rocks, mountain slopes, and water. Until quite recently its soft, pliable boughs were favored for camp beds.

To the naturalist and the wilderness visitor, balsam fir in its native undisturbed habitat, whether viewed as a freshly germinated seedling, a fast-growing young sapling, or a dying, broken old tree, gives an esthetic satisfaction that is derived from its harmonious role in an integrated, dynamic forest community.

The colorful display of foliage in the fall is most effective in southeastern Canada and in the northeastern United States (Wyman, 1951), an area coinciding with the optimum range of balsam fir. There its deep green foliage stands out in striking contrast to the brilliant autumnal colors of the northern hardwoods. From June to October, good cone crops increase interest and add to the natural beauty of balsam fir. Its somber green adds pleasing color and contrast to the wintry, snow-covered landscape and its fragrance is unmatched among northern conifers. In fact, balsam fir would be sorely missed by the visitor viewing the northern forested landscape were it not there.

Two great national parks are located within the natural range of balsam fir in the United States, Acadia on the coast of Maine and Isle Royale in Lake Superior. Many fine state parks — including Itasca in Minnesota, the Porcupine Mountains in Michigan, the Adirondacks and Catskills in New York, and Baxter in Maine — are beautified by the presence of balsam fir. Canada, too, has equally fine reservations in such areas as Cape Breton in Nova Scotia, Laurentides and Mount Tremblant in Quebec, Algonquin in Ontario, and Riding Mountain in Manitoba. Unique among natural reservations is the two-million-acre Quetico-Superior wilderness area, shared jointly by the United States and Canada. There the northern forest with its pleasing vistas of rocks, trees, and water is a major attraction to the canoeist and hiker.

Recreational demands upon the forests and lakes are doubling each decade. Especially attractive roadside and lakeshore stands are being preserved for the visitor. It is probable, however, that not only natural, relatively undisturbed stands will be sought by the public. Managed forests will also provide recreational resources, and forest management aimed at meeting such objectives will become more common. The esthetic values of balsam fir combined with its other desirable qualities will assure it an important place in the multiple-use forests of the future.

Conclusions

1. Much of the available data relating to the physical and mechanical properties of wood and its chemical composition are based upon relatively few measurements, often made under unspecified conditions. Frequently, neither the number of observations nor the variability is indicated. One of the objects of the increased research by university and governmental laboratories located in the commercial range of balsam fir should be the development of more complete information on its properties.

2. Little information is available on the effect of growing conditions on the physical, mechanical, and chemical properties of balsam fir. Studies of this type throughout the wide range of the species are needed to provide information for the guidance of forest managers. Of special interest would be a comparison between the properties of balsam fir in old-growth and second-growth stands.

3. Because of its technical properties balsam fir frequently does not rank among the superior species. Experience has shown that high quality pulp and paper products can be produced from high quality balsam fir pulpwood by several pulping processes. Moreover, experience has also shown that balsam fir lumber can compete successfully in local markets with lumber from other species of similar density and log size. The problem is to determine conditions under which this species has the greatest economic utility.

4. Several early studies indicated that selective logging might be as economical as clearcutting in mature stands. Because of changing logging practices and mechanization, new studies should be initiated to determine the present situation relative to clearcutting and selective cutting practices.

5. The reported slow deterioration and possible economic salvage of fire-killed balsam fir many years after killing, as contrasted with the rapid deterioration and poor salvage prospects in budworm-killed stands, need further study to determine the reasons for these reported differences. Added study of the utilization of knotty top wood and partially decayed wood likewise is needed.

6. Tree, log, and lumber grading rules are available for most timber species used for lumber. There exist only local specifications for pulpwood. Establish-

ment of pulpwood grades and grading rules for balsam fir could be of benefit to both the growers and users of this species and improve its acceptability.

7. The published statistical information is not adequate for evaluating properly the role of balsam fir and other species in the supply of such vital material as lumber, pulpwood, and other forest products. Not only is there a lack of specialized information, but even summary data — such as were available in earlier years — are no longer reported. With little added effort and cost, valuable information on the production and consumption of balsam fir could be made available. Particularly in the United States the statistical information on production and utilization of the species is less complete today than in earlier years.

8. Christmas tree consumption in the United States has in recent years shifted appreciably from balsam fir to the pines, largely as a result of the production of high quality plantation-grown pine. If balsam fir is to meet this competition, methods of handling young wild stands and of establishing balsam fir plantations that will yield a high percentage of acceptable and readily salable trees must be found. Balsam fir has inherently desirable properties for Christmas trees. In the United States, however, forest conditions are such that fewer trees of acceptable quality are produced annually in wild stands and little research is being done on production of trees from plantations.

9. Increased recreational use of both natural and managed forest lands will require investigations aimed toward better manipulation of balsam fir on campgrounds and recreation areas.

10. Balsam fir will continue to be of considerable importance in the economy of eastern North America. Its esthetic values will become more appreciated and further developed, and it will remain an interesting subject for intensive scientific investigations.

PLATES

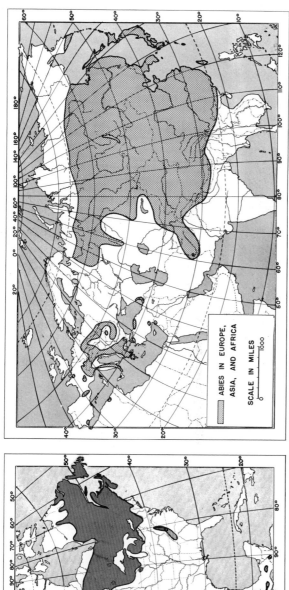

Plate 1. Distribution of *Abies balsamea* in relation to *Abies fraseri, Abies lasiocarpa,* and other *Abies* species. (Sources: after Department of Forestry, Canada; U.S. Forest Service; Mattfeld (1925); Schmucker (1942); Martinez (1953); and others.)

331

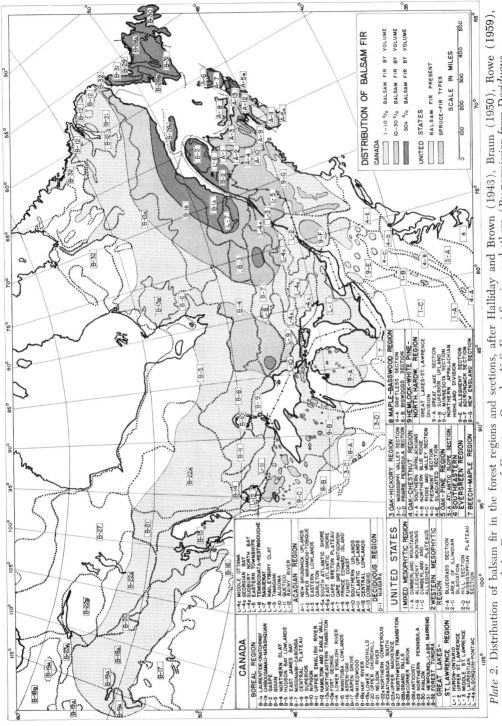

Plate 2. Distribution of balsam fir in the forest regions and sections, after Halliday and Brown (1943), Braun (1950), Rowe (1959), various sources of the Department of Forestry, Canada, U.S. Forest Service, and others. (By permission from *Deciduous Forests of Eastern North America* by E. L. Braun. Copyright 1950 by the McGraw-Hill Book Company, Inc.)

NORTHERN HARDWOODS   OAK   PINES   WHITE SPRUCE BALSAM FIR   CONIFEROUS SWAMP   BOTTOMLAND HARDWOODS   NOT SAMPLED

NORTHERN HARDWOOD CLIMAX REGION

		SURFACE SOIL				
	SAND	LIGHT LOAM AND LOAMY SAND	LOAM AND HEAVY LOAM	VARIABLE	MUCK	PEAT AND MUCK
ROCK	✕	◯	◯	◯ ⬡		
SAND	⬡ ⬡	◯ ◯	◯	◯ ◯	◯	◆
CLAY	⬡	⬡ ◯	◯	◯		
SOIL MOISTURE	DRY	FRESH	MOIST	PERIODICALLY WET	WET	SATURATED

SUBSOIL

WHITE SPRUCE–BALSAM FIR CLIMAX REGION

		SURFACE SOIL				
	SAND	LIGHT LOAM AND LOAMY SAND	LOAM AND HEAVY LOAM	VARIABLE	MUCK	PEAT AND MUCK
ROCK	✕	✕	✕	▲		
SAND	⬡	⬡	◯ ⬡ ▲	◯	▲ ◆	◆
CLAY	✕	✕	◯ ▲	◯ ⬡ ▲		
SOIL MOISTURE	DRY	FRESH	MOIST	PERIODICALLY WET	WET	SATURATED

SUBSOIL

Plate 3. Relation of the potential forest cover to different soil groups in Michigan, Wisconsin, and Minnesota. (Redrawn from Roe, 1935. Courtesy of the Lake States Forest Experiment Station, U.S. Forest Service.)

333

I. SITE REGIONS

2. SOIL DEVELOPMENT

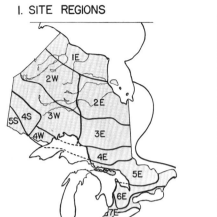

LEACHING DEPTH OF LIME IN SILTY CLAY SOILS, INCHES	FREQUENCY PERCENT OF A₁-HORIZON IN DRY AND MOIST SOILS	MICROORGANIC ACTIVITY ON RICH-FRESH SOILS	WET SOILS	PEAT DEPTH ON CLAY, INCHES
0 – 1	0–1			
0 – 6	0–1	VERY LOW	NIL	8–36
2 – 8		LOW	NIL	18–36
3 – 12	1–3	MOD. LOW	VERY LOW	6–24
3 – 12	3–10	MODERATE	LOW	3–12
10 – 60		MOD. HIGH	MOD. LOW	0–6
6 – 18	60 – 80	HIGH	MOD	0
12 – 24	80 – 100	VERY HIGH	MOD. HIGH	0

3. BALSAM FIR AND OTHER SPECIES IN PHYSIOGRAPHIC SITE CLASSES UNDER LEAST DISTURBED CONDITIONS IN HUMID EASTERN REGIONS

ECOCLIMATE SOIL	HOTTER DRIER	HOTTER FRESH	HOTTER WETTER	NORMAL DRIER	NORMAL FRESH	NORMAL WETTER	COLDER DRIER	COLDER FRESH	COLDER WETTER
IE HUDSON BAY	BETULA PAPYR., POPULUS TREM., PICEA MARIANA, PINUS BANKSIANA	PICEA GLAUCA, PICEA MARIANA, BETULA PAPYRIF., POPULUS TREM., PINUS BANKSIANA	OPEN, PICEA MARIANA, LARIX LARICINA, ALNUS SPP.	PICEA MARIANA, PICEA GLAUCA, PINUS BANKSIANA	OPEN, PICEA MARIANA, LARIX LARICINA	OPEN, PICEA MARIANA, LARIX LARICINA	MOSSES AND LICHENS		
2E JAMES BAY	LOCALLY OPEN, PINUS BANKSIANA, BETULA PAPYR.	PINUS BANKSIANA, PICEA MARIANA, PICEA GLAUCA	OPEN, PICEA MARIANA, PICEA GLAUCA	PINUS BANKSIANA, PICEA MARIANA, PICEA GLAUCA	ABIES BALSAMEA, PICEA GLAUCA, PICEA MARIANA, POPULUS TREM.	OPEN, PICEA MARIANA, LARIX LARICINA	OPEN, PICEA MARIANA, LARIX LARICINA	MOSSES AND LICHENS	
3E LAKE ABITIBI	PINUS BANKSIANA, PICEA MARIANA	PINUS BANKSIANA, PICEA GLAUCA, PINUS STROBUS, PINUS RESINOSA	POPULUS BALSAM., THUJA OCCIDENT., ULMUS AMERICANA	PINUS BANKSIANA, BETULA PAPYRIF, POPULUS TREM.	ABIES BALSAMEA, PICEA GLAUCA, POPULUS TREM., POPULUS BALSAM., BETULA PAPYRIF.	PICEA MARIANA, ABIES BALSAMEA	PICEA MARIANA, PINUS BANKSIANA, LARIX LARICINA	LOCALLY OPEN, PICEA MARIANA, LARIX LARICINA	MOSSES AND LICHENS
4E LAKE TIMAGAMI	PINUS STROBUS, PINUS RESINOSA, PINUS BANKSIANA, BETULA PAPYR., POPULUS TREM., POPULUS GRAND.	ACER SACCHARUM, ACER RUBRUM, BETULA ALLEGHAN, PINUS STROBUS, ABIES BALSAMEA, PICEA GLAUCA	BETULA ALLEGH., THUJA OCCIDENT., ULMUS AMERICANA	PINUS STROBUS, PINUS RESINOSA, PINUS BANKSIANA, ACER SACCHARUM, ACER RUBRUM	ABIES BALSAMEA, PICEA GLAUCA, PINUS STROBUS, PINUS RESINOSA, BETULA PAPYR., POPULUS TREM.	ABIES BALSAMEA, PICEA GLAUCA, PICEA MARIANA	ABIES BALSAMEA, THUJA OCCIDENT.	PICEA MARIANA, LARIX LARICINA	LOCALLY OPEN, PICEA MARIANA, LARIX LARICINA
5E GEORGIAN BAY	QUERCUS RUBRA, PINUS STROBUS, PINUS RESINOSA, QUERCUS MACROC.	ACER SACCHARUM, QUERCUS RUBRA, FAGUS AMERICANA, TSUGA CANADENSIS	ULMUS AMERICANA, FRAXINUS NIGRA, ACER RUBRUM, PICEA GLAUCA, PICEA RUBRA, TSUGA CANADENSIS	PINUS STROBUS, PINUS RESINOSA, TSUGA CANADENSIS, FAGUS AMERICANA, FRAXINUS AMERIC.	ACER SACCHARUM, BETULA ALLEGHAN, TSUGA CANADENSIS, PINUS STROBUS, PICEA GLAUCA, ABIES BALSAMEA	FRAXINUS NIGRA, ACER RUBRUM, ABIES BALSAMEA, PICEA GLAUCA, ABIES BALSAMEA	PICEA GLAUCA, ABIES BALSAMEA, PINUS STROBUS	ABIES BALSAMEA, PICEA GLAUCA, PINUS STROBUS, THUJA OCCIDENT., BETULA PAPYRIF.	ABIES BALSAMEA, PICEA MARIANA, PICEA GLAUCA, LARIX LARICINA
6E LAKES SIMKOE-RIDEAU	FAGUS AMERICANA, TSUGA CANADENSIS, QUERCUS RUBRA, ULMUS THOMASII, POPULUS GRAND., POPULUS TREM., JUNIPERUS VIRG.	QUERCUS ALBA, CARYA OVATA, ULMUS THOMASII, JUGLANS CINEREA	ACER SACCHARUM, ACER SACCHARIN., ACER RUBRUM, BETULA ALLEGH., JUGLANS NIGRA, PRUNUS SEROTINA	ACER SACCHARUM, ACER RUBRUM, QUERCUS ALBA, QUERCUS RUBRA, FRAXINUS AMER., ULMUS AMERICANA	FAGUS AMERICANA, ACER SACCHARUM, QUERCUS RUBRA, TSUGA CANADENS.	TSUGA CANADENS., PINUS STROBUS, ULMUS AMERICANA, FRAXINUS AMER., ULMUS RUBRA, THUJA OCCIDENT., ABIES BALSAMEA	PINUS STROBUS, PINUS RESINOSA, FRAXINUS AMER., ULMUS AMERICANA, ULMUS THOMASII, THUJA OCCIDENT.	PICEA GLAUCA, ABIES BALSAMEA	ABIES BALSAMEA, PICEA MARIANA, PICEA GLAUCA, LARIX LARICINA
7E LAKES ERIE-ONTARIO	QUERCUS MACR., QUERCUS RUBRA, QUERCUS COCCIN., QUERCUS BICOL., CASTANE A DENT., CARYA GLABRA, CARYA OVATA, JUGLANS CINEREA, PINUS STROBUS, JUNIPERUS VIRG.	LIRIODENDR. TUL., JUGLANS NIGRA, QUERCUS ALBA, QUERCUS RUBRA, CARYA TOMENTOSA, CARYA GLABRA, FRAXINUS QUADR., FRAXINUS AMER., JUGLANS CINEREA	PLATANUS OCC., LIRIODENDR. TULIP., ULMUS AMERICANA, ULMUS RUBRA, FRAXINUS SPP., BETULA ALLEGH.	QUERCUS RUBRA, QUERCUS ALBA, QUERCUS RUBRA, CARYA OVATA, CARYA GLABRA, ULMUS AMERICANA, ULMUS THOMASII	ACER SACCHARUM, FAGUS AMERICANA, QUERCUS ALBA, QUERCUS RUBRA, CARYA OVATA	QUERCUS BICOLOR, QUERCUS PALUST., FRAXINUS PENNS., FRAXINUS NIGRA, ULMUS RUBRA, ULMUS AMERICANA, CARYA CORDIFORM.	TSUGA CANADENS., FAGUS AMERICANA, BETULA ALLEGH., PINUS STROBUS	ULMUS AMERICANA, ULMUS RUBRA, FRAXINUS NIGRA, QUERCUS BICOLOR, ACER RUBRUM	PICEA MARIANA, PICEA GLAUCA, ACER SACCHARINUM, ACER RUBRUM, BETULA ALLEGH.

GENERIC NAMES WRITTEN WITH SMALL LETTERS INDICATE COMMON BUT NOT CHARACTERISTIC SPECIES
☐ BALSAM FIR PRESENT

Plate 4. Site regions (1), soil development (2), and changes in species composition (3) in the humid eastern part of Ontario. The humid and subhumid western sections are indicated by the letters "w" and "s," respectively. The characteristic species are listed first. Micro-organic activity is shown, as indicated, by the humification of the organic matter. (Redrawn from Hills, 1960a. Courtesy of Dr. G. A. Hills, Ontario Department of Lands and Forests.)

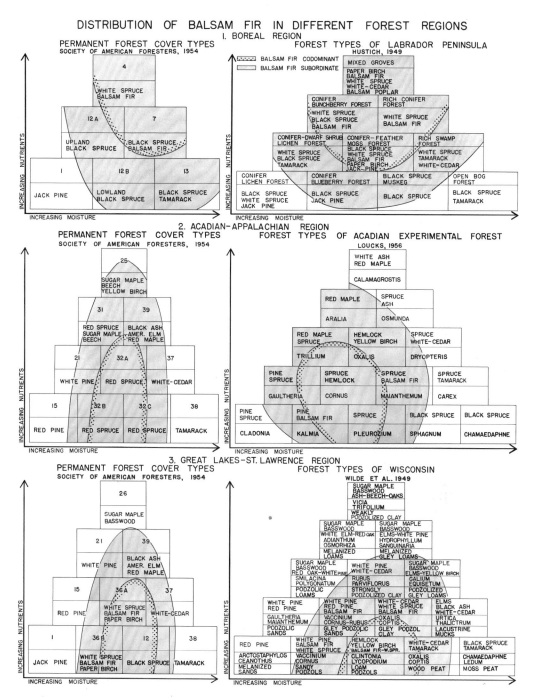

Plate 5. The position of balsam fir in regional edaphic fields as outlined by a modified forest cover type system (Society of American Foresters, 1954) and local forest classification systems for the Labrador Peninsula (Hustich, 1949), the Acadian Experimental Forest (Loucks, 1956), and for Wisconsin (Wilde et al., 1949).

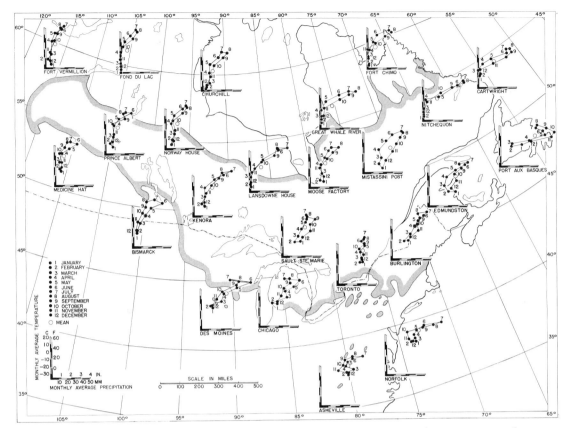

Plate 6. Climographs characterizing the range of balsam fir. Average monthly temperature and precipitation from the Meteorological Division of Canada and the United States Weather Bureau. Intersections of scales mark the locations of weather stations.

336

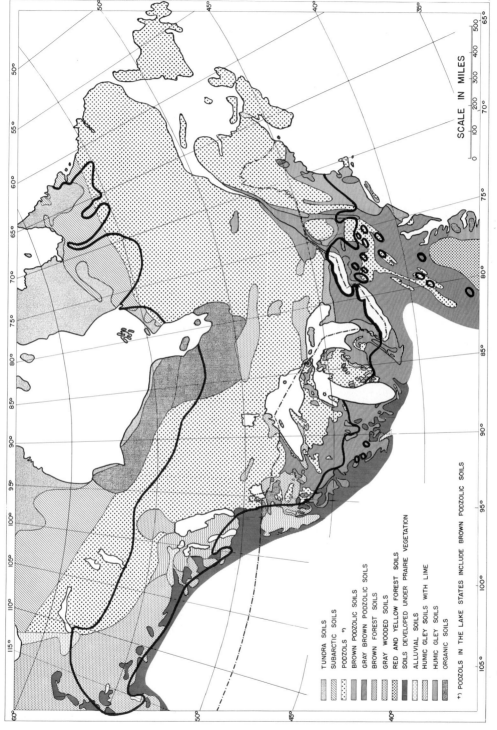

Plate 7. Major soil groups in the natural range of balsam fir. (After U.S. Department of Agriculture, 1938; Canada Geographical Branch, 1957; and North Central Agricultural Experiment Stations, 1960.)

SCALE IN MILES

TUNDRA SOILS

SUBARCTIC SOILS

PODZOLS *)

BROWN PODZOLIC SOILS

GRAY BROWN PODZOLIC SOILS

BROWN FOREST SOILS

GRAY WOODED SOILS

RED AND YELLOW FOREST SOILS

SOILS DEVELOPED UNDER PRAIRIE VEGETATION

ALLUVIAL SOILS

HUMIC GLEY SOILS WITH LIME

HUMIC GLEY SOILS

ORGANIC SOILS

*) PODZOLS IN THE LAKE STATES INCLUDE BROWN PODZOLIC SOILS

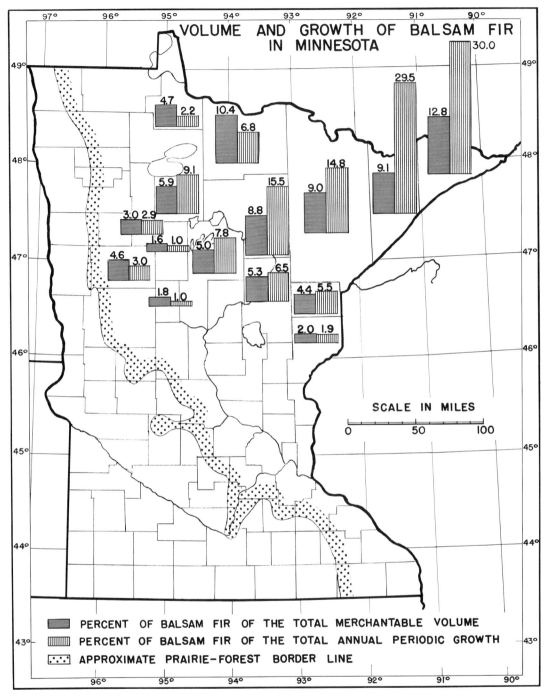

Plate 8. Distribution of balsam fir in Minnesota. (Data from publications
of the Iron Range Resources Rehabilitation Office, 1950–1963.)

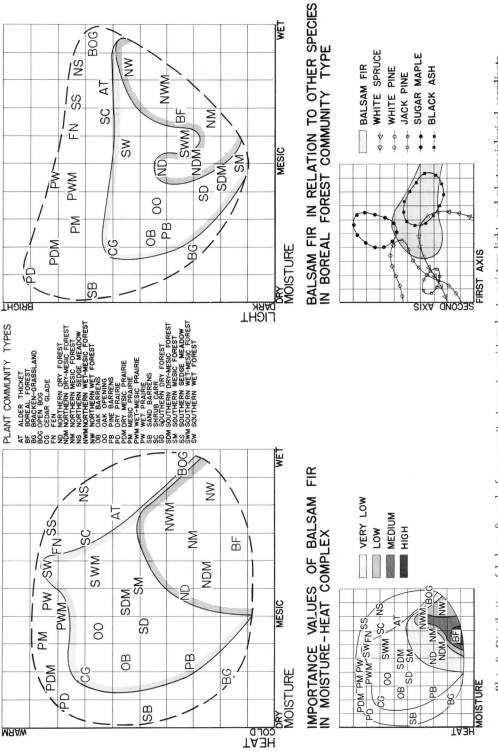

DISTRIBUTION OF BALSAM FIR ▨   DISTRIBUTION OF FOREST VEGETATION ▬   DISTRIBUTION OF TERRESTRIAL VEGETATION —

PLANT COMMUNITY TYPES

AT ALDER THICKET
BF BOREAL FOREST
BG BRACKEN-GRASSLAND
BOG OPEN BOG
CG CEDAR GLADE
FN FEN
ND NORTHERN DRY FOREST
NDM NORTHERN DRY-MESIC FOREST
NM NORTHERN MESIC FOREST
NS NORTHERN SEDGE MEADOW
NWM NORTHERN WET-MESIC FOREST
NW NORTHERN WET FOREST
OB OAK BARRENS
OO OAK OPENING
PB PINE BARRENS
PD DRY PRAIRIE
PDM DRY MESIC PRAIRIE
PM MESIC PRAIRIE
PWM WET-MESIC PRAIRIE
PW WET PRAIRIE
SB SAND BARRENS
SC SHRUB CARR
SD SOUTHERN DRY FOREST
SDM SOUTHERN DRY-MESIC FOREST
SM SOUTHERN MESIC FOREST
SS SOUTHERN SEDGE MEADOW
SWM SOUTHERN WET-MESIC FOREST
SW SOUTHERN WET FOREST

IMPORTANCE VALUES OF BALSAM FIR
IN MOISTURE-HEAT COMPLEX

▢ VERY LOW
▨ LOW
▨ MEDIUM
■ HIGH

BALSAM FIR IN RELATION TO OTHER SPECIES
IN BOREAL FOREST COMMUNITY TYPE

▨ BALSAM FIR
▽ WHITE SPRUCE
○ WHITE PINE
□ JACK PINE
● SUGAR MAPLE
■ BLACK ASH

Plate 9. Distribution of balsam fir and of community types in moisture-heat, moisture-light, and phytosociological coordinate axes in Wisconsin. (Redrawn with permission of the Regents of the University of Wisconsin, from J. T. Curtis, *The Vegetation of Wisconsin.* University of Wisconsin Press, 1959.)

339

PHYTOSOCIOLOGICAL   ORDINATION

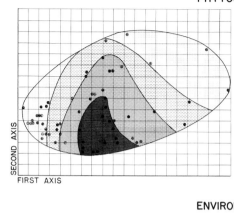

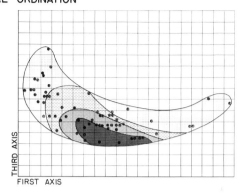

ENVIRONMENTAL   ORDINATION

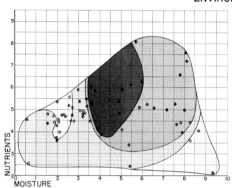

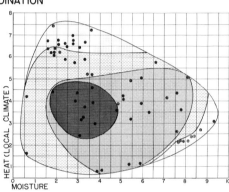

Plate 10. Importance values of balsam fir in phytosociological and environmental coordinate systems in northwestern New Brunswick. Importance values computed and phytosociological ordination follow the techniques used by the Wisconsin school (Curtis, 1959). The environmental coordinate systems are based on synthetic scalars of physiographic measurements and indicate increasing intensity of the factor complexes. (Redrawn from Loucks, 1962. Courtesy of Dr. O. L. Loucks, University of Wisconsin.)

FREQUENCY DISTRIBUTION OF BALSAM FIR IN TWO-DIMENSIONAL SCALE

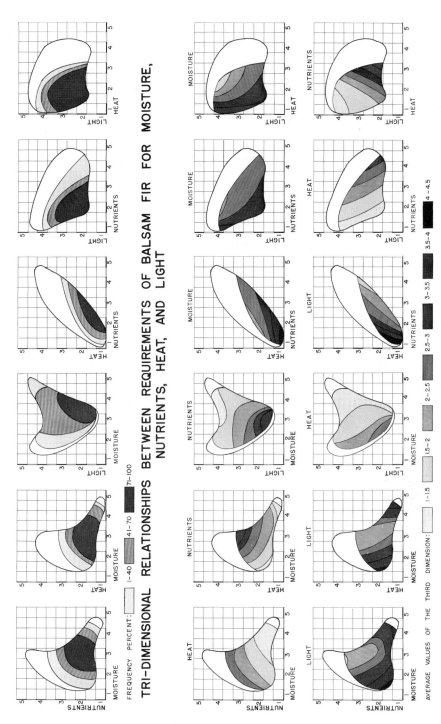

FREQUENCY DISTRIBUTION OF BALSAM FIR IN TWO-DIMENSIONAL SCALE

TRI-DIMENSIONAL RELATIONSHIPS BETWEEN REQUIREMENTS OF BALSAM FIR FOR MOISTURE, NUTRIENTS, HEAT, AND LIGHT

FREQUENCY PERCENT: 1-40 41-70 71-100

AVERAGE VALUES OF THE THIRD DIMENSION: 1-1.5 1.5-2 2-2.5 2.5-3 3-3.5 3.5-4 4-4.5

Plate 11. Frequency distribution of balsam fir in bivariate combinations of synecological moisture, nutrient, heat, and light coordinates in Minnesota. Balsam fir is represented in the shaded areas of the synecological fields, which are shown in outline on a relative scale of from 1 to 5, for all Minnesota forest communities. Frequency percentages are computed from the total number of forest communities in a bivariate unit of the scale. The tri-dimensional graphs in the lower two series of diagrams give the average coordinate for a third axis when two coordinates are given.

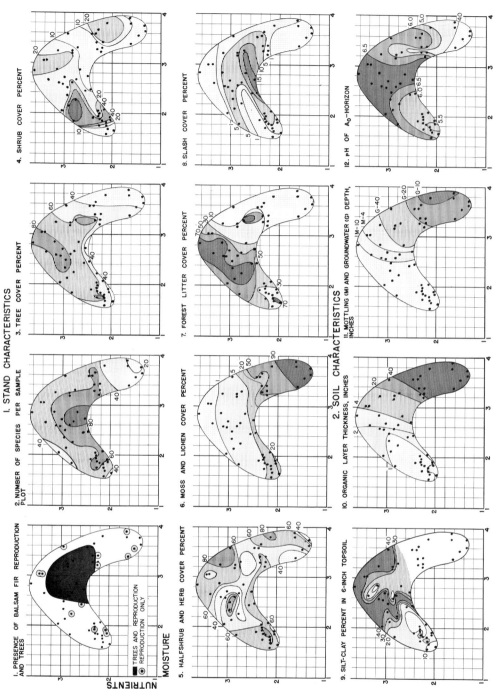

Plate 12. Distribution of balsam fir in the edaphic field of the Central Minnesota Pine Section. (1) *Stand characteristics:* occurrence of reproduction and trees; shrub, herb, moss, and lichen cover; litter and slash (dead and down trees) cover. (2) *Soil characteristics:* silt-clay percentage, thickness of organic layer, mottling and groundwater depth, and pH of the A_0 horizon.

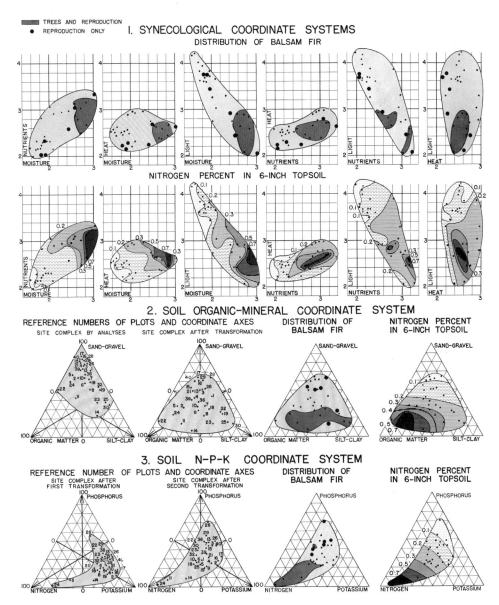

Plate 13. Distribution of balsam fir, and soil nitrogen percentages, in upland forest communities of the Central Minnesota Pine Section in synecological, soil organic-mineral, and soil N-P-K coordinate systems. Reproduction includes first-year seedlings. Trees are considered from 1-inch dbh. Soil data refer to the 6-inch (15 cm) topsoil. (1) Data are shown in all six bivariate combinations of the synecological moisture, nutrient, heat, and light coordinate systems on a relative scale of from 1 to 5. (2) Data are shown in a soil organic-mineral (organic matter, silt-clay, and sand-gravel) triangular coordinate system. In the first outline the coordinate percentages are computed following the normal procedure of triangular coordinate systems. In the second transformed or working outline the original data of organic matter, silt-clay, and sand-gravel percentages are divided by the corresponding mean values from all 30 plots, and the coordinate percentages are computed from these relative values and their sums. (3) Data are shown in a soil N-P-K (total nitrogen, available phosphorus, and exchangeable potassium) triangular coordinate system. In the first transformation the original N figures are divided by 100 before computing the sums of N+P+K and the coordinate percentages. In the second transformation the original figures for N, P, and K are divided by the corresponding means of all 30 plots, and the coordinate percentages computed from these relative figures and their sums.

343

Plate 14. Degrees of frost injury sustained by balsam fir during late spring freeze in Minnesota, May, 1946: (1) laterals killed, terminal leader unharmed; (2) terminal and five laterals killed, one lateral surviving to become the leader; (3) terminal killed, lateral partially developed at time of frost with only tips killed, with some damage evident on immature needles back from tip, growth of laterals entirely arrested, no dormant buds at time of frost and no laterals developing into leaders (recurved laterals); (4) three laterals killed, one lateral and terminal developed after frost; (5) all laterals and terminal killed, five new terminals developing adventitiously; (6) all laterals and terminals killed and numerous laterals and terminals developing adventitiously. Examples 1–4 from Itasca State Park; 5 and 6 from St. Paul campus, University of Minnesota. (Source: Hansen and Rees, 1947.)

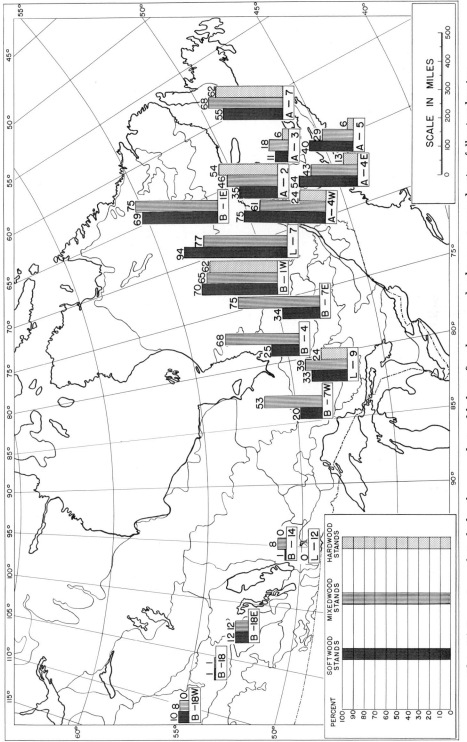

Plate 15. Percentage of stocked milacre quadrats of balsam fir advance and subsequent reproduction following logging in softwood, mixed-wood, and hardwood stands in different forest regions and sections of Canada. (After Candy, 1951.)

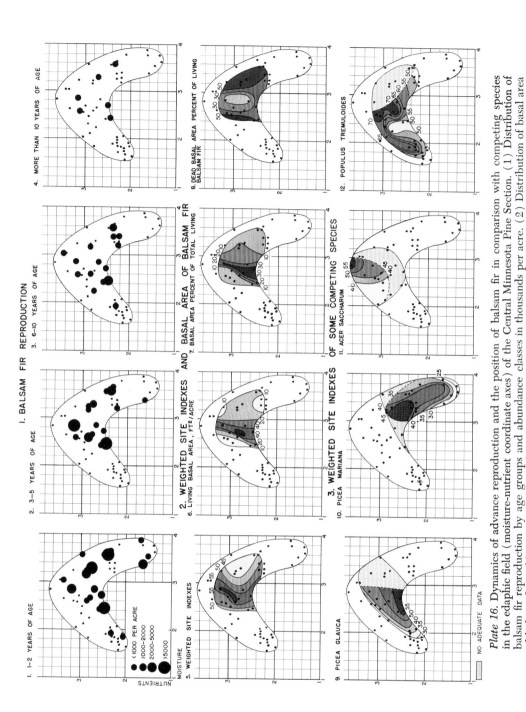

Plate 16. Dynamics of advance reproduction and the position of balsam fir in comparison with competing species in the edaphic field (moisture-nutrient coordinate axes) of the Central Minnesota Pine Section. (1) Distribution of balsam fir reproduction by age groups and abundance classes in thousands per acre. (2) Distribution of basal area of balsam fir in ft² per acre, expressed as percentage of the total living stand basal area; dead-standing or down balsam fir basal area, as percentage of living balsam fir basal area. (3) Weighted site indexes of four competing species, expressed as height in feet according to Lake States site-index curves (see Chapter 8 for details).

APPENDIXES, LITERATURE CITED, AND INDEX

APPENDIX I

A TENTATIVE LIST OF MYXOMYCETES AND FUNGI ASSOCIATED WITH BALSAM FIR

THIS list contains the names and synonyms of fungi reported to have been found on balsam fir (*Abies balsamea* (L.) Mill.). The fungi are grouped by families according to the arrangement by Clements and Shear (1954, reprinted from 1931). The major sources of synonyms are the classical works of Saccardo (1882–1931) and Oudemans (1919) and more recent studies by specialists on separate groups. The main host lists are taken from Farlow and Seymour (1888), Seymour (1929), *North American Flora* (New York Botanical Garden, 1906–37), Bisby et al. (1929, 1938), Wehmeyer (1950), disease surveys by the U.S. Department of Agriculture (Anderson et al., 1926; Weiss and O'Brien, 1953), and the annual reports of the forest insect and disease survey of Canada (Can. For. Biol. Div., 1942–59; Can. Dep. For., 1960–62). The list by Weiss and O'Brien was republished by the U.S. Agricultural Research Service in 1960. The Forest Biology Division of the Canada Department of Agriculture was reorganized recently into the Forest Entomology and Pathology Branch of the Department of Forestry. In the compilation of the list, the *Dictionary of the Fungi* (Ainsworth and Bisby, 1954, 1961) was very helpful.

Although an attempt was made to present a complete list of fungi (including slime molds) associated with balsam fir, omissions may occur. Considerable difficulty was experienced in establishing the identity of a number of fungi, especially those listed only by early workers, and the synonyms listed may not be true synonyms in all cases. As mycologists are well aware, the taxonomic status of many groups of fungi is imperfectly known, and there is considerable disagreement among specialists on the treatment of critical families and genera. The primary purpose of this list is to review the historical record and to establish to some degree the identity of the organisms that have received a diversity of names from the various workers who have studied them at different times. The list contains the names of the principal workers who have investigated the various species. References to species found in standard textbooks of forest pathology (Hartig, 1894; Hubert, 1931; Boyce, 1948, 1961; Baxter, 1952a), and in forest tree disease surveys of the United States and Canada are also included.

MYXOMYCETES

Comatricha nigra (Pers.) Schröt. — Moore (1906, 1913), Wehmeyer (1950).
Cribraria argillacea Pers. — Moore (1913), Wehmeyer (1950).
Lamproderma arcyrionema Rost. — Seymour (1929), Wehmeyer (1950).
Leocarpus fragilis (Dicks.) Rostaf. — *L. spermoides* Link., *L. vernicosus* Link. — Saccardo (1901), Oudemans (1919), Seymour (1929).
Physarum confertum Macbr. — Wehmeyer (1935, 1950).
Physarum contextum Pers. — *Chondrioderma contextum* Rostaf., *Diderma contextum* P., *D. granulatum* Fckl. — Oudemans (1919), Bisby et al. (1938).
Physarum flavidum Pk. — *Didymium flavidum* Pk., *P. flavidum* (Pk.) Berl. — Farlow and Seymour (1888), Saccardo (1898), Seymour (1929).

349

Physarum galbeum Wingate. — Moore (1906), Seymour (1929), Wehmeyer (1950).
Physarum globuliferum (Bull.) Pers. — *Cytidium globuliferum* Morg. — Saccardo (1901), Moore (1906, 1913), Oudemans (1919), Seymour (1929), Wehmeyer (1935, 1950).

ASCOMYCETES
CAPNODIACEAE

Adelopus nudus (Pk.) Höhn. — *A. balsamicola* (Pk.) Th., *Asterella nuda* (Pk.) Sacc., *Asterina nuda* Pk., *Cryptopus balsamicola* (Pk.) Th., *C. nudus* (Pk.) Th., *Dimerosporium balsamicola* (Pk.) Ell. & Ev., *Meliola balsamicola* Pk., *Phaeocryptopus nudus* (Pk.) Petr., *Zukalia balsamicola* (Pk.) Sacc. — Farlow and Seymour (1888), Saccardo (1898), Seymour (1929), Schenck (1939), Hahn (1947), Wehmeyer (1942, 1950), Weiss and O'Brien (1953), Can. For. Biol. Div. (1951–56), Spaulding (1961).
Apiosporium pinophilum Fckl. (*Apiosporium = Sclerotium*). — Oudemans (1919), Kujala (1930).
Scorias spongiosa (Fr.) Schw. — MacKay (1908), Wehmeyer (1950).

SPHAERIACEAE

Acanthostigma parasiticum (Hart.) Sacc. — *Trichosphaeria parasitica* Hart. — Oudemans (1919), Hubert (1931), Weiss and O'Brien (1953), Spaulding (1956), Schwerdtfeger (1957), Can. For. Biol. Div. (1958), Spaulding (1961), Smerlis (1961), Can. For. Ent. Path. Br. (1961).
Amphisphaeria incustrans Ell. & Ev. — Bisby et al. (1938).
Amphisphaeria juniperi Tracy & Earle. — Wehmeyer (1942, 1950).
Bertia moriformis (Tode ex Fr.) de Not. — *Sphaeria moriformis* Tode. — Oudemans (1919), Seymour (1929).
Ceratocystis bicolor (Davidson & Wells) Davidson. — *Ophiostoma bicolor* Davidson & Wells. — Davidson (1955, 1958), Basham (1959).
Ceratocystis brunneocrinita Wright & Cain. — Wright and Cain (1961).
Herpotrichia nigra Hart. — Hubert (1931), Schwerdtfeger (1957), Can. For. Biol. Div. (1958).
Kirschteiniella thujina (Pk.) Pomerleau & Etheridge. — *Amphisphaeria thujina* (Pk.) Sacc. — Etheridge and Pomerleau (1961), Can. For. Ent. Path. Br. (1961).
Melogramma boreale Ell. & Ev. — Oudemans (1919), Seymour (1929).
Rehmiellopsis abietis (E. Rostr.) O. Rostr. — Waterman (1937, 1945), Waterman and Aldrich (1940), Spaulding (1956, 1961).
Rehmiellopsis balsameae Wat. — *R. bohemica* Bulb. & Kab. — Oudemans (1919), Waterman (1937, 1945), Waterman and Aldrich (1940), Wehmeyer (1950), Baxter (1952a), Weiss and O'Brien (1953), Westcott (1955), Can. For. Biol. Div. (1951–59), Boyce (1961), Can. For. Ent. Path. Br. (1961).
Valsa abietis Fr. — *Cytospora abietis* Sacc., *Sphaeria abietis* Fr. — Oudemans (1919), Wehmeyer (1950), Weiss and O'Brien (1953), Can. For. Biol. Div. (1959), Raymond and Reid (1961), Can. For. Ent. Path. Br. (1961), Spaulding (1961).
Valsa brevis Pk. — Saccardo (1898), Seymour (1929).
Valsa colliculus (Wormsk. ex Fr.) B. — *Sphaeria colliculus* Wormsk. — Oudemans (1919), Seymour (1929).
Valsa friesii Duby. — *Cytospora friesii* Sacc., *C. pinastri* Fr., *Valsa friesii* (Duby) Fckl., *V. friesii* Fckl. — Farlow and Seymour (1888), Saccardo (1898), Oudemans (1919), Faull (1930a), Weiss and O'Brien (1953), Can. For. Biol. Div. (1951–59), Can. For. Ent. Path. Br. (1960–61), Spaulding (1961).
Valsa kunzei Fr. — *Cytospora kunzei* Sacc., *Sphaeria kunzei* Fr. — Saccardo (1901), Oudemans (1919), Wehmeyer (1942, 1950), Can. For. Biol. Div. (1951–55), Spaulding (1961).

HYPOCREACEAE

Calonectria balsamea (Cke. & Pk.) Sacc. — Saccardo (1898).
Nectria balsamea Cke. & Pk. — *Chilonectria balsamea* (Cke. & Pk.) Sacc., *C. rosellini* (Carsest.) Sacc., *Scoleconectria balsamea* (Cke. & Pk.) Seav., *Thyronectria balsamea* (Cke. & Pk.) Seeler. — Saccardo (1898), Seymour (1929), Bisby et al. (1938), Wehmeyer (1942, 1950), Can. For. Biol. Div. (1951–59), Weiss and O'Brien (1953), Westcott (1955), Raymond and Reid (1961), Can. For. Ent. Path. Br. (1961).
Nectria cucurbitula Sacc. (non (Tode) Fr.). — *Calonectria cucurbitula* Sacc., *Chilonectria cucurbitula* Sacc., *C. cucurbitula* (Curr.) Sacc., *Creonectria cucurbitula* (Sacc.) Seav., *Nectria cucurbitula* Fr., *N. cucurbitula* (Curr.) Cke., *N. cucurbitula* f. *abietis* Roum. —

Farlow and Seymour (1888), Anderson et al. (1926), Robak (1951), Weiss and O'Brien (1953), Can. For. Biol. Div. (1951–55), Spaulding (1956, 1961).

Nectria cucurbitula var. *macrospora* Wr. — *Cylindrocarpon cylindroides* Wr. — Roll-Hansen (1962).

Passerinula candida Sacc. — Wehmeyer (1942, 1950).

Ophionectria scolecospora Bref. — *Scoleconectria scolecospora* (Bref.) Seav. — Oudemans (1919), Seymour (1929), Wehmeyer (1942, 1950), Weiss and O'Brien (1953), Westcott (1955).

DOTHIDEACEAE

Montagnella abietina Ell. & Ev. — Seymour (1929).

MICROPELTACEAE

Micropeltis pitya Sacc. — Seymour (1929).

HYSTERIACEAE

Glonium stellatum Mühl. — Oudemans (1919), Bisby et al. (1938).

Hypodermella mirabilis Darker. — Darker (1932), Boyce (1948, 1952), Weiss and O'Brien (1953), Westcott (1955), Smerlis (1961), Can. For. Ent. Path. Br. (1961).

Hypodermella nervata Darker. — Darker (1932), Boyce (1948, 1952), Wehmeyer (1940, 1950), Weiss and O'Brien (1953), Can. For. Biol. Div. (1951–59), Westcott (1955), Smerlis (1961).

Hypodermella punctata Darker. — Smerlis (1961).

Lophium mytilinum (Pers.) Fr. — *Mytilidion karstenii* Sacc. — Oudemans (1919), Seymour (1929), Can. For. Biol. Div. (1951–55).

Lophodermium autumnale Darker. — Darker (1932), Wehmeyer (1950), Weiss and O'Brien (1953), Westcott (1955).

Lophodermium lacerum Darker. — Darker (1932), Weiss and O'Brien (1953), Westcott (1955), Smerlis (1961).

Lophodermium nervisequum (DC. ex Fr.) Rehm. — *Hypoderma nervisequum* DC., *Hypodermella nervisequia* (DC. ex Fr.) Lagerb., *Hysterium nervisequum* Fr., *L. nervisequum* (DC.) Rehm. — Spaulding (1912), Oudemans (1919), Anderson et al. (1926), Seymour (1929).

Lophodermium piceae (Fckl.) Höhn. — *Coccomyces piceae* Sacc., *Hypodermina abietis* Hilitzer, *Lophodermium abietis* Rostr. — Darker (1932), Jacot (1939), Bisby et al. (1938), Westcott (1955), Spaulding (1961), Smerlis (1961).

GRAPHIDACEAE

Graphis scripta (L.) Ach. — Wehmeyer (1940, 1950).

PHACIDIACEAE

Bifusella faullii Darker. — Boyce (1948, 1952), Darker (1932), Wehmeyer (1950), Weiss and O'Brien (1953), Can. For. Biol. Div. (1951–59), Westcott (1955), Boyce (1961), Smerlis (1961).

Clithris graphis Rehm (*Clithris = Colpoma*). — Seymour (1929).

Phacidium abietinellum Dearn. — *P. abietinum* Schm. ex Fr., *P. abietinum* Kze. & Schm., *P. pinastri* Lib. — Farlow and Seymour (1888), Saccardo (1898), Oudemans (1919), Seymour (1929), Boyce (1948), Weiss and O'Brien (1953), Can. For. Biol. Div. (1951–55), Westcott (1955).

Phacidium balsameae J. J. Davis. — *Stegopezizella balsameae* Syd. — Dearness (1926), Seymour (1929), Darker (1932), Boyce (1948), Weiss and O'Brien (1953), Can. For. Biol. Div. (1951–55), Westcott (1955), Boyce (1961).

Phacidium infestans Karst. — Faull (1930b), Weiss and O'Brien (1953), Westcott (1955), Spaulding (1956, 1961), Can. For. Biol. Div. (1956), Smerlis (1961).

Potebniamyces balsamicola Smerlis. — Smerlis (1961).

DERMATACEAE

Cenangella abietina Ell. & Ev. (?). — Oudemans (1919), Seymour (1929), Seaver (1951).

Cenangium abietis (Pers.) Rehm. — *C. abietis* (Pers.) Duby, *C. abietis* Rehm, *C. ferruginosum* Fr., *Dermatea pini* Phill. & Hark., *Peziza abietis* Pers. — Farlow and Seymour (1888), Saccardo (1898), Long (1924), Seymour (1929), Seaver (1951), Weiss and O'Brien (1953), Can. For. Biol. Div. (1951–55), Westcott (1955), Spaulding (1961).

Dermea balsamea (Pk.) Seav. — *Cenangium balsameum* Pk., *C. balsameum* var. *abietinum* Pk., *Gelatinosporium abietinum* Pk., *Micropera erumpens* Ell. & Ev. — Farlow and Seymour (1888), Seymour (1929), Wehmeyer (1940, 1950), Seaver (1951), Can. For. Biol. Div. (1957–59), Can. For. Ent. Path. Br. (1961), Raymond and Reid (1961).

Godronia abietina (Ell. & Ev.) Seav. — *Scleroderris abietina* Ell. & Ev. — Oudemans (1919), Seymour (1929), Bisby et al. (1938), Wehmeyer (1940, 1950), Seaver (1951), Can. For. Ent. Path. Br. (1961).

Godronia abietis (Naum.) Seav. — *Ascocalyx abietis* Naum., *Bothrodiscus pinicola* Shear, *Cenangium pityum* B. & C., *C. pityum* Fr., *Fusisporium berenice* B. & C., *Pycnocalyx abietis* Naum., *Scleroderris pitya* (B. & C.) Sacc. — Farlow and Seymour (1888), Saccardo (1898), Oudemans (1919), Seymour (1929), Wehmeyer (1950), Seaver (1951), Can. For. Biol. Div. (1951–56), Can. For. Ent. Path. Br. (1960–61).

Nothophacidium abietinellum (Dearn.) Reid & Cain. — *Phacidium abietinellum* Dearn. — Dearness (1926), Reid and Cain (1962).

Pezicula phyllophila (Pk.) Seav. — *Dermatea phyllophila* Pk. — Farlow and Seymour (1888), Seymour (1929), Seaver (1951).

Tympanis abietina Groves. — Groves (1952).

Tympanis pinastri (Pers.) Tul. — *Cenangella pinastri* (P. ex Fr.) Sacc., *C. pinastri* Sacc., *Cenangium laricinum* Fckl., *C. pinastri* Fr., *C. pinastri* Rehm, *C. pinastri* P. ex Fr., *Micropera pinastri* Sacc., *Peziza pinastri* P., *Phacidium pinastri* Fr., *Tryblidiopsis pinastri* Karst., *Tryblidium pinastri* Fr., *Tympanis laricina* (Fckl.) Sacc., *T. pinastri* Rehm, *T. pinastri* f. *laricinum* Rehm, *T. pitya* (Fr.) Karst. — Farlow and Seymour (1888), Saccardo (1898), Oudemans (1919), Seymour (1929), Bisby et al. (1929), Wehmeyer (1940, 1950), Seaver (1951), Weiss and O'Brien (1953).

Tympanis truncatula (Pers. ex Fr.) Rehm. — *Cycledum truncatulum* Wallr., *Lecanidion truncatulum* Sacc., *Peziza truncatula* Fr., *Peziza truncatula* Pers., *Tympanis pinastri* sensu Tul. (not *Cenangium pinastri* Fr.). — Oudemans (1919), Groves (1952).

PATELARIACEAE

Patellaria patinelloides (Sacc. & Roum.) Sacc. — *Karschia patinelloides* Sacc. — Saccardo (1901), Oudemans (1919), Seymour (1929).

Patinella punctiformis Rehm. — Wehmeyer (1940, 1950).

BULGARIACEAE

Coryne sarcoides (Jacq.) Tul. — Etheridge and Carmichael (1957).

Gloeocalyx rufa (S.) Sacc. var. *magna* Pk. (*Gloeocalyx* = *Sarcosoma* = *Bulgaria*). — Seymour (1929).

HELOTIACEAE

Helotium pallescens (Pers.) Fr. (?). — *Calycella pallescens* Quél., *Peziza pallescens* Quél. — Wehmeyer (1950), Seaver (1951).

Lachnella agassizii (Berk. & Curt.) Seav. — *Dasyscypha agassizii* Sacc., *D. agassizii* (Berk. & Curt.) Sacc., *D. incarnata* Clem., *Elvela calyciformis* Batch (?), *Helotium calyciforme* Wettst., *Lachnella subtilissima* Phill., *Octospora calyciformis* Hedw. (?), *Peziza agassizii* Berk. & Curt., *P. calyciformis* Willd. (?), *P. subtilissima* Cke. — Farlow and Seymour (1888), Saccardo (1898), Seymour (1929), Bisby et al. (1938), Wehmeyer (1940, 1950), Seaver (1951), Stillwell (1959).

Lachnella arida (Phill.) Seav. — *Dasyscypha arida* (Phill.) Sacc., *D. flavovirens* Bres., *Lachnum engelmannii* Earle, *Peziza arida* Phill. — Oudemans (1919), Seaver (1951), Weiss and O'Brien (1953).

Lachnella hahniana Seav. — *Dasyscypha calyciformis* (Will.) Rehm, *D. calycina* Fckl., *D. calycina* (Schum. ex Fr.) Fckl., *D. calycina* (Schum.) Fckl., *Peziza calycina* Fr. — Oudemans (1919), Seymour (1929), Seaver (1951), Weiss and O'Brien (1953), Spaulding (1956, 1961).

Lachnella resinaria Phill. — *Dasyscypha resinaria* (Cke. & Phill.) Rehm, *D. resinaria* Rehm, *Lachnellula resinaria* Rehm, *Peziza resinaria* Cke. & Phill. — Anderson (1902), Oudemans (1919), Seymour (1929), Seaver (1951), Weiss and O'Brien (1953), Westcott (1955), Boyce (1961).

Sclerotinia kerneri Wettst. — Cash (1941).

Tapesia balsamicola (Pk.) Sacc. — *Peziza balsamicola* Pk. — Farlow and Seymour (1888), Saccardo (1898), Seymour (1929), Seaver (1951).

PEZIZACEAE

Barlaea lacunosa Ell. & Ev. (*Barlaea* = *Barlaeina* = *Lamprospora*). — Seymour (1929).

Patella irregularis (Clem.) Seav. — *Lachnea irregularis* (Clem.) Sacc. & D. Sacc. — Seaver (1928), Seymour (1929), Wehmeyer (1950).

AGYRIACEAE

Agyrium rufum (Pers.) Fr. — Farlow and Seymour (1888), Saccardo (1898), Seymour (1929).

BASIDIOMYCETES

PUCCINIACEAE

Caeoma arcticum Kauffm. — Seymour (1929).

MELAMPSORACEAE

Calyptospora goeppertiana Kühn. — *Aecidium columnare* A. & S., *A. ornamentale* Farl., *Caeoma columnare* Lk., *Melampsora goeppertiana* Wint., *Peridermium columnare* Schm. & Kze., *P. holwayi* Syd., *P. ornamentale* Arth., *Pucciniastrum goeppertianum* (Kühn) Kleb., *Uredo columnaris* Spreng. — Oudemans (1919), Anderson et al. (1926), Seymour (1929), New York Bot. Gard. (1931), Bartholomew (1933), Arthur (1934), Faull (1939), Hunter (1948), Boyce (1948), Wehmeyer (1950), Westcott (1955), Spaulding (1961).

Hyalopsora aspidiotus (Pk.) Magn. — *Caeoma aspidiotus* Pk., *Hyalopsora aspidiotus* Magn., *H. polypodii-dryopteridis* Magn., *Peridermium pycnoconspicuum* Bell, *Pucciniastrum aspidiotus* Karst., *Uredinopsis polypodii-dryopteridis* Liro, *Uredo aspidiotus* Pk. — Bell (1924), Seymour (1929), Faull (1930a), Hubert (1931), Bartholomew (1933), Arthur (1934), Boyce (1948), Hunter (1948), Wehmeyer (1950), Weiss and O'Brien (1953), Westcott (1955), Spaulding (1961).

Melampsora abieti-capraearum Tub. — *Caeoma abietis-pectinatae* Riess, *C. americana* (?), *Melampsora americana* Arth., *M. arctica* Rostr., *M. capraearum* Thim., *M. farinosa* (P.) Schröt., *M. humboldtiana* Arth., *M. humboldtiana* Speg., *Uredo abietis-pectinatae* (Riess) Sacc., *U. alpina* Arth., *U. rostrupiana* Arth. — Fraser (1913), Weir and Hubert (1918), Oudemans (1919), Anderson et al. (1926), Boyce (1928), Seymour (1929), Hubert (1931), New York Bot. Gard. (1931), Bartholomew (1933), Arthur (1934), Bisby et al. (1938), Boyce (1948), Hunter (1948), Wehmeyer (1950), Weiss and O'Brien (1953), Westcott (1955), Spaulding (1961).

Melampsora epitea Thüm. — Can. For. Ent. Path. Br. (1961).

Melampsorella caryophyllacearum Schröt. — *Aecidium coloradense* Diet., *A. elatinum* A. & S., *Caeoma caryophyllacearum* Lk., *C. cerastii* Schröt., *C. elatina* Lk., *Exobasidium stellaria* Syd., *Hypodermium* (*Uredo*) *stellarium* Lk., *Melampsora cerastii* (Pers.) Schröt., *M. cerastii* Wint., *Melampsorella cerastii* Wint., *M. cerastii* (Wint.) Schröt., *M. elatina* (A. & S.) Schröt., *Peridermium boreale* A. & K., *P. coloradense* A. & K., *P. elatinum* (A. & S.) Kze. & Schm., *Uredo cerastii* Mart., *U. pustulata cerastii* Pers. — Anderson (1897), Saccardo (1898), Freeman (1905), Oudemans (1919), Stone (1920), Anderson et al. (1926), Boyce (1928), Seymour (1929), New York Bot. Gard. (1931), Conners (1931–41), Bartholomew (1933), Arthur (1934), Wehmeyer (1935, 1950), Boyce (1948), Weiss and O'Brien (1953), Can. For. Biol. Div. (1951–59), Westcott (1955), Can. For. Ent. Path. Br. (1960–61), Spaulding (1961), Smerlis (1961).

Milesia kriegeriana (Magn.) Arth. — *Melampsora kriegeriana* Magn., *Melampsorella kriegeriana* Magn., *M. kriegeriana* P., *Milesia fructuosa* Faull, *M. intermedia* Faull, *Milesina kriegeriana* Magn., *Peridermium balsameum* Pk., *Uredo polypodii* Mart., *Uredinopsis filicinus* (Niess) Magn. — Oudemans (1919), Boyce (1928), Hubert (1931), New York Bot. Gard. (1931), Faull (1932), Bartholomew (1933), Arthur (1934), Boyce (1948), Hunter (1948), Wehmeyer (1950), Weiss and O'Brien (1953), Westcott (1955), Can. For. Biol. Div. (1956), Can. For. Ent. Path. Br. (1961).

Milesia marginalis Faull & Wats. — *M. kriegeriana* Arth., *Milesina marginalis* Faull & Wats. — Faull (1932), Arthur (1934), Boyce (1948), Hunter (1948), Westcott (1955).

Milesia polypodophila (Bell) Faull. — *M. polypodophila* Faull, *M. pycnograndis* (Bell) Arth., *Milesina polypodophila* (Bell) Faull, *Peridermium pycnogrande* Bell, *Uredinopsis polypodophila* Bell. — Boyce (1928), Seymour (1929), Hubert (1931), New York Bot. Gard. (1931), Bartholomew (1933), Arthur (1934), Boyce (1948), Hunter (1948), Wehmeyer (1950), Westcott (1955).

Pucciniastrum epilobii Otth. — *Melampsora chamaenerii* Rostr., *M. epilobii* Fckl., *M. herbarum* Desm., *M. pustulata* Schröt., *Peridermium pustulatum* (P.) Kauffm., *Phragmospora epilobii* Magn., *Pucciniastrum abieti-chamaenerii* Kleb., *P. chamaenerii* Rostr., *P. pustulatum* (Pers.) Diet., *Uredo pustulata* Pers., *U. epilobii* DC. — Oudemans (1919), Anderson et al. (1926), Boyce (1928), Hubert (1931), Bartholomew (1933), Arthur (1934), Faull (1938b), Hunter (1948), Wehmeyer (1950), Westcott (1955), Can. For. Ent. Path. Br. (1961), Boyce (1961), Spaulding (1961).

Uredinopsis atkinsonii Magn. — *Milesia atkinsonii* Arth., *M. copelandi* Arth., *Milesina atkinsonii* (Magn.) Arth., *Peridermium balsameum* Auct. Am., *Uredinopsis atkinsonii* Magn. ex Arth., *U. copelandi* Syd., *U. copelandi* Syd. ex Jacks., *U. mirabilis* (Pk.) Magn. ex Arth. (in part), *U. struthiopteridis* Störm. ex Arth. — Fraser (1913), Oudemans (1919), Boyce (1928), Seymour (1929), New York Bot. Gard. (1931), Bartholomew (1933), Faull (1938c), Hunter (1948).

Uredinopsis ceratophora Faull. — *U. atkinsonii* Magn. ex Davis, *U. copelandi* Syd. ex Arth., *U. struthiopteridis* Störm. ex Arth. — Faull (1938a,c), Boyce (1948).

Uredinopsis longimucronata Faull. — *U. atkinsonii* Magn. ex Arth., *U. copelandi* Syd. ex Jacks., *U. mirabilis* (Pk.) Magn., *U. struthiopteridis* Störm. ex Arth. — Faull (1938a,c), Boyce (1948).

Uredinopsis mirabilis (Pk.) Magn. — *Aecidium balsameum* Diet., *Gloeosporium mirabile* Pk., *Milesia mirabilis* Arth., *Peridermium balsameum* Pk., *Rhabdospora mirabilis* (Pk.) O. Kuntze, *Septoria mirabilis* Pk., *Uredinopsis americana* Syd., *U. mirabilis* (Pk.) Arth. — Saccardo (1898), Fraser (1913), Bell (1924), Boyce (1928, 1948), Seymour (1929), Hubert (1931), Arthur (1934), Wehmeyer (1935, 1950), Bisby et al. (1938), Faull (1938a,c), Hunter (1948), Weiss and O'Brien (1953), Westcott (1955).

Uredinopsis osmundae Magn. — *Milesia osmundae* Arth., *M. osmundae* (Magn.) Arth., *Peridermium balsameum* Auct. Am., *Uredinopsis mirabilis* (Pk.) Magn. ex Rhoads (in part). — Peck (1875), Fraser (1913), Bell (1924), Boyce (1928), Seymour (1929), New York Bot. Gard. (1931), Bartholomew (1933), Arthur (1934), Faull (1938a,c), Hunter (1948), Weiss and O'Brien (1953), Can. For. Biol. Div. (1951–55), Westcott (1955).

Uredinopsis phegopteridis Arth. — *Peridermium balsameum* Pk., *Uredinopsis mirabilis* (Pk.) Magn. ex Rhoads (in part). — Fraser (1913), Oudemans (1919), Bell (1924), Seymour (1929), Bartholomew (1933), Arthur (1934), Faull (1938a,c), Hunter (1948), Boyce (1948), Wehmeyer (1950), Weiss and O'Brien (1953), Westcott (1955).

Uredinopsis struthiopteridis Störm. — *Gloeosporium struthiopteridis* Störm., *Peridermium balsameum* Auct. Am. — Fraser (1913), Oudemans (1919), Boyce (1928), Seymour (1929), Bartholomew (1933), Arthur (1934), Faull (1938a,c), Bisby et al. (1938), Hunter (1948), Weiss and O'Brien (1953), Spaulding (1961).

AURICULACEAE

Auricularia auricula (Hook.) Underw. — *A. auricularis* (S. F. Gray) Martin, *Hirneola auricula-judae* (L. ex Fr.) Berk., *H. indiformis* Lév. — Farlow and Seymour (1888), Saccardo (1898), Oudemans (1919), Seymour (1929), Can. For. Biol. Div. (1951–55), Stillwell (1959), Spaulding (1961).

TREMELLACEAE

Tremella encephala Willd. — *Naematelia encephala* (Willd.) ex Fr. — Oudemans (1919), Seymour (1929), Wehmeyer (1935, 1950).

Tremella saccharina Fr. var. *foliacea* (Bref.) Bres. — *T. saccharina* Bon. — Oudemans (1919), Bisby et al. (1938).

Tulasnella fuscoviolacea Bres. — Oudemans (1919), Seymour (1929).

Ulocolla foliacea (P.) Bref. (*Ulocolla = Exidia*). — *Tremella foliacea* P. — Oudemans (1919), Seymour (1929), Wehmeyer (1935, 1950).

DACRYOMYCETACEAE

Calocera viscosa Fr. — *C. palmata* (Schum.) Fr. — Wehmeyer (1950).

Dacryomyces minor Pk. — Wehmeyer (1935, 1950).
Ditiola radicata Fr. — Wehmeyer (1935, 1950).

HYPOCHNACEAE

Hypochnus avellaneus Burt (*Hypochnus = Tomentella*). — Seymour (1929).
Hypochnus fumosus Fr. — *Coniophora fumosa* Sacc., *Thelephora fumosa* P., *T. menieri* Pat. —
Saccardo (1898), Oudemans (1919), Bisby et al. (1938).
Hypochnus umbrinus (Fr.) Quél. — *H. umbrinus* Wallr., *Thelephora umbrina* Fr. — Oudemans
(1919), Bisby et al. (1938).

THELEPHORACEAE

Aleurodiscus amorphus (Pers.) Rabh. — *A. amorphus* Rabh., *Corticium amorphum* Fr., *C. amor-*
phum (P.) Fr., *Nodularia balsamicola* Pk., *Peziza amorpha* P., *P. balsamicola* Cke. nec. Pk.,
Thelephora amorpha Fr., *T. amorpha* (P.) Fr. — Farlow and Seymour (1888), Saccardo
(1898), Farlow (1905), Seymour (1929), Bisby et al. (1929), Hansbrough (1934), Weh-
meyer (1940, 1950), Boyce (1948), Stillwell (1959), Spaulding (1961), Can. For. Ent.
Path. Br. (1961).
Aleurodiscus farlowii Burt. — Seymour (1929).
Coniophora olivacea (Fr.) Karst. — *Coniophorella olivacea* (Fr.) Karst. — Oudemans (1919),
Bisby et al. (1938).
Coniophora puteana (Schum. ex Fr.) Karst. — *C. cerebella* Pers., *Corticium puteanum* Fr. —
Oudemans (1919), Spaulding and Hansbrough (1944), Boyce (1948), Basham (1951a),
Baxter (1952a), Basham et al. (1953), Can. For. Biol. Div. (1956), Stillwell (1959),
Spaulding (1961).
Coniophora suffocata (Pk.) Massee. — Wehmeyer (1950).
Corticium bicolor Pk. — *C. bicolor* (P.) Fr. — Oudemans (1919), Bisby et al. (1938).
Corticium confluens Fr. — *C. caesio-album* Karst., *Thelephora confluens* Fr. — Saccardo (1901),
Oudemans (1919), Seymour (1929).
Corticium furfuraceum Bres. — *C. calceum* Fr., *C. subpallidulum* Litsch. — Wehmeyer (1950).
Corticium fuscostratum Burt. — Can. For. Biol. Div. (1959).
Corticium galactinum (Fr.) Burt. — *C. galactinum* var. *alni* (Fr.) Bres., *Stereum alneum* Fr.
(?), *S. suaveolens* Fr. (?), *Thelephora alni* Fr. (?), *T. galactina* Fr., *T. suaveolens* Fr. —
Schrenk (1902), Long (1917), Kress et al. (1925), Burt (1926), Bisby et al. (1938), Cooley
and Davidson (1940), Cooley (1948, 1951), Marshall (1948), Nobles (1948), Davidson
(1951), White (1951), Basham et al. (1953), Can. For. Biol. Div. (1951–56), Stillwell
(1959), Spaulding (1961), Can. For. Ent. Path. Br. (1961).
Corticium hydnans (Schw.) Burt. — *Radulum orbiculare* Fr. (?). — Wehmeyer (1950).
Corticium lacteum Fr. — *Peniophora subcremea* Höhn — Oudemans (1919), Seymour (1929).
Corticium laeve Pers. — *C. pelliculare* Karst. — Saccardo (1901), Oudemans (1919), Bisby et
al. (1929), Smerlis (1961).
Corticium lividum Pers. — Wehmeyer (1940, 1950).
Corticium odoratum (Fr.) Bourd. & Galz. — *Stereum odoratum* Fr., *Thelephora odorata* Fr. —
White (1951), Spaulding (1961).
Corticium radiosum Fr. — Oudemans (1919), Can. For. Biol. Div. (1951–55).
Corticium roseo-cremeum Bres. (?). — Wehmeyer (1950).
Corticium subcoronatum Höhn. & Litsch. — Oudemans (1919), Bisby et al. (1938).
Corticium sulphureo-isabellinum Litsch. — Jackson (1948).
Corticium sulphureum Fr. — Saccardo (1898), Oudemans (1919), Seymour (1929).
Hymenochaete agglutinans Ell. — Can. For. Biol. Div. (1959).
Hymenochaete badio-ferruginea (Mont.) Lev. — Wehmeyer (1950).
Hymenochaete tabacina (Sow. ex Fr.) Lev. — *Thelephora variegata* Schrad. — Oudemans (1919),
Bisby et al. (1938).
Hymenochaete tenuis Pk. — Oudemans (1919), Bisby et al. (1938).
Pellicularia vaga (Berk. & Curt.) Rog. ex Linder. — *Corticium botryosum* Bres., *C. vagum* B. &
C. — Rogers (1943), Wehmeyer (1950).
Peniophora aspera (Pers.) Sacc. — *Odontia setigera* (Fr.) L. W. Miller, *Peniophora setigera*
(Fr.) Höhn. & Litsch., *Thelephora aspera* Pers. — Can. For. Biol. Div. (1951–55), Weh-
meyer (1940, 1950), Slysh (1960).
Peniophora byssoides (Pers. ex Fr.) Bres. — *Coniophora byssoides* (Pers. ex Fr.) Karst. — Rogers
and Jackson (1943), Wehmeyer (1950), Slysh (1960).

Peniophora carnosa Burt. — Wehmeyer (1940, 1950), Slysh (1960).
Peniophora gigantea (Fr.) Massee. — *Corticium giganteum* Fr., *C. thelephoroides* Ell. & Ev., *Hypochnus pallescens* (S.) Burt, *H. thelephoroides* (Ell. & Ev.) Burt, *Peniophora globifera* Ell. & Ev. — Farlow and Seymour (1888), Saccardo (1898), Oudemans (1919), Seymour (1929), Basham et al. (1953), Slysh (1960).
Peniophora gracillima Ell. & Ev. — *P. glebulosa* sensu Bres. — Wehmeyer (1950), Slysh (1960).
Peniophora greschikii (Bres.) Bourd. & Galz. — *P. subcremea* Höhn. & Litsch., *P. rudis* (Karst.) Bourd. & Galz., *P. alba* Burt. — Wehmeyer (1950), Slysh (1960).
Peniophora hamata Jacks. — Jackson (1948), Slysh (1960).
Peniophora inornata Jacks. & Rogers. — Jackson (1948), Slysh (1960).
Peniophora pallidula (Bres.) Bres. — *P. alutaria* Burt. — Bisby et al. (1938), Slysh (1960).
Peniophora perexigua Jacks. — Jackson (1948), Slysh (1960).
Peniophora phlebioides Jacks. & Dearn. — Stillwell (1959), Slysh (1960).
Peniophora piceina Overh. — Bisby et al. (1938).
Peniophora ralla Jacks. — Jackson (1948), Slysh (1960).
Stereum abietinum Pers. — *S. abietinum* (Fr.) Fr. — Wehmeyer (1940, 1950), Spaulding (1961).
Stereum chailletii Pers. — *Thelephora chailletii* P. — Oudemans (1919), Basham (1951a, 1959), Basham et al. (1953), Stillwell (1959).
Stereum purpureum Pers. — Smerlis (1961).
Stereum sanguinolentum A. & S. ex Fr. — *S. crispum* Schröt. — Oudemans (1919), Schierbeck (1922), Faull (1923), McCallum (1928), Seymour (1929), Spaulding et al. (1932), Kaufert (1935), Spaulding and Hansbrough (1944), Boyce (1948), Pomerleau (1948), Wehmeyer (1940, 1950), Davidson (1951), Baxter (1952a), Basham et al. (1953), Weiss and O'Brien (1953), Can. For. Biol. Div. (1951–58), Westcott (1955), Stillwell (1959), Can. For. Ent. Path. Br. (1960), Spaulding (1961).
Thelephora laciniata (P.) ex Fr. — *Elvela pineti* (L.) Schrank. — Oudemans (1919), Seymour (1929).
Trechispora raduloides (Karst.) Rog. — Basham et al. (1953).

CLAVARIACEAE

Sparassis radicata Weir. — Seymour (1929), Hubert (1931).

HYDNACEAE

Caldesiella viridis (A. & S.) Pat. (*Caldesiella* = *Odontia* = *Tomentella*). — Bisby et al. (1938).
Echinodontium tinctorium E. & E. — Boyce (1948), Westcott (1955).
Grandinia crustosa (P.) Fr. — *Hydnum crustosum* P. — Oudemans (1919), Seymour (1929).
Hydnum auriscalpium L. ex Fr. — Seymour (1929).
Hydnum balsameum Pk. — Seymour (1929), Brown (1935), Weiss and O'Brien (1953), Westcott (1955).
Hydnum farinaceum P. ex Fr. — Seymour (1929).
Mucronella aggregata Fr. — Wehmeyer (1950).
Odontia barba-jovis Fr. — Wehmeyer (1950).
Odontia bicolor (A. & S. ex Fr.) Bres. — *Grandinia mucida* Fr., *Hydnum bicolor* A. & S., *H. subtile* Fr. — Oudemans (1919), Fritz (1923), Miller (1934), Brown (1935), Basham et al. (1953), Nobles (1953), Can. For. Biol. Div. (1951–55), Stillwell (1959).
Odontia crustosa (Pers.) Quél. — Wehmeyer (1950).
Odontia lactea Karst. — Wehmeyer (1950).
Odontia spathulata (Fr.) Litsch. — Wehmeyer (1950).
Oxydontia alboviride (Morg.) L. W. Miller (*Oxydontia* = *Sarcodontia*). — Bisby et al. (1938).
Radulum orbiculare Fr. — *R. hornotium* Karst., *R. sitaneum* Karst. — Oudemans (1919), Wehmeyer (1950), Can. For. Biol. Div. (1951–55), Stillwell (1959).

POLYPORACEAE

Fomes annosus (Fr.) Cke. — *F. annosus* (Fr.) Karst., *Polyporus annosus* Fr., *P. irregularis* Underw. — Schrenk (1900), Rankin (1918), Oudemans (1919), Seymour (1929), Overholts (1953), Can. For. Biol. Div. (1951–55), Spaulding (1956, 1961), Lowe (1957).
Fomes fuliginosus (Scop. ex Fr.) Cke. — *Ischnoderma fuliginosum* (Scop. ex Fr.) Murr. — Oudemans (1919), Seymour (1929).

Fomes pini (Thore ex Fr.) Karst. — *Boletus pini* Brot., *B. pini* Thore, *Cryptoderma yamanoi* Imazeki, *Daedalea pini* Brot. ex Fr., *D. vorax* Harkn., *Fomes abietis* Karst., *F. pini* (Fr.) Karst., *F. pini* (Thore) Lloyd, *Polyporum pini* Thore ex Pers., *Polyporus piceinus* Pk., *Polystictus piceinus* Pk., *Porodaedalea pini* (Thore ex Fr.) Murr., *Trametes pini* Thore ex Fr. — Schrenk (1900), Hedgcock (1912), Rankin (1918), Faull (1919), Oudemans (1919), Anderson et al. (1926), Bisby et al. (1929), Seymour (1929), Wehmeyer (1950), Basham et al. (1953), Weiss and O'Brien (1953), Can. For. Biol. Div. (1951–56), Lowe (1957), Spaulding (1961).

Fomes pini var. *abietis* (Karst.) Overh. — *F. abietis* (Karst.) Overh., *Polyporus piceinus* Pk., *Trametes abietis* Karst., *T. pini* var. *abietis* (Karst.) Schrenk. — Seymour (1929), Baxter (1952a), Overholts (1953).

Fomes pinicola (Sw.) Cke. — *Boletus marginatus* Pers., *B. pinicola* (Sw. ex Fr.), *Fomes ungulatus* Schaeff. ex Sacc., *Polyporus pinicola* Sw. ex Fr., *P.* (*Fomes*) *ponderosus* Schrenk. — Schrenk (1900), Rankin (1918), Faull (1919), Oudemans (1919), Schierbeck (1922), Anderson et al. (1926), Seymour (1929), Bisby et al. (1938), Hirt and Eliason (1938), Basham (1951a), Basham et al. (1953), Overholts (1953), Can. For. Biol. Div. (1951–59), Lowe (1957), Stillwell (1959), Spaulding (1961).

Fomes robustus f. *abietis* Baxter. — Baxter (1952b).

Fomes roseus (A. & S. ex Fr.) Cke. — *Boletus roseus* A. & S., *Fomes roseus* (A. & S. ex Fr.) Karst., *Polyporus carneus* Nees, *Trametes arctica* Berk. — Hedgcock (1914), Rankin (1918), Oudemans (1919), Anderson et al. (1926), Seymour (1929), Wehmeyer (1950), Overholts (1953), Can. For. Biol. Div. (1951–55), Lowe (1957), Spaulding (1961).

Fomes subroseus (Weir) Overh. — *Trametes carnea* Auct. Am. (not *T. carnea* Nees), *T. subrosea* Weir. — Bisby et al. (1938), Baxter (1948), Wehmeyer (1950), Overholts (1953), Can. For. Biol. Div. (1951–55).

Ganoderma applanatum (Pers.) Pat. — *G. applanatum* (S. F. Gray) Pat. — Nobles (1948), Spaulding (1961).

Lenzites saepiaria Wulf. ex Fr. — *Agaricus saepiarius* P., *A. saepiarius* Wulf., *Daedalea saepiaria* Wulf. ex Fr., *D. saepiaria* Sw., *Gloeophyllum hirsutum* Schaeff. ex Murr., *G. saepiarium* Schröt., *Lenzites saepiaria* (Wulf. ex Fr.) Fr., *Merulius saepiarius* P., *Sesia hirsuta* Schaeff. ex. Murr. — Spaulding (1911), Oudemans (1919), Schierbeck (1922), Seymour (1929), Baxter (1950), Wehmeyer (1950), Davidson and Newell (1953), Overholts (1953), Weiss and O'Brien (1953), Can. For. Biol. Div. (1951–55), Stillwell (1959).

Merulius aureus Fr. — *M. vastator* Fr. (non Tode), *M. vastator* Gmel. — Farlow and Seymour (1888), Saccardo (1898), Burt (1917), Oudemans (1919), Seymour (1929).

Merulius americanus Burt. — *Boletus arboreus* Sow., *Gyrophana himantioides* (Fr.) Bourd. & Galz., *Hydnum pinastri* Fr., *Merulius himantioides* Fr., *M. lacrymans* var. *tenuissimum* Berk., *M. papyraceus* Fr., *M. silvester* Falck, *M. tenuis* Pk., *Serpula americana* (Burt) W. B. Cooke, *S. lacrimans* var. *himantioides* W. B. Cooke. — Burt (1917), Oudemans (1919), Seymour (1929), Cooke (1943, 1957), Basham et al. (1953), Davidson and Lombard (1953), Prielipp (1957).

Merulius subaurantiacus Pk. — *M. ceracellus* de Thumen, *M. fugax* Fr., *M. molluscus* Fr. — Burt (1917), Oudemans (1919), Seymour (1929).

Polyporus abietinus Dicks. ex Fr. — *Boletus abietinus* Dicks., *Coriolus abietinus* (Dicks. ex Fr.) Quél., *Daedalea unicolorviolacea* Clem., *Polyporus abietinus* Fr., *Polystictus abietinus* (Fr.) Sacc., *P. abietinus* (Dicks. ex Fr.) Cke., *P. abietinus* (Dicks.) Sacc. & Cub. — Farlow and Seymour (1888), Saccardo (1898), Oudemans (1919), Faull (1919), Schierbeck (1922), Boyce (1948), Wehmeyer (1950), Basham (1951a,b), Baxter (1952a), Basham et al. (1953), Overholts (1953), Can. For. Biol. Div. (1951–55), Findlay (1956), Stillwell (1959).

Polyporus adustus (Willd.) Fr. — *Boletus adustus* Willd., *B. isabellinus* Schw., *Leptoporus adustus* (Willd.) ex Fr., *Polyporus caerulus* Fr. — Baxter (1947), Spaulding (1961).

Polyporus amorphus Fr. — *Boletus abietinus* DC., *B. irregularis* Sow., *B.* (*Poria*) *nitida* A. & S., *Polyporus aureolus* P., *P. irregularis* Berk., *P. roseoporis* Rostk. — Oudemans (1919), Seymour (1929).

Polyporus anceps Pk. — *P. ellisianus* (Murr.) Sacc. & Trott., *Tyromyces anceps* (Pk.) Murr., *T. ellisianus* Murr. — Seymour (1929), Overholts (1953), Can. For. Biol. Div. (1951–1955).

Polyporus balsameus Pk. — *Coriolus balsameus* (Pk.) Murr., *P. crispellus* Pk., *P. floriformis* Quél., *P. tephroleucus* Fr., *Polystictus balsameus* Cke., *P. balsameus* Pk., *Tyromyces balsameus* (Pk.) Murr. — Peck (1878), Farlow and Seymour (1888), Oudemans (1919), Faull (1919, 1922), Schierbeck (1922), McCallum (1925, 1928), Seymour (1929), Hubert (1929,

1931), Spaulding et al. (1932), Kaufert (1935), Heimburger and MacCallum (1940), Spaulding and Hansbrough (1944), Boyce (1948), Baxter (1947, 1952a), Wehmeyer (1950), Basham et al. (1953), Overholts (1953), Can. For. Biol. Div. (1951–55), Spaulding (1961).

Polyporus borealis Fr. — *Spongipellis borealis* (Fr.) Pat. — Oudemans (1919), Seymour (1929), Wehmeyer (1950), Spaulding (1961).

Polyporus caesius (Schrad.) ex Fr. — Oudemans (1919), Seymour (1929), Baxter (1948), Wehmeyer (1950).

Polyporus circinatus Fr. — *P. dualis* Pk., *P. tomentosus* var. *circinatus* (Fr.) Sartory & Maire, *Polystictus circinatus* (Fr.) Sacc., *Trametes circinatus* Fr. — Oudemans (1919), Christensen (1940), Basham et al. (1953), Overholts (1953), Weiss and O'Brien (1953), Can. For. Biol. Div. (1951–55), Spaulding (1961).

Polyporus cutifractus Murr. — *Tyromyces cutifractus* Murr. — Seymour (1929).

Polyporus destructor Fr. (?). — Wehmeyer (1950).

Polyporus fibrillosus Karst. — Wehmeyer (1940, 1950), Baxter (1953).

Polyporus fragilis Fr. — *Spongipellis fragilis* (Fr.) Murr., *S. sensibilis* Murr. — Oudemans (1919), Overholts (1953), Weiss and O'Brien (1953).

Polyporus guttulatus Pk. — *P. maculatus* Pk., *Tyromyces guttulatus* (Pk.) Murr., *T. substipitatus* Murr., *T. tiliophila* Murr. — Seymour (1929), Wehmeyer (1950), Basham et al. (1953), Weiss and O'Brien (1953).

Polyporus hirtus Quél. — *P. hispidellus* Pk., *Scutiger hispidellus* (Pk.) Murr. — Oudemans (1919), Overholts (1953), Weiss and O'Brien (1953).

Polyporus picipes Fr. — *P. fissus* Berk., *P. varius* Fr. — Oudemans (1919), Rhoads (1921), Seymour (1929), Overholts (1953).

Polyporus resinosus Schrad. ex Fr. — *Ochroporus resinosus* Schröt., *Polyporus benzoinus* Wahlenb., *P. resinosus* (Schrad.) Fr. — Oudemans (1919), Seymour (1929), Basham et al. (1953).

Polyporus schweinitzii Fr. — *Phaeolus sistotremoides* (A. & S.) Murr., *Polyporus hispidioides* Pk., *Trametes schweinitzii* Thüm. — Schrenk (1900), Hedgcock (1912), Rankin (1918), Oudemans (1919), Faull (1922), Schierbeck (1922), Anderson et al. (1926), Seymour (1929), Bisby et al. (1938), Wehmeyer (1940, 1950), Pomerleau (1948), Baxter (1952a), Basham et al. (1953), Overholts (1953), Can. For. Biol. Div. (1951–55).

Polyporus tsugae (Murr.) Overh. — *Ganoderma lucidum* (Leys.) Karst., (*Ganoderma = Fomes*), *G. tsugae* Murr. — Oudemans (1919), Boyce (1948), Overholts (1953), Can. For. Biol. Div. (1951–55).

Polyporus violaceus Fr. — *Poria violacea* (Fr.) Cke., *P. violacea* (Fr.) Sacc. — Farlow and Seymour (1888), Oudemans (1919), Seymour (1929).

Polyporus volvatus Pk. — *Cryptoporus volvatus* (Pk.) Hubbard, *C. volvatus* (Pk.) Shear, *C. volvatus* Shear, *Fomes volvatus* Cke., *F. volvatus* Pk., *Polyporus inflatus* Ell. & Martindale, *P. obvolutus* Berk. & Cke., *P. volvatus* Berk. & Cke. — Farlow and Seymour (1888), Saccardo (1898), Murrill (1903), Schmitz (1923), Seymour (1929), Basham et al. (1953), Overholts (1953).

Poria asiatica (Pilát) Overh. — Smerlis (1961).

Poria candidissima (Schw.) Cke. — Wehmeyer (1950).

Poria cocos (Schw.) Wolf. — *Lentinus tuber regium* Fr., *Lycoperdon cervinum* Walt., *L. solidum* Clayton, *Pachyma cocos* (Schw.) Fr., *P. coniferarum* Horaninow, *P. pinetorum* Horaninow, *P. solidum* Oken., *Sclerotium cocos* (Schw.) Fr., *S. giganteum* MacBride, *Tuckhaus rugosus* Rafin. — Wolf (1922, 1926), Baxter (1947), Gilbertson (1956), Prielipp (1957), Lowe (1958).

Poria crustulina Bres. — *P. flavicans* Karst. — Wehmeyer (1950).

Poria ferruginosa (Schrad. ex Fr.) Karst. — *Polyporus ferruginosus* Fr. — Wehmeyer (1950), Spaulding (1961).

Poria lenis Karst. — *Physiosporus lenis* Karst., *Polyporus vulgaris* Fr. — Baxter (1935).

Poria monticola Murr. — *P. microspora* Overh. — Stillwell (1959), Boyce (1961).

Poria subacida (Pk.) Sacc. — *Polyporus subacidus* Pk. — Schrenk (1900), Oudemans (1919), Schierbeck (1922), Faull (1922), McCallum (1928), Seymour (1929), Spaulding et al. (1932), Kaufert (1935), Heimburger and McCallum (1940), Spaulding and Hansbrough (1944), Pomerleau (1948), Baxter (1952a), Basham et al. (1953), Weiss and O'Brien (1953), Can. For. Biol. Div. (1951–58), Westcott (1955), Spaulding (1961).

Poria subincarnata (Pk.) Murr. — Wehmeyer (1940, 1950).

Poria taxicola (Pers.) Bres. — *Merulius ravenellii* Berk., *Polyporus haematodes* Rostk. — Baxter (1937).

Poria vaporaria Fr. — *Polyporus vaporarius* Fr., *P. vaporarius* (P.) ex Fr., *Poria vaporaria* P., *P. vaporaria* (P. ex Fr.) Cke. — Oudemans (1919), Seymour (1929), Weiss and O'Brien (1953), Spaulding (1961).

Poria vulgaris Fr. sensu Romell. — Wehmeyer (1950).

Poria weirrii Murr. — Seymour (1929).

Poria xantha (Fr.) Lind. — *P. luteo-alba* Karst. — Baxter (1935).

Trametes heteromorpha (Fr.) Bres. — *Coriolus hexagoniformis* Murr., *Daedalea heteromorpha* Fr., *Lenzites heteromorpha* Fr., *Trametes lacerata* Lloyd. — Seymour (1929), Overholts (1953), Weiss and O'Brien (1953).

Trametes sepium Berk. — Wehmeyer (1950).

Trametes serialis Fr. — Oudemans (1919), Seymour (1929).

Trametes variiformis Pk. — *Polyporus variiformis* Pk. — Baxter (1935).

AGARICACEAE

Agaricus mitis Pers. — *Pleurotus mitis* Fr., *P. mitis* Quél., *P. mitis* (P.) Sacc. — Farlow and Seymour (1888), Saccardo (1901), Oudemans (1919), Seymour (1929).

Armillaria mellea (Vahl. ex Fr.) Quél. — *Agaricus melleus* Vahl., *A. polymyces* Pers., *A. putridus* Scop., *Armilaria mellea* Vahl., *Armillaria solipides* Pk., *A. melleorubens* Ber. & Curt. — Schrenk (1900), New York Bot. Gar. (1914), Oudemans (1919), Faull (1919, 1922), Schierbeck (1922), Whitney (1952), Basham et al. (1953), Weiss and O'Brien (1953), Christensen and Hodson (1954), Can. For. Biol. Div. (1951–59), Westcott (1955), Spaulding (1961), Can. For. Ent. Path. Br. (1961).

Clitocybe sulphurea Pk. — Farlow and Seymour (1888), Saccardo (1898), Oudemans (1919), Seymour (1929).

Collybia stipitaria Fr. — *C. stipitaria* (Fr.) Sacc. — Oudemans (1919), Bisby et al. (1929).

Crepidotus herbarum Pk. — Wehmeyer (1950).

Entoloma adirondackense Murr. — Seymour (1929).

Flammula subviridis Murr. (*Flammula* = *Pholiota*). — *Gymnopilus subviridis* Murr. — Seymour (1929).

Hypholoma capnoides (Fr.) Sacc. — *Agaricus capnoides* Fr. — Oudemans (1919), Seymour (1929).

Lactarius subdulcis Fr. var. *oculatus* Pk. — Seymour (1929).

Lentinus strigosus (S.) Fr. — *L. strigosus* Fr. — Farlow and Seymour (1888), Saccardo (1898), Oudemans (1919), Seymour (1929).

Marasmius androsaceus Fr. — *Agaricus androsaceus* Scop. — Oudemans (1919), Seymour (1929).

Marasmius campanellus (Pk.) Atk. & House. — Bisby et al. (1938).

Marasmius filipes Pk. — Farlow and Seymour (1888), Saccardo (1898), Seymour (1929).

Mycena avellana Murr. — *Prunulus avellaneus* Murr. — Seymour (1929).

Mycena elegantula Pk. — Smith and Wehmeyer (1936), Wehmeyer (1950).

Mycena flavifolia Pk. — Seymour (1929).

Mycena fuliginosa Murr. — *Prunulus fuliginosus* Murr. — Seymour (1929).

Omphalia aurantiaca Pk. (*Omphalia* = *Omphalina*). — *Omphalopsis aurantiaca* (Pk.) Murr. — Seymour (1929).

Omphalia rugosodisca Pk. — Wehmeyer (1950).

Paxillus curtisii B. — Farlow and Seymour (1929), Seymour (1929).

Pholiota adiposa Fr. — Oudemans (1919), Lorenz and Christensen (1937), Can. For. Biol. Div. (1951–55).

Pholiota flammans Fr. — *Agaricus flammans* Fr., *Geopetalum blakei* (B. & C.) Murr., *Hypodendrum flammans* (Batsch ex Fr.) Murr., *Pleurotus blakei* B. & C. — Oudemans (1919), Seymour (1929).

Pholiota spectabilis (Fr.) Quél. — Wehmeyer (1950).

Pleurotus serotinus (Schräd.) Fr. — Stillwell (1959).

Schizophyllum commune Fr. — *S. communis* Fr. — Oudemans (1919), Can. For. Biol. Div. (1951–55), Stillwell (1959).

Xeromphalina campanella (Fr.) Kühner & Maire — *Agaricus campanella* Batsch, *A. fragilis* Schaeff., *Omphalia campanella* Fr., *O. campanella* (Batsch ex Fr.) Sacc. — Oudemans (1919), Nobles (1948), Singer (1949), Basham et al. (1953), Smith (1954), Can. For. Biol. Div. (1951–55).

DEUTEROMYCETES (FUNGI IMPERFECTI)

PHOMACEAE

Ceutosphora abietina Ell. & Ev. — Seymour (1929).

Diplodia lignicola Pk. — *Diplodiella lignicola* Sacc., *D. lignicola* (Pk.) Sacc. — Farlow and Seymour (1888), Seymour (1929).

Diplodina parasitica (Horst.) Prill. — Wehmeyer (1950).

Fusicoccum abietinum Prill. & Delacr. — *Dothiorella pitya* Prill. & Delacr., *Phoma abietina* Htg., *Phomopsis abietina* (Htg.) Wils. & Hahn. — Hartig (1894), Saccardo (1898), Oudemans (1919), Spaulding (1956), Can. For. Biol. Div. (1956–59), Can. For. Ent. Path. Br. (1960–61).

Hendersonia falcata Ell. & Ev. (*Hendersonia* = *Stagnospora*). — Oudemans (1919).

Phoma balsamea Brem. — Saccardo (1898).

Phomopsis pseudotsugae M. Wilson. — Seymour (1929).

Rhizosphaera abietis Magn. & Har. — Oudemans (1919), Can. For. Biol. Div. (1955).

Sphaeropsis abietis Povah. — Weiss and O'Brien (1953).

LEPTOSTROMACEAE

Discosia artocreas (Tode) ex Fr. — *Diplodina aquilina* Fr. (?), *Discosia faginea* Lib. — Saccardo (1901), Oudemans (1919), Seymour (1929).

Leptostromella conigena Dearn. — Seymour (1929).

Rhizothyrium abietis Naum. — Can. For. Biol. Div. (1955).

Sacidium pini (Cda.) Fr. — *Leptothyrium pini* Sacc. — Farlow and Seymour (1888), Oudemans (1919), Seymour (1929).

ZYTHIACEAE

Zythia resinae (Ehr. ex Fr.) Karst. — Can. For. Ent. Path. Br. (1960).

MELANCONIACEAE

Coryneum abietinum Ell. & Ev. — Oudemans (1919), Seymour (1929).

Cryptosporium falcatum Dearn. (*Cryptosporium* = *Cryptosporiopsis*). — Seymour (1929).

Cryptosporium macrospermum Pk. — Weiss and O'Brien (1953), Westcott (1955).

Cryptosporium noveboracense B. & C. — Farlow and Seymour (1888), Saccardo (1898), Seymour (1929).

Gloeosporium balsameae J. J. Davis. — Seymour (1929), Weiss and O'Brien (1953), Westcott (1955).

Kabietella balsameae (J. J. Davis) Arx. — Can. For. Ent. Path. Br. (1961).

Pestalotia camptosperma Pk. (*Pestalotia* = *Pestalozzia*). — *Coryneum bicorne* Rostr., *Monochaetia camptosperma* (Pk.) Sacc., *Scolecosporium camptosperma* (Pk.) Höhn., *Toxosporium abietinum* Vuill., *T. camptospermum* (Pk.) Maubl. & Vuill. — Farlow and Seymour (1888), Saccardo (1898), Seymour (1929), Wehmeyer (1950), Guba (1961).

Pestalotia scirrofaciens N. A. Brown. — Brown (1920), Guba (1961).

Pestalotia hartigii Tub. — *P. tumefaciens* Henn. — Saccardo (1898), Oudemans (1919), Brown (1920), Guba (1961), Spaulding (1961).

Steganosporium cenangioides Ell. & Rothr. — *Myxocyclus cenangioides* (Ell. & Rothr.) Petrak. — Farlow and Seymour (1888), Saccardo (1898), Seymour (1929).

MONILIACEAE

Cephalosporium album Preuss. — Christensen (1937), Baxter (1952a), Christensen and Hodson (1954).

Fusoma parasiticum Tub. — *F. pini* Htg. — Manshard (1927), Spaulding (1956).

DEMATIACEAE

Phragmocephala minima Mason & Hughes. — Can. For. Biol. Div. (1951–55).

TUBERCULARIACEAE

Epicoccum diversisporum Preuss. — Oudemans (1919), Seymour (1929).

Trimmatostroma abietina Doherty. — Oudemans (1919), Seymour (1929).

STILBACEAE

Stilbum glomerulisporum Ell. & Ev. — Seymour (1929).
Stilbum resinarium Pk. — Oudemans (1919), Seymour (1929).

(PSEUDOSACCHAROMYCETES)

Pullularia pullulans (de Bary) Berk. — Can. For. Biol. Div. (1951–55).

APPENDIX II

A TENTATIVE LIST OF INSECTS ASSOCIATED
WITH BALSAM FIR

This list contains the names and synonyms of species of insects found on balsam fir (*Abies balsamea* (L.) Mill.) by various authors, who have often followed different taxonomic rules. No attempt has been made to examine the validity of reported occurrences, to evaluate their significance in relation to balsam fir, or to clarify the numerous taxonomic problems involved. In general, species names have been used as reported in Craighead (1950) or in the most recently published reference. The arrangement follows the classification system by Brues et al. (1954). Synonyms are taken largely from the checklist of North American Lepidoptera by McDunnough (1938–39), and the *Coleopterum Catalogus* by Junk and Schenkling (1910–40). Following each species' name and its synonyms is a list of references which treat the taxonomy of the species or relate the insect to balsam fir.

HEMIPTERA

APHIDIDAE

Cinara curvipes (Patch). — Doane et al. (1936) Craighead (1950).
Mindarus abietinus (Koch), balsam twig aphid. — Peirson (1927), Craighead (1950), Anderson (1960).

PHYLLOXERIDAE

Adelges piceae (Ratz.), balsam woolly aphid. — *Chermes piceae* Ratz., *Dreyfusia piceae* (Ratz.). — Patch (1909), Balch (1934), MacAloney (1935), Doane et al. (1936), Brower (1947), Craighead (1950), Balch (1952), Can. For. Biol. Div. (1950–59), Varty (1956), Anderson (1960), Can. For. Ent. Path. Br. (1960–61).

DIASPIDIDAE

Aspidiotus ithacae (Ferris), hemlock scale. — *A. abietis* Schr. — Felt (1905–6), Craighead (1950).

LEPIDOPTERA

GELECHIIDAE

Recurvaria obliquistrigella Cham. — *Gelechia obliquistrigella* Cham. — Packard (1890), McDunnough (1938–39).

BLASTOBASIDAE

Holcocera chalcofrontella Clem. — Felt (1905–6), McDunnough (1938–39).

TORTRICIDAE

Acleris variana (Fern.) black-headed budworm. — *Peronea variana* Fern. — Balch (1932), Doane et al. (1936), McDunnough (1938–39), Craighead (1950), Raizenne (1952), Can. For. Biol. Div. (1950–59), Beckwith and Ewan (1956), Anderson (1960), Can. For. Ent. Path. Br. (1960–61).

Argyrotaenia velutinana (Wlk.). — *A. incertana* Clem., *A. triferana* (Wlk.), *Lophoderus velutinana* (Wlk.). — Packard (1890), Leonard (1928), McDunnough (1938–39), Craighead (1950).

Choristoneura fumiferana (Clem.), spruce budworm. — *Archips fumiferana* Clem., *A. nigridia* Rob., *Cacoecia fumiferana* Clem., *Harmologa fumiferana* Clem., *Tortrix fumiferana* Clem., *T. nigridia* Robinson. — Packard (1890), Felt (1905–6), Meyrick (1913), Forbes (1923), Peirson (1923, 1927), Graham (1928), Doane et al. (1936), McDunnough (1938–39), Craighead (1950), Raizenne (1952), Can. For. Biol. Div. (1950–55), Beckwith and Ewan (1956), Waters (1956), Freeman (1958), Anderson (1960), Can. For. Ent. Path. Br. (1960–61).

Tortrix afflictana Wlk. — *Archips afflictana* Wlk., *Loxotoemia afflictana* Wlk., *T. fuscolineana* Clem. — Packard (1890), Felt (1905–6), McDunnough (1938–39).

Tortrix packardiana Fern. — Packard (1890), Felt (1905–6), Doane et al. (1936), McDunnough (1938–39), Raizenne (1952).

PYRALIDIDAE

Dioryctria abietivorella Grt., cone pyralid. — *D. abietella* D. & S., *D. abietivorella* D. & S. — Doane et al. (1936), McDunnough (1938–39).

Dioryctria reniculella (Grt.), spruce needle worm. — McDunnough (1938–39), Can. For. Biol. Div. (1950–59), Can. For. Ent. Path. Br. (1960–61).

GEOMETRIDAE

Abbottana clemataria (J. E. Smith). — *A. clemataria* A. & S., *A. transducens* Wlk., *A. transferens* Wlk., *A. transfingens* Wlk. — McDunnough (1938–39), Raizenne (1952).

Anacamptodes vellivolata Hlst. — McDunnough (1938–39), Raizenne (1952).

Caripeta divisata Wlk. — *C. albopunctata* Morr. — McDunnough (1938–39), McGuffin (1955b).

Cingilia catenaria (Dru.), chain spotted geometer. — *C. devinctaria* Gn., *C. humeralis* Wlk. — Leonard (1928), McDunnough (1938–39), Craighead (1950).

Ectropis crepuscularia (Schiff.). — *E. cineraria* Wlk., *E. cunearia* D'Urban, *E. intrataria* Wlk., *E. occiduaria* Gn., *E. signaria* Wlk., *E. spatiosaria* Wlk. — McDunnough (1938–39), Raizenne (1952).

Eufidonia notataria (Wlk.). — *E. quadripunctata* Morr. — Leonard (1928), McDunnough (1938–39), Raizenne (1952).

Eupithecia filmata Pears. — McDunnough (1938–39), Raizenne (1952).

Eupithecia luteata Pack. — *E. catskillata* Pears., *E. fasciata* Tayl., *Tephroclystis luteata* Pack. — Packard (1890), Felt (1905–6), Peirson (1927), McDunnough (1938–39), Raizenne (1952).

Hydriomena divisaria Wlk. — McDunnough (1938–39), Raizenne (1952).

Lambdina fiscellaria fiscellaria (Gn.), hemlock looper. — *Ellopia fiscellaria* Gn. — McDunnough (1938–39), Craighead (1950), Raizenne (1952), Schaffner (1952), Can. For. Biol. Div. (1950–55), Anderson (1960), Can. For. Ent. Path. Br. (1960–61).

Melanolophia canadaria (Gn.). — McDunnough (1938–39), McGuffin (1955b).

Nemoria mimosaria Gn. — *Aplodes coniferaria* Pack., *N. approximaria* Pack., *N. coniferaria* Pack., *N. latiaria* Pack., *N. tractaria* Wlk., *N. venustus* Walsh. — Packard (1890), McDunnough (1938–39), Raizenne (1952).

Nepytia canosaria (Wlk.). — *N. pulchraria* Pack. — McDunnough (1938–39), Craighead (1950), Raizenne (1952), Anderson (1960).

Nepytia semiclusaria Wlk., whitish black-marked spanworm. — *N. fumosaria* Stkr. — Packard (1890), Felt (1906), Doane et al. (1936), McDunnough (1938–39).

Nyctobia limitaria Wlk. — *N. lobophorata* Wlk., *N. fusifasciata* Wlk., *N. longipennis* Wlk. — McDunnough (1938–39), Raizenne (1952).

Paraphia piniata Pack. — *P. guttata* Hlst. — McDunnough (1938–39), Raizenne (1952).

Paraphia subatomaria Wood, brown spanworm. — *P. deplanaria* Gn., *P. exsuperata* Wlk., *P. fidoniaria* Wlk., *P. impropriata* Wlk., *P. mammuraria* Gn. — Packard (1890), Felt (1905–6), McDunnough (1938–39).

Paraphia unipunctata Haw. — Doane et al. (1936), McDunnough (1938–39).

Pero morrisonarius Hy. Edw. — McDunnough (1938–39), Raizenne (1952).

Prochoerodes transversata (Dru.). — *P. contingens* Wlk., *P. goniata* Gn. — McDunnough (1938–39), Raizenne (1952).

Protoboarmia porcelaria Gn. — *P. filaria* Wlk., *P. maestosa* Hlst. — McDunnough (1938–39), Raizenne (1952).

Semiothisa bicolorata Fabr. — *Macaria praeatomata* Haw. var. *bisignata* Wlk., *S. consepta* Wlk., *S. consimilata* Zell., *S. grassata* Hlst. — Felt (1906), McDunnough (1938–39).

Semiothisa bisignata Wlk. — *S. galbineata* Zell. — Packard (1890), Peirson (1927), McDunnough (1938–39).

Semiothisa granitata (Gn.). — McDunnough (1938–39), Forbes (1948), Craighead (1950), Raizenne (1952), McGuffin (1955b), Can. For. Ent. Path. Br. (1960), Anderson (1960).

LASIOCAMPIDAE

Tolype laricis (Fitch). — *T. minuta* Grt. — McDunnough (1938–39), Raizenne (1952).

SPHINGIDAE

Sphinx gordius Cram. — *S. borealis* Clark, *S. campestris* McD., *S. coxeyi* Cad., *S. poecila* Steph. — McDunnough (1938–39), McGugan, chmn. (1958).

NOCTUIDAE (PHALAENIDAE, AGROTIDAE)

Anomogyna elimata Gn. — *A. badicollis* Grt., *A. grisatra* Sm. — McDunnough (1938–39), Raizenne (1952), Prentice, chmn. (1962).

Anomogyna perquiritata Morr. — *A. baileyana* Grt., *A. beddeki* Hamp., *A. clarkei* Benj., *A. partita* McD. — McDunnough (1938–39), Prentice, chmn. (1962).

Aplectoides condita Gn. — *A. trabalis* Grt. — McDunnough (1938–39), Prentice, chmn. (1962).

Elaphria versicolor Grt. — *Oligia* (*Olygia*) *versicolor* Grt. — Packard (1890), Felt (1905–6), Raizenne (1952), Prentice, chmn. (1962).

Epizeuxis aemula (Hbn.). — *Camptylochila aemula* Hbn., *C. concisa* Wlk., *C. herminioides* Wlk., *C. mollifera* Wlk., *C. undulalis* Steph. — McDunnough (1938–39), Raizenne (1952), Prentice, chmn. (1962).

Feralia jocosa Gn. — *F. furtiva* Sm. — McDunnough (1938–39), Raizenne (1952), Prentice, chmn. (1962).

Lithophane innominata Sm. — *Graptolitha innominata* Sm. — McDunnough (1938–39), Prentice, chmn. (1962).

Palthis angulalis Hbn. — *P. aracinthusalis* Wlk. — McDunnough (1938–39), Raizenne (1952), Prentice, chmn. (1962).

Panthea acronyctoides Wlk. — *P. albosuffusa* McD., *P. leucomelana* Morr. — Felt (1905–6), McDunnough (1938–39), Prentice, chmn. (1962).

Phlogophora periculosa Gn. — McDunnough (1938–39), Prentice, chmn. (1962).

Orthosia hibisci Gn. — *O. brucei* Sm., *O. confluens* Morr., *O. inflava* Sm., *O. inherita* Sm., *O. latirena* Dod., *O. malora* Sm., *O. nubilata* Sm., *O. proba* Sm., *O. quinquefasciata* Sm. — McDunnough (1938–39), Prentice, chmn. (1962).

Syngrapha alias Ottol. — *Autographa alias* Ottol., *A. interalia* Ottol. — McDunnough (1938–39), McGuffin (1955a), Prentice, chmn. (1962).

Syngrapha rectangula Kby. — *Autographa rectangula* Kby., *A. demaculata* Strand, *A. mortuorum* Gn., *A. nargenta* Ottol. — McDunnough (1938–39), Prentice, chmn. (1962).

Syngrapha selecta Wlk. — *Autographa selecta* Wlk., *A. viridisigma* Grt., *A. viridisignata* Grt. — McDunnough (1938–39), McGuffin (1955a), Prentice, chmn. (1962).

Zanclognatha minoralis Sm. — *Epizeuxis minoralis* Sm. — McDunnough (1938–39), Prentice, chmn. (1962).

ARCTIIDAE

Clemensia albata Pack. — *C. albida* Wlk., *C. cana* Wlk., *C. irrorata* Hy. Edw. — McDunnough (1938–39), McGugan, chmn. (1958).

Hypoprepia fucosa Hbn. — *H. inornata* Ottol., *H. subornata* N. & D., *H. tricolor* Fitch. — McDunnough (1938–39), Raizenne (1952), McGugan, chmn. (1958).

Lexis bicolor Grt. — *L. argillacea* Pack. — McDunnough (1938–39), Raizenne (1952), McGugan, chmn. (1958).

LYMANTRIIDAE (LIPARIDAE)

Hemerocampa leucostigma (J. E. Smith), white-marked tussock moth. — *H. borealis* Fitch, *H. intermedia* Fitch, *H. leucographa* Gey., *H. leucostigma* (A. & S.), *H. obliviosa* Hy. Edw., *Orgyia leucostigma* A. & S. — Packard (1890), Felt (1905–6), McDunnough (1938–39), Craighead (1950), Can. For. Biol. Div. (1950–59), Can. For. Ent. Path. Br. (1960), Prentice, chmn. (1962).

Olene plagiata Wlk. — *O. pini* Dyar, *O. pinicola* Dyar, *Parorgyia plagiata* Wlk. — McDunnough (1938–39), Craighead (1950), Anderson (1960), Prentice, chmn. (1962).

Olene vagans B. & McD. — *O. grisea* B. & McD., *O. willingi* B. & McD., *Parorgia vagans* B. & McD. — McDunnough (1938–39), Raizenne (1952), Prentice, chmn. (1962).

Orgyia antiqua L., rusty tussock moth. — *Notolophus antiqua* L., *N. antiqua nova* Fitch, *N. nova* Fitch. — McDunnough (1938–39), Raizenne (1952), Can. For. Ent. Path. Br. (1961), Prentice, chmn. (1962).

Porthetria dispar (L.), gypsy moth (*Porthetria = Ocneria = Liparis*). — McDunnough (1938–39), Conklin (1952).

DIPTERA

CECIDOMYIIDAE (ITONIDIDAE)

Dasyneura balsamicola (Lint.), balsam gall midge. — *Cecidomyia balsamicola* Lint., *Itonida balsamicola* (Lint.). — Felt (1905–6, 1924), Peirson (1927), Doane et al. (1936), Craighead (1950), Giese and Benjamin (1959), Waters and Mook (1959), Can. For. Ent. Path. Br. (1960–61).

COLEOPTERA

BUPRESTIDAE

Buprestis aurulenta L. — Schenck (1909).

Buprestis maculativentris Say. — *B. maculiventris* Gemm. & Haw., *B. maura* Cast. & Gory, *B. sexnotata* Cast. & Gory. — Felt (1906, 1924), Peirson (1927), Doane et al. (1936), Junk and Schenkling (1910–40), Wilson (1961b).

Chrysobothris pusila Cast. — *C. aegrota* Dej., *C. biguttata* Cast. & Gory, *C. strangulata* Melsh. — Doane et al. (1936), Junk and Schenkling (1910–40).

Chrysobothris scabripennis Cast. — *C. consimilis* Dej., *C. proxima* Kby., *C. scabra* Gory, *C. scabripennis* Cast. & Gory. — Junk and Schenkling (1910–40), Wilson (1961b).

Dicerca tenebrosa (Kby.), flatheaded borer. — Doane et al. (1936).

Dicerca tuberculata (Chev.), flatheaded borer. — *D. hilaris* Lec. (?), *D. manca* Lec. (?). — Doane et al. (1936), Junk and Schenkling (1910–40).

Melanophila acuminata (DeG.), flatheaded borer. — Doane et al. (1936), Craighead (1950).

Melanophila drummondi (Kby.), flatheaded borer. — *M. guttulata* Mannerh. (non Gabler), *M. umbellatarum* Kby. — Doane et al. (1936), Junk and Schenkling (1910–40), Craighead (1950), Anderson (1960).

Melanophila drummondi abies C. & K., flatheaded borer. — Doane et al. (1936).

Melanophila fulvoguttata (Harr.), hemlock borer. — *M. actospilota* Cast. & Gory, *M. cauta* Dej., *M. guttulata* Hamilton, *M. subguttata* Dej. l. c. — Doane et al (1936), Junk and Schenkling (1910–40), Craighead (1950).

MELANDRYIDAE (SERROPALPIDAE)

Serropalpus barbatus (Schall.), blazed tree borer. — *S. biguttatus* Schellenberg, *S. striatus* Hellenius. — Felt (1905–6, 1924), Schenck (1909), Peirson (1927), Junk and Schenkling (1910–40), Craighead (1950).

Serropalpus substriatus (Hald.). — Junk and Schenkling (1910–40), Wilson (1961b).

CERAMBYCIDAE

Anoplodera canadensis (Oliv.). — *Leptura canadensis* Ol., *L. erythroptera* Kby., *L. tenuicornis* Hald. — Junk and Schenkling (1910–40), Wilson (1961b).

Evodinus monticola (Rand.). — *Pachyta monticola* Rand. — Schenck (1909), Junk and Schenkling (1910–40).

Leptura tibialis Lec. (*Leptura = Anoplodera*). — Doane et al. (1936).

Monochamus marmorator Kby., balsam fir sawyer. — *M. acutus* Lacord, *M. fautor* Lec. — Felt (1906), Peirson (1923, 1927), Swaine et al. (1924), Doane et al. (1936), Junk and Schenkling (1910–40), Craighead (1950), Belyea (1952a,b), Anderson (1960), Wilson (1961b).

Monochamus notatus (Dru.), northeastern sawyer. — *M. confusor* Kby., *M. dentator* Westw., *M. varius* Fröhl. — Packard (1890), Felt (1905–6), Doane et al. (1936), Junk and Schenkling (1910–40), Wilson (1961b).

Monochamus scutellatus (Say), white-spotted sawyer. — *M. mutator* Lec., *M. resutor* Kby. — Felt (1906), Peirson (1923, 1927), Junk and Schenkling (1910–40), Belyea (1952a,b), Anderson (1960), Wilson (1961b).

Monochamus titillator (Fab.), southern pine sawyer. — *M. dentator* Haw. — Felt (1905–6), Peirson (1923, 1927), Junk and Schenkling (1910–40), Craighead (1950).

Stenocorus lineatus Oliv., the ribbed pine borer. — *Rhagium cephalotes minor* Voet., *R. exile* Gmel., *R. indagator* F., *R. indagatrix* Latr., *R. inquisitor* L., *R. inquisitor* var. *lineatum* Ol., *R. lineatum* Oliv., *R. minutum* F., *R. nubecula* Bergstr. — Packard (1890), Felt (1906), Junk and Schenkling (1910–40), Craighead (1950), Anderson (1960).

Semanotus ligneus (Fab.), the cedar tree borer. — *Hylotrupes ligneus* Fab. — Felt (1906), Junk and Schenkling (1910–40), Craighead (1950).

Tetropium cinnamopterum Kby., the eastern large borer. — Craighead (1950), Belyea (1952a,b).

Xylotrechus undulatus (Say). — *X. undatus* Kby. — Junk and Schenkling (1910–40), Wilson (1961b).

CURCULIONIDAE

Hylobius pales (Hbst.). — *H. assimilis* Roel, *H. macellus* Germ. — Carter (1916), Doane et al. (1936), Craighead (1950).

Hylobius pinicola (Couper). — *H. confusus* Payk., *H. excavatus* Laich., *H. inaccessus* Schrank, *H. picatus* Ol., *H. piceus* De Geer, *H. pineti* F. — Junk and Schenkling (1910–40), Can. For. Biol. Div. (1957), Smerlis (1957, 1961).

Pissodes dubius Rand., balsam weevil. — Hopkins (1904), Schenck (1909), Swaine et al. (1924), Peirson (1927), Junk and Schenkling (1910–40), Craighead (1950), Belyea (1952a,b).

Pissodes rotundatus Lec. — Craighead (1950).

Pissodes similis Hopk. — Peirson (1927), Craighead (1950).

Pissodes strobi (Peck), white pine weevil. — Packard (1890), Junk and Schenkling (1910–40), Craighead (1950).

SCOLYTIDAE (IPIDAE)

Cryphalus balsameus Hopk. — Swaine (1917–18), Doane et al. (1936), Craighead (1950).

Crypturgus atomus Lec. — Packard (1890), Swaine (1917–18), Felt (1924), Doane et al. (1936).

Dryocoetes confusus Sw. — *D. abietis* Hopk. — Swaine (1917–18), Doane et al. (1936).

Dryocoetes pseudotsugae Sw. — Felt (1924).

Ips borealis Sw. — Swaine (1917–18), Doane et al. (1936).

Orthotomicus caelatus (Eichh.). — *Ips caelatus* Eichh., *Tomicus caelatus* Eichh., *T. decretus* Eichh., *T. xylographus* Fitch, *Xyleborus caelatus* Eichh., *X. caelatus* Zimm., *X. vicinus* Lec. — Packard (1890), Felt (1906), Junk and Schenkling (1910–40), Chamberlin (1939), Craighead (1950).

Pityophthorus augustus Blkm. — Craighead (1950).

Pityophthorus balsameus Blkm. — Craighead (1950).

Pityophthorus cariniceps Lec. — *P. canadensis* Sw. — Doane et al. (1936), Chamberlin (1939), Craighead (1950).

Pityophthorus consimilis Lec. — *P. canadensis* Sw., *P. granulatus* Sw. — Swaine (1917–18), Doane et al. (1936), Chamberlin (1939), Craighead (1950).

Pityophthorus nudus Sw. — Craighead (1950).

Pityophthorus opaculus Lec. — Swaine (1917–18), Doane et al. (1936), Craighead (1950).

Pityophthorus patchi Blkm. — Craighead (1950).

Pityophthorus puberulus Lec. — *Cryphalus puberulus* L., *P. infans* Eichh. — Swaine (1917–18), Doane et al. (1936), Junk and Schenkling (1910–40), Craighead (1950).

Pityophthorus pulchellus Eichh. — *P. hirticeps* Lec., *P. pusio* Lec. — Doane et al. (1936), Junk and Schenkling (1910–40), Craighead (1950).

Pityokteines sparsus (Lec.), balsam bark beetle. — *Ips sparsus* Lec., *Pityogenes sparsus* Eichh. & Schwarz, *P. balsameus* Lec., *Pityophthorus sparsus* Lec., *Tomicus balsameus* Lec., *Xyleborus balsameus* Lec. — Hopkins (1904), Felt (1905, 1924), Schenck (1909), Swaine (1918), Peirson (1923, 1927), Doane et al. (1936), Chamberlin (1939), Junk and Schenkling (1910–40), Craighead (1950), Belyea (1952a,b), Anderson (1960).

Polygraphus rufipennis (Kby.), spruce bark beetle. — *Apate* (*Lepisomus*) *nigriceps* Kby.,

Hylesinus rufipennis Mann., *P. rufipennis* Lec., *P. saginatus* Mann. — Felt (1906), Junk and Schenkling (1910–40), Craighead (1950).

Scolytus piceae (Sw.) — *Eccoptogaster piceae* Sw. — Swaine (1917–18), Felt (1924), Blackman (1934), Doane et al. (1936), Craighead (1950).

Scolytus subscaber Lec. — *Eccoptogaster subscaber* Lec. — Hopkins (1904), Swaine (1917–18).

Trypodendron bivittatum (Kby.), ambrosia beetle. — *Apate bivittata* Kby., *A. rufitarsus* Kby., *Bostrichus cavifrons* Mann., *B. melanocephalus* F., *T. bivittatum* Provancher, *T. lineatum* (Oliv.), *T. lineatum* Eichh., *T. vittiger* Eichh., *Xyloterus bivittatus* Kby., *X. bivittatus* Mann. — Packard (1890), Felt (1905–6), Swaine (1917–18), Doane et al. (1936), Junk and Schenkling (1910–40), Chamberlin (1939), Belyea (1952a,b).

Xylechinus americanus Blkm. — Schenck (1909).

HYMENOPTERA

XYELIDAE

Pleroneura borealis Felt, balsam bud-mining sawfly. — Can. For. Biol. Div. (1950–59), Can. For. Ent. Path. Br. (1960–61).

DIPRIONIDAE

Diprion hercyniae (Htg.), European spruce sawfly. — *Gilpinia hercyniae* Htg., *Lophyrus hercyniae* Htg. — Balch (1942), Muesebeck et al. (1951), MacAloney and Dowden (1952).

Neodiprion abietis Harr., balsam fir sawfly. — *Lophyrus abietis* Harr. — Packard (1890), Felt (1906), Doane et al. (1936), Craighead (1950), Muesebeck et al. (1951), Can. For. Biol. Div. (1950–59), Waters and Mook (1958), Anderson (1960), Can. For. Ent. Path. Br. (1960–61).

SIRICIDAE

Sirex cyaneus Fab., blue horntail. — *Paururus cyaneus* Fabr., *S. duplex* Shuckard, *S. hirsutus* Kby., *S. juvencus* L. (?), *S. varipes* Walker. — Felt (1906), Muesebeck et al. (1951), Belyea (1952a,b), Stillwell (1960), Wilson (1961b).

Urocerus albicornis (Fab.), white-horned horntail. — *Sirex albicornis* Fab., *S. stephensi* Kby., *U. abdominalis* Harr. — Muesebeck et al. (1951), Belyea (1952a,b), Wilson (1961b).

Urocerus flavicornis (F.). — Craighead (1950), Wilson (1961b).

Xeris spectrum (L.). — *Ichneumon spectrum* L., *Sirex melancholicus* Westwood, *Urocerus caudatus* Cresson. — Muesebeck et al. (1951), Stillwell (1960).

TORYMIDAE

Megastigmus specularis Walley, balsam fir seed chalcid. — Walley (1932), Muesebeck et al. (1951), Hedlin (1956).

FORMICIDAE

Camponotus herculeanus L., large carpenter ant. — *C. herculeanus pennsylvanicus* (DeG.). — Felt (1905), Peirson (1927), Graham (1939), Craighead (1950), Schread (1952).

LITERATURE CITED

Chapter 1. Botanical Foundations

Ahlgren, C. E. 1957. Phenological observations of nineteen native tree species. Ecology 38: 622–628.

Anderson, A. P. 1897. Comparative anatomy of the normal and diseased organs of *Abies balsamea* affected with *Aecidium elatinum*. Bot. Gaz. 24:309–344.

Anderson, C. H. 1932. Root development in seedlings of Norway pine, balsam fir, and white spruce with relation to the texture and moisture of sand soils. Master's thesis. Univ. Minn. 78 pp.

Bailey, I. W. 1954. Contributions to plant anatomy. Chronica Botanica. Waltham, Mass. 259 pp.

Bailey, L. H. 1933. The cultivated conifers in North America. Macmillan. New York, N.Y. 404 pp.

——. 1950. The standard cyclopedia of horticulture. Vol. 1. Macmillan. New York, N.Y. 1200 pp.

——, and E. Z. Bailey. 1941. Hortus second. Macmillan. New York, N.Y. 778 pp.

Balch, R. E. 1935. Cultural practices and forest insects. Ent. Soc. Ont. 65th Annual Rep. (1934): 43–49.

——. 1946. "Staminate trees" and spruce budworm abundance. Bi-m. Progr. Rep. For. Insect Invest. Dep. Agric. Can. 2(3):1.

——. 1952. Studies on the balsam woolly aphid, *Adelges piceae* (Ratz.) and its effects on balsam fir *Abies balsamea* (L.) Mill. Publ. Dep. Agric. Can. 867. 76 pp.

Baldwin, H. I. 1931. The period of height growth in some northeastern conifers. Ecology 12:665–689.

Bannan, M. W. 1936. A comparison of the distribution of albuminous and tracheary cells in the gymnosperms. Amer. J. Bot. 23:36–40.

——. 1940. The root systems of northern Ontario conifers growing in sand. Amer. J. Bot. 27:108–114.

——. 1941. Variability in wood structure in roots of native Ontario conifers. Bull. Torrey Bot. Cl. 68:173–194.

——. 1942. Notes on the origin of adventitious roots in the native Ontario conifers. Amer. J. Bot. 29:593–598.

Barghoorn, E. S. 1940. Origin and development of the uniserate ray in the conifers. Bull. Torrey Bot. Cl. 67:303–343.

Beal, J. M. 1934. Chromosome behavior in *Pinus banksiana* following fertilization. Bot. Gaz. 95:660–666.

Beissner, L. 1894. Mitteilungen über Coniferen. Mitt. Dtsch. Dendrol. Ges. 3:51–60.

——. 1897. Neues und interessantes über Coniferen. Mitt. Dtsch. Dendrol. Ges. 6 (repr. 1909): 291–305.

——. 1901. Mitteilungen über Coniferen. Mitt. Dtsch. Dendrol. Ges. 10:72–87.

——. 1903. Mitteilungen über Coniferen. Mitt. Dtsch. Dendrol. Ges. 12:17, 20.

——. 1906. Mitteilungen über Coniferen. Mitt. Dtsch. Dendrol. Ges. 15:89–100.

Belyea, R. M., D. A. Fraser, and A. H. Rose. 1951. Seasonal growth of some trees in Ontario. For. Chron. 27:300–305.

Boivin, B. 1959. *Abies balsamea* (Linné) Miller et ses variations. Naturaliste Canadien 86:219–223.

Bonner, E. 1941. Balsam fir in the Clay Belt of northern Ontario. Master's thesis. Univ. Toronto. 102 pp.

Boucher, P. 1664. Histoire véritable et naturelle de Moeurs et Prod. du pays de la Nouvelle France le Canada. 3rd ed. P. 49.

Brown, H. P., and A. J. Panshin. 1934. Identification of the commercial timbers of the United States. McGraw-Hill. New York, N.Y. 223 pp.

———, and C. C. Forsaith. 1949. Textbook of wood technology. Vol. I. McGraw-Hill. New York, N.Y. 652 pp.

Browne, D. J. 1832. The Silva Americana. Hyde & Co. Boston, Mass. 408 pp.

Buchholz, J. T. 1920. Polyembryony among Abietineae. Bot. Gaz. 69:153–167.

———. 1942. A comparison of the embryogeny of *Picea* and *Abies*. Madrono 6:156–167.

Canada Forest Research Branch. 1961. Annual report on forest research. Year ended March 31, 1961. For. Dep. Can. 79 pp.

Canada Forestry Branch. 1955. Sixth annual report on active forest research projects. Mimeo. For. Br. Can. 144 pp.

———. 1956. Native trees of Canada. 5th ed. Bull. For. Br. Can. 61. 293 pp.

Carpenter, C. H., and L. Leney. 1952. 382 photomicrographs of 91 papermaking fibers. Tech. Publ. New York St. Coll. For. 74. 152 pp.

Cash, E. K. 1941. An abnormality of *Abies balsamea*. Plant Dis. Rep. U.S. Dep. Agric. 25: 548.

Chamberlain, C. J. 1935. Gymnosperms, structure and evolution. Univ. Chicago Press. Chicago, Ill. 484 pp.

Chang, Ying-Pe. 1954a. Bark structure of North American conifers. Tech. Bull. U.S. Dep. Agric. 1095. 86 pp.

———. 1954b. Anatomy of common North American pulpwood barks. Tappi Monogr. 14. 249 pp.

Chittenden, A. K. 1905. Forest conditions of northern New Hampshire. Bull. U.S. Bur. For. 55. 100 pp.

Chouinard, L., and L. Parrot. 1958. The callusing and rooting of air-layers in *Betula papyrifera*, *Populus tremuloides*, *Larix laricina* and *Abies balsamea*. Contr. Laval Univ. For. Res. Found. 2. 16 pp.

Clark, J. 1956. Photosynthesis of white spruce and balsam fir. Bi-m. Progr. Rep. Div. For. Biol. Dep. Agric. Can. 12(5):1–2.

———. 1961. Photosynthesis and respiration in white spruce and balsam fir. Tech. Bull. N.Y. St. Coll. For. 85. 72 pp.

———, and J. M. Bonga. 1961. An indole compound in the needles of balsam fir. Bi-m. Progr. Rep. Br. For. Ent. Path. Dep. For. Can. 17(4):2.

———. 1963. Evidence for indole-3-acetic acid in balsam fir, *Abies balsamea* (L.) Mill. Canad. J. Bot. 41:165–173.

Codman, H. S. 1889. The gardens of Petit-Trianon. Garden and Forest. 2:2.

Cook, D. B. 1941. The period of growth in some northeastern trees. J. For. 39:956–959.

———. 1945. An abnormal balsam fir. Torreya 45:13–14.

Cooper, W. S. 1911. Reproduction by layering among conifers. Bot. Gaz. 52:369–379.

Dallimore, W., and A. B. Jackson. 1931 (3rd ed., 1948; repr. 1954). Coniferae including Ginkgoaceae. Arnold & Co. London. 582 pp.

Davis, J. H. 1946. The peat deposits of Florida. Bull. Geol. Surv. Florida 30. 247 pp.

Dorner, H. B. 1899. The resin ducts and strengthening cells of *Abies* and *Picea*. Proc. Indiana Acad. Sci. 116–129.

Edlin, H. E. 1956. Gaelic and Norse words in the Scots forester's vocabulary. Scot. For. 10:198–200.

Emerson, G. B. 1846. Report on the trees and shrubs growing naturally in the forests of Massachusetts. Dutton and Wentworth. Boston, Mass. 547 pp.

Endlicher, S. F. L. 1847. Synopsis Coniferarum. Scheitlin & Zollikofer. Sangalli. 368 pp.

Engelmann, G. 1878. A synopsis of the American firs, *Abies* Link. Trans. Acad. Sci. St. Louis 3:593–602.

Evelyn, J. 1664. Sylva. Vol. 2. Repr. of 1706 (4th) ed. Doubleday & Co. London. 215 pp.

Facey, V. 1956. Abscission of leaves in *Picea glauca* (Moench) Voss and *Abies balsamea* (L.) Mill. Proc. North Dakota Acad. Sci. 10:38–43.

Fegel, A. C. 1941. Comparative anatomy and varying physical properties of trunk, branch, and root wood in certain northeastern trees. Bull. New York St. Coll. For. 55. 20 pp.

Fernald, M. L. 1909. A new variety of *Abies balsamea*. Rhodora 11:201–203.

———. 1950. Gray's manual of botany. 8th ed. American Book Co. New York, N.Y. 1632 pp.

———, and C. A. Weatherby. 1932. *Abies balsamea* (L.) Mill., forma *hudsonia* (Bosc.) comb. nov. Rhodora 34:190–191.

Florin, R. 1931. Untersuchungen zur Stammesgeschichte der Coniferales und Cordaitales. I. Teil: Morphologie und Epidermisstruktur der Assimilationsorgane bei den rezenten Koniferen. Almqvist & Wiksells. Stockholm. 588 pp.

Foster, A. S., and E. M. Gifford. 1959. Comparative morphology of vascular plants. Freeman and Co. San Francisco, Calif. 555 pp.

Franco, J. do Amaral. 1950. Abetos [Firs]. An. Inst. Agron. (Lisboa) 17:1–260.

Fulling, E. H. 1934. Identification, by leaf structure, of the species of *Abies* cultivated in the United States. Bull. Torrey Bot. Cl. 61:497–524.

———. 1936. *Abies intermedia*, the Blue Ridge fir, a new species. Jour. South. Appal. Bot. Club (Castanea) 1:91–94.

Gordon, G. 1875 (1st ed., 1858). The pinetum. Bohn. London. 484 pp.

Gosse, P. H. 1840. The Canadian naturalist. van Voorst. London. 372 pp.

Greguss, P. 1955. Identification of living gymnosperms on the basis of xylotomy. Akademiai Kiado. Budapest. 263 pp. + 350 pl.

Hahn, G. G., C. Hartley, and A. S. Rhoads. 1920. Hypertrophied lenticels in the roots of conifers and their relation to moisture and aeration. J. Agric. Res. 20:253–265.

Heimburger, C. C. 1945. Comment on the budworm outbreak in Ontario and Quebec. For. Chron. 21:114–126.

———, and M. Holst. 1955. Notes from a trip to the southern United States, January, 1953. For. Chron. 31:60–73.

Hereman, S. 1868. Paxton's botanical dictionary. Evans & Co. London. 623 pp.

Hofmeister, W. 1848. Über die Entwicklung des Pollens. Bot. Ztg. 6:425–434; 649–658; 670–674.

Holdheide, W. 1951. Anatomie mitteleuropäischen Gehölzrinden. *In* H. Freund, Handbuch der Mikroskopie in der Technik. Umschau. Frankfurt-am-Main. Vol. 5, Pt. 1, pp. 193–367.

Hooker, W. J., and G. Spratt. 1832. *Abies balsamea*. Woodville Med. Bot. London 5:1–4.

Hopkins, H. T., and R. L. Donahue. 1939. Forest tree root development as related to soil morphology. Proc. Soil Sci. Soc. Amer. 4:353.

Hort, A. 1916. Theophrastus. Enquiry into plants. Vol. 1–2. Heinemann, London. Putnam's Sons, New York, N.Y.

Huntington, A. O. 1904. Balsam fir. New England Mag. 31:225.

Hutchinson, A. H. 1914. The male gametophyte of *Abies*. Bot. Gaz. 57:148–153.

———. 1915. Fertilization in *Abies balsamea*. Bot. Gaz. 60:457–472.

———. 1924. Embryogeny of *Abies*. Bot. Gaz. 77:280–289.

International Union of Biological Sciences. 1961. International code of nomenclature for cultivated plants. Int. Assoc. Plant Taxonomy. Utrecht, Netherlands. 30 pp.

Jacombe, F. W. H. 1920. What is a "fir" tree and why? Canad. For. Journ. 16:5–6.

Jeffrey, E. C. 1917. The anatomy of woody plants. Univ. Chicago Press. Chicago, Ill. 478 pp.

Johansen, D. A. 1950. Plant embryology. Chronica Botanica. Waltham, Mass. 305 pp.

Johnson, L. P. V. 1939. A descriptive list of natural and artificial interspecific hybrids in North American forest-tree genera. Canad. J. Res. 17(C):411–444.

Josselyn, J. 1674. An account of two voyages to New England. London. 279 pp. Repr. 1833. Mass. Hist. Soc. Coll. Ser. 3. III. Pp. 211–396.

Kelsey, H. P., and W. A. Dayton, eds. 1942. Standardized plant names. 2nd ed. McFarland. Harrisburg, Pa. 675 pp.

Kent, A. H. 1900. Veitch's manual of the Coniferae. 2nd ed. Veitch & Sons. London. 562 pp.

Kienholz, R. 1934. Leader, needle, cambial, and root growth of certain conifers and their interrelations. Bot. Gaz. 96:73–92.

Klaehn, F. N., and J. A. Winieski. 1962. Interspecific hybridization in the genus *Abies*. Silv. Genet. 11:130–142.

Kozlowski, T. T. 1961. Shoot elongation characteristics of forest trees. For. Sci. 7:357–368.

———, and J. H. Cooley. 1960. Natural root grafting in forest trees. For. Res. Notes Wis. Coll. Agric. 56. 2 pp.

———. 1961. Natural root grafting in northern Wisconsin. J. For. 59:105–107.

Kozlowski, T. T., and R. C. Ward. 1957. Seasonal height growth of conifers. For. Sci. 3:61–66.

Kramer, P. J. 1943. Amount and duration of growth of various species of tree seedlings. Plant Physiol. 18:239–251.

Krüssmann, G. 1960 (1st ed., 1955). Die Nadelgehölze. 2nd ed. Parey. Berlin u. Hamburg. 335 pp.

Kukachka, B. F. 1960. Identification of coniferous woods. Tappi 43:887–896.

Laing, E. V. 1956. The genus *Abies* and recognition of species. Scot. For. 10:20–25.

Lamb, G. N. 1915. A calendar of the leafing, flowering, and seeding of the common trees of the eastern United States. Weather Bureau. Monthly Weather Rev. Supplement 2(1): 5–19.

Lamb, W. H. 1914. A conspectus of North American firs (exclusive of Mexico). Proc. Soc. Amer. For. 9:528–538.

Lanjouw, J., chmn. 1956. International code of botanical nomenclature. Int. Assoc. Plant Taxonomy. Utrecht, Netherlands. 338 pp.

Lawson, P., and Son. 1836. List No. 10, *Abietineae*, p. 11.

Linné, C. von. 1753. Species Plantarum. Holmiae impensis L. Salvii. 1200 pp.

Little, E. L. 1953. Check list of native and naturalized trees of the United States (including Alaska). Agric. Handb. U.S. Dep. Agric. 41. 472 pp.

Loudon, J. C. 1838. Arboretum et fruticetum britannicum. Vol. 4:2339. Longman, Orme, Brown, Green and Longmans. London.

McAdams, T. 1909. The human interest in firs. Gdn. Mag. 10(1):12–14.

MacGillivray, H. G. 1957. Rooting balsam fir cuttings under intermittent mist. For. Chron. 33:353–354.

McLeod, C. H. 1924. The pathological anatomy of tissue produced in *Abies balsamea* following an attack of the spruce budworm. Phytopathology 14:345.

McNab, W. R. 1877. A revision of the species of *Abies*. Proc. Roy. Irish Acad. 2. Ser. II:673–704.

Marriott, F. G., and C. Greaves. 1947. Canada balsam, its preparation and uses. Mimeo. For. Prod. Lab., Can. 123. 5 pp.

Masters, M. T. 1891. Review of some points in the comparative morphology, anatomy and life-history of the Coniferae. Jour. Linnean Soc. Bot. 27:226–332.

———. 1895. The Balm of Gilead. Gdn. Chron. Ser. 3:422–423.

Mattfeld, J. 1925. Die in Europa und dem Mittelmeergebiet wildwachsenden Tannen. Mitt. Dtsch. Dendrol. Ges. 35:1–37.

Melchior, H., and E. Werdermann. 1954. A. Engler's Syllabus der Pflanzenfamilien. 12th ed. Vol. I. Gebr. Borntraeger. Berlin-Nikolasee. 367 pp.

Mergen, F., and D. T. Lester. 1961a. Microsporogenesis in *Abies*. Silvae Genet. 10:146–156.

———. 1961b. Colchicine-induced polyploidy in *Abies*. For. Sci. 7:314–319.

Meyer, E. A. 1914. Die Nadelhölzer im Arboretum des landwirtsch. Instituts in Moskou. Mitt. Dtsch. Dendrol. Ges. 23:188–200.

Michaux, A. 1803. Flora boreali-americana. Paris. 2 vol.

Michaux, F. A. 1810–13. Histoire des arbres forestiers de l'Amerique septentrionale. Paris. 3 vol.

———. 1818–19. The North American Sylva or a description of the forest trees of the United States, Canada and Nova Scotia. Philadelphia, Pa. 3 vol.

Miller, Ph. 1768. The gardener's dictionary. 8th ed. Francis Rivington. London. 1334 pp.

Miyake, K. 1903. Contribution to the fertilization and embryogeny of *Abies balsamea*. Beih. Bot. Centralbl. 14:134–144.

Moore, B. 1917. Some factors influencing the reproduction of red spruce, balsam fir, and white pine. J. For. 15:827–853.

———. 1922. Humus and root systems in certain northeastern forests in relation to reproduction and competition. J. For. 20:235–254.

Morris, R. F. 1948. How old is a balsam tree? For. Chron. 24:106–110.

———. 1951. The effects of flowering on the foliage production and growth of balsam fir. For. Chron. 27:40–57.

———, F. E. Webb, and C. W. Bennet. 1956. A method of phenological survey for uses in forest insect studies. Canad. J. Zool. 34:533–540.

Moss, E. H. 1959. Flora of Alberta. Univ. Toronto Press. Toronto, Ont. 546 pp.

Münchhausen, O. 1770. Verzeichnis aller Bäume und Stauden welche in Deutschland vorkommen. Hausvater 5:93–368.

Myers, J. E. 1922. Ray volumes of the commercial woods of the United States and their significance. J. For. 20:337–351.

Myers, O., and E. H. Bormann. 1963. Phenotypic variation in *Abies balsamea* in response to altitudinal and geographic gradients. Ecology 44:429–436.

Otis, C. H. 1913. Michigan trees. Univ. Mich. Ann Arbor, Mich. 246 pp.

Parlatore, P. 1868. Coniferae. *In* A. de Candolle, Prodromus systematis naturalis regni vegetabilis. Treuttel and Wurtz. Paris. Vol. 16(2), pp. 361–521.

Penhallow, D. P. 1907. A manual of North American gymnosperms. Ginn & Co. Boston, Mass. 374 pp.

Pepin, P. D. 1860. *Abies balsamea denudata.* Revue Horticole. Paris. P. 10.

Pickering, Ch. 1879. Chronological history of plants. Little, Brown & Co. Boston, Mass. 1222 pp.

Pilger, R. 1926. Pinaceae. *In* A. Engler and K. Prantl, Die natürlichen Pflanzenfamilien. Engelmann. Leipzig. 2nd ed. Vol. 13, pp. 271–342.

Pitcher, J. A. 1960. Heteroplastic grafting in the genera *Acer, Fraxinus, Picea,* and *Abies.* Proc. Ntheast. For. Tree Impr. Conf. 7:52–57.

Potzger, J. E. 1937. Vegetative reproduction in conifers. Am. Midl. Nat. 18:1001–1004.

Pritzel, G. A. 1866. Iconum Botanicarum. Nikolaische Verlagsbuchhandl. Berlin. Vol. 1–2.

Provancher, L. 1862. Floré canadienne. Quebec, P.Q. Vol. 2, p. 556.

Raup, H. M. 1946. Phytogeographic studies in the Athabaska–Great Slave Lake region. II. J. Arnold Arbor. 27:1–85.

Ray, J. 1704. Arboris Balsamum Gileadense fundens. Historiae Plantarum, Lib. XXV. Dendr. Secundus. Londini. Vol. III, p. 8.

Regel, E. 1884. *Abies balsamea* Ait. in Park zu Ropscha bei Petersburg. Gartenfl. 33:300–301.

Rehder, A. 1928. New species, varieties and combinations from the herbarium and the collections of the Arnold Arboretum. J. Arnold Arbor. 9:29–31.

———. 1940. Manual of cultivated trees and shrubs. 2nd ed. Macmillan. New York, N.Y. 996 pp.

———. 1949. Bibliography of cultivated trees and shrubs. Arnold Arboretum. Harvard Univ. 825 pp.

Richard, C. L. 1810. *Peuce balsamea* C. L. Richard. Ann. Mus. Hist. Nat. Paris 16:298.

Richens, R. H. 1945. Forest tree breeding and genetics. Joint Publ. Imp. Agric. Bur. 8. 79 pp.

Roe, E. I. 1946. Extended periods of seed fall of white spruce and balsam fir. Tech. Note Lake St. For. Exp. Sta. 261. 1 p.

———. 1948a. Balsam fir seed, its characteristics and germination. Sta. Pap. Lake St. For. Exp. Sta. 11. 13 pp.

———. 1948b. Early seed production by balsam fir and white spruce. J. For. 46:529.

Rosendahl, C. O. 1955. Trees and shrubs of the Upper Midwest. Univ. Minnesota Press. Minneapolis, Minn. 411 pp.

Rothrock, J. T. 1910. Balsam fir. For. Leaves 12(7):105.

Roy, H. 1940. Spruce regeneration in Canada: Quebec. For. Chron. 16:10–20.

Rudolf, P. O. 1956. Guide for selecting superior forest trees and stands in the Lake States. Sta. Pap. Lake St. For. Exp. Sta. 40. 32 pp.

Salisbury, R. A. 1796. Prodromus stirpium in horto ad Chapel Allerton vingentium. O. Londini. 422 pp.

Sargent, C. S. 1884. Forests of North America. U.S. Dept. Int. Census Off. 612 pp.

——— (C. S. S.). 1889. *Abies fraseri.* Garden and Forest. 2:472.

——— (S.). 1892. The Waukegan nurseries. Garden and Forest. 5:274–275.

——— (C. S. S.). 1897. Notes on cultivated conifers. Garden and Forest. 10:509–512.

———. 1898. The Silva of North America. Vol. 12. Houghton Mifflin. Boston, Mass. 144 pp.

———. 1926. Manual of the trees of North America. Houghton Mifflin. Boston, Mass. 910 pp.

Sax, K. 1918. The behavior of the chromosomes in fertilization. Genetics 3:309–327.

———, and H. J. Sax. 1933. Chromosome number and morphology in the conifers. J. Arnold Arbor. 14:356–375.

Schenck, C. A. 1939. Fremdländische Wald- und Parkbäume. Parey. Berlin. Vol. 1–2.

Schwerin, F. 1903. Die Augsburger Forstgärten in Diedorf. Mitt. Dtsch. Dendrol. Ges. 12:91–95.

———. 1929. Botanische und forstliche Mitteilungen über Koniferen. Mitt. Dtsch. Dendrol. Ges. 41:159–174.

Seitz, F. W. 1951. Chromosomenzahlenverhältnisse bei Holzpflanzen. Z. Forstgenet. 1:22–32.

Stone, E. L. 1953. The origin of epicormic branches in fir. J. For. 51:366.

Sudworth, G. B. 1897. Nomenclature of the arborescent flora of the United States. Bull. U.S.D.A. For. Div. 14. 419 pp.

———. 1898. Check list of the forest trees of the United States, their names and ranges. Bull. U.S. Div. For. 17. 144 pp.

———. 1916. The spruce and balsam fir trees of the Rocky Mountain region. Bull. U.S. Dep. Agric. 327. 43 pp.

———. 1927. Check list of the forest trees of the United States, their names and ranges. Misc. Circ. U.S. Dep. Agric. 92. 295 pp.

Taubert, F. 1926. Beiträge zur äusseren und inneren Morphologie der Licht- und Schattenadeln bei der Gattung *Abies* Juss. Mitt. Dtsch. Dendrol. Ges. 37(2):206–252.

Thompson, W. P. 1912. Ray tracheids in *Abies*. Bot. Gaz. 53:331–338.

Trelease, W., and A. Gray. 1887. The botanical works of the late George Engelmann. Wilson & Son. Cambridge, Mass. 548 pp.

United States Forest Service. 1948. Woody-plant seed manual. Misc. Publ. U.S. Dep. Agric. 654. 416 pp.

Viguié, M., and H. Gaussen. 1928–29. Revision du genre *Abies*. Bull. Soc. Hist. Nat. Toulouse 57:369–434; 58:245–564.

Wardlaw, C. W. 1955. Embryogenesis of plants. Wiley & Sons. New York, N.Y. 381 pp.

Wesmael, A. 1890. Naturalization of American conifers in Belgium. Garden and Forest 3:494.

White, L. T. 1951. Studies of Canadian *Thelephoraceae*. VIII. *Corticium galactinum* (Fr.) Burt. Canad. J. Bot. 29:279–296.

Wiesehuegel, E. G. 1932. Diagnostic characteristics of the xylem of the North American *Abies*. Bot. Gaz. 93:55–70.

Willkomm, M. 1863. *Abies balsamea* var. *phanerolepis* Willkomm. *In* Delectus Semium Hortus Dorpatensis.

———. 1869. *Abies balsamea* var. *brachylepis*. Ind. Sem. Hort. Acad. Dorpat. 1868. P. 7.

Wodehouse, R. P. 1935. Pollen grains. McGraw-Hill. New York, N.Y. 574 pp.

Wright, H. E., T. C. Winter, and H. L. Patten. 1963. Two pollen diagrams from southeastern Minnesota: problems in the regional late-glacial and postglacial vegetational history. Bull. Geol. Soc. Amer. 74:1371–1396.

Wright, J. W. 1953. Notes on flowering and fruiting of northeastern trees. Sta. Pap. Ntheast. For. Exp. Sta. 60. 38 pp.

Wyman, D. 1943. A simple foliage key to the firs. Arnoldia 3:65–71.

Zon, R. 1914. Balsam fir. Bull. U.S. For. Serv. 55. 68 pp.

Chapter 2. Geography and Synecology

Alexander, R. R. 1958. Silvical characteristics of subalpine fir. Sta. Pap. Rocky Mt. For. Range Exp. Sta. 32. 15 pp.

American Forestry Association. 1945. Report on American big trees. Amer. Forests 51:30–36.

Antevs, E. 1932. Alpine zone of the Mt. Washington range. Merrill & Webber. Auburn, Me. 119 pp.

Artist, R. C. 1939. Pollen spectrum studies on the Anoka sand plain in Minnesota. Ecol. Monogr. 9:494–535.

Auer, V. 1927. Stratigraphical and morphological investigations of peat bogs of southern Canada. Commun. Inst. Quest. For. Finl. 12:1–32.

Aughanbaugh, J. E., H. R. Muckley, and O. D. Diller. 1958. Performance records of woody plants in the Secrest Arboretum. Dep. Ser. Ohio Agric. Exp. Sta. 41. 90 pp.

Bakuzis, E. V., H. L. Hansen, and V. Kurmis. 1962. Weighted site indices in relation to some characteristics of the edaphic field of the Central Pine Section of Minnesota forests. Minn. For. Notes 119. 2 pp.

Baskerville, G. L. 1960. Conversion to periodic selection management in a fir, spruce, and birch forest. Tech. Note For. Br. Can. 86. 19 pp.

Bergman, F. H., and H. Stallard. 1916. The development of climax formations in northern Minnesota. Minn. Bot. Studies 4:333–378.

Betts, H. S. 1945. Balsam fir (*Abies balsamea*). Amer. Woods. U.S. For. Serv. 8 pp.

Blais, J. R. 1961. Spruce budworm outbreaks in the lower St. Lawrence and Gaspé regions. For. Chron. 37:192–202.

Braun, E. L. 1950. Deciduous forests of eastern North America. Blakiston. Philadelphia, Pa. 596 pp.

———. 1955. The phytogeography of unglaciated eastern United States and its interpretation. Bot. Rev. 21:297–375.

Braun-Blanquet, J. 1951. Pflanzensoziologie. Springer. Wien. 631 pp.

Bray, W. L. 1930. The development of the vegetation of New York State. Bull. N.Y. St. Coll. For. 29. 189 pp.

Brown, D. M. 1941. Vegetation of Roan Mountain: A phytosociological and successional study. Ecol. Monogr. 11:61–97.

Brown, R. T., and J. T. Curtis. 1952. The upland conifer-hardwood forests of northern Wisconsin. Ecol. Monogr. 22:217–234.

Browne, D. J. 1832. The Silva Americana. Hyde & Co. Boston, Mass. 408 pp.

Buell, M. F. 1945. Lake Pleistocene forests of southeastern North Carolina. Torreya 45:117–118.

———. 1956. Spruce-fir, maple-basswood competition in Itasca Park, Minnesota. Ecology 37:606.

———, and W. E. Gordon. 1945. Hardwood-conifer forest contact zone in Itasca Park, Minnesota. Amer. Midl. Nat. 34:433–439.

Buell, M. F., and E. W. Martin. 1961. Competition between maple-basswood and fir-spruce communities in Itasca Park, Minnesota. Ecology 42:428–491.

Buell, M. F., and W. A. Niering. 1953. Vegetation of a raised bog near Itasca Park, Minnesota. Bull. Torrey Bot. Cl. 80:123–130.

Butters, F. K., and E. C. Abbe. 1953. A floristic study of Cook County, Minnesota. Rhodora 55:21–55, 63–101, 161–201.

Cain, S. A. 1944. Pollen analysis of some buried soils, Spartanburg County, South Carolina. Bull. Torrey Bot. Cl. 71:11–22.

Cajander, A. K. 1923. Der Anbau ausländischer Holzarten als forstliches und pflanzengeographisches Problem. Acta For. Fenn. 24(1):1–15.

Canada Geographical Branch. 1957. Atlas of Canada. Dep. Mines Tech. Surv. Can. 110 pl.

Cheyney, E. G. 1942. American silvics and silviculture. Univ. Minnesota Press. Minneapolis, Minn. 472 pp.

Christensen, C. M., J. J. Clausen, and J. T. Curtis. 1959. Phytosociology of the lowland forests of northern Wisconsin. Amer. Midl. Nat. 62:232–247.

Churchill, J. R., chmn. 1933. Reports on the flora of Massachusetts. Rhodora 35:351–359.

Clausen, J. J. 1957. A phytosociological ordination of the conifer swamps of Wisconsin. Ecology 38:638–646.

Clements, F. E. 1916. Plant succession, an analysis of development of vegetation. Publ. Carnegie Inst. Washington, D.C. 512 pp.

Cline, M. G. 1953. Major kinds of profiles and their relationships in New York. Proc. Soil Sci. Soc. Amer. 17:123–127.

Coker, W. C., and H. R. Totten. 1934. The trees of the southeastern states. Univ. North Carolina Press. Chapel Hill, N.C. 399 pp.

Conard, H. S. 1939. The fir forests of Iowa. Proc. Iowa Acad. Sci. (1938) 45:69–72.

Conway, V. M. 1949. The bogs of central Minnesota. Ecol. Monogr. 19:173–206.

Cooper, W. S. 1913. The climax forest of Isle Royale, Lake Superior, and its development. Bot. Gaz. 55:1–44, 115–140, 189–235.

———. 1932. Reconstruction of a late Pleistocene biotic community in Minneapolis, Minnesota. Ecology 13:63–73.

Core, E. L. 1940. New plant records for West Virginia. Torreya 40:5–9.

Curtis, J. T. 1959. The vegetation of Wisconsin. Univ. Wisconsin Press. Madison, Wis. 657 pp.

Cushing, E. J. 1963. Late-Wisconsin pollen stratigraphy in east-central Minnesota. Univ. Minn. Ph.D. thesis, Minneapolis. 165 pp.

Dansereau, P., and F. Segadas-Vianna. 1952. Ecological study of the peat bogs of eastern North America. Canad. J. Bot. 30:490–520.

Darlington, H. T. 1931. Vegetation of the Porcupine Mountains, northern Michigan. Pap. Mich. Acad. Sci. 13:9–65.

Davis, M. B. 1958. Three pollen diagrams from central Massachusetts. Amer. J. Sci. 256:540–570.

———. 1960. A late glacial pollen diagram from Taunton, Massachusetts. Bull. Torrey Bot. Cl. 87:258–269.

———, and J. C. Goodlett. 1960. Comparison of the present vegetation with pollen-spectra in surface samples from Brownington Pond, Vermont. Ecology 41:346–357.

Deam, C. C. 1921. Trees of Indiana. Publ. Ind. Dep. Conserv. 13. 317 pp.

Deevey, E. S. 1951. Late-glacial and postglacial pollen diagrams from Maine. Amer. J. Sci. 249:177–204.

Deters, M. E., and H. Schmitz. 1936. Drought damage to prairie shelter-belts in Minnesota. Bull. Minn. Agric. Exp. Sta. 329. 28 pp.

Dixon, D. 1961. These are the champs. Amer. Forests 67:41–46.

Donahue, R. L. 1940. Forest-site quality studies in the Adirondacks. I. Tree growth as related to soil morphology. Mem. Cornell Agric. Exp. Sta. 229. 44 pp.

Döring, von. 1927. Neuere Erfahrungen über den Anbau fremdländischer Forstarten. Mitt. Dtsch. Dendrol. Ges. 38:341–363.

———. 1930. Exoten in der Schildfelder Forst. Mitt. Dtsch. Dendrol. Ges. 41:261–264.

Ellis, R. C. 1960. An observation on the invasion of black spruce sites by balsam fir in central Newfoundland. Pulp Paper Mag. Can. 61(6):164, 166.

Emberger, L. 1944. Les plantes fossiles. Mason et Cie. Paris. 492 pp.

Engler, A. 1879. Versuch einer Entwicklungsgeschichte der extratropischen Florengebiete der nördlichen Hemisphäre. Engelmann. Leipzig. 202 pp.

Essex, B. L., C. D. Chase, and A. G. Horn. 1955. Michigan forest survey. Timber resources, southeastern block Lower Peninsula, Michigan. Mich. Dep. Conserv. Lansing, Mich. 49 pp.

Fassett, N. C. 1929. Preliminary reports on the flora of Wisconsin. V. Coniferales. Trans. Wis. Acad. Sci. 24:257–268.

Fenneman, N. M. 1938. Physiography of eastern United States. McGraw-Hill. New York, N.Y. 691 pp.

Fernald, M. L. 1925. Persistence of plants in unglaciated areas of boreal America. Mem. Amer. Acad. Arts Sci. 15:241–342.

Fernow, B. E. 1912. Forest resources and problems of Canada. Proc. Soc. Amer. For. 7:133–144.

Flaccus, E. 1959. Revegetation of landslides in the White Mountains of New Hampshire. Ecology 40:692–703.

Flint, R. F. 1957. Glacial and Pleistocene geology. Wiley & Sons. New York, N.Y. 553 pp.

Fontaine, W. M. 1889. The Potomac or Younger Mesozoic Flora. Monogr. U.S. Geol. Surv. 15. 375 pp.

Forster, H. 1905. Über ausländische Coniferen. Mitt. Dtsch. Dendrol. Ges. 14:157–168.

Franco, J. do Amaral. 1950. Abetos [Firs]. An. Inst. Agron. (Lisboa) 17:1–260.

Fries, M. 1962. Pollen profiles of Lake Pleistocene and recent deposits from Weber Lake, Minnesota. Ecology 43:295–308.

Fuller, G. D. 1919. Vegetation of Cape Breton. Bot. Gaz. 67:370–373.

———. 1939. Interglacial and postglacial vegetation of Illinois. Trans. Ill. Acad. Sci. 32(1):5–15.

Gates, F. C. 1926. Plant successions about Douglas Lake, Cheboygan County, Michigan. Bot. Gaz. 82:170–182.

Gegelsky, I. N. 1953. Pikhta balsamicheskaya v Chernigovskoi oblasti. Lesn. Khoz. 6(7):38–40.

George, J. 1953. Tree and shrub species for the northern great plains. Circ. U.S. Dep. Agric. 912. 46 pp.

Gleason, H. A. 1926. The individualistic concept of the plant association. Bull. Torrey Bot. Cl. 53:7–26.

———. 1939. The individualistic concept of the plant association. Amer. Midl. Nat. 21:92–110.

Graham, A., and C. Heimsch. 1960. Pollen studies of some Texas peat deposits. Ecology 41:751–763.

Graham, S. A. 1941. Climax forests of the Upper Peninsula of Michigan. Ecology 22:355–362.

Grant, M. L. 1929. The burn succession in Itasca County, Minnesota. Master's thesis. Univ. Minnesota. 63 pp.

———. 1934. The climax forest community in Itasca County, Minnesota, and its bearing upon the successional status of the pine community. Ecology 15:243–257.

Graves, H. S. 1899. Practical forestry in the Adirondacks. Bull. U.S. Dep. Agric. 26. 84 pp.

Great Britain Forestry Commission. 1955. Guide to the national Pinetum and forest plots at Bedgebury. HMSO. London. 65 pp.

Griggs, R. F. 1946. The timberlines of northern America and their interpretation. Ecology 27:275–289.

Grimm, W. C. 1950. The trees of Pennsylvania. Stackpole and Heck. New York, N.Y. 363 pp.

Grundner, F. 1921. Die Anbauversuche mit fremdländischen Holzarten in den braunschweigischen Statsforsten. Mitt. Dtsch. Dendrol. Ges. 31:19–63.

Halliday, W. E. D. 1937. A forest classification for Canada. Bull. For. Serv. Can. 89. 50 pp.

———, and A. W. A. Brown. 1943. The distribution of some important forest trees in Canada. Ecology 24:353–373.

Hansen, H. L. 1946. An analysis of jack pine sites in Minnesota with special reference to site index, plant indicators, and certain soil features. Ph.D. thesis, Univ. Minnesota. 117 pp.

Hansen, H. P. 1939. Postglacial vegetation of the driftless area of Wisconsin. Amer. Midl. Nat. 21:752–762.

———. 1949a. Postglacial forests in south-central Alberta, Canada. Amer. J. Bot. 36:54–65.

———. 1949b. Postglacial forests in west-central Alberta, Canada. Bull. Torrey Bot. Cl. 76:278–289.

———. 1952. Postglacial forests in the Grande Prairie–Lesser Slave Lake region of Alberta, Canada. Ecology 33:31–40.

Hare, F. K. 1950. Climate and zonal divisions of the Boreal forest formations in eastern Canada. Geogr. Rev. 40:615–635.

Harper, E. B., chmn. 1922. Addition to the flora of Connecticut. Rhodora 24:111–121.

Harshberger, J. 1911. Phytogeographic survey of North America. Engelmann. Leipzig. 790 pp.

Hart, A. C. 1959. Silvical characteristics of balsam fir (*Abies balsamea*). Sta. Pap. Ntheast. For. Exp. Sta. 122. 22 pp.

Hartley, T. G. 1960. Plant communities of the La Crosse area in western Wisconsin. Proc. Iowa Acad. Sci. 67:174–188.

Hawley, R. C., and A. F. Hawes. 1912. Forestry in New England. Wiley & Sons. New York, N.Y. 479 pp.

Heimburger, C. C. 1934. Forest-type studies in the Adirondack region. Mem. Cornell Agric. Exp. Sta. 165. 122 pp.

———. 1941. Forest-site classification and soil investigation on Lake Edward Forest Experimental Area. Silv. Res. Note For. Serv. Can. 66. 60 pp.

Heinselman, M. L. 1954. The extent of natural conversion to other species in the Lake States aspen-birch type. J. For. 52:737–738.

Heusser, C. J. 1960. Late-Pleistocene environments of North Pacific North America. Spec. Publ. Amer. Geogr. Soc. 35. 308 pp.

Hills, G. A. 1960a. Regional site research. For. Chron. 36:401–423.

———. 1960b. Comparison of forest ecosystems (vegetation and soil) in different climatic zones. Silva Fenn. 105:33–35.

———, and W. G. E. Brown. 1955. The sites of the University of Toronto forest. Adv. Rep. 8th Ntheast. For. Soils Conf. Ont., 1955. 141 pp.

Hoff, A. 1931. Bemerkenswerte fremdländische Nadelhölzer in den Harburger Waldungen. Mitt. Dtsch. Dendrol. Ges. 43:423–425.

Hosie, R. C. 1937. Botanical investigations in Batchawana Bay region, Lake Superior. Bull. Nat. Mus. Can. 88:1–65.

Hough, R. B. 1907. The trees of the northern states and Canada east of the Rocky Mountains. Publ. author. Lowville, N.Y. 470 pp.

Hultén, E. 1937. Outline of the history of arctic and boreal biota during the Quaternary Period. Bokförlags Aktiebolaget Thule. Stockholm. 168 pp. + 43 pl.

Hustich, I. 1949. On the forest geography of the Labrador Peninsula. Acta Geogr. 10(2):1–63.

———. 1950. Notes on the forests on the east coast of Hudson Bay and James Bay. Acta Geogr. 11(1):1–71.

———. 1952. Barrträdsarternas polara gräns på norra halvklotet. (The polar limits of the coniferous species.) Commun. Inst. For. Fenn. 40(29):1–20.

———. 1954. On forests and tree growth in the Knob Lake area, Quebec, Labrador Peninsula. Acta Geogr. 13(1):3–60.

———. 1955. Forest-botanical notes from the Moose River area, Ontario, Canada. Acta Geogr. 13(2):1–50.

Illick, J. S. 1914. Pennsylvania trees. Pa. Dep. For. Harrisburg, Pa. 231 pp.

———. 1928. Pennsylvania trees. Bull Pa. Dep. For. Waters 11. 237 pp.

Ilvessalo, L. 1926. Über die Anbaumöglichkeit ausländischer Holzarten. Mitt. Dtsch. Dendrol. Ges. 36:96–132.

Iron Range Resources Rehabilitation Office. 1950–63. (Forest resources of Minnesota counties.) Office of Iron Range Resources and Rehabilitation. St. Paul, Minn.

Jarvis, J. M. 1960. Forty-five years' growth on the Goulais River watershed. Tech. Note For. Br. Can. 84. 31 pp.

Jelgersma, S. 1962. A late-glacial pollen diagram from Madelia, southcentral Minnesota. Amer. J. Sci. 260:522–529.

Kapper, O. G. 1954. Khvoinie porodi. Goslesbumizdat. Moscow. 303 pp.

Kell, L. L. 1938. The effect of the moisture-retaining capacity of soils on forest succession in Itasca Park, Minnesota. Amer. Midl. Nat. 20:682–694.

Keller, C. O. 1943. A comparative study of three Indiana bogs. Butler Univ. Bot. Stud. 6:65–80.

Kerner, A. von M. 1863 (tr. 1950). The plant life of the Danube basin. Transl. by H. S. Conard (The background of plant ecology). Iowa State College Press. Ames, Iowa. 238 pp.

King, J. E., and R. O. Kapp. 1963. Modern pollen rain studies in eastern Ontario. Can. J. Bot. 41:243–252.

Kittredge, J. 1938. The interrelations of habitat, growth rate, and associated vegetation in the aspen community of Minnesota and Wisconsin. Ecol. Monogr. 8:151–246.

———, and S. R. Gevorkiantz. 1929. Forest possibilities of aspen lands in the Lake States. Tech. Bull. Minn. Agric. Exp. Sta. 60. 84 pp.

Knowlton, C. H. 1900. Further notes on the flora of Worcester County, Massachusetts. Rhodora 2:201.

Knowlton, F. H. 1898. A catalogue of Cretaceous and Tertiary plants of North America. Bull. U.S. Geol. Surv. 152. 247 pp.

Köppen, W. 1936. Das geographische System der Klimate. *In* W. Köppen and R. Geiger, Handbuch der Klimatologie. Gebr. Borntraeger. Berlin. Vol. 1(C). 44 pp.

Lake States Forest Experiment Station. 1942. Jack pine reverts to balsam fir in northeastern Minnesota. Tech. Note Lake St. For. Exp. Sta. 188. 1 p.

Lakela, O. 1948. Ferns and flowering plants of Beaver Island, Lake Superior, Minnesota. Bull. Torrey Bot. Cl. 75:265–271.

LaMotte, R. S. 1952. Catalogue of the Cenozoic plants of North America through 1950. Mem. Geol. Soc. Amer. 51. 381 pp.

Lee, S. C. 1924. Factors controlling forest successions at Lake Itasca, Minnesota. Bot. Gaz. 78:129–174.

Leisman, G. A. 1957. A vegetation and soil chronosequence on the Mesabi Iron Range soil banks, Minnesota. Ecol. Monogr. 27:221–245.

Linteau, A. 1955. Forest site classification of the northeastern section Boreal forest region, Quebec. Bull. For. Br. Can. 118. 85 pp.

Lipa, I. L. 1940. Supplement to the gymnosperms of gardens and parks in the Ukrainian SSR. Bot. Zh. 1:119–126.

Little, E. L. 1949. Important forest trees of the United States. Pp. 763–814. *In* Trees, the yearbook of agriculture, 1949. U.S. Dep. Agric. Washington, D.C.

———. 1953. Check list of native and naturalized trees of the United States. Agric. Handb. U.S. Dep. Agric. 41. 472 pp.

Livingston, B. E. 1921. The distribution of vegetation in the United States, as related to climatic conditions. Publ. Carnegie Inst. Washington 284. 585 pp.

Long, H. D. 1952. Forest types and sites of the Acadia Forest Experiment Station. Master's thesis. Univ. of New Brunswick.

Loomis, F. B. 1938. Physiography of the United States. Doubleday, Doran & Co. New York, N.Y. 350 pp.

Loucks, O. L. 1956. Site classification, Acadia Forest Experiment Station. Progr. Rep. For. Br. Maritime Distr. M-226. 48 pp.

———. 1960. Environmental and phytosociological ordination of a regional forest vegetation. Ph.D. thesis. Univ. Wisconsin. Micr. Xerox Publ. Univ. Microfilms. Ann Arbor, Mich. 213 pp.

———. 1962a. Ordinating forest communities by means of environmental scalars and phytosociological indices. Ecol. Monogr. 32:137–166.

———. 1962b. A forest classification for the Maritime Provinces. Proc. Nova Scotian Inst. Sci. 25:85–167.

Loudon, C. 1838. Arboretum et fruticetum Britannicum. Vol. 4:2339. Longman, Orme, Brown, Green and Longmans. London.

Macbride, T. H. 1895. Forest trees of Allamakee County, Iowa. Iowa Geol. Surv. (1894) 4:112–120.

MacLean, D. W. 1949. Forest development on the Goulais River watershed, 1910–1946. Silv. Res. Note For. Serv. Can. 94. 54 pp.

———. 1960. Some aspects of the aspen-birch-spruce-fir type in Ontario. Tech. Note For. Res. Br. Can. 94. 24 pp.

Mägdefrau, K. 1953. Paläobiologie der Pflanzen. 2nd ed. Fischer. Jena. 438 pp.

Marr, J. W. 1948. Ecology of the forest-tundra ecotone on the East Coast of Hudson Bay. Ecol. Monogr. 18:117–144.

Martin, N. D. 1959. An analysis of forest succession in Algonquin Park, Ontario. Ecol. Monogr. 29:187–218.

Martinez, M. 1953. Las Pinaceas Mexicanas. Secretaria de Agricultura y Ganaderia. Mexico, D.F. 362 pp.

Mattfeld, J. 1925. Die in Europa und dem Mittelmeergebiet wildwachsenden Tannen. Mitt. Dtsch. Dendrol. Ges. 35:1–37.

Maycock, P. F. 1961. The spruce-fir forests of the Keweenaw Peninsula, northern Michigan. Ecology 42:357–365.

———, and J. T. Curtis. 1960. The phytosociology of boreal-conifer-hardwood forests of the Great Lakes region. Ecol. Monogr. 30:1–35.

Moss, E. H. 1953. Forest communities in northwestern Alberta. Canad. J. Bot. 31:212–252.

Moyle, J. B. 1946. Relict boreal plants in southern Minnesota. Rhodora 48:163.

Müller, K. M. 1938. *Abies grandis* und ihre Klimarassen. Neumann. Neudamm. 118 pp.

Munger, T. T. 1947. The Wind River Arboretum from 1937 to 1946. Prog. Rep. Pacif. Nthwest. For. Range Exp. Sta. 3. 25 pp.

———, and E. L. Kolbe. 1932. The Wind River Arboretum from 1912 to 1932. Rep. Pacif. Nthwest. For. Range Exp. Sta. 24 pp.

———. 1937. The Wind River Arboretum from 1932 to 1937. Rep. Pacif. Nthwest. For. Range Exp. Sta. 15 pp.

Munns, E. N. 1938. The distribution of important forest trees of the United States. Misc. Publ. U.S. Dep. Agric. 287. 176 pp.

Nachtigall. 1943. Über Bedeutung und Anbau fremdländischer Holzarten. Thar. Forstl. Jb. 94:141–186.

Newberry, J. S. 1898. The later extinct floras of North America. U.S. Surv. Monogr. 35:151.

Nichols, G. E. 1913. The vegetation of Connecticut. Torreya 13:87–112.

———. 1935. The hemlock–white pine–northern hardwood region of eastern North America. Ecology 16:394–422.

Nielsen, E. L. 1935. A study of a pre-Kansan peat deposit. Torreya 35:53–56.

North Central Agricultural Experiment Stations. 1960. Soils of the North Central region of the United States. Bull. Wis. Agric. Exp. Sta. 544. 192 pp.

Oosting, H. J. 1956. The study of plant communities — an introduction to plant ecology. Freeman and Co. San Francisco, Calif. 440 pp.

———, and W. D. Billings. 1951. A comparison of virgin spruce-fir forest in the northern and southern Appalachian system. Ecology 32:84–103.

Pammel, L. H. 1920. Yellow river region in Allamakee county. Rep. Iowa St. Rd. Conserv. 1919:163.

Peattie, D. C. 1950. A natural history of trees of eastern and central North America. Houghton Mifflin. Boston, Mass. 606 pp.

Penhallow, D. P. 1907. A manual of North American gymnosperms. Ginn and Co. Boston, Mass. 374 pp.

Place, I. C. M. 1955. The influence of seed-bed conditions on the regeneration of spruce and balsam fir. Bull. For. Br. Can. 117. 87 pp.

Plochmann, R. 1956. Bestockungsaufbau und Baumartenwändel nordischer Urwälder. Beih. Forstwiss. Cbl. Forstwiss. Forsch. 6:1–96.

Pogrebniak, P. S. 1930. Über die Methodik der Standortsuntersuchungen in Verbindung mit den Waldtypen. Proc. Congr. Int. Union For. Res. Org. Stockholm, 1929. Pp. 455–471.

———. 1955. Osnovi lesnoi tipologii. Akad. Nauk Ukrainskoi SSR, Kiev. 455 pp.

Pool, R. J. 1939. Some reactions of the vegetation in the towns and cities of Nebraska to the great drought. Bull. Torrey Bot. Cl. 66:457–464.

Potter, L. D. 1947. Postglacial forest sequence of north-central Ohio. Ecology 28:396–417.

Potzger, J. E. 1941. The vegetation of Mackinac Island, Michigan: An ecological survey. Amer. Midl. Nat. 25:298–323.

———. 1942. Pollen spectra from four bogs on the Gillen Nature Reserve along the Michigan-Wisconsin state line. Amer. Midl. Nat. 28:501, 511.

———. 1944. Pollen frequency of *Abies* and *Picea* in peat: a correction on some published records from Indiana bogs and lakes. Butler Univ. Bot. Studies 6:123–130.

———. 1949. Pollen study in the Quetico-Superior Memorial Forest. Amer. J. Bot. 36:819.

———. 1950. Bogs of the Quetico-Superior country tell its forest history. The President's Quetico-Superior Committee. 26 pp.

———. 1953a. History of forests in the Quetico-Superior country from fossil pollen studies. J. For. 51:560–565.

———. 1953b. Nineteen bogs from southern Quebec. Canad. J. Bot. 31:383–401.

———, and A. Courtemanche. 1954. Bog and lake studies on the Laurentian Shield in Mont Tremblant Park, Quebec. Canad. J. Bot. 32:549–560.

Potzger, J. E., and B. C. Tharp. 1947. Pollen profile from a Texas bog. Ecology 28:274–280.

Potzger, J. E., and H. B. Wales. 1950. Trees of the Quetico-Superior country. The President's Quetico-Superior Committee. 37 pp.

Preston, R. J. 1948. North American trees. 1st ed. (2nd ed., 1961). Iowa St. Coll. Press. Ames, Ia. 371 pp.

Raup, H. M. 1935. Botanical investigations in Wood Buffalo Park. Bull. Nat. Mus. Can. 74. 174 pp.

———. 1946. Phytogeographic studies in the Athabaska–Great Slave Lake region II. J. Arnold Arbor. 27:1–85.

Ray, R. G. 1941. Site-types and rate of growth, Lake Edward, Champlain County, P.Q., 1915–1936. Silv. Res. Note For. Serv. Can. 65. 56 pp.

————. 1956. Site-types, growth and yield at the Lake Edward Forest Experiment Area, Quebec. Tech. Note For. Br. Can. 27. 53 pp.

Rehder, A. 1939. The firs of Mexico and Guatemala. J. Arnold Arboretum 20:281–287.

Roe, E. I. 1935. Forest soils — the basis of forest management. Lake St. For. Exp. Sta. 9 pp.

Rosendahl, C. O. 1948. A contribution to the knowledge of the Pleistocene flora of Minnesota. Ecology 29:284–315.

————. 1955. Trees and shrubs of the Upper Midwest. Univ. Minnesota Press. Minneapolis, Minn. 411 pp.

Roth, F. 1898. Forestry conditions and interests of Wisconsin. Bull. U.S. Dep. Agric. 16. 76 pp.

Roth, J. 1920. Maifrostschäden an Exoten. Cbl. Ges. Forstw. 46:151–161.

Rowe, J. S. 1955. Factors influencing white spruce reproduction in Manitoba and Saskatchewan. Tech. Note For. Br. Can. 3. 27 pp.

————. 1959. Forest regions of Canada. Bull. For. Br. Can. 123. 71 pp.

————. 1961. Critique of some vegetational concepts as applied to forests of northwestern Alberta. Can. J. Bot. 39:1007–1017.

Russell, N. H. 1953. Plant communities of the Apple-River Canyon, Wisconsin. Proc. Iowa Acad. Sci. 60:238–242.

Rydberg, P. A. 1906. Flora of Colorado. Bull. Agric. Exp. Sta. Colorado Agric. College 100. 447 pp.

————. 1912. Phytogeography and its relation to taxonomy and other branches of science. Torreya 12:73–85.

Sargent, C. S. 1884. Forests of North America. U.S. Dep. Int. Census Off. Washington, D.C. 612 pp.

————. 1898. The Silva of North America. Vol. 12. Houghton Mifflin. Boston, Mass. 144 pp.

Schelle, E. 1920. Mitteilungen über Koniferen. Mitt. Dtsch. Dendrol. Ges. 29:37–52.

Schenck, C. A. 1939. Fremdländische Wald- und Parkbäume. Parey. Berlin. Vol. 1–2.

Schimper, A. F. W. 1903. Plant-geography upon a physiological basis. English translation by W. R. Fisher, revised and edited by Percy Groom and I. B. Balfour. Clarendon Press. Oxford. 839 pp.

Schmucker, O. 1942. Die Baumarten der nördlich gemässigter Zone und ihre Verbreitung. Silvae Orbis 4:1–156 + 250 distribution maps.

Schwerin, F. 1929. Botanische und forstliche Mitteilungen über Koniferen. Mitt. Dtsch. Dendrol. Ges. 41:159–174.

Sears, P. B. 1930. Common fossil pollen of the Erie Basin. Bot. Gaz. 87:95–106.

————. 1938. Climatic interpretation of postglacial pollen deposits in North America. Bull. Amer. Meteor. Soc. 19:177–181.

————. 1941. Postglacial vegetation in the Erie-Ohio area. Ohio J. Sci. 41:225–234.

————. 1942. Forest sequences in the north-central states. Bot. Gaz. 103:751–761.

Shantz, H. L., and R. Zon. 1924. Natural vegetation. Atlas of Amer. Agric. U.S. Dep. Agric. 24 pp.

Shimek, B. 1906. Flora of Winneshiek County. Iowa Geol. Surv. (1905) 16:165.

Silen, R. R., and L. R. Woike. 1959. The Wind River Arboretum. Res. Pap. Pacif. Nthwest. For. Range Exp. Sta. 33. 50 pp.

Sisam, J. W. B. 1938a. The correlation of tree species and growth with site-types. Silv. Res. Note For. Serv. Can. 53. 19 pp.

————. 1938b. Site as a factor in silviculture — its determination with special reference to the use of plant indicators. Silv. Res. Note For. Serv. Can. 54. 81 pp.

Sivers, M. 1911. Dendrologische Mitteilungen aus der baltischen Provinzen. Mitt. Dtsch. Dendrol. Ges. 20:150–164.

Society of American Foresters. 1940. Forest cover types of the eastern United States. Society of Amer. For. Washington, D.C. 39 pp.

————. 1954. Forest cover types of North America (exclusive of Mexico). Soc. Amer. For. Washington, D.C. 67 pp.

Soper, J. H., and P. F. Maycock. 1963. A community of arctic-a pine plants on the east shore of Lake Superior. Canad. J. Bot. 41:183–198.

Spaulding, P. 1956. Diseases of North American forest trees planted abroad. Agr. Handb. U.S. Dep. Agric. 100. 144 pp.

Stallard, H. 1929. Secondary succession in the climax forest formations of northern Minnesota. Ecology 10:476–547.

Stott, C. B., W. W. Barton, and J. H. Stone. 1942. Results of forest cutting practices in various Lake States timber types. U.S. For. Serv. Reg. 9. Milwaukee, Wis. 56 pp.

Sudia, T. W. 1952. A statistical analysis of twenty pollen spectra from a single stratum of Amanda bog (Ohio). Ohio J. Sci. 52:213–215.

Sudworth, G. B. 1898. Check list of the forest trees of the United States, their names and regions. Bull. U.S. Div. For. 17. 144 pp.

———. 1916. The spruce and balsam fir trees of the Rocky Mountain region. Bull. U.S. Dep. Agric. 327. 43 pp.

———. 1927. Check list of the forest trees of the United States, their names and ranges. Misc. Circ. U.S. Dep. Agric. 92. 295 pp.

Sukachev, V. N. 1932. Die Untersuchungen der Waldtypen des osteuropäischen Flachlandes. Abderhalden's Handb. Biol. Arbeitsmethoden 11(6):191–250.

Sylvain, C. 1940. Compte rendu préliminaire du travail forestier à la station de Kenscoff, Haiti. Carib. For. 1(2):16–22.

Thornthwaite, C. W. 1941. Atlas of the climatic types in the United States 1900–1939. Misc. Publ. U.S. Dep. Agric. 421. 7 pp. + 96 pl.

———. 1948. An approach toward a rational classification of climate. Geogr. Rev. 38:55–94.

———, and J. R. Mather. 1955. The water balance. Drexel Inst. Tech. Publ. in Climatology 8(1):1–104.

Tryon, E. H., A. W. Goodspeed, R. P. True, and C. J. Johnson. 1951. Christmas trees. Their profitable production in West Virginia. Circ. W. Va. Agric. Exp. Sta. 82. 28 pp.

United States Department of Agriculture. 1938. Soils and men. Yearb. Agric. U.S. Dep. Agric. 1232 pp.

———. 1941. Climate and man. Yearb. Agric. U.S. Dep. Agric. 1248 pp.

United States Forest Service. 1949. Areas characterized by major forest types in the United States. (A map.) U.S. For. Serv. Washington, D.C.

———. 1958. Timber resources for America's future. For. Res. Rep. U.S. Dep. Agric. 14. 713 pp.

Upham, W. 1884. Catalogue of the flora of Minnesota. Johnson, Smith, and Harrison. Minneapolis, Minn. 193 pp.

Viguié, M. T., and H. Gaussen. 1928–29. Revision du genre *Abies*. Bull. Soc. Hist. Nat. Toulouse 57:369–434; 58:245–564.

Voss, J. 1933. Pleistocene forests of central Illinois. Bot. Gaz. 94:808–814.

———. 1934. Postglacial migration of forests in Illinois, Wisconsin, and Minnesota. Bot. Gaz. 96:3–43.

Wagner, H. 1954. Gedanken zur Berücksichtigung der Mehrdimensionalen Beziehungen der Pflanzengesellschaften in der Vegetationssystematik. Proc. Int. Bot. Congr. (Paris) 8(7):9–11.

———. 1958. Grundfragen der Systematik der Waldgesellschaften. *In* Festschrift Werner Lüdi. Veröfftl. Inst. Rübel. (Zürich) 33:241–252.

Wang, C. W. 1961. The forests of China with a survey of grasslands and desert vegetation. Publ. Maria Moors Cabot Found. Harvard Univ. 5. 313 pp.

Waring, R. H. 1959. Some characteristics of the upland forest types in the rock outcrop area of northeastern Minnesota. Master's thesis. Univ. Minnesota. 49 pp.

Wein, K. 1931. Die erste Einführung nordamerikanischer Gehölze in Europe. II. Mitt. Dtsch. Dendrol. Ges. 43:95–160.

Wenner, C. 1948. Pollen diagrams from Labrador. Geografiska Annaler 29:137–374.

Westveld, M. 1931. Reproduction on pulpwood lands in the Northeast. Tech. Bull. U.S. Dep. Agric. 223. 52 pp.

———. 1952. A method of evaluating forest site quality from soil, forest cover, and indicator plants. Sta. Pap. Ntheast. For. Exp. Sta. 48. 12 pp.

———. 1953. Ecology and silviculture of the spruce-fir forests of eastern North America. J. For. 51:422–430.

———, chmn. 1956. Natural forest vegetation zones of New England. J. For. 54:332–338.

Whitehead, D. R., and E. S. Barghoorn. 1962. Pollen analytical investigations of Pleistocene deposits from western North Carolina and South Carolina. Ecol. Monogr. 32:347–369.

Wilde, S. A. 1958. Forest soils, their properties and relation to silviculture. Ronald Press. New York, N.Y. 537 pp.

———, F. G. Wilson, and D. P. White. 1949. Soils of Wisconsin in relation to silviculture. Publ. Wis. Conserv. Dep. 525-49. 171 pp.

Wilson, L. R. 1938. The postglacial history of vegetation in northwestern Wisconsin. Rhodora 40:137–145.

———, and R. M. Kosanke. 1940. The microfossils in a pre-Kansan peat deposit near Belle Plaine, Iowa. Torreya 40:1–5.

Wilson, L. R., and R. M. Webster. 1942. Fossil evidence of wider post-Pleistocene range for butternut and hickory in Wisconsin. Rhodora 44:409–414.

Wilton, W. C. 1956. Forest resources of Avalon Peninsula, Newfoundland. Tech. Note For. Br. Can. 50. 33 pp.

———. 1959. Forest types of the Grand Lake and Northwestern Lake areas of Labrador. Tech. Note For. Br. Can. 83. 30 pp.

Winchell, N.H. 1884. The geology of Mower County. Geol. Survey of Minnesota. Final Report. 1:347–366.

Winter, T. C. 1962. Pollen sequence of Kirchner Marsh, Minnesota. Science 138:526–528.

Wodehouse, R. P. 1933. Tertiary Pollen. II. The oil shales of the Eocene Green River formation. Bull. Torrey Bot. Cl. 60:470–524.

Zehetmayr, J. W. L. 1954. Experiments in tree planting on peat. Bull. Great Brit. For. Comm. 22. 110 pp.

Zon, R. 1914. Balsam fir. Bull. U.S. For. Serv. 55. 68 pp.

Zonn, S. W. 1955. Die biogeozönotische Methode und ihre Bedeutung für die Erforschung der Rolle der biologischen Faktoren in der Bodengenese unter Wald. Arch. Forstw. 4:578–587.

Chapter 3. Ecological Factors

Abbott, H. G. 1962. Tree seed preferences of mice and voles in the Northeast. J. For. 60:97–99.

———, and A. C. Hart. 1960. Mice and voles prefer spruce seeds. Sta. Pap. Ntheast. For. Exp. Sta. 153. 12 pp.

Ahlgren, C. E., and H. L. Hansen. 1957. Some effects of temporary flooding on coniferous trees. J. For. 55:647–650.

Aldous, S. E. 1942. Winter deer food problems. Conserv. 5(27):37–39.

———. 1948. Fall and winter food habits of deer in northeastern Minnesota. Wildl. Leafl. U.S. Dep. Int. 310. 8 pp.

———, and L. W. Krefting. 1946. The present status of moose on Isle Royale. Trans. North Amer. Wildlife Conf. 11:296–308.

American Ornithologists' Union. 1957. Check-list of North American birds. 5th ed. Lord Baltimore Press. Baltimore, Md. 691 pp.

Arnold, C. I. 1958. Frost damage. Tree Plant. Notes 31:8–9.

Averell, J. L., and P. C. McGrew. 1929. The reaction of swamp forests to drainage in northern Minnesota. Minn. Comm. Drainage and Waters. 66 pp.

Baker, F. S. 1950. Principles of silviculture. McGraw-Hill. New York, N.Y. 414 pp.

Bakuzis, E. V. 1959. Synecological coordinates in forest classification and in reproduction studies. Ph.D. thesis. Univ. Minnesota. Micr. Xerox Publ. Univ. Microfilms. Ann Arbor, Mich. 242 pp.

———. 1961. Synecological coordinates and investigation of forest ecosystems. Pap. Congr. Int. Union For. Res. Org. 13. Vienna. 10 pp.

Balch, R. E. 1942. A note to squirrel damage to conifers. For. Chron. 18:42.

Bartlett, I. H. 1950. Michigan deer. Mich. Dep. Conserv. 50 pp.

Behre, C. E. 1921. A study of windfall in the Adirondacks. J. For. 19:632–637.

Bent, A. C. 1932. Life histories of the North American gallinaceous birds. Bull. U.S. Nat. Mus. 162. 490 pp.

Bonner, E. 1941. Balsam fir in the Clay Belt of northern Ontario. Master's thesis. Univ. Toronto. 102 pp.

Bowman, A. B. 1944. Growth and occurrence of spruce and fir on pulpwood lands in northern Michigan. Tech. Bull. Mich. Agric. Exp. Sta. 188. 82 pp.

Boyko, H. 1947. On the role of plants as quantitative climate indicators and the geo-ecological law of distribution. J. Ecol. 35:138–157.

Brooks, M. 1955. An isolated population of the Virginia varying hare. J. Wildlife Mgmt. 19:54–61.

Bump, G., R. W. Darrow, F. C. Edminster, and W. F. Crissey. 1947. The ruffed grouse. N.Y. St. Conserv. Dep. Albany, N.Y. 915 pp.

Cahalane, V. H. 1947. Mammals of North America. Macmillan. New York, N.Y. 682 pp.

Canada Forest Biology Division. 1942–59. Annual report of the forest insect and disease survey. Dep. Agric. Can.

Canada Forest Entomology and Pathology Branch. 1960–62. Annual report of the forest insect and disease survey. Dep. For. Can.

Canada Geographical Branch. 1957. Atlas of Canada. Dep. Mines Tech. Surv. Can. 110 pl.

Cayford, J. H., and R. A. Haig. 1961. Glaze damage in forest stands in southeastern Manitoba. Tech. Note For. Res. Br. Can. 102. 16 pp.

Chandler, R. F. 1944. Amount of mineral nutrients of freshly fallen needle litter of some northeastern conifers. Proc. Soil Sci. Soc. Amer. 8:409–411.

Cheyney, E. G. 1942. American silvics and silviculture. Univ. Minnesota Press. Minneapolis, Minn. 472 pp.

Cook, D. B., and S. B. Robeson. 1945. Varying hare and forest succession. Ecology 26:406–410.

Cringan, A. T. 1957. History, food habits, and range requirements of woodland caribou of continental North America. Trans. North Amer. Wildlife Conf. 22:485–501.

Curry, J. R., and T. W. Church. 1952. Observations on winter drying of conifers in the Adirondacks. J. For. 50:114–116.

Curtis, J. T. 1955. A prairie continuum in Wisconsin. Ecology 36:558–566.

———. 1959. The vegetation of Wisconsin. Univ. Wisconsin Press. Madison, Wis. 657 pp.

Dahlberg, B. L., and R. C. Guettinger. 1956. The white-tailed deer of Wisconsin. Tech. Wildl. Bull. Wis. Conserv. Dep. 14. 282 pp.

Davenport, L. A. 1939. Results of deer feeding experiments at Cusino, Michigan. Trans. North Amer. Wildlife Conf. 4:268–274.

Davis, K. P. 1959. Forest fire, control and use. McGraw-Hill. New York, N.Y. 584 pp.

Dell, V. 1952–53. Snowshoe rabbits. New homes for displaced animals. N.Y. St. Conserv. (Dec.–Jan.):23.

Deters, M. E., and H. Schmitz. 1936. Drought damage to prairie shelter-belts in Minnesota. Bull. Minn. Agric. Exp. Sta. 329. 28 pp.

Dice, L. R. 1952. Natural Communities. Univ. Michigan Press. Ann Arbor, Mich. 547 pp.

Diebold, G. H. 1941. Effect of fire and logging upon the depth of the forest floor in the Adirondack region. Proc. Soil Sci. Soc. Amer. 6:409–413.

Donahue, R. L. 1940. Forest-site quality studies in the Adirondacks. I. Tree growth as related to soil morphology. Mem. Cornell Agric. Exp. Sta. 229. 44 pp.

Dyer, H. J. 1948. Preliminary plan for wildlife management on Baxter State Park. Master's thesis. Univ. Maine. 79 pp.

Edminster, F. C. 1947. The ruffed grouse. Macmillan. New York, N.Y. 385 pp.

Ellenberg, H. 1950. Unkrautgemeinschaften als Zeiger für Klima und Boden. Ulmer. Stuttgart. 141 pp.

Erickson, A. B., V. E. Gunvalson, D. W. Burcalow, M. H. Stenlund, and L. H. Blankenship. 1961. The white-tailed deer of Minnesota. Tech. Bull. Minn. Dep. Conserv. 5. 64 pp.

Etheridge, D. E., and L. A. Morin. 1961. Decay associated with bear damage to balsam fir in the Gaspé Peninsula. Bi-m. Progr. Rep. For. Ent. Path. Br. Can. 17(6):1.

Flaccus, E. 1959. Revegetation of landslides in the White Mountains of New Hampshire. Ecology 40:692–703.

Gates, F. C., and E. C. Woollett. 1926. The effect of inundation above a beaver dam upon upland vegetation. Torreya 26:45–50.

Gill, J. D. 1957. Review of deer yard management. Bull. Maine Dep. Inl. Fish. Game Div. 5. 61 pp.

Graham, S. A. 1954a. Scoring tolerance of forest trees. Mich. For., 4. 2 pp.

———. 1954b. Changes in northern Michigan forests from browsing from deer. Trans. North Amer. Wildlife Conf. 19:526–533.

Grange, W. B. 1932. Observations on the snowshoe hare *Lepus americanus phaenotus*. Allen. J. Mammal. 13:1–19.

———. 1948. Wisconsin grouse problems. Publ. Wis. Conserv. Dep. 328. 318 pp.

———. 1949. The way to game abundance. Scribner's Sons. New York, N.Y. 365 pp.

Grant, M. L. 1934. The climax forest community in Itasca County, Minnesota, and its bearing upon the successional status of the pine community. Ecology 15:243–257.

Grisez, T. J. 1954. Results of a mouse census in a spruce-fir-hemlock forest in Maine. For. Res. Note Ntheast. For. Exp. Sta. 28. 2 pp.

———. 1955. 1954 hurricane damage on Penobscot Experiment Forest. J. For. 53:207.

Hall, C. W. 1880. Report of Professor C. W. Hall. *In* Geological and natural history survey of Minnesota. 8th Ann. Rep. for the Year 1879. Chapt. VI, p. 134.

Halliday, W. E. D. 1937. A forest classification for Canada. Bull. For. Serv. Can. 89. 50 pp.

Hamilton, W. V. 1941. The food of some small forest mammals in eastern United States. J. Mammal. 22:250–263.

———, and D. B. Cook. 1940. Small mammals and the forest. J. For. 36:468–473.

Hansen, H. L., and L. W. Rees. 1947. Late spring frost damage to forest trees in Minnesota, May, 1947. Unpubl. Rep. Minn. For. Sch. 10 pp.

Hart, A. C. 1956. Changes after partial cutting of a spruce-fir stand in Maine. Sta. Pap. Ntheast. For. Exp. Sta. 86. 8 pp.

Hawley, R. C., and P. W. Stickel. 1948. Forest protection. Wiley & Sons. New York, N.Y. 355 pp.

Heimburger, C. C. 1941. Forest-site classification and soil investigation on Lake Edward Forest Experimental Area. Silv. Res. Note For. Serv. Can. 66. 60 pp.

Hickie, P. F. (Undated.) Michigan moose. Game Div. Mich. Dep. Conserv. 57 pp.

Hoffman, R. S., R. G. Janson, and F. Hartkorn. 1958. Effect on grouse populations of DDT spraying for spruce budworm. J. Wildlife Mgmt. 22:92–93.

Hosley, N. W. 1949. The moose and its ecology. Wildlife Leafl. U.S. Dep. Int. 312. 51 pp.

Howard, W. J. 1937. Notes on winter foods of Michigan deer. J. Mammal. 18:77–80.

Hutchinson, A. H. 1918. Limiting factors in relation to specific ranges of tolerance of forest trees. Bot. Gaz. 66:465–493.

Jenkins, D. H., and I. H. Bartlett. 1959. Michigan whitetails. Game Div. Mich. Dep. Conserv. 80 pp.

Jenny, H. 1941. Factors of soil formation. McGraw-Hill. New York, N.Y. 281 pp.

Johnson, C. E. 1927. The beaver in the Adirondacks. Roosevelt Wildlife Bull. 4:501–641.

Kell, L. L. 1938. The effect of the moisture-retaining capacity of soils on forest succession in Itasca Park, Minnesota. Amer. Midl. Nat. 20:682–694.

Kellogg, R. 1939. Annotated list of Tennessee mammals. Proc. U.S. Nat. Mus. 86:245–303.

Kelly, J. A. B., and I. C. M. Place. 1950. Windfirmness of residual spruce and fir. Pulp Paper Mag. Can. 51(6):124.

Kendeigh, S. C. 1947. Bird population studies in the coniferous forest biome during a spruce budworm outbreak. Biol. Bull. Dep. Lds. For. Ont. 1. 100 pp.

King, R. T. 1937. Ruffed grouse management. J. For. 35:523–532.

Krefting, L. W. 1941. Methods of increasing deer browse. J. Wildlife Mgmt. 5:95–102.

———. 1951. What is the future of the Isle Royale moose herd? Trans. North Amer. Wildl. Conf. 16:461–72.

———. 1956. Meadow mouse damage to a mixed plantation of red pine and balsam fir. Unpubl. Rep. U.S. Fish Wildl. Serv. 3 pp.

———. 1959. Survival and growth of some wildlife cover plantings in Minnesota. Minn. For. Notes 79. 2 pp.

———. 1963. The beaver of Isle Royale, Lake Superior. Naturalist 14(2):2–11.

———, A. B. Erickson, and V. E. Gunvalson. 1955. Results of deer hunts on the Tamarac National Wildlife Refuge. J. Wildlife Mgmt. 19:346–352.

Krefting, L. W., H. L. Hansen, and M. H. Stenlund. 1955. Use of herbicides in inducing regrowth of mountain maple for deer browse. Minn. For. Notes 42. 2 pp.

———. 1956. Stimulating regrowth of mountain maple for deer browse by herbicides, cutting, and fire. J. Wildlife Mgmt. 20:434–441.

Krefting, L. W., and E. I. Roe. 1949. The role of some birds and mammals in seed germination. Ecol. Monogr. 19:269–286.

Krefting, L. W., and M. H. Stenlund. 1951. Poor winter yards — fewer deer. Conserv. 14(80):16–20.

Krefting, L. W., J. H. Stoeckeler, B. J. Bradle, and W. D. Fitzwater. 1962. Porcupine-timber relationships in the Lake States. J. For. 60:325–330.

Lake States Forest Experiment Station. 1936. Woody food preferences of the snowshoe rabbit in the Lake States. Tech. Note Lake St. For. Exp. Sta. 109. 1 p.

———. 1938. Food habits of Minnesota deer. Tech. Note. Lake St. For. Exp. Sta. 113. 1 p.

———. 1939. Comparative resistance of native Wisconsin trees to snow breakage. Tech. Note Lake St. For. Exp. Sta. 152. 1 p.

———. 1959. Deer browse surveys in northern hardwoods and aspen-balsam fir. Annual Report 1958. Lake St. For. Exp. Sta. 64 pp.

Lee, S. C. 1924. Factors controlling forest successions at Lake Itasca, Minnesota. Bot. Gaz. 78:129–174.

Leinfelder, R. P., and H. L. Hansen. 1954. Tolerance of conifers to various applications of 2,4-D and 2,4,5-T in Minnesota. Res. Rep. Nth. Centr. Weed Contr. Conf. Pp. 133–134.

Leisman, G. A. 1957. A vegetation and soil chronosequence on the Mesabi Iron Range spoil banks. Ecol. Monogr. 27:221–245.

Lemieux, G. J. 1961. An evaluation of Paterson's CVP-index in eastern Canada. Tech. Note Dep. For. Can. 112. 10 pp.

Lemon, P. C. 1961. Forest ecology of ice storms. Bull. Torrey Bot. Cl. 88:21–29.

Leopold, A. 1933. Game management. Scribner's Sons. New York, N.Y. 481 pp.

———. 1947. The distribution of Wisconsin hares. Trans. Wis. Acad. Sci. 37:1–14.

Lieth, H., and R. Quellette. 1962. Studies on the vegetation of the Gaspé Peninsula. II. The soil respiration of some plant communities. Canad. J. Bot. 40:127–140.

Littlefield, E. W., W. S. Schoonmaker, and D. B. Cook. 1946. Field mouse damage to coniferous plantations. J. For. 44:745–749.

Loucks, O. L. 1956. Site classification, Acadia Forest Experiment Station. Progr. Rep. For. Br. Can. Maritime Distr. M-266. 48 pp.

———. 1962. Ordinating forest communities by means of environmental scalars and phyto-sociological indices. Ecol. Monogr. 32:137–166.

Lundgren, A. L. 1954. An investigation of the 1953 blowdown in Itasca State Park. Unpubl. Rep. Univ. Minnesota. 47 pp.

Lutz, H. J., and R. F. Chandler. 1951. Forest soils. Wiley & Sons. New York, N.Y. 514 pp.

McLintock, T. F. 1954. Factors affecting wind damage in selectively cut stands of spruce and fir in Maine and Northern New Hampshire. Sta. Pap. Ntheast. For. Exp. Sta. 70. 17 pp.

———. 1958. Hardpans in the spruce-fir forests of Maine. Proc. Soc. Amer. For. (1957):65–66.

MacLulich, D. A. 1937. Fluctuations in the numbers of the varying hare (*Lepus americanus*). Biol. Ser. Toronto Univ. Studies 43. 136 pp.

Magnus, L. T. 1949. Cover type use of the ruffed grouse in relation to forest management on the Cloquet Experiment Station. Flicker 21:29–44, 73–85.

Maki, T. E. 1950. Some factors affecting the composition and development of forest floor. Ph.D. thesis. Univ. Minnesota. 85 pp.

Marshall, W. H., and K. E. Winsness. 1953. Ruffed grouse on the Cloquet experimental forest. Minn. For. Notes 20. 2 pp.

Martin, A. C., H. S. Zim, and A. L. Nelson. 1951. American wildlife and plants. McGraw-Hill. New York, N.Y. 500 pp.

Martin, N. D. 1960. An analysis of bird populations in relation to forest succession in Algonquin Provincial Park, Ontario. Ecology 41:126–140.

Maycock, P. F. 1957. The phytosociology of boreal conifer-hardwood forests of the Great Lakes region. Diss. Abstr. 17(10):2132.

Melin, E. 1930. Biological decomposition of some types of litter from North American forests. Ecology 11:72–101.

Merriam, C. H. 1898. Life zones and crop zones of the United States. Biol. Surv. Bull. U.S. Dep. Agric. 10. 79 pp.

Millar, J. B. 1936. The silvicultural characteristics of black spruce in the Clay Belt of northern Ontario. Master's thesis. Univ. Toronto. 81 pp.

Miller, G. S., and R. Kellogg. 1955. List of North American recent mammals. Bull. U.S. Nat. Mus. 205. 954 pp.

Morris, R. F. 1955. Population studies on some small forest mammals in eastern Canada. J. Mammal. 36:21–35.

Murie, A. 1934. The moose of Isle Royale. Misc. Publ. Mich. Univ. Mus. Zool. 25. 44 pp.

Newsom, W. M. 1937. Mammals on Anticosti Island. J. Mammal. 18:435–442.

Paterson, S. S. 1961. Introduction to phyochorology of Norden. Medd. Skogsforskn. Inst. Stockholm. 50(5):1–145.

Patrick, E. F., and W. L. Webb. 1954. A preliminary report on intensive beaver management. J. For. 52:31–32.

Pearce, J. 1937. The effect of deer browsing on certain western Adirondack forest types. Roosevelt Wildlife Bull. 7:1–61.

Peterson, R. I. 1955. North American moose. Univ. Toronto Press. Toronto, Ont. 280 pp.

Pierce, R. S. 1953. Oxidation-reduction potential and specific conductance of ground water: their influence on natural forest distribution. Proc. Soil Sci. Soc. Amer. 17:61–65.

Pimlott, D. H. 1953. A survey of moose damage to forest reproduction. Rep. Newfoundld. For. Prot. Ass. 1952/53:19–23.

———. 1955. Moose and the Newfoundland forests. A report to the Nfld. Royal Commission on Forestry. Wildlife Div. Nfld. Dep. Mines Res. 26 pp.

———. 1961. The ecology and management of moose in North America. Terre et Vie 2:246–265.

Place, I. C. M. 1955. The influence of seed-bed conditions on the regeneration of spruce and balsam fir. Bull. For. Br. Can. 117. 87 pp.

Pogrebniak, P. S. 1930. Über die Methodik der Standortsuntersuchungen in Verbindung der Waldtypen. Proc. Intern. Congr. For. Res. Org. Stockholm, 1929. Pp. 455–471.

———. 1955. Osnovi lesnoj tipologii. Akad. Nauk Ukrainskoi SSR. Kiev. 455 pp.

Pomerleau, R. 1944. Observations sur quelques maladies non parasitaires des arbres dans le Québec. Canada J. Res. 22C:171–189.

Pool, R. J. 1939. Some reactions of the vegetation in the towns and cities of Nebraska to the great drought. Bull. Torrey Bot. Cl. 66:457–464.

Pospelov, S. M. 1957. (Birds and mammals in various age classes of a spruce forest.) In Russian. Zoologicesky Zhurnal 26:603–607.

Potzger, J. E. 1941. The vegetation of Mackinac Island, Michigan: an ecological survey. Amer. Midl. Nat. 25:298–323.

Pulling, A. V. S. 1924. Small rodents and northeastern conifers. J. For. 22:813–814.

Quick, H. F. 1954. Small mammal populations on the University Forest. Tech. Note Univ. Maine For. Dep. 28. 2 pp.

——. 1955. Rodent populations on the University Forest. Tech. Note Univ. Maine For. Dep. 41. 3 pp.

Ray, R. G. 1941. Site types and rate of growth, Lake Edward, Champlain County, P.Q., 1915–1936. Silv. Res. Note For. Serv. Can. 65. 56 pp.

Redmond, D. R. 1954. Conditions of tree rootlets in Nova Scotia following a hurricane. Bi-m. Progr. Rep. Div. For. Biol. Dep. Agric. Can. 10(5):1.

Reeks, W. A. 1942. Notes on the Canada porcupine damage in the Maritime Provinces. For. Chron. 18:182–187.

——. 1958. Weather injuries to trees in Manitoba. Bi-m. Progr. Rep. Div. For. Biol. Dep. Agric. Can. 14(5):1–2.

Rowe, J. S. 1955. Factors influencing white spruce reproduction in Manitoba and Saskatchewan. Tech. Note For. Br. Can. 3. 27 pp.

——. 1956. Uses of undergrowth plant species in forestry. Ecology 37:461–473.

——. 1959. Forest regions of Canada. Bull. For. Br. Can. 123. 71 pp.

Satterlund, D. R., and S. A. Graham. 1957. Effect of drainage on the growth in stagnant sphagnum bogs. Mich. For. 19. 2 pp.

Schantz-Hansen, T. 1923. Second growth on cutover lands in St. Louis County. Bull. Minn. Agric. Exp. Sta. 203. 50 pp.

——. 1934. The cutover lands of Lake County. Bull. Minn. Agric. Exp. Sta. 304. 23 pp.

——. 1937. Storm damage on the Cloquet Forest. J. For. 35:463–465.

——, and P. N. Joranson. 1939. Some effects of the 1936 drought on the forest at the Cloquet Forest Experiment Station. J. For. 37:635–639.

Seed, G. K. 1951. A study of wind loss of residual spruce and fir in the Patricia district of Ontario. Unpubl. Rep. Univ. Minnesota. 17 pp.

Severaid, J. H. 1942. The snowshoe hare, its life history and artificial propagation. Maine Dep. Inl. Fish. Game. 95 pp.

Shelford, V. E. 1945. The relative merits of the life zone and biome concepts. Wilson Bull. 57:248–252.

——. 1963. The ecology of North America. Univ. Illinois Press. Urbana, Ill. 610 pp.

Sisam, J. W. B. 1938a. The correlation of tree species and growth with site-types. Silv. Res. Note For. Serv. Can. 53. 19 pp.

——. 1938b. Site as a factor in silviculture — its determination with special reference to the use of plant indicators. Silv. Res. Note For. Serv. Can. 54. 81 pp.

Skolko, A. J. 1947. Deterioration of fire-killed pulpwood stands in eastern Canada. For. Chron. 23:128–145.

Smerlis, E. 1961. Pathological condition of immature balsam fir stands of Hylocomium-Oxalis type in the Laurentide Park, Quebec. For. Chron. 37:109–115.

Smith, C. F., and S. E. Aldous. 1947. The influence of mammals and birds in retarding artificial and natural reseeding of coniferous forests in the United States. J. For. 45:361–369.

Society of American Foresters. 1954. Forest cover types of North America (exclusive of Mexico). 2nd ed. Soc. Amer. For. Washington, D.C. 69 pp.

Spiker, C. J. 1933. Some late winter and early spring observations on the white-tailed deer of Adirondacks. Roosevelt Wildlife Bull. 6:328–385.

Starker, T. J. 1934. Fire resistance in the forest. J. For. 32:462–467.

Stewart, D. C. 1951. Burning and natural vegetation in the United States. Geogr. Rev. 41:317–320.

Stewart, G. 1925. Forest cover types of northern swamps. J. For. 23:160–172.

Stickel, P. W. 1941. On the relation between bark character and resistance to fire. Tech. Note Ntheast. For. Exp. Sta. 39. 2 pp.

Stoeckeler, J. H., and Carl Arbogast, Jr. 1955. Forest management lessons from a 1949 windstorm in northern Wisconsin and upper Michigan. Sta. Pap. Lake St. For. Exp. Sta. 34. 11 pp.

Stoeckeler, J. H., and P. O. Rudolf. 1949. Winter injury and recovery of conifers in the Upper Midwest. Sta. Pap. Lake St. For. Exp. Sta. 18. 20 pp.

Stoeckeler, J. H., R. O. Strothmann, and L. W. Krefting. 1957. Effect of deer browsing on reproduction in the northern hardwood-hemlock type in northeastern Wisconsin. J. Wildlife Mgmt. 21:75–80.

Swift, E. 1948. Wisconsin deer damage to forest reproduction survey-final report. Publ. Wis. Conserv. Dep. 347. 24 pp.

Taylor, W. P., ed. 1956. The deer of North America. Stackpole, Harrisburg, Pa. and Wildlife Mgmt. Inst. Washington, D.C. 668 pp.

Thomas, J. B. 1956. Hail damage in the Lake Nipigon area. Bi-m. Progr. Rep. Div. For. Biol. Dep. Agric. Can. 12(3):2.

Trimble, G. R. 1959. A problem analysis and program for watershed-management research in the White Mountains of New Hampshire. Sta. Pap. Ntheast. For. Exp. Sta. 116. 46 pp.

Trippensee, R. E. 1948. Wildlife management. McGraw-Hill. New York, N.Y. 479 pp.

United States Department of Agriculture. 1941. Climate and man. Yearb. Agric. U.S. Dep. Agric. Washington, D.C. 1248 pp.

United States Forest Service. 1958. Manual for forest fire control. Fire fighting methods and techniques. U.S. For. Serv. Reg. 7. Ntheast. For. Fire Prot. Comm. Chatham, N.Y. 268 pp.

Van Dersal, W. R. 1938. Native woody plants of the United States, their erosion control and wildlife value. Misc. Publ. U.S. Dep. Agric. 303. 362 pp.

Vesall, D. B., R. W. Nyman, and R. H. Gensch. 1948. Beaver management. Proc. Amer. Soc. For. (1947):195–200.

Vos, A. de, and R. L. Peterson. 1951. A review of the status of woodland caribou (*Rangifer caribou*) in Ontario. J. Mammal. 32:329–337.

Waldron, R. M. 1959. Experimental cutting in a mixedwood stand in Saskatchewan, 1924. Tech. Note For. Br. Can. 74. 14 pp.

Walter, H. 1947–1956. Einführung in die Phytologie. Vol. 1–4. Ulmer. Stuttgart.

Ward, R. de C., C. F. Brooks, and A. J. Connor. 1938. The climates of North America. *In* W. Köppen and R. Geiger, Handbuch der Klimatologie. Gebr. Borntraeger. Berlin. Vol. 2. 424 pp.

Webb, W. L. 1948. Environmental analysis of a winter deer range. Trans. North Amer. Wildl. Conf. 13:442–450.

Weetman, G. F. 1957. The chief causes of blow-down in pulpwood stands in eastern Canada. Woodl. Res. Index Pulp Paper Res. Inst. Can. 99. 8 pp.

Westveld, M. 1953. Ecology and silviculture of the spruce-fir forests of eastern North America. J. For. 51:422–430.

Westveld, R. H. 1949. Applied silviculture in the United States. Wiley and Sons. New York, N.Y. 590 pp.

Wilde, S. A. 1933. The relation of soils and forest vegetation of the Lake States region. Ecology 14:94–105.

———. 1953. Trees of Wisconsin, their ecological and silvicultural silhouettes. Misc. Publ. Soils Dep. Univ. Wis. and Wis. Conserv. Dep. 118. 44 pp.

———. 1954. Reaction of soils: facts and fallacies. Ecology 35:89–92.

———, F. G. Wilson, and D. P. White. 1949. Soils of Wisconsin in relation to silviculture. Publ. Wis. Conserv. Dep. 525-49. 171 pp.

Winchell, N. H. 1879. Sketch of the work of the season of 1878. Chapt. II, pages 24–25, in Geol. and Nat. History Survey of Minnesota, Seventh Ann. Rep. for the year 1878.

Woodward, C. W. 1917. Tree growth and climate in the United States. J. For. 15:521–531.

Wright, B. S. 1952. A report to the Minister of Lands and Mines on the moose of New Brunswick. Ntheast. Wildlife Sta. Fredericton, N.B. 43 pp.

Zederbauer, E. 1919. Über Anbauversuche mit fremdländischen Holzarten in Österreich. Cbl. Ges. Forstw. 45:153–169.

Zeedyk, W. D. 1957. Why do bears girdle balsam fir in Maine? J. For. 55:731–732.

Zentgraf, E. 1949. Die Edeltanne. (*Abies alba*) Allg. Forst- u. Jagdztg. 121:7–16.

Zon, R. 1914. Balsam fir. Bull. U.S. For. Serv. 55. 68 pp.

———, and H. S. Graves. 1911. Light relations to tree growth. Bull. U.S. For. Serv. 92. 59 pp.

Chapter 4. Microbiology

Ainsworth, G. C., and G. R. Bisby. 1954 (2nd ed., 1961). A dictionary of the fungi. Commonw. Mycol. Inst. Kew, Surrey. 475 pp.

Anderson, A. P. 1897. Comparative anatomy of the normal and diseased organs of *Abies balsamea* affected with *Aecidium elatinum*. Bot. Gaz. 24:309–344.

———. 1902. *Dasyscypha resinaria* causing canker growth in *Abies balsamea* in Minnesota. Bull. Torrey Bot. Cl. 29:23–34.

Anderson, P. J., R. J. Haskell, W. C. Muenscher, C. J. Weld, J. J. Wood, and G. H. Martin. 1926. Check list of diseases of economic plants in the United States. Bull. U.S. Dep. Agric. 1366. 112 pp.

Arthur, J. C. 1934. Manual of rusts in the United States and Canada. Purdue Res. Found. Lafayette, Ind. 438 pp.

Bartholomew, E. 1933. Handbook of the North American Uredinales. Spurier. Morland, Kans. 238 pp.

Basham, J. T. 1951a. The pathological deterioration of balsam fir killed by the spruce budworm. Pulp Paper Mag. Can. 52(9):120–134.

———. 1951b. Deterioration of balsam fir killed by the spruce budworm. Bi-m. Progr. Rep. Div. For. Biol. Dep. Agric. Can. 7(2):2.

———. 1959. Studies in forest pathology. XX. Investigations of the pathological distribution in killed balsam fir. Canad. J. Bot. 37:291–326.

———, and R. M. Belyea. 1960. Death and deterioration of balsam fir weakened by spruce budworm defoliation in Ontario. For. Sci. 6:78–96.

Basham, J. T., P. V. Mook, and A. G. Davidson. 1953. New information concerning balsam fir decays in eastern North America. Canad. J. Bot. 31:334–360.

Baxter, D. V. 1935. Some resupinate polypores from the region of the Great Lakes. VII. Pap. Mich. Acad. Sci. 21:243–267.

———. 1937. Some resupinate polypores from the region of the Great Lakes. IX. Pap. Mich. Acad. Sci. 23:285–305.

———. 1947. Some resupinate polypores from the region of the Great Lakes. XIX. Pap. Mich. Acad. Sci. 33:9–30.

———. 1948. Some resupinate polypores from the region of the Great Lakes. XX. Pap. Mich. Acad. Sci. 34:41–56.

———. 1950. Some resupinate polypores from the region of the Great Lakes. XXII. Pap. Mich. Acad. Sci. 36:69–83.

———. 1952a. Pathology in forest practice. Wiley & Sons. New York, N.Y. 601 pp.

———. 1952b. Some resupinate polypores from the region of the Great Lakes. XXIII. Pap. Mich. Acad. Sci. 37:93–110.

———. 1953. Some resupinate polypores from the region of the Great Lakes. XXV. Pap. Mich. Acad. Sci. 39:125–138.

Bell, H. P. 1924. Fern rusts of *Abies*. Bot. Gaz. 77:1–31.

Bisby, G. R., A. H. R. Buller, and J. Dearness. 1929. The fungi of Manitoba. Longmans, Green and Co. New York, N.Y. 194 pp.

Bisby, G. R., A. H. R. Buller, J. Dearness, W. P. Fraser, R. C. Russell, and H. T. Güssow. 1938. The fungi of Manitoba and Saskatchewan. Mem. Canad. fungi. Nat. Res. Counc. Can. 189 pp.

Boyce, J. S. 1928. A conspectus of needle rusts on balsam fir in North America. Phytopathology 18:705–708.

———. 1948 (3rd ed., 1961). Forest pathology. McGraw-Hill. New York. 550 pp.

———. 1952. Needle cast disease of conifers (caused by *Bifusella, Hypoderma, Hypodermella, Lophodermium*). Pp. 87–89. *In* Important tree pests of the Northeast. Soc. Amer. For. New Engl. Sec.

Brooks, F. T. 1953. Plant diseases. Oxford Univ. Press. New York. 457 pp.

Brown, C. A. 1935. Morphology and biology of some species of *Odontia*. Bot. Gaz. 96:640–675.

Brown, N. A. 1920. A *Pestalozzia* producing a tumor on the Sapodilla tree. Phytopathology 10:383–394.

Burt, E. A. 1917. *Merulius* in North America. Ann. Mo. Bot. Gdn. 4:305–361.

———. 1926. The Thelephoraceae of North America. XV. Ann. Mo. Bot. Gdn. 13:173–354.

Canada Forest Biology Division. 1942–59. Annual report of the forest insect and disease survey. Dep. Agric. Can.

Canada Forest Entomology and Pathology Branch. 1960–62. Annual report of the forest insect and disease survey. Dep. For. Can.

Cartwright, K. S. G. 1938. A further note on fungus association in the Siricidae. Ann. Appl. Biol. 25:430.

———, and W. P. K. Findlay. 1944. Timber decay. For. Abstr. 5(4):217–228.

Cash, E. K. 1941. An abnormality of *Abies balsamea*. U.S. Dep. Agric. Plant Dis. Reptr. 25:548.

Christensen, C. M. 1937. *Cephalosporium* canker of balsam fir. Phytopathology 27:788–791.

———. 1940. Observations on *Polyporus circinatus*. Phytopathology 30:957–963.

———, and A. C. Hodson. 1954. Artificially induced senescence of forest trees. J. For. 52:126–129.

Christensen, C. M., and F. H. Kaufert. 1942. A blue-staining fungus inhabiting the heartwood of certain species of conifers. Phytopathology 32:735–736.

Clements, F. E., and C. L. Shear. 1931 (repr. 1954). The genera of fungi. Hafner. New York, N.Y. 496 pp.

Conners, I. L. 1931–41. Annual report of the Canadian Plant Disease Survey (11th to 21st rep.). Dep. Agric. Can.

Cooke, W. B. 1943. Some Basidiomycetes from Mount Shasta. Mycologia 35:277–303.

———. 1957. The genera *Serpula* and *Meruliporia*. Mycologia 49:197–225.

Cooley, J. S. 1948. Natural infection of replanted apple trees by white root rot fungus. Phytopathology 38:110–113.

———. 1951. Control of white root rot with chlorpicrin in a border of ornamental plants. Phytopathology 41:379–380.

———, and R. W. Davidson. 1940. A white root rot of apple trees caused by *Corticium galactinum*. Phytopathology 30:139–148.

Crowell, I. H. 1940a. Heart bluestain of white spruce and balsam fir. Woodl. Rev. 11(7):3–4.

———. 1940b. Heart bluestain of white spruce and balsam fir. Pulp Paper Mag. Can. 41(7):451–452.

Darker, G. D. 1932. The Hypodermataceae of conifers. Contr. Arnold Arboretum. Harvard Univ. 1. 131 pp.

Davidson, A. G. 1951. Decay of balsam fir in New Brunswick. Bi-m. Progr. Rep. Div. For. Biol. Dep. Agric. Can. 7(5):2–3.

———. 1952. Forest pathological studies initiated in Newfoundland. Bi-m. Progr. Rep. Div. For. Biol. Dep. Agric. Can. 8(4):1.

———. 1957. Studies in forest pathology. XVI. Decay of balsam fir, *Abies balsamea* (L.) Mill., in the Atlantic Provinces. Canad. J. Bot. 35:857–874.

———, and D. E. Etheridge. 1963. Infection of balsam fir, *Abies balsamea* (L.) Mill., by *Stereum sanguinolentum* (Alb. and Schw. ex Fr.). Canad. J. Bot. 41:759–765.

Davidson, A. G., and W. R. Newell. 1953. Pathological deterioration in wind-thrown balsam fir in Newfoundland. For. Chron. 29:100–107.

———. 1962. Infection courts of red heart rot in balsam fir. Bi-m. Progr. Rep. For. Ent. Path. Br. Can. 18(1):1.

Davidson, R. W. 1955. Wood-staining fungi associated with bark beetles in Engelmann spruce in Colorado. Mycologia 47:58–67.

———. 1958. Additional species of Ophiostomaceae from Colorado. Mycologia 50:661–670.

———, and F. K. Lombard. 1953. Large-brown-spored house-rot fungi in the United States. Mycologia 45:88–100.

Dearness, J. 1926. New and noteworthy fungi. Mycologia 18:236–255.

Etheridge, D. E. 1962. Conditions influencing the entry of *Stereum sanguinolentum* Alb. & Schw. ex Fr. in balsam fir. Bi-m. Progr. Rep. For. Ent. Path. Br. Can. 18(1):2–3.

———, and E. C. Carmichael. 1957. Observations on *Coryne sarcoides* (Jacq.) Tul. Bi-m. Progr. Rep. Div. For. Biol. Dep. Agric. Can. 13(4):3.

Etheridge, D. E., and L. A. Morin. 1962. Wetwood formation in balsam fir. Canad. J. Bot. 40:1335–1345.

Etheridge, D. E., and R. Pomerleau. 1961. Identity of a fungus causing the blue stain in balsam fir. Science 133:2062–2063.

Farlow, W. G. 1905. Bibliographical index of North American fungi. Carnegie Inst. Washington, D.C. 1(1), 312 pp.

———, and A. B. Seymour. 1888. A provisional host-index of the fungi of the United States. Cambridge, Mass. 219 pp.

Faull, J. H. 1919–1924. Forest pathology. Ann. Rep. Dep. Lds. For., Ontario.

———. 1929a. Studies being made and progress in control of forest diseases. A fungous disease of conifers related to the snow cover. For. Chron. 5:29–34.

———. 1929b. A fungous disease of conifers related to snow cover. J. Arnold Arbor. 10(1):3–8.

———. 1930a. Notes on forest diseases of Nova Scotia. J. Arnold Arbor. 11:55–58.

———. 1930b. The spread and the control of *Phacidium* blight in spruce plantations. J. Arnold Arbor. 11:136–147.

———. 1932. Taxonomy and geographical distribution of the genus *Milesia*. Contr. Arnold Arboretum. Harvard Univ. 2. 138 pp.

———. 1938a. Taxonomy and geographical distribution of the genus *Uredinopsis*. Contr. Arnold Arboretum. Harvard Univ. 11. 120 pp.

———. 1938b. *Pucciniastrum* on *Epilobium* and *Abies*. J. Arnold Arbor. 19:163–173.

————. 1938c. The biology of rusts of the genus *Uredinopsis*. J. Arnold Arbor. 19:402–436.

————. 1939. A review and extension of our knowledge of *Calyptospora goeppertiana* Kuehn. J. Arnold Arbor. 20:104–113.

————, and J. Mounce. 1924. *Stereum sanguinolentum* as the cause of "sapin rouge" or red heart rot of balsam. Phytopathology 14:349–350.

Findlay, W. P. K. 1956. Timber decay — a survey of recent work. For. Abstr. 17(3):317–327.

Fraser, W. P. 1913. Further cultures of heteroecious rusts. Phytopathology 3:73.

Freeman, E. M. 1905. Minnesota plant diseases. Univ. Minn. St. Paul, Minn. 432 pp.

Fritz, C. W. 1923. Cultural criteria for the distribution of wood-destroying fungi. Proc. Trans. Roy. Soc. Can. Ser. 3, 17(5):191–288.

Gevorkiantz, S. R., and L. P. Olsen. 1950. Growth and yield of upland balsam fir in the Lake States. Sta. Pap. Lake St. For. Exp. Sta. 22. 24 pp.

Gilbertson, R. L. 1956. The genus *Poria* in the Central Rocky Mountains and Pacific Northwest. Lloydia 19:65–85.

Groves, J. W. 1952. The genus *Tympanis*. Canad. J. Bot. 30:571–651.

Guba, E. F. 1961. Monograph on *Monochaetia* and *Pestalotia*. Harvard Univ. Press. Cambridge, Mass. 342 pp.

Hahn, G. C. 1947. Analysis of Peck's types of *Meliola balsamicola* and *Asterina nuda*. Mycologia 39:479–490.

Hansbrough, J. R. 1934. Occurrence and parasitism of *Aleurodiscus amorphus* in North America. J. For. 32:452–458.

Hartig, R. 1894. Textbook of the diseases of trees. Macmillan and Co. New York. 331 pp.

Hartley, C., R. W. Davidson, and B. S. Crandall. 1961. Wetwood, bacteria, and increased pH in trees. Rep. U.S. For. Prod. Lab. 2215. 36 pp.

Hedgcock, G. G. 1912. Notes of some diseases of trees in our national forests. II. Phytopathology 2:73–80.

————. 1914. Notes on some diseases of trees in our national forests. IV. Phytopathology 4:181–188.

Heimburger, C. C., and A. W. McCallum. 1940. Balsam fir butt-rot in relation to some site factors. Pulp Paper Mag. Can. 41(4):301–303.

Hirt, R. R., and E. J. Eliason. 1938. The development of decay in living trees inoculated with *Fomes pinicola*. J. For. 36:705–709.

Hubert, E. E. 1929. A butt-rot of balsam fir caused by *Polyporus balsameus* Pk. Phytopathology 19:725–732.

————. 1931. An outline of forest pathology. Wiley & Sons. New York, N.Y. 543 pp.

Hunter, L. M. 1927. Comparative study of spermogonia of rusts of *Abies*. Bot. Gaz. 83:1–23.

————. 1948. A study of the mycelium and haustoria of the rusts of *Abies*. Canad. J. Res. 26(C):219–234.

Jackson, H. S. 1947. *Trichomonascus*, a new genus among simple Ascomycetes. Mycologia 39:709–715.

————. 1948. Studies of Canadian Thelephoraceae. Canad. J. Res. 26(C):128–157.

Jacot, A. P. 1939. Reduction of spruce and fir litter by minute animals. J. For. 37:858–860.

Kamei, S. 1940. Studies on the cultural experiments of the fern rusts of *Abies* in Japan. II. J. Fac. Agric. Hokkaido Univ. 47:93–191.

Kaufert, F. 1935. Heart rot of balsam fir in the Lake States with special reference to forest management. Tech. Bull. Minn. Agric. Exp. Sta. 110. 27 pp.

Kelley, A. P. 1950. Mycotrophy in plants. Chronica Botanica. Waltham, Mass. 223 pp.

Kress, O., C. J. Humphrey, C. A. Richards, M. W. Bray, and J. A. Staidl. 1925. Control of decay in pulp and pulpwood. Bull. U.S. Dep. Agric. 1298. 80 pp.

Krstic, M. 1956. Prospects of application of biological control in forest pathology. Bot. Rev. 22:38–44.

Kujala, V. 1930. Über die Krankheiten der *Abies sibirica* in Finnland. Mitt. Dtsch. Dendrol. Ges. 42:380–383.

Long, W. H. 1917. Investigations of the rotting of slash in Arkansas. Bull. U.S. Dep. Agric. 496. 15 pp.

————. 1924. Self-pruning of western yellow pine. Phytopathology 14:326–337.

Lorenz, R. C., and C. M. Christensen. 1937. A survey of forest tree diseases in their relation to stand improvement in the Lake States and Central States. Unpubl. Rep. Univ. Minnesota. 52 pp.

Lowe, J. L. 1957. Polyporaceae of North America. The genus *Fomes*. Tech. Bull. N.Y. St. Coll. For. 80. 97 pp.

———. 1958. The genus *Poria* in North America. Lloydia 21:100–114.

McCallum, A. W. 1925. A study of decay in balsam fir. Phytopathology 15:302.

———. 1928. Studies in forest pathology. I. Decay in balsam fir (*Abies balsamea* (L.) Mill.). Bull. Dep. Agric. Can. 104. 25 pp.

McDougall, W. B. 1922. Mycorrhizae of coniferous trees. J. For. 20:255–260.

MacKay, A. H. 1908. Fungi of Nova Scotia: a provisional list. Proc. Trans. Nova Scotian Inst. Sci. 11:122–143.

Manshard, E. 1927. Krankheiten und Schädlinge im Saatbeet der forstlich wichtigsten Holzarten. Mitt. Dtsch. Dendrol. Ges. 38:198–229.

Marshall, R. P. 1948. A white root rot which merits study as a possible cause of white pine blight. Sci. Tree Topics 1(9):66–68. Bartlett Tree Res. Lab. Stamford, Conn.

Miller, L. W. 1934. The Hydnaceae of Iowa. II. The genus *Odontia*. Mycologia 26:13–32.

Moore, B. 1922. Humus and root systems in certain northeastern forests in relation to reproduction and competition. J. For. 20:235–254.

Moore, C. L. 1906. The Myxomycetes of Pictou Co. Proc. Trans. Nova Scotian Acad. Sci. Ass. 1:11–16.

———. 1913. The Myxomycetes of Pictou Co. Proc. Trans. Nova Scotian Inst. Sci. 12:165–206.

Morawski, Z. J. R., J. T. Basham, and K. B. Turner. 1958. A survey of pathological conditions in the forests of Ontario. Dep. Lds. For. Ont. 96 pp.

Morris, R. F. 1948. How old is a balsam tree? For. Chron. 24:106–110.

Morris, R. F., D. R. Redmond, A. B. Vincent, E. L. Howie, and D. W. Hudson. 1955. The Green River project — a decade of forestry research. Pulp Paper Mag. Can. 56(9):147–163.

Murrill, W. A. 1903. The Polyporaceae of North America V. The genera *Cryptoporus, Piptoporus, Scutiger* and *Porodiscus*. Bull. Torrey Bot. Cl. 30:423–434.

New York Bot. Garden. 1906–37. North American flora. N.Y. Bot. Gdn. New York, N.Y.

Nissen, T. V. 1956. Actinomycetes antagonistic to *Polyporus annosus* Fr. Experientia 12:229–230.

Nobles, M. K. 1948. Studies in forest pathology. VI. Identification of cultures of wood rotting fungi. Canad. J. Res. 26(C):281–341.

———. 1953. Studies in wood-inhabiting Hymenomycetes. I. *Odontia bicolor*. Canad. J. Bot. 31:745–749.

Oudemans, C. A. J. A. 1919. Enumeratio Systematica Fungorum Vol. 1–5. Editum auspicis Societatis hollandicae disciplinarum harlemensis. Hague Comitum, apud M. Nijhoff.

Overholts, L. O. 1953. The Polyporaceae of the United States, Alaska and Canada. Univ. Mich. Press. Ann Arbor, Mich. 466 pp.

Peace, T. R. 1939. Forest pathology in North America. Forestry 13:36–45.

Peck, C. H. 1875. Annual report. N.Y. St. Mus. 27:104.

———. 1878. Annual report. N.Y. St. Mus. 30:46.

Peirson, H. B. 1923. Insects attacking forest and shade trees. Bull. Maine For. Serv. 1:8–25.

Pomerleau, R. 1948. Les caries du sapin. For. Québec 13:537–546.

———. 1958. Relation entre development des caries des conifères et le milieu. Abstr. in Annales de l'Association Canadienne-Française pour l'Advancement des Sciences 24:90.

———, and D. E. Etheridge. 1961. A bluestain in balsam fir. Mycologia 53:155–170.

Prielipp, D. O. 1952. Decay in balsam fir in the Upper Peninsula of Michigan. Unpubl. Rep. Univ. Minnesota. 58 pp.

———. 1957. Balsam fir pathology for Upper Michigan. Publ. Kimberly-Clark of Michigan. 69 pp.

Quirke, D. A., and H. H. V. Hord. 1955. A canker and dieback of balsam fir in Ontario. Bi-m. Progr. Rep. Div. For. Biol. Dep. Agric. Can. 11(6):2.

Rankin, W. H. 1918. Manual of tree diseases. Macmillan, New York, N.Y. 398 pp.

———. 1920. Butt rots of the balsam fir in Quebec Province (Abstract). Phytopathology 10:314–315.

Raymond, F. L., and J. Reid. 1957. Comments on the agents responsible for the cankering and killing of balsam fir in eastern Canada. Bi-m. Progr. Rep. Div. For. Biol. Dep. Agric. Can. 13(6):1–2.

———. 1961. Dieback of balsam fir in Ontario. Canad. J. Bot. 39:233–251.

Redmond, D. R. 1957. Infection courts of butt-rotting fungi in balsam fir. For. Sci. 3:15–21.

Reid, J., and R. F. Cain. 1962. Studies on the organisms associated with "snow-blight" of conifers in North America. I. A new genus of the Helotiales. Mycologia 54:194–200.

Rhoads, A. S. 1921. Some new or little known hosts for wood destroying fungi. III. Phytopathology 11:319–326.

Robak, H. 1951. Noen iakttakelser til belysning av forholdet mellom klimatiske skader og soppangrep på naletraer. (Some observations to elucidate the connection between climatic injuries and fungal infection of conifers.) Medd. Vestl. forstl. Forsøkssta. 27(8,2):43 pp.

Rogers, D. P. 1943. The genus *Pellicularia* (Thelephoraceae). Farlowia 1:95–118.

——, and H. S. Jackson. 1943. Notes on the synonyms of North American Thelephoraceae and other resupinates. Farlowia 1:263–336.

Roll-Hansen, F. 1962. *Nectria cucurbitula* sensu Wollenweber, its *Cephalosporium* state and some *Cephalosporium* spp. from stems of conifers. Rep. Norwegian For. Res. Inst. 61:291–312.

Saccardo, P. A. 1882–1931. Sylloge fungorum hucusque cognitorum. 25 vol. Paria.

Schaeffer, A. 1926. Branches, noeuds et mode de traitèment. Bull. Trimest. Soc. Forest. Franche-Comte 16:495–496.

Schenck, C. A. 1939. Fremdländische Wald- und Parkbäume. Parey. Berlin. Vol. 1–2.

Schierbeck, O. 1922. Treatise on the spruce budworm, bark beetle and borer. The causes of the dying of the balsam fir in Quebec and suggestions for preventive measures. Barnjum. Montreal. 38 pp.

Schmitz, M. 1923. Notes on wood decay. I. The wood destroying properties of *Polyporus volvatus*. J. For. 21:502–503.

Schrenk, H. 1900. Some diseases of New England conifers. Bull. U.S. Dep. Agric. Div. Veg. Physiol. Path. 25. 54 pp.

——. 1902. A root rot of apple trees caused by *Thelephora galactina* Fr. Bot. Gaz. 34:65.

——. 1905. Glassy fir. Rep. Mo. Bot. Gdn. 16:117–120.

——, and P. Spaulding. 1909. Diseases of deciduous forest trees. Bull. U.S. Dep. Agric. Bur. Plant Ind. 149. 85 pp.

Schwerdtfeger, F. 1957. Die Waldkrankheiten. Paul Parey. Hamburg u. Berlin. 485 pp.

Seaver, F. J. 1928. The North American cup-fungi (Operculates). Publ. author. New York, N.Y. 284 pp.

——. 1951. The North American cup-fungi (Inoperculates). Publ. author. New York. 428 pp.

Seymour, A. B. 1929. Host index of the fungi of North America. Harvard Univ. Press. Cambridge, Mass. 732 pp.

Singer, R. 1949. The "Agaricales" (mushrooms) in modern taxonomy. Lilloa, Revista de Botanica 22:1–832.

Skolko, A. J. 1947. Deterioration of fire-killed pulpwood stands in eastern Canada. For. Chron. 23:128–145.

Slysh, A. R. 1960. The genus *Peniophora* in New York State and adjacent regions. Tech. Publ. N.Y. St. Coll. For. 83. 95 pp.

Smerlis, E. 1961. Pathological conditions of immature balsam fir stands of Hylocomium-Oxalis type in the Laurentide Park. For. Chron. 37:108–115.

——. 1962. Taxonomy and morphology of *Potebniamyces balsamicola* sp. nov. associated with a twig and branch blight of balsam fir in Canada. Canad. J. Bot. 40:351–359.

Smith, A. H. 1954. Mushrooms in their natural habitats, Vol. I. Sawyers, Portland, Ore. 626 pp.

——, and L. E. Wehmeyer. 1936. Contributions to a study of the fungous flora of Nova Scotia. II. Agaricaceae and Boletaceae. Pap. Mich. Acad. Sci. 21:163–197.

Spaulding, P. 1911. The timber rot caused by *Lenzites sepiaria*. Bull. U.S. Dep. Agric. Bur. Plant Ind. 214. 46 pp.

——. 1912. Notes upon tree diseases in the eastern states. Phytopathology 2:93.

——. 1956. Diseases of North American forest trees planted abroad. Agric. Handb. U.S. Dep. Agric. 100. 144 pp.

——. 1961. Foreign diseases of forest trees of the world. Agric. Handb. U.S. Dep. Agric. 197. 361 pp.

——, and J. R. Hansbrough. 1944. Decay in balsam fir in New England and New York. Tech. Bull. U.S. Dep. Agric. 872. 30 pp.

Spaulding, P., G. H. Hepting, and H. J. MacAloney. 1932. Investigations in decays of balsam fir. I. Gale River Exp. Forest, New Hampshire. Tech. Note Ntheast. For. Exp. Sta. 11. 3 pp.

Stephenson, J. N. 1950. Preparation and treatment of wood pulp. Vol. 1. McGraw-Hill. New York, N.Y. 1043 pp.

Stickel, P. W., and F. Marcott. 1936. Forest fire studies in Northeast. III. Relation between fire injury and fungal infection. J. For. 34:420–423.

Stillwell, M. A. 1956. Pathological aspects of spruce budworm attack. For. Sci. 2:174–180.

——. 1958. Deterioration of fire-killed balsam fir and spruce stands in Newfoundland and New Brunswick. Bi-m. Progr. Rep. Div. For. Biol. Dep. Agric. Can. 14(4):1.

——. 1959. Further studies of the pathological deterioration in wind-thrown balsam fir in Newfoundland. For. Chron. 35:212–218.

————. 1960. Decay associated with woodwasps in balsam fir weakened by insect attack. For. Sci. 6:225–231.

————. 1962. Deterioration of balsam fir killed by the eastern hemlock looper. Bi-m. Progr. Rep. For. Ent. Path. Br. Can. 18(27):1.

————, and D. R. Redmond. 1956. Decay of balsam fir arising through spruce budworm injury. Bi-m. Progr. Rep. Div. For. Biol. Dep. Agric. Can. 12(4):1–2.

Stone, R. E. 1920. Witches' broom of the Canada balsam and the alternate hosts of the causal organism. Phytopathology 10:315.

Trimble, G. R. 1942. Logging damage in partial cutting of spruce-fir stands. Tech. Note Ntheast. For. Exp. Sta. 51. 2 pp.

Wagener, W. W., and R. W. Davidson. 1954. Heart rots in living trees. Bot. Rev. 20:61–134.

Waterman, A. M. 1937. New hosts and distribution of *Rehmiellopsis bohemica*. Phytopathology 27:734–736.

————. 1945. Tip blight of species of *Abies* caused by a new species of *Rehmiellopsis*. J. Agric. Res. 70:315–337.

————, and K. F. Aldrich. 1940. *Rehmiellopsis* needle blight of balsam fir in Maine. Plant Dis. Reptr. 24:201–205.

Wehmeyer, L. E. 1935. Contributions to a study of the fungous flora of Nova Scotia. I. Pap. Mich. Acad. Sci. 20:233–266.

————. 1940. Contributions to a study of the fungous flora of Nova Scotia. IV. Additional Basidiomycetes. Canad. J. Res. 18(C):92–110.

————. 1942. Contributions to a study of the fungous flora of Nova Scotia. VI. Pyrenomycetes. Canad. J. Res. 20(C):572–594.

————. 1950. Fungi of New Brunswick, Nova Scotia, Prince Edward Island. Nat. Res. Counc. Can. 1890. 150 pp.

Weir, J. R., and E. E. Hubert. 1918. Notes on forest tree rusts. Phytopathology 8:114–118.

Weiss, F., and M. J. O'Brien. 1953. Index of plant diseases in the United States. U.S. Dep. Agric. Plant Dis. Surv. Part 5:809–811.

Westcott, C. 1955. Plant disease handbook. Van Nostrand, New York. 746 pp.

White, L. T. 1951. Studies of Canadian Thelephoraceae. VIII. *Corticium galactinum* (Fr.) Burt. Canad. J. Bot. 29:279–296.

Whitney, R. D. 1952. Root disease of balsam fir. Bi-m. Progr. Rep. Div. For. Biol. Dep. Agric. Can. 8(5):2.

Wolf, F. A. 1922. The fruiting stage of the tuckahoe *Pachyma cocos*. Elisha Mitchell Sci. Soc. 38:127–137.

————. 1926. Tuckahoe on maize. Elisha Mitchell Sci. Soc. 42:288–290.

Wright, E. F., and R. F. Cain. 1961. New species of the genus *Ceratocystis*. Canad. J. Bot. 39:1215–1230.

Zon, R. 1914. Balsam fir. Bull. U.S. For. Serv. 55. 68 pp.

Chapter 5. Entomology

Anderson, G. W., and D. C. Schmiege. 1959. The forest insect and disease situation, Lake States, 1958. Sta. Pap. Lake St. For. Exp. Sta. 70. 18 pp.

————. 1961. The forest insect and disease situation, Lake States, 1960. Sta. Pap. Lake St. For. Exp. Sta. 88. 18 pp.

Anderson, R. F. 1960. Forest and shade tree entomology. Wiley and Sons. New York, N.Y. 428 pp.

Angus, T. A., A. M. Heimpel, and R. A. Fisher. 1961. Tests of a microbial insecticide against forest defoliators. Bi-m. Rep. For. Ent. Path. Br. Can. 17(3):1–2.

Annand, P. N. 1928. A contribution toward a monograph of the Adelginae (Phylloxeridae) of North America. Stanford Univ. Biol. Sci. Ser. 6(1):146 pp.

Anonymous. 1945. Spraying experiments for spruce budworm control. Bi-m. Progr. Rep. For. Insect Invest. Dep. Agric. Can. 1(3):1.

Atwood, C. E. 1945a. Spruce budworm — investigations on the susceptibility and vulnerability of forests. Bi-m. Progr. Rep. For. Insect Invest. Dep. Agric. Can. 1(2):2.

————. 1945b. Spruce budworm studies. Bi-m. Progr. Rep. For. Insect Invest. Dep. Agric. Can. 1(3):2.

Baird, A. B. 1947. Biological control of the spruce budworm in eastern Canada by parasite introduction. Bi-m. Progr. Rep. For. Insect Invest. Dep. Agric. Can. 3(2):1.

Balch, R. E. 1932. The black-headed budworm. Spec. Circ. Can. Dep. Agric. Ent. Br. 2 pp.

————. 1934. The balsam woolly aphid, *Adelges piceae* (Ratz.) in Canada. Sci. Agric. (Ottawa) 14(7):374–383.

————. 1942. European spruce sawfly in 1942. Pulp Paper Mag. Can. 44(3):279–282.

————. 1946a. Spruce budworm hazard map. Bi-m. Progr. Rep. For. Insect Invest. Dep. Agric. Can. 2(2):1.

————. 1946b. "Staminate trees" and spruce budworm abundance. Bi-m. Progr. Rep. For. Insect Invest. Dep. Agric. Can. 2(3):1.

————. 1950. Flights of balsam twig aphid detected from aeroplanes. Bi-m. Progr. Rep. For. Insect Invest. Dep. Agric. Can. 6(5):1.

————. 1952. Studies of the balsam woolly aphid, *Adelges piceae* (Ratz.) and its effects on balsam fir *Abies balsamea* (L.) Mill. Publ. Dep. Agric. Can. 867. 76 pp.

————. 1953a. Current practices and future trends in forest insect control. For. Chron. 29:6–13.

————. 1953b. Susceptibility of species of fir to balsam woolly aphid *Adelges piceae* Ratz. Bi-m. Progr. Rep. Div. For. Biol. Dep. Agric. Can. 9(4):1.

————. 1956. Damage caused by balsam woolly aphid. Bi-m. Progr. Rep. Div. For. Biol. Dep. Agric. Can. 12(5):1.

————. 1962. The morphology and host relationships of adelgids on fir in North America. *In* Annual Report For. Ent. Path. Br. Dep. For. Can., p. 45.

————, F. E. Webb, and J. J. Fettes. 1955–56. The use of aircraft in forest insect control. For. Abstr. 16:453–465; 17:3–9, 149–159.

Balch, R. E., J. Clark, and J. M. Bonga. 1964. Hormonal action in production of tumours and compression wood by an aphid. Nature 202:121–122.

Baldwin, H. I., R. C. Brown, H. B. Peirson, H. A. Reynolds, T. R. H. Hansbrough, P. M. Reed, and L. W. Rathburn. 1952. Important tree pests of the Northeast. New Engl. Sect. Soc. Amer. For. 191 pp.

Barker, R. B., and H. A. Fyfe. 1947. A mixed infestation in the Spruce Woods Forest Reserve, Manitoba. Bi-m. Progr. Rep. For. Insect Invest. Dep. Agric. Can. 3(4):3.

Baskerville, G. L. 1960. Mortality in immature balsam fir following severe budworm defoliation. For. Chron. 36:342–345.

Bean, J. L. 1960. Frass size as an indicator of spruce budworm larval instars. Ann. Ent. Soc. Amer. 52:605–608.

————. 1961a. Predicting emergence of second-instar spruce budworm larvae from hibernation under field conditions in Minnesota. Ann. Ent. Soc. Amer. 54:175–177.

————. 1961b. The use of balsam fir shoot elongation for timing spruce budworm aerial spray programs in the Lake States. J. Econ. Ent. 54:996–1000.

————, and H. O. Batzer. 1956. A spruce budworm risk-rating and spruce-fir types in the Lake States. Tech. Note Lake St. For. Exp. Sta. 453. 2 pp.

Bean, J. L., and H. Graeber. 1957. Spruce budworm increasing in Minnesota. Tech. Note Lake St. For. Exp. Sta. 479. 2 pp.

Bean, J. L., and W. E. Waters. 1961. Spruce budworm in eastern United States. For. Pest Leafl. U.S. For. Serv. 58. 8 pp.

Beckwith, L. C., and H. G. Ewan. 1956. The forest insect situation in the Lake States. Sta. Pap. Lake St. For. Exp. Sta. 35. 15 pp.

Beckwith, L. C., and H. L. MacAloney. 1955. The more important forest insects in the Lake States in 1954. Misc. Rep. Lake St. For. Exp. Sta. 36. 9 pp.

Belyea, R. M. 1948. Role of beetles in the dying of defoliated balsam fir. Bi-m. Progr. Rep. For. Insect Invest. Dep. Agric. Can. 4(6):1.

————. 1952a. Death and deterioration of balsam fir weakened by spruce budworm defoliation in Ontario. J. For. 50:729–738.

————. 1952b. Death and deterioration of balsam fir weakened by spruce budworm defoliation in Ontario. Part I. Notes on the seasonal history and habits of insects breeding in severely weakened and dead trees. Canad. Ent. 84:325–335.

Bergold, G. H. 1949. The polyhedral disease of the spruce budworm, *Choristoneura fumiferana* (Clem.). Bi-m. Progr. Rep. For. Insect Invest. Dep. Agric. Can. 5(3):2.

Bess, H. A. 1945. Ecological studies of the spruce budworm. Bi-m. Progr. Rep. For. Insect Invest. Dep. Agric. Can. 1(3):2.

————. 1946. Staminate flowers and spruce budworm abundance. Bi-m. Progr. Rep. For. Insect Invest. Dep. Agric. Can. 2(2):3–4.

Blackman, M. W. 1934. A revisional study of the genus *Scolytus* Geoffrey in North America. Tech. Bull. U.S. Dep. Agric. 431. 30 pp.

Blais, J. R. 1952. The relationship of the spruce budworm (*Choristoneura fumiferana* (Clem.)) to the flowering condition of balsam fir (*Abies balsamea* (L.) Mill.). Canad. J. Zool. 30:1–29.

———. 1953a. Effects of the destruction of the current year's foliage of balsam fir on the fecundity and habits of flight of the spruce budworm. Canad. Ent. 85:446–448.

———. 1953b. Borer control in balsam fir, spruce and jack pine logs. Bi-m. Progr. Rep. Div. For. Biol. Dep. Agric. Can. 9(2):2–3.

———. 1954. The recurrence of spruce budworm infestation in the past century in the Lac Seul area of northwest Ontario. Ecology 35:62–71.

———. 1956. Aerial spraying against spruce budworm in Quebec – 1955. Bi-m. Progr. Rep. Div. For. Biol. Dep. Agric. Can. 12(1):1.

———. 1957. Some relationships of the spruce budworm, *Choristoneura fumiferana* (Clem.) to black spruce, *Picea mariana*. For. Chron. 33:364–372.

———. 1958a. Effects of defoliation by spruce budworm (*Choristoneura fumiferana* (Clem.)) on radial growth at breast height of balsam fir (*Abies balsamea* (L.) Mill.) and white spruce (*Picea glauca* (Moench) Voss). For. Chron. 34:39–47.

———. 1958b. The vulnerability of balsam fir to spruce budworm attack in northwestern Ontario with special reference to the physiological age of the tree. For. Chron. 34:405–422.

———. 1960a. A residual spruce budworm outbreak and aerial spraying operations in the lower St. Lawrence region in 1960. Bi-m. Progr. Rep. Div. For. Biol. Dep. Agric. Can. 16(5):2.

———. 1960b. Spruce budworm outbreaks and the climax of the boreal forest in eastern North America. Rep. Quebec Soc. Prot. Pl. 1959:69–75.

———. 1961a. Spruce budworm outbreaks in the lower St. Lawrence and Gaspé regions. For. Chron. 37:192–202.

———. 1961b. Aerial application of insecticides and the suppression of incipient spruce budworm outbreaks. For. Chron. 37:203–210.

———. 1963. Control of spruce budworm outbreak in Quebec through aerial spraying operations. Canad. Ent. 95:821–827.

———, and R. Martineau. 1956. Spruce budworm defoliation and egg-mass survey in lower St. Lawrence and Gaspé Peninsula – 1955. Bi-m. Progr. Rep. Div. For. Biol. Dep. Agric. Can. 12(2):3.

———. 1960. A recent spruce budworm outbreak in the lower St. Lawrence and Gaspé Peninsula with reference to aerial spraying operations. For. Chron. 36:209–224.

Blais, J. R., and A. J. Thorsteinson. 1948. The spruce budworm in relation to flowering conditions of the host tree. Bi-m. Progr. Rep. For. Insect Invest. Dep. Agric. Can. 4(5):1.

Brower, A. E. 1947. The balsam woolly aphid in Maine. J. Econ. Ent. 40: 689–694.

Brown, N. R. 1947. Spread of imported predator of the balsam woolly aphid. Bi-m. Progr. Rep. For. Insect Invest. Dep. Agric. Can. 3(6):6.

———, and R. C. Clark. 1956. Studies of predators of the balsam woolly aphid, *Adelges piceae* (Ratz.) (Homoptera: Adelgidae) II. An annotated list of the predators associated with the balsam woolly aphid in eastern Canada. Canad. Ent. 88:678–683.

———. 1959. Studies of predators of the balsam woolly aphid, *Adelges piceae* (Ratz.) (Homoptera: Adelgidae) VI. *Aphidecta obliterata* (L.) (Coleoptera: Coccinellidae) an introduced predator in eastern Canada. Canad. Ent. 91:596–599.

———. 1960. Studies of predators of the balsam woolly aphid, *Adelges piceae* (Ratz.) (Homoptera: Adelgidae) VIII. Syrphidae (Diptera). Canad. Ent. 92:801–811.

Brown, R. C. 1952. The spruce budworm. Pp. 683–688. *In* Insects. Yearb. Agric. U.S. Dep. Agric.

———, H. J. MacAloney, and P. B. Dowden. 1949. The spruce budworm. Pp. 423–427. *In* Trees. Yearb. Agric. U.S. Dep. Agric.

Brues, C. T., A. L. Melander, and F. M. Carpenter. 1954. Classification of insects. Bull. Harvard Univ. Mus. Compar. Zool. 108. 917 pp.

Campbell, I. M. 1953. Morphological differences between the pupae and the egg clusters of *Choristoneura fumiferana* (Clem.) and *C. pinus* Free. (Lepidoptera: Tortricidae). Canad. Ent. 85:134–135.

Canada Forest Biology Division. 1942–59. Annual report of the forest insect and disease survey. Dep. Agric. Can.

Canada Forest Entomology and Pathology Branch. 1960–62. Annual report of the forest insect and disease survey. Dep. For. Can.

Carroll, W. J. 1962. The *Neodiprion abietis* complex in Newfoundland. *In* Annual Report For. Ent. Path. Br. Dep. For. Can., pp. 30–31.

———, and D. G. Bryant. 1960. A review of balsam woolly aphid in Newfoundland. For. Chron. 36:278–290.

Carter, E. E. 1916. *Hylobius pales* as a factor in the reproduction of conifers in New England. Proc. Soc. Amer. For. 11:297–307.

Cartwright, K. S. G. 1938. A further note on fungus association in the Siricidae. Ann. Appl. Biol. 25:430.

Chamberlin, W. J. 1939. The bark and timber beetles of North America north of Mexico. OSC Cooperative Assoc. Corvallis, Ore. 513 pp.

Clark, R. C., and N. R. Brown. 1957. Studies of predators of the balsam woolly aphid, *Adelges piceae* (Ratz.) (Homoptera: Adelgidae) III. Field identification and some notes on the biology of *Neoleucopis pinicola* Mall. (Diptera: Chamaeyiidae). Canad. Ent. 89:40–49.

———. 1958. Studies of predators of the balsam woolly aphid, *Adelges piceae* (Ratz.) (Homoptera: Adelgidae). V. *Laricobius erichsonii* Rosen. (Coleoptera: Derodontidae), an introduced predator in eastern Canada. Canad. Ent. 90:657–672.

———. 1959. Predator introduction for balsam woolly aphid control. Bi-m. Progr. Rep. Div. For. Biol. Dep. Agric. Can. 15(6):1.

———. 1960. Studies of predators of the balsam woolly aphid, *Adelges piceae* (Ratz.) (Homoptera: Adelgidae). VII. *Laricobius rubidus* Lec. (Coleoptera: Derodontidae), a predator on *Pineus strobi* (Htg.) (Homoptera: Adelgidae). Canad. Ent. 92:237–240.

Conklin, J. G. 1952. Gypsy moth (*Porthetria dispar*). Pp. 33–35. *In* Important tree pests of the Northeast. New Engl. Sect. Soc. Amer. For.

Coppel, H. C., and M. G. Maw. 1954. Studies on dipterous parasites of the spruce budworm *Choristoneura fumiferana* (Clem.) (Lepidoptera: Tortricidae). Canad. J. Zool. 32:144–156, 314–323.

Coppel, H. C., and B. C. Smith. 1957. Studies on dipterous parasites of the spruce budworm, *Choristoneura fumiferana* (Clem.) (Lepidoptera: Tortricidae). V. *Omotoma fumiferana* (Tot.) (Diptera: Tachinidae). Canad. J. Zool. 35:581–592.

Cox, C. E. 1953. Analysis of frequency distribution of adults and larvae of *Choristoneura fumiferana* (Clem.) and *C. pinus* Free. (Lepidoptera: Tortricidae). Canad. Ent. 85:136–141.

Craighead, F. C. 1925. Relation between mortality of the trees attacked by the spruce budworm (*Cacoecia fumiferana* Clem.) and previous growth. J. Agric. Res. 30(6):541–555.

———. 1950. Insect enemies of eastern forests. Misc. Publ. U.S. Dep. Agric. 657. 679 pp.

Daviault, L. 1950. Les parasites de la tordeuse des bourgeons de l'épinette (*Choristoneura fumiferana* Clem.) dans la province de Québec. Rep. Québec Soc. Prot. Pl. 1948–49:41–45.

———. 1954. Aerial spraying against the spruce budworm in Quebec in 1954. Bi-m. Progr. Rep. Div. For. Biol. Dep. Agric. Can. 11(1):2–3.

Doane, R. W., E. C. van Dyke, W. J. Chamberlin, and H. E. Burke. 1936. Forest insects. McGraw-Hill. New York, N.Y. 463 pp.

Dowden, P. B. 1962. Parasites and predators of forest insects liberated in the United States through 1960. Agric. Handb. U.S. For. Serv. 226. 70 pp.

———, and V. M. Carolin. 1947. Studies on the biological control factors affecting spruce budworm in the United States. Bi-m. Progr. Rep. For. Insect Invest. Dep. Agric. Can. 3(2):6–7.

Dowden, P. B., H. A. Jaynes, and V. M. Carolin. 1953. The role of birds in a spruce budworm outbreak in Maine. J. Econ. Ent. 46:307–312.

Eichhorn, O. 1957. Eine neue Tannenlaus der Gattung *Dreyfusia* (*Dreyfusia merkeri* nov. spec.). Z. Angew. Zool. 44:303–348.

Elliot, K. R. 1960. A history of recent infestations of the spruce budworm in northwestern Ontario and an estimate of resultant timber losses. For. Chron. 36:61–82.

———. 1962. Important infestations. *In* Annual Report For. Ent. Path. Br. Dep. For. Can., p. 93.

Felt, E. P. 1905–6. Insects affecting park and woodland trees. Mem. N.Y. St. Mus. 8. 875 pp.

———. 1924. Manual of tree and shrub insects. Macmillan. New York, N.Y. 382 pp.

Fettes, J. J. 1949. Spruce budworm population trends in a section of Lake Nipigon infestation. Bi-m. Progr. Rep. For. Insect Invest. Dep. Agric. Can. 5(1):2.

———. 1960. Control of the spruce budworm by aircraft spraying and the hazard to aquatic fauna. Bi-m. Progr. Rep. Div. For. Biol. Dep. Agric. Can. 16(1):1–2.

Forbes, W. T. M. 1923. The Lepidoptera of New York and neighboring states. Mem. Cornell Agric. Exp. Sta. 68:489.

———. 1948. Lepidoptera of New York and neighboring states. Mem. Cornell Agric. Exp. Sta. 274(2):128–175.

Franz, J. 1957. Ein Vergleich des europäischen und des nord-amerikanischen Tannentriebwicklers (*Choristoneura muriana* (Hb.) und *C. fumiferana* (Clem.)). Z. Pfl.-Krankh. 64:578–584.

Franz, J. M. 1952. Observations on collecting parasites of *Cacoecia histrionana* (Froel.) (Lep., Tortricidae). Bull. Ent. Res. 43:1–19.

Freeman, T. N. 1953. The spruce budworm, *Choristoneura fumiferana* (Clem.) and an allied new species on pine (Lepidoptera: Tortricidae). Canad. Ent. 85:121–127.

———. 1958. The Archipinae of North America (Lepidoptera: Tortricidae). Suppl. Canad. Ent. 7. 89 pp.

Frost, F. 1958. The balsam woolly aphid and prescribed burning. Publ. Newfoundland Res. Comm. St. John's. 1:19–35, 64–65.

George, J. L., and R. T. Mitchell. 1948. Calculations on the extent of spruce budworm control by insectivorous birds. J. For. 46:454–455.

Ghent, A. W. 1958a. Mortality of overstory aspen in relation to outbreaks of the forest tent caterpillar and the spruce budworm. Ecology 39:222–232.

———. 1958b. Studies of regeneration in forest stands devastated by the spruce budworm II. For. Sci. 4:135–146.

———, D. A. Fraser, and J. B. Thomas. 1957. Studies of regeneration in forest stands devastated by the spruce budworm. For. Sci. 3:184–208.

Giese, R. L., and D. M. Benjamin. 1959. The biology and ecology of the balsam gall midge in Wisconsin. For. Sci. 5:193–208.

Graham, S. A. 1923. The dying balsam fir and spruce in Minnesota. Spec. Bull. Minn. Agric. Ext. Serv. 68. 12 pp.

———. 1928. Host selection by the spruce budworm. Pap. Mich. Acad. Sci. 8:517–523.

———. 1935. The spruce budworm in Michigan pine. Bull. Sch. For. Conserv. Univ. Mich. 6. 56 pp.

———. 1939. Principles of forest entomology. 2nd ed. McGraw-Hill. New York, N.Y. 410 pp.

———. 1956a. Ecology of forest insects. Ann. Rev. Ent. 1:261–280.

———. 1956b. Hazard rating of stands containing balsam fir according to expected injury by spruce budworm. Mich. For. 13. 2 pp.

———. 1956c. Forest insects and the law of natural compensation. Canad. Ent. 88:44–55.

———, and L. W. Orr. 1940. The spruce budworm in Minnesota. Tech. Bull. Minn. Agric. Exp. Sta. 142. 27 pp.

Greenbank, D. O. 1956. The role of climate and dispersal in the initiation of outbreaks of the spruce budworm in New Brunswick. Canad. J. Zool. 34:453–476.

———. 1957. The role of climate and dispersal in the initiation of outbreaks of the spruce budworm in New Brunswick. II. The role of dispersal. Canad. J. Zool. 35:385–403.

———. 1963a. The development of the outbreak. *In* R. F. Morris, ed., The dynamics of epidemic spruce budworm populations. Mem. Ent. Soc. Can. 31:19–23.

———. 1963b. Climate and the spruce budworm. *In* R. F. Morris, ed., The dynamics of epidemic spruce budworm populations. Mem. Ent. Soc. Can. 31:174–180.

———. 1963c. Staminate flowers and the spruce budworm. *In* R. F. Morris, ed., The dynamics of epidemic spruce budworm populations. Mem. Ent. Soc. Can. 31:202–218.

———. 1963d. Host species and the spruce budworm. *In* R. F. Morris, ed., The dynamics of epidemic spruce budworm populations. Mem. Ent. Soc. Can. 31:219–223.

Gryse, J. J. de. 1944. Spruce budworm threatens Ontario forests. Pulp Paper Mag. Can. 45:542–544.

Hart, A. C. 1956. Changes after partial cutting of a spruce-fir stand in Maine. Sta. Pap. Ntheast. For. Exp. Sta. 86. 8 pp.

Hawley, R. C., and P. W. Stickel. 1948. Forest protection. Wiley & Sons. New York, N.Y. 355 pp.

Haynes, D. L. 1962. Invertebrate predators of forest insects. *In* Annual Report For. Ent. Path. Br. Dep. For. Can., pp. 39–40.

Hedlin, A. F. 1956. Studies on the balsam fir seed chalcid, *Megastigmus specularis* Walley (Hymenoptera: Chalcididae). Canad. Ent. 88:691–697.

Heller, R. C., J. L. Bean, and J. W. Marsh. 1952. Aerial survey of spruce budworm damage in Maine in 1950. J. For. 50:8–11.

Henson, W. R. 1951. Mass flights of the spruce budworm. Canad. Ent. 83:240.

Hopkins, A. D. 1904. Catalogues of exhibits of insect enemies of forests and forest products at the Louisiana purchase exposition. St. Louis, Missouri, 1904. Bull. U.S. Bur. Ent. 48. 56 pp.

Jacot, A. P. 1939. Reduction of spruce and fir litter by minute animals. J. For. 37:858–860.

Jahn, E. 1961. Forstschutzprobleme in Tirol. *In* Exkursionsführer "Forstschutz," Int. Verb. Forstl. Forsch. (IUFRO) 13. Kongress, Wien. Sept. 10–29, 1961. Pp. 89–96.

Jaynes, H. A. 1954. Parasitization of spruce budworm larvae at different crown heights by *Apanteles* and *Glypta*. J. Econ. Ent. 47:355–356.

———, and V. M. Carolin. 1952. Spruce budworm (*Choristoneura fumiferana*). *In* Important tree pests of the Northeast. Soc. Amer. For., New Engl. Sect., pp. 51–54.

Jaynes, H. A., and C. F. Speers. 1949. Biological and ecological studies of the spruce budworm. J. Econ. Ent. 42:221–225.

Junk, W., and S. Schenkling. 1910–40. Coleopterorum Catalogus. Verl. f. Naturwiss. Berlin and S'-Gravenhage.

Kendeigh, S. C. 1947. Bird population studies in the coniferous forest biome during a spruce budworm outbreak. Biol. Bull. Div. Res. Dep. Lds. For. Ont. 1. 100 pp.

Kloft, W. 1955. Untersuchungen an der Rinde von Weisstannen (*Abies pectinata*) bei Befall durch *Dreyfusia (Adelges) piceae* Ratz. Z. Angew. Ent. 37:340–348.

———. 1957. Further investigations concerning the interrelationship between bark condition of *Abies alba* and infestation by *Adelges piceae typica* and *A. nüsslini schneideria*. Z. Angew. Ent. 41:438–442.

Kotinsky, J. 1916. The European fir trunk bark louse (*Chermes (Dreyfusia) piceae* Ratz.) apparently long established in the United States. Proc. Ent. Soc. Washington 18:14–16.

Kotschy, K. 1960. Untersuchungen über die Tannentrieblaus *Dreyfusia nüsslini*. Cbl. Ges. Forstw. 77:127–155.

Krieg, A. 1961. *Bacillus thuringiensis* Berliner. Über seine Biologie, Pathogenie und Anwendung in der biologischen Schädlingsbekämpfung. Mitt. Biol. Bundesanst. Land- u. Forstw. Berlin-Dahlem 103. 79 pp.

Leonard, M. D. 1928. A list of the insects of New York. Mem. Cornell Univ. Agric. Exp. Sta. 101. 1121 pp.

Lindquist, O. H., and W. L. Sippell. 1961. Bud-miners. *Argyresthia* spp. on spruce and balsam fir in Ontario. Bi-m. Progr. Rep. For. Biol. Div. Dep. For. Can. 17(1):2.

Loughton, B. G., C. Derry, and A. S. West. 1963. Spiders and the spruce budworm. *In* R. F. Morris, ed., The dynamics of epidemic spruce budworm populations. Mem. Ent. Soc. Can. 31:249–268.

MacAloney, H. J. 1935. The balsam woolly aphid in the Northeast. J. For. 33:481–484.

———. 1952. Balsam woolly aphid (*Chermes piceae*). *In* Important tree pests of the Northeast. Soc. Amer. For. New Engl. Sect., pp. 61–63.

———, D. Crosby, and J. L. Brown. 1947. The entomological aspects of forest management in spruce budworm in the northeastern United States. Bi-m. Progr. Rep. For. Insect Invest. Dep. Agric. Can. 3(2):4–5.

MacAloney, H. J., and P. B. Dowden. 1952. European spruce sawfly (*Diprion hercyniae*). *In* Important tree pests of the Northeast. Soc. Amer. For. New Engl. Sect., pp. 7–9.

McCambridge, W. F., and G. L. Downing. 1960. Black-headed budworm. For. Pest Leafl. U.S. For. Ser. 45. 4 pp.

Macdonald, D. R., and F. E. Webb. 1963. Insecticides and the spruce budworm. *In* R. F. Morris, ed., The dynamics of epidemic spruce budworm populations. Mem. Ent. Soc. Can. 31:288–310.

McDunnough, J. 1938–39. Check list of the Lepidoptera of Canada and the United States of America. Mem. South. California Acad. Sci. Los Angeles, Calif. Vol. 1, 275 pp. Vol. 2, 171 pp.

McGuffin, W. C. 1955a. Descriptions of larvae of forest insects: *Syngrapha*, *Autographa* (Lepidoptera: Phalaenidae). Canad. Ent. 86:39.

———. 1955b. Notes on life histories of some Eunominae (Lepidoptera: Geometridae). Canad. Ent. 87:41–44.

McGugan, B. M. 1954. Needle mining habits and larval instars of the spruce budworm. Canad. Ent. 86:439–454.

———, chmn. 1958. Forest Lepidoptera of Canada recorded by the forest insect survey. Vol. 1. Papilionidae to Arctiidae. Publ. Div. For. Biol. Dep. Agric. Can. 1034. 76 pp.

———, and J. R. Blais. 1959. Spruce budworm parasite studies in northwestern Ontario. Canad. Ent. 91:758–783.

MacKay, M. R. 1953. The larvae of *Choristoneura fumiferana* (Clem.) and *C. pinus* Free. (Lepidoptera: Tortricidae). Canad. Ent. 85:128–133.

McLeod, C. H. 1949. Fungi associated with spruce budworm. Bi-m. Progr. Rep. Div. For. Insect Invest. Dep. Agric. Can. 5(1):1.

MacLeod, D. M., and T. C. Lougheed. 1956. Utilization of carbohydrates by a *Hirsutella* sp. from *Choristoneura fumiferana* (Clem.). Bi-m. Progr. Rep. Div. For. Biol. Dep. Agric. Can. 12(5):2.

McLintock, T. F. 1947. Silvicultural practices of spruce budworm. J. For. 45:655–658.

———. 1949. Mapping vulnerability of spruce-fir stands in the Northeast to spruce budworm attack. Sta. Pap. Ntheast. For. Exp. Sta. 21. 20 pp.

———. 1955. How damage to balsam fir develops after a spruce budworm epidemic. Sta. Pap. Ntheast. For. Exp. Sta. 75. 18 pp.

Merker, E. 1962. Zur Biologie und Systematik der mittel-europäischen Tannenläuse (Gattung *Dreyfusia*). Allg. Forst-u. Jagdztg. 133:149–199.

Meyrick, E. 1913. Lepidoptera: Heterocera, Fam. Tortricidae. Genera Insectorum 149:41.

Miller, C. A. 1957. A technique for estimating the fecundity of natural populations of the spruce budworm. J. Zool. 35:1–13.

———. 1958. The measurement of spruce budworm populations and mortality during the first and second larval instars. Canad. J. Zool. 36:409–422.

———. 1960. The interaction of the spruce budworm, *Choristoneura fumiferana* (Clem.), and the parasite *Glypta fumiferanae* (Vier.). Canad. Ent. 92:839–850.

———. 1962. The population dynamics of the black-headed budworm. In Annual Report For. Ent. Path. Br. Dep. For. Can., p. 40.

———. 1963a. The spruce budworm. In R. F. Morris, ed., The dynamics of epidemic spruce budworm populations. Mem. Ent. Soc. Can. 31:12–19.

———. 1963b. Parasites and the spruce budworm. In R. F. Morris, ed., The dynamics of epidemic spruce budworm populations. Mem. Ent. Soc. Can. 31:228–244.

———, and D. R. Macdonald. 1961. The Green River project — spruce budworm studies. Bi-m. Progr. Rep. For. Ent. Path. Br. Can. 17(4):1.

Mitchell, R. T. 1952. Consumption of spruce budworms by birds in Maine spruce-fir forest. J. For. 50:387–389.

Morris, R. F. 1948. How old is a balsam tree? For. Chron. 24:106–110.

———. 1950. Technique for population sampling on standing trees. Bi-m. Progr. Rep. For. Insect Invest. Dep. Agric. Can. 6(6):1–2.

———. 1951. The effects of flowering on the foliage production and growth of balsam fir. For. Chron. 27:40–57.

———. 1954. A sequential sampling technique for spruce budworm egg surveys. Canad. J. Zool. 32:302–313.

———. 1955. The development of sampling techniques for forest insect defoliators, with particular reference to the spruce budworm. Canad. J. Zool. 33:225–294.

———. 1958. A review of the important insects affecting the spruce-fir forest in the Maritime Provinces. For. Chron. 34:159–189.

———. 1960. Sampling insect populations. Ann. Rev. Ent. 5:243–264.

———. 1963a. Introduction. In R. F. Morris, ed., The dynamics of epidemic spruce budworm populations. Mem. Ent. Soc. Can. 31:7–12.

———. 1963b. Foliage depletion and the spruce budworm. In R. F. Morris, ed., The dynamics of epidemic spruce budworm populations. Mem. Ent. Soc. Can. 31:223–228.

———. 1963c. Résumé. In R. F. Morris, ed., The dynamics of epidemic spruce budworm populations. Mem. Ent. Soc. Can. 31:311–320.

———, ed. 1963. The dynamics of epidemic spruce budworm populations. Mem. Ent. Soc. Can. 31. 332 pp.

———, and C. A. Miller. 1954. The development of life tables for the spruce budworm. Canad. J. Zool. 32:283–301.

Morris, R. F., W. F. Chesire, C. A. Miller, and D. G. Mott. 1958. The numerical response of avian and mammalian predators during a gradation of the spruce budworm. Ecology 39:487–494.

Morris, R. F., F. E. Webb, and C. W. Bennet. 1956. A method of phenological survey for uses in forest insect studies. Canad. J. Zool. 34:533–540.

Morris, R. F., and D. G. Mott. 1963. Dispersal and the spruce budworm. In R. F. Morris, ed., The dynamics of epidemic spruce budworm populations. Mem. Ent. Soc. Can. 31:180–189.

Mott, D. B. 1954. Secondary insects in chemically debarked trees. Bi-m. Progr. Rep. Div. For. Biol. Dep. Agric. 10(2):1.

———. 1963a. The analysis of the survival of small larvae in the unsprayed area. In R. F. Morris, ed., The dynamics of epidemic spruce budworm populations. Mem. Ent. Soc. Can. 31:42–52.

———. 1963b. The forest and the spruce budworm. In R. F. Morris, ed., The dynamics of epidemic spruce budworm populations. Mem. Ent. Soc. Can. 31:189–202.

———, L. D. Nairn, and J. A. Cook. 1957. Radial growth in forest trees and effects of insect defoliation. For. Sci. 3:286–304.

Muesebeck, C. F. W., K. V. Krombein, and H. K. Townes. 1951. Hymenoptera of America north of Mexico. Agric. Monogr. U.S. Dep. Agric. 2. 142 pp.

Neatby, K. W. 1955. Research trends in Canadian agriculture and forestry. Ann. Appl. Biol. 42:54–64.

Neilson, M. M. 1963. Disease and the spruce budworm. *In* R. F. Morris, ed., The dynamics of epidemic spruce budworm populations. Mem. Ent. Soc. Can. 31:272–287.

Nordin, V. J., and B. M. McGugan. 1962. Recent developments in forest insect and disease investigations in Canada. Rep. 8th Brit. Commonw. For. Conf. East Africa. 14 pp.

Orr, L. W. 1954. The role of surveys in forest insect control. J. For. 52:250–252.

Packard, A. S. 1890. Insects injurious to forest trees. Rep. U.S. Ent. Comm. 5. 928 pp.

Patch, E. 1909. Chermes of Maine conifers. Bull. Maine Agric. Exp. Sta. 173:277–308.

Peirson, H. B. 1923. Insects attacking forest and shade trees. Bull. Maine For. Serv. 1:8–25.

———. 1927. Manual of forest insects. Bull. Maine For. Serv. 5. 130 pp.

———. 1950. Spruce budworm control a cooperative project in Maine. Circ. Maine For. Serv. 8 8 pp.

Peterson, L. O. 1947. Balsam fir sawfly, *Neodiprion abietis* Harr. Bi-m. Progr. Rep. For. Insect Invest. Dep. Agric. Can. 3(4):3.

Pilon, J. G., and J. R. Blais. 1961. Weather and outbreaks of the spruce budworm in Province Quebec from 1939 to 1956. Canad. J. Ent. 93:118–123.

Pope, R. B. 1957. The role of aerial photography in the current balsam woolly aphid outbreak. For. Chron. 33:263–264.

Prebble, M. L. 1961. Tests of a microbial insecticide against forest defoliation. Bi-m. Progr. Rep. For. Biol. Div. Dep. Agric. Can. 17(3):1.

———, and J. E. Bier. 1954. The situation with respect to forest entomology and pathology in Canada, 1943 to 1953. For. Chron. 30:25–29.

Prentice, R. M., chmn. 1962. Forest Lepidoptera of Canada recorded by the forest insect survey. Vol. 2. Nycteolidae, Notodontidae, Noctuidae, Liparidae. Bull. For. Ent. Path. Br. Dep. For. Can. 128. 281 pp.

Raizenne, H. 1952. Forest Lepidoptera of southern Ontario and their parasites. Div. For. Biol. Dep. Agric. Can. 277 pp.

Randall, A. P. 1963. Evidence of DDT-resistance in spruce budworm (*Choristoneura fumiferana* Clem.) from forest populations subjected to repeated large-scale DDT sprays. Bi-m. Progr. Rep. For. Ent. Path. Br. Dep. For. Can. 19(4):1.

Raymond, F. L., and J. Reid. 1961. Dieback of balsam fir in Ontario. Canad. J. Bot. 39:233–251.

Redmond, D. R. 1957. Infection courts of butt-rotting fungi in balsam fir. For. Sci. 3:15–20.

———. 1959. Mortality of rootlets in balsam fir defoliated by the spruce budworm. For. Sci. 5:64–69.

Reeks, W. A., and R. S. Forbes. 1950. Native parasites of the spruce budworm in the Maritime Provinces. Bi-m. Progr. Rep. For. Insect Invest. Dep. Agric. Can. 6(5):1.

Roe, E. I. 1950. Balsam fir in Minnesota — a summary of present knowledge. Misc. Rep. Lake St. For. Exp. Sta. 13. 25 pp.

Rose, A. H., and J. R. Blais. 1954. A relation between April and May temperatures and spruce budworm larval emergence. Canad. Ent. 86:174–177.

Schaffner, J. V. 1952. Hemlock loopers (*Lambdina fiscellaria fiscellaria* and *Lambdina athasaria athasaria*). *In* Important tree pests of the Northeast. Soc. Amer. For., New Engl. Sect., pp. 54–56.

Schenck, C. A. 1909. Forest protection. Inland Press. Asheville, N.C. 159 pp.

Schmiege, D. C., and G. W. Anderson. 1960. The forest insect and disease situation, Lake States, 1959. Sta. Pap. Lake St. For. Exp. Sta. 79. 18 pp.

Schmiege, D. C., and R. L. Anderson. 1958. The forest insect and disease situation, Lake States, 1957. Sta. Pap. Lake St. For. Exp. Sta. 60. 22 pp.

Schread, J. C. 1952. Ants injurious to the trees and wood products. *In* Important tree pests in the Northeast. Soc. Amer. For., New Engl. Sect., pp. 148–154.

Secrest, J. P., and D. G. Thornton. 1959. A comparison of the toxicity of various insecticides to the spruce budworm. J. Econ. Ent. 52:212–214.

Shepherd, R. F. 1958. Factors controlling the internal temperatures of spruce budworm larvae, *Choristoneura fumiferana* (Clem.) Canad. J. Zool. 36:779–786.

Silen, R. R., and L. R. Woike. 1959. The Wind River Arboretum. Res. Pap. Pacif. Nthwest. For. Range Exp. Sta. 33. 50 pp.

Smerlis, E. 1957. *Hylobius* injuries of infection of root and butt-rots in immature balsam fir stands. Bi-m. Progr. Rep. Div. For. Biol. Dep. Agric. Can. 13(2):1.

————. 1961. Pathological condition of immature balsam fir stands of Hylocomium-Oxalis type in the Laurentide Park, Quebec. For. Chron. 37:108–115.

Smirnoff, W. A. 1963. Tests of *Bacillus thuringiensis* var. *thuringiensis* Berliner and *B. cereus* Frankland and Frankland on larvae of *Choristoneura fumiferana* (Clemens). Canad. Ent. 95:127–133.

Smith, S. G. 1954. A partial breakdown of temporal and ecological isolation between *Choristoneura* species (Lepidoptera: Tortricidae). Evolution 8:206–224.

Stairs, G. R. 1960. On the embryology of the spruce budworm, *Choristoneura fumiferana* (Clem.) (Lepidoptera: Tortricidae). Canad. Ent. 92:147–154.

Stillwell, M. A. 1960. Rootlet recovery in balsam fir defoliated by the spruce budworm. Bi-m. Progr. Rep. Div. For. Biol. Dep. Agric. Can. 16(5):1.

Swaine, J. M. 1917–18. Canadian bark beetles. Bull. Ent. Dep. Agric. Can. 14(1–2):143 + 32 pp.

————, F. C. Craighead, and J. W. Bailey. 1924. Studies on spruce budworm (*Cacoecia fumiferana* Clem.). Bull. Can. Dep. Agric. 37. 91 pp.

Thomson, H. M. 1955a. Microsporidian disease of spruce budworm. Bi-m. Progr. Rep. Div. For. Biol. Dep. Agric. Can. 11(5):2.

————. 1955b. *Perezia fumiferana* n. sp., a new species of Microsporidia from the spruce budworm *Choristoneura fumiferana* (Clem.). J. Parasitology 41(4):8 pp.

————. 1957a. The effect of a microsporidian disease on the rate of development of the spruce budworm. Bi-m. Progr. Rep. Div. For. Biol. Dep. Agric. Can. 13(3):3.

————. 1957b. A note on the predation of spruce budworm pupae. Bi-m. Progr. Rep. Div. For. Biol. Dep. Agric. Can. 13(4):2.

————. 1958. Some aspects of the epidemiology of a microsporidian parasite of the spruce budworm, *Choristoneura fumiferana* (Clem.). Canad. J. Zool. 36:309–316.

————. 1960. The possible control of a budworm infestation by a microsporidian disease. Bi-m. Progr. Rep. Div. For. Biol. Dep. Agric. Can. 16(4):1.

Tothill, J. D. 1923. Notes on the outbreaks of spruce budworm, forest tent caterpillar, and larch sawfly in New Brunswick. Proc. Acadian Ent. Soc. (1922) 8:172–182.

Turner, K. B. 1952. The relation of mortality of balsam fir (*Abies balsamea* (L.) Mill.), caused by the spruce budworm (*Choristoneura fumiferana* (Clem.)), to forest composition in the Algoma Forest of Ontario. Publ. Dep. Agric. Canad. 875. 107 pp.

United States Bureau of Entomology. 1952. Cooperative economic insect report. Res. Adm. Bur. Ent. 1(4):97–99.

Varty, J. W. 1956. *Adelges* insects of silver firs. Bull. For. Comm. Lond. 26. 75 pp.

Vincent, A. B. 1962. Development of balsam fir thickets in the Green River watershed following the spruce budworm outbreak of 1913–1919. Tech. Note For. Res. Br. Can. 119. 20 pp.

Walley, G. S. 1932. Host records and new species of Canadian Hymenoptera. Canad. Ent. 64:181–189.

Warren, G. L. 1954. The spruce needle worm, *Dioryctria reniculella* Grt. as a predator of spruce budworm. Bi-m. Progr. Rep. Div. For. Biol. Dep. Agric. Can. 10(3):2–3.

Waters, W. E. 1954. Forest insect conditions in the Northeast, 1954. Sta. Pap. Ntheast. For. Exp. Sta. 76. 22 pp.

————. 1955. Sequential sampling in forest insect surveys. For. Sci. 1:68–79.

————. 1956. Forest insect conditions in the Northeast, 1955. Sta. Pap. Ntheast. For. Exp. Sta. 79. 19 pp.

————, and T. McIntyre. 1954. Forest insect conditions in the Northeast, 1953. Sta. Pap. Ntheast. For. Exp. Sta. 72. 17 pp.

Waters, W. E., and P. V. Mook. 1958. Forest insect and disease conditions in the Northeast, 1957. Sta. Pap. Ntheast. For. Exp. Sta. 107. 31 pp.

Waters, W. E., and R. C. Heller, and J. L. Bean. 1958. Aerial appraisal of damage by the spruce budworm. J. For. 56:269–276.

Waters, W. E., and A. M. Waterman. 1957. Forest insect and disease conditions in the Northeast – 1956. Sta. Pap. Ntheast. For. Exp. Sta. 94. 23 pp.

Watt, K. E. F. 1961. Mathematical models for use in insect pest control. Suppl. Canad. Ent. 19. 62 pp.

Webb, F. E. 1954. Aerial spraying against spruce budworm in New Brunswick. Bi-m. Progr. Rep. Div. For. Biol. Dep. Agric. Can. 11(1):1–2.

————. 1955. Four years of aerial spraying against spruce budworm in New Brunswick. Pulp Paper Mag. Can. 56(13):132–135.

————. 1956. Aerial spraying against spruce budworm in New Brunswick, 1955. Bi-m. Progr. Rep. Div. For. Biol. Dep. Agric. Can. 12(2):1–2.

———. 1958. Developments in forest spraying against spruce budworm in New Brunswick. Pulp Paper Mag. Can. 59 (Convent. No.):301–308.

———, D. R. Macdonald, and D. G. Cameron. 1959. Aerial spraying against spruce budworm in New Brunswick — 1958. Bi-m. Progr. Rep. For. Biol. Div. Dep. Agric. Can. 15(1):1–2.

Webb, F. E., J. R. Blais, and R. W. Nash. 1961. A cartographic history of spruce budworm outbreaks and aerial forest spraying in the Atlantic region of North America, 1949–1959. Canad. Ent. 93:160–166.

Wellington, W. G. 1948. Measurements of the physical environment (of spruce budworm). Bi-m. Progr. Rep. For. Insect Invest. Dep. Agric. Can. 4(5):1–2.

———. 1950. Effects of radiation on the temperatures of insect habitats. Sci. Agric. 30:209–234.

———, and W. R. Henson. 1947. Notes on the effects of physical factors on the spruce budworm, *Choristoneura fumiferana* (Clem.). Canad. Ent. 79:168–195.

Wellington, W. G., J. J. Fettes, K. B. Turner, and R. M. Belyea. 1950. Physical and biological indicators of the spruce budworm, *Choristoneura fumiferana* (Clem.) (Lepidoptera: Tortricidae). Canad. J. Res. 28(D):308–331.

Wellington, W. G., C. R. Sullivan, and G. W. Green. 1951. Polarized light as orientation factors in the light reaction of some hymenopterous larvae. Canad. J. Zool. 29:339–351.

Westveld, M. 1945. A suggested method for rating the vulnerability of spruce-fir stands to budworm damage. Ntheast. For. Exp. Sta. 4 pp.

———. 1946. Forest management as a means of controlling the spruce budworm. J. For. 44: 949–953.

———. 1954. A budworm vigor-resistance classification for spruce and balsam fir. J. For. 52: 11–24.

Wilkes, A., H. C. Coppel, and W. G. Mathers. 1948. Notes on the insect parasites of the spruce budworm, *Choristoneura fumiferana* (Clem.) in British Columbia. Canad. Ent. 80:138–155.

Wilson, L. F. 1960. Reducing pulpwood losses from borer attacks by shading conventional pulp piles. Tech. Notes Lake St. For. Exp. Sta. 533. 2 pp.

———. 1961a. Calculating volume loss in balsam fir pulpwood from wood-boring insects. Tech. Notes Lake St. For. Exp. Sta. 599. 2 pp.

———. 1961b. Attraction of wood-boring insects to freshly cut pulpwood. Tech. Notes Lake St. For. Exp. Sta. 610. 2 pp.

Chapter 6. Reproduction

Anonymous. 1948. Growth of balsam fir in Newfoundland. Pulp Paper Mag. Can. 49(3):280, 284, 286.

Baldwin, H. I. 1933. The density of spruce and fir reproduction related to the direction of exposure. Ecology 14:152–156.

———, and A. Pleasonton. 1952. Cold storage of nursery stock. Fox For. Note 48. 2 pp.

Baskerville, G. L. 1960. Conversion to periodic selection management in a fir, spruce, and birch forest. Tech. Note For. Br. Can. 86. 19 pp.

———, E. L. Hughes, and O. L. Loucks. 1960. Research by the Federal Forestry Branch in the Green River project. For. Chron. 36:265–277.

Bell, G. W. 1962. National regeneration resolution committee report, Northern Ontario Section. For. Chron. 38:79–85.

Bonner, E. 1941. Balsam fir in the Clay Belt of northern Ontario. Master's thesis. Univ. Toronto. 102 pp.

Bowman, A. B. 1944. Growth and occurrence of spruce and fir on pulpwood lands in northern Michigan. Tech. Bull. Mich. Agric. Exp. Sta. 188. 82 pp.

Buell, M. F. 1956. Spruce-fir, maple-basswood competition in Itasca Park, Minnesota. Ecology 37:606.

Candy, R. H. 1951. Reproduction on cut-over and burned-over land in Canada (east of the Rocky Mountains). Silv. Res. Note For. Br. Can. 92. 224 pp.

Carman, R. S. 1953. How tree seed is procured. Pulp Paper Mag. Can. 54(1):100–107.

Cheyney, E. G. 1946. Forest plantation in northern Minnesota. J. For. 44:39–40.

Deters, M. E., and H. Schmitz. 1936. Drought damage to prairie shelterbelts in Minnesota. Bull. Minn. Agric. Exp. Sta. 329. 28 pp.

Dixon, A. C., and I. C. M. Place. 1952. The influence of microtopography on the survival of spruce and fir reproduction. Silv. Leafl. For. Br. Can. 68. 4 pp.

Eichel, G. H. 1957. Management of spruce-balsam stands towards natural regeneration. For. Chron. 33:233–237.

George, E. J. 1953. Tree and shrub species for the northern great plains. Circ. U.S. Dep. Agric. 912. 46 pp.

Ghent, A. W. 1958a. Mortality of overstory aspen in relation to outbreaks of the forest tent caterpillar and the spruce budworm. Ecology 39:222–232.

———. 1958b. Studies in forest stands devastated by the spruce budworm. II. For. Sci. 4:135–146.

———. 1963. Studies of regeneration in forest stands devastated by the spruce budworm. III. For. Sci. 9:295–310.

———, D. A. Fraser, and J. B. Thomas. 1957. Studies of regeneration in forest stands devastated by the spruce budworm. For. Sci. 3:184–208.

Halliday, W. E. D. 1937. A forest classification for Canada. Bull. For. Serv. Can. 89. 50 pp.

Hart, A. C. 1958. Report on forest tree seed crop in New England in 1957. For. Res. Note Ntheast. For. Exp. Sta. 79. 2 pp.

———. 1959. Report on forest tree seed crop in New England. For. Res. Note Ntheast. For. Exp. Sta. 86. 2 pp.

Hatcher, R. J. 1960. Development of balsam fir following clear-cut in Quebec. Tech. Note For. Br. Can. 87. 21 pp.

———. 1961. Partial cutting balsam fir stands on the Epaule River watershed, Quebec. Tech. Note For. Br. Can. 105. 29 pp.

Hawley, R. C., and A. F. Hawes. 1912. Forestry in New England. Wiley & Sons. New York, N.Y. 479 pp.

———, and D. M. Smith. 1954. The practice of silviculture. Wiley & Sons. New York, N.Y. 525 pp.

Heit, C. E., and E. J. Eliason. 1940. Coniferous tree seed testing and factors affecting germination and seed quality. Tech. Bull. N.Y. Agric. Exp. Sta. 255. 45 pp.

Holt, L. 1950. Cutting methods in pulpwood operations of eastern Canada. Woodl. Res. Index Pulp Pap. Res. Inst. Can. 84. 58 pp.

Horton, K. W. 1962. Regenerating white pine with seed trees and ground scarification. Tech. Note Dep. For. Can. 118. 19 pp.

Hosie, R. C. 1947. Report on regeneration studies. Ontario Res. Rep. Dep. Lds. For. Ont. 10. 56 pp.

———. 1953. Forest regeneration in Ontario. For. Bull. Univ. Toronto 2. 134 pp.

Hsiung, W. W. Y. 1951. An ecological study of the beaked hazel (*Corylus cornuta* Marsh.) in the Cloquet Experimental Forest, Minnesota. Ph.D. thesis. Univ. Minnesota.

Illick, J. S. 1928. Pennsylvania trees. Bull. Pa. Dep. For. Wat. 11. 237 pp.

International Seed Testing Association. 1959. International rules for seed testing. Proc. Int. Seed Test. Ass. 24:475–584.

Jones, J. R. 1956. Logging mortality and first year post-logging mortality in advance reproduction in a fir-spruce-intolerant hardwood stand. Unpubl. Rep. Univ. Minnesota. 26 pp.

Kabzems, A. 1952. Stand dynamics and development in the mixed forest. For. Chron. 28:7–22.

Kittredge, J. 1929. Forest planting in the Lake States. Bull. U.S. Dep. Agric. 1097. 88 pp.

———, and H. C. Belyea. 1923. Reproduction with fire protection in the Adirondacks. J. For. 21:784–787.

———, and S. R. Gevorkiantz. 1929. Forest possibilities of aspen lands in the Lake States. Tech. Bull. Minn. Agric. Exp. Sta. 60. 84 pp.

Koroleff, A. 1954. Full-tree logging – a challenge to research. Woodl. Res. Ind. Pulp Pap. Res. Inst. Can. 93. 101 pp.

Kozlowski, T. T. 1960. Effect of moist stratification and storage in polyethylene bags on germination of forest tree seed. For. Res. Notes Univ. Wisconsin 59. 5 pp.

Krefting, L. W. 1953. Effect of cutting mountain maple on the production of deer browse. Minn. For. Notes 21. 2 pp.

———, H. L. Hansen, and M. H. Stenlund. 1955. Use of herbicides in inducing regrowth of mountain maple for deer browse. Minn. For. Notes 42. 2 pp.

———. 1956. Stimulating regrowth of mountain maple for deer browse by herbicides, cutting and fire. J. Wildlife Mgmt. 20:434–441.

Lake States Forest Experiment Station. 1950. Annual report for 1949. Lake St. For. Exp. Sta. 32 pp.

———. 1953. Annual report for 1952. Lake St. Exp. Sta. 43 pp.

LeBarron, R. K. 1945. Mineral soil is favorable seed bed for spruce and fir. Tech. Note Lake St. For. Exp. Sta. 237. 1 p.

Linteau, A. 1955. Forest site classification of the northeastern Boreal forest region, Quebec. Bull. For. Br. Can. 118. 85 pp.

Long, H. D. 1940. Spruce regeneration in Canada: the Maritimes. For. Chron. 16:6–9.

McArthur, J. D. 1963. Effect of mechanized logging on the composition of balsam fir stand in the Gaspé Peninsula. Pulp Paper Mag. Can. 64: WR208–210.

McConkey, T. W. 1960. Report on 1959 forest tree seed crop in New England. For. Res. Notes Ntheast. For. Exp. Sta. 96. 3 pp.

———. 1961. Report on 1960 forest tree seed crop in New England. For. Res. Notes Ntheast. For. Exp. Sta. 115. 3 pp.

McCullough, H. A. 1948. Plant succession on fallen logs in a virgin spruce-fir forest. Ecology 29:508–513.

MacGillivray, H. G. 1955. Germination of spruce and fir seed following different stratification periods. For. Chron. 31:365.

MacLean, D. W. 1955. The Black River experimental area. Second ecological report. Mimeo. For. Br. Can. 7 pp.

———. 1959. Five-year progress report on project RC-17. Woodl. Res. Ind. Pulp Paper Res. Inst. Can. 112. 142 pp.

———. 1960. Some aspects of aspen-birch-spruce-fir type in Ontario. Tech. Note For. Br. Can. 94. 24 pp.

Morris, R. F. 1951. The effect of flowering on the foliage production and growth of balsam fir. For. Chron. 27: 40–57.

Moss, A. 1962. Spruce silvicultural management for the future. For. Chron. 36:156–162.

Murphy, L. W. 1917. Red spruce: its growth and management. Tech. Bull. U.S. Dep. Agric. 544. 100 pp.

Nickerson, D. E. 1958. Studies of regeneration on burned forest land in Newfoundland. Publ. Newfoundl. Res. Comm. St. John's 1:11–18, 62–64.

Northeastern Forest Soils Conference. 1961. Planting sites in the Northeast. Sta. Pap. Ntheast. For. Exp. Sta. 157. 24 pp.

Olmsted, N. W., and J. D. Curtis. 1947. Seeds of the forest floor. Ecology 28:49–52.

Olson, C. E., and R. R. Bay. 1955. Reproduction of upland cutover spruce-balsam-hardwood stands in Minnesota. J. For. 53:833–835.

Oosting, H. J., and J. F. Reed. 1944. Ecological composition of pulpwood forests in northwestern Maine. Amer. Midl. Nat. 31:182–210.

Paton, R. R., E. Secrest, and H. A. Ezri. 1944. A survey of forest plantations in Ohio. Bull. Ohio Agric. Exp. Sta. 647. 77 pp.

Pike, R. T. 1955. Yield of white spruce and balsam fir in an undisturbed stand, Duck Mountain, Manitoba. Tech. Note For. Br. Can. 11. 3 pp.

Place, I. C. M. 1952a. Comparative growth of spruce and fir seedlings in sandflats. Silv. Leafl. For. Br. Can. 64. 4 pp.

———. 1952b. The influence of bracken fern on establishment of spruce and fir seedlings. Silv. Leafl. For. Br. Can. 70. 4 pp.

———. 1955. The influence of seed-bed conditions on the regeneration of spruce and balsam fir. Bull. For. Br. Can. 117. 87 pp.

Pool, R. J. 1939. Some reactions of the vegetation in the towns and cities of Nebraska to the great drought. Bull. Torrey Bot. Cl. 66:457–464.

Prebble, M. L. 1949. Regeneration in budworm-devastated stands. Bi-m. Progr. Rep. For. Insect Invest. Dep. Agric. Can. 5(2):2.

Ray, R. G. 1941. Site-types and rate of growth, Lake Edward, Champlain County, P.Q., 1915–1936. Silv. Res. Note For. Serv. Can. 65. 56 pp.

———. 1956. Site-types, growth and yield at the Lake Edward Forest Experiment Area, Quebec. Tech. Note For. Br. Can. 27. 53 pp.

———. 1957. Patch cutting in second-growth balsam fir at Matane, Quebec. Pulp Paper Mag. Can. 58(10):120–205.

Recknagel, A. B., and M. Westveld. 1942. Results of second remeasurement of Adirondack cutting plots. J. For. 40:837–840.

———, H. L. Churchill, C. Heimburger, and M. Westveld. 1933. Experimental cutting of spruce and fir in the Adirondacks. J. For. 31:680–688.

Robbins, P. W. 1930. Need mineral soil for spruce and balsam seedlings. Quart. Bull. Mich. Agric. Exp. Sta. 12:79–81.

Roe, E. I. 1946. Extended periods of seedfall of white spruce and balsam fir. Tech. Notes Lake St. For. Exp. Sta. 261. 1 p.

———. 1948. Balsam fir seed, its characteristics and germination. Sta. Pap. Lake St. For. Exp. Sta. 11. 13 pp.

———. 1953. Regeneration of balsam fir guaranteed by continuous reserve of small seedlings. Tech. Note Lake St. For. Exp. Sta. 404. 1 p.

———. 1957. Growth of established balsam fir reproduction greatly stimulated by cutting. Tech. Note Lake St. For. Exp. Sta. 512. 2 pp.

Rohmeder, E. 1938. Neuzeitliche Geräte und Arbeitsverfahren zur Prüfung des Forstsaatgutes. Forstw. Cbl. 60:218–231, 244–254, 265–278.

———. 1951. Beiträge zur Keimungsphysiologie der Forstpflanzen. Bayer. Landw. Verl. München. 148 pp.

———. 1960. Forstliche Samenkunde, Genetik und Züchtung. *In* F. Bauer, ed., Fortschritte in der Forstwirtschaft. BLV Verlagsgesellschaft. München-Bonn-Wien. Pp. 31–46.

Rowe, J. S. 1955. Factors influencing white spruce reproduction in Manitoba and Saskatchewan. Tech. Note For. Br. Can. 3. 27 pp.

———. 1959. Forest regions of Canada. Bull. For. Br. Can. 123. 71 pp.

———. 1961. Critique of some vegetational concepts as applied to forests of northwestern Alberta. Can. J. Bot. 39:1007–1017.

Rudolf, P. O. 1946. The reforestation job in the Lake States — a new estimate. Sta. Pap. Lake St. For. Exp. Sta. 4. 9 pp.

———. 1950a. Forest plantations in the Lake States. Tech. Bull. U.S. Dep. Agric. 1010. 171 pp.

———. 1950–63. Forest tree seed crop in the Lake States. Annual Report. Tech. Notes Lake St. For. Exp. Sta.

Schantz-Hansen, T. 1923. Second growth on cut-over lands in St. Louis County. Bull. Minn. Agric. Exp. Sta. 203. 50 pp.

———. 1934. The cut-over lands of Lake County. Bull. Minn. Agric. Exp. Sta. 304. 23 pp.

———, and O. F. Hall. 1952. Results of testing exotic trees and shrubs for hardiness in northern Minnesota. Minn. For. Notes 5. 2 pp.

Schenck, C. A. 1912. The art of the second growth, or American sylviculture. Brandow. Albany, N.Y. 206 pp.

Stoeckeler, J. H. 1952. Disking to regenerate pulpwood species. Tech. Note Lake St. For. Exp. Sta. 372. 1 p.

———. 1955. Establishment of natural regeneration by disking in Wisconsin. Rep. Lake States For. Exp. Sta. 33 pp.

———, and D. D. Skilling. 1959. Direct seeding and planting of balsam fir in northern Wisconsin. Sta. Pap. Lake St. For. Exp. Sta. 72. 22 pp.

Stott, C. B., W. W. Barton, and J. H. Stone. 1942. Results of forest cutting practices in various Lake States timber types. U.S. For. Serv. Reg. 9. Milwaukee, Wis. 56 pp.

Toumey, J. W., and E. J. Neethling. 1923. Some effects of cover over coniferous seed beds in southern New England. Bull. Yale Sch. For. 9. 39 pp.

———, and C. L. Stevens. 1928. The testing of coniferous tree seeds at the School of Forestry, Yale University, 1906–1926. Bull. Yale Sch. For. 21:1–14.

Tryon, E. H., A. W. Goodspeed, R. P. True, and G. J. Johnson. 1951. Christmas trees. Their profitable production in West Virginia. Circ. W. Va. Agric. Exp. Sta. 82. 28 pp.

United States Forest Service. 1948. Woody-plant seed manual. Misc. Publ. U.S. Dep. Agric. 654. 416 pp.

Vincent, A. B. 1952. Logging damage to spruce (*Picea glauca*) and fir (*Abies balsamea*) advance growth. Silv. Leafl. For. Br. Can. 69. 4 pp.

———. 1953. Mountain maple. Silv. Leafl. For. Br. Can. 80. 3 pp.

———. 1954. Release of balsam fir and white spruce reproduction from shrub competition. Silv. Leafl. For. Br. Can. 100. 4 pp.

———. 1955. Development of balsam fir and white spruce forest in northwestern New Brunswick. Tech. Note For. Br. Can. 6. 27 pp.

———. 1956. Balsam fir and white spruce reproduction on the Green River watershed. Tech. Note For. Br. Can. 40. 24 pp.

Waldron, R. M. 1959. Experimental cutting in a mixedwood stand in Saskatchewan, 1924. Tech. Note For. Br. Can. 74. 14 pp.

Webb, L. S. 1957. The growth and development of balsam fir in Gaspé. Pulp Paper Mag. Can. 58(10):206–214.

Westveld, M. 1931. Reproduction on pulpwood lands in the Northeast. Tech. Bull. U.S. Dep. Agric. 223. 52 pp.

———. 1953. Ecology and silviculture of the spruce-fir forests of eastern North America. J. For. 51:422–430.

Westveld, R. H. 1933. The relation of certain soil characteristics to forest growth and composition in the northern hardwood forest of northern Michigan. Tech. Bull. Mich. Agric. Exp. Sta. 135. 52 pp.

Zasada, Z. A. 1952. Reproduction on cut-over swamplands in the Upper Peninsula of Michigan. Sta. Pap. Lake St. For. Exp. Sta. 27. 15 pp.

Zon, R. 1914. Balsam fir. Bull. U.S. For. Serv. 55. 68 pp.

Chapter 7. Stand Development

Anonymous. 1948. Growth of balsam fir in Newfoundland. Pulp Paper Mag. Can. 49(3):280, 284, 286.

Atkins, E. S. 1956. Susceptibility of certain trees of eastern Ontario to basal bark sprays. Tech. Note For. Br. Can. 38. 5 pp.

Baskerville, G. L. 1959. Softwoods respond to weeding. Pulp Paper Mag. Can. 60(8):140, 144.

———. 1961a. Response of young fir and spruce to release from shrub competition. Tech. Note Dep. For. Can. 98. 14 pp.

———. 1961b. Development of immature balsam fir following crown release. Tech. Note Dep. For. Can. 101. 15 pp.

———, E. L. Hughes, and O. L. Loucks. 1960. Research by the Federal Forestry Branch in the Green River project. For. Chron. 36:265–277.

Blanchard, K. L., J. Hall, and R. B. Anderson. 1957. Effect of aminotriazole on various coniferous seedlings. (Abstr.) Res. Rep. 14th Ann. Nth. Cent. Weed Control Conf., pp. 143–144.

Bonner, E. 1941. Balsam fir in the Clay Belt of northern Ontario. Master's thesis. Univ. Toronto. 102 pp.

Bowman, A. B. 1944. Growth and occurrence of spruce and fir on pulpwood lands in northern Michigan. Tech. Bull. Mich. Agric. Exp. Sta. 188. 82 pp.

Candy, R. H. 1938. Growth and regeneration surveys in Canada. Bull. For. Serv. Can. 90. 50 pp.

———. 1942. Forest growth on the upper Lièvre valley, P.Q. Silv. Res. Note For. Serv. Can. 71. 25 pp.

Cary, A. 1928. The hardwood problem of the Northeast. J. For. 26:864–870.

Chapman, C. L. 1954. The problem of over stocking in an intensively managed softwood stand. Tech. Note Univ. Maine 24. 3 pp.

Churchill, H. L. 1927. Girdling of hardwoods to release young conifers. J. For. 25:708–714.

Clarke, W. B. M. 1940. Experimental girdling in mixedwood stands in New Brunswick. Silv. Res. Note For. Serv. Can. 62. 24 pp.

Daly, E. G. 1950. Improvement cutting in a mixedwood stand. Silv. Leafl. For. Br. Can. 46. 3 pp.

Day, M. W. 1942. Selective cutting well adapted to spruce-fir forests. Quart. Bull. Mich. Agric. Exp. Sta. 24:238–240.

———. 1945. Spruce-fir silviculture. Quart. Bull. Mich. Agric. Exp. Sta. 28:59–65.

———. 1950. Basal-stem sprays effective on alder and red maple. (Abstr.) Res. Rep. 7th Ann. Nth. Cent. Weed Control Conf., p. 235.

———. 1954a. Results of thinning a spruce-fir stand. Quart. Bull. Mich. Agric. Exp. Sta. 36: 270–274.

———. 1954b. Foliage sprays for maple sprout clumps. (Abstr.) Res. Rep. 11th Ann. Nth. Cent. Weed Control Conf., p. 131.

———. 1961. Further results from thinning a spruce-fir stand. Quart. Bull. Mich. Agric. Exp. Sta. 48:21–26.

———, and R. E. Dils. 1951. The effect of basal stem sprays on red maple and alder. (Abstr.) Res. Rep. 8th Ann. Nth. Cent. Weed Control Conf., p. 150.

Duncan, D. P., and A. C. Hodson. 1958. Influence of the forest tent caterpillar upon the aspen forests of Minnesota. For. Sci. 4:71–93.

Ghent, A. W. 1958. Mortality of overstory aspen in relation to outbreaks of the forest tent caterpillar and the spruce budworm. Ecology 39:222–232.

———, D. A. Fraser, and J. B. Thomas. 1957. Studies of regeneration in forest stands devastated by the spruce budworm. For. Sci. 3:184–208.

Graham, S. A. 1941. Climax forests of the Upper Peninsula of Michigan. Ecology 22:355–362.

Graves, H. S. 1899. Practical forestry in the Adirondacks. Bull. U.S. Dep. Agric. 26. 84 pp.

Halliday, W. E. D. 1937. A forest classification for Canada. Bull. For. Serv. Can. 89. 50 pp.

Hansen, H. L. 1955. Changes in plant communities following spraying, and release effects on conifers. Misc. Rep. Lake St. For. Exp. Sta. 39:12–17.

———, and C. E. Ahlgren. 1950a. Effects of various foliage sprays on beaked hazel (*Corylus cornuta*) at the Quetico-Superior Wilderness Research Center. (Abstr.) Res. Rep. 7th Ann. Nth. Cent. Weed Control Conf., p. 239.

———. 1950b. Effects of various sprays on mountain maple (*Acer spicatum*) at the Quetico-Superior Wilderness Research Center. (Abstr.) Res. Rep. 7th Ann. Nth. Cent. Weed Control Conf., p. 240.

Hart, A. C. 1956. Changes after partial cutting of a spruce-fir stand in Maine. Sta. Pap. Ntheast. For. Exp. Sta. 86. 8 pp.

———. 1961. Thinning balsam fir thickets with soil sterilants. Sta. Pap. Ntheast. For. Exp. Sta. 152. 8 pp.

Hatcher, R. J. 1960. Development of balsam fir following a clearcut in Quebec. Tech. Note For. Br. Can. 87. 21 pp.

———. 1961. Partial cutting in balsam fir stands on the Epaule River watershed, Quebec. Tech. Note For. Dep. Can. 105. 29 pp.

Heimburger, C. C. 1934. Forest-type studies in the Adirondack region. Mem. Cornell Agric. Exp. Sta. 165. 122 pp.

———. 1941. Forest-site classification and soil investigation of Lake Edward Forest Experimental Area. Silv. Res. Note For. Serv. Can. 66. 60 pp.

Heinselman, M. L. 1954. The extent of natural conversion to other species in the Lake States aspen-birch type. J. For. 52:737–738.

Holt, L. 1949. Progress report on a thinning experiment in young spruce–balsam fir stands on the limits of Fraser Companies, Ltd. Woodl. Res. Index Pulp Pap. Res. Inst. Can. 58. 16 pp.

Jankowski, E. J. 1955. Effectiveness of chemical sprays on resistant species. Misc. Rep. Lake St. For. Exp. Sta. 39:7–10.

Jarvis, J. M. 1960. Forty-five years' growth on the Goulais River watershed. Tech. Note For. Br. Can. 84. 31 pp.

Jensen, R. A. 1948. Results of 2,4-D foliage spray on hazel brush (*Corylus rostrata*). (Abstr.) Res. Rep. 5th Ann. Nth. Cent. Weed Control Conf. Sect. 6, p. 8.

Kabzems, A. 1952. Stand dynamics and development in the mixed forest. For. Chron. 28:7–22.

Kagis, A. I. 1954. The story of a cut-over. For. Chron. 30:158–182.

Kittredge, J., and S. R. Gevorkiantz. 1929. Forest possibilities of aspen lands in the Lake States. Tech. Bull. Minn. Agric. Exp. Sta. 60. 84 pp.

Klug, H. A., and H. L. Hansen. 1960. Relative effectiveness of various concentrations of 2,4-D in basal, dormant-season applications. Minn. For. Notes 94. 2 pp.

Krefting, L. W. 1953. Effect of cutting mountain maple on the production of deer browse. Minn. For. Notes 21. 2 pp.

———, H. L. Hansen, and M. H. Stenlund. 1955. Use of herbicides in inducing regrowth of mountain maple for deer browse. Minn. For. Notes 42. 2 pp.

———. 1956. Stimulating regrowth of mountain maple for deer browse by herbicides, cutting and fire. J. Wildlife Mgmt. 20:434–441.

Leinfelder, R. P., and H. L. Hansen. 1954. Tolerance of conifers to various applications of 2,4-D and 2,4,5-T in Minnesota. (Abstr.) Res. Rep. 11th Ann. Nth. Cent. Weed Control Conf., pp. 133–135.

Linteau, A. 1955. Forest site classification of northeastern section, Boreal forest region, Quebec. Bull. For. Br. Can. 118. 85 pp.

McCarthy, E. F. 1918. Accelerated growth of balsam fir in the Adirondacks. J. For. 16:304–307.

MacConnell, W. P., and R. S. Bond. 1961. Hardwood control with mist blowers. Proc. Ntheast. Weed Control Conf. 15:506–515.

McCormack, R. J. 1953. Forest growth in the upper Lièvre River valley, 1930 to 1951. Silv. Leafl. For. Br. Can. 92. 4 pp.

MacLean, D. W. 1949. Forest development on the Goulais River watershed, 1910–1946. Silv. Res. Note For. Serv. Can. 94. 54 pp.

———. 1960. Some aspects of aspen-birch-spruce-fir type in Ontario. Tech. Note For. Br. Can. 94. 24 pp.

McLintock, T. F. 1951. Budworm damage in Canada: some observations on mortality. Res. Notes Ntheast. For. Exp. Sta. 4:1–2.

———. 1954. Factors affecting wind damage in selectively cut stands of spruce and fir in Maine and northern New Hampshire. Sta. Pap. Ntheast. For. Exp. Sta. 70. 17 pp.

———. 1955. How damage to balsam fir develops after a spruce budworm epidemic. Sta. Pap. Ntheast. For. Exp. Sta. 75. 18 pp.

Meagher, G. S., and A. B. Recknagel. 1935. The growth of spruce and fir on the Whitney Park in the Adirondacks. J. For. 33:499–502.

Melander, L. W., chmn. 1948. Investigations of methods of control of undesirable woody plants. Res. Rep. 5th Ann. Nth. Cent. Weed Control Conf. Sect. 6. 38 pp.

———, chmn. 1949. Investigations of methods of control of undesirable woody plants. Res. Rep. 6th Ann. Nth. Cent. Weed Control Conf., pp. 137–178.

Morris, R. F. 1948. How old is a balsam tree? For. Chron. 24:106–110.

Mott, D. G., L. D. Nair, and J. A. Cook. 1957. Radial growth in forest trees and effects of insect defoliation. For. Sci. 3:286–304.

Mulloy, G. A. 1931. Growth on cut-over pulpwood forest in Ontario. Silv. Res. Note For. Serv. Can. 36. 20 pp.

———. 1941a. Cleaning of scattered young balsam fir and spruce in cut-over hardwood stands. Silv. Res. Note For. Serv. Can. 67. 19 pp.

———. 1941b. Improvement cuttings in intolerant hardwood conifer type. Silv. Res. Note For. Serv. Can. 68. 25 pp.

Northeast Pulpwood Research Center. 1952. Forest practice survey report. Ntheast. Pulpw. Res. Cent. Gorham, N.H. 191 pp.

Pike, R. T. 1955. Yield of white spruce and balsam fir in an undisturbed stand, Duck Mountain, Manitoba. Tech. Note For. Br. Can. 11. 2 pp.

Playfair, L. 1956a. 2,3,6-trichlorobenzoic acid as a foliar spray. (Abstr.) Res. Rep. 13th Ann. Nth. Cent. Weed Control Conf., p. 165.

———. 1956b. 2,3,6-trichlorobenzoic acid as a dormant over-all spray. (Abstr.) Res. Rep. 13th Ann. Nth. Cent. Weed Control Conf., p. 165.

Plice, M. J., and G. W. Hedden. 1931. Selective girdling of hardwoods to release young growth conifers. J. For. 29:32–40.

Plochmann, R. 1956. Bestockungsaufbau und Baumartenwandel nordischer Urwälder. Beih. Forstwiss. Centrbl. Forstwiss. Forsch. 6:1–96.

Ray, R. G. 1941. Site-types and rate of growth, Lake Edward, Champlain County, P.Q., 1915–1936. Silv. Res. Note For. Serv. Can. 65. 56 pp.

———. 1956. Site-types, growth and yield at the Lake Edward Forest Experiment Area, Quebec. Tech. Note For. Br. Can. 27. 53 pp.

Recknagel, A. B. 1933. Sustained yield of Adirondack spruce and fir. J. For. 31:343–344.

———. 1936. Five-year remeasurement of sample plots. J. For. 34:994–995.

———. 1942. Ten-year remeasurement of sample plots. J. For. 40:265.

———, and G. W. Abel. 1942. Components of spruce and fir volume as influenced by cutting. J. For. 40:962.

Recknagel, A. B., and M. Westveld. 1942. Results of second remeasurement of Adirondack cutting plots. J. For. 40:837–840.

Robertson, W. M. 1942. Effect of mature yellow birch upon spruce and fir understory. Silv. Leafl. For. Serv. Can. 13. 2 pp.

———. 1949. Partial cutting pays. Woodl. Res. Index Pulp Pap. Res. Inst. Can. 42. 3 pp.

———. 1951. Silviculture — cost and production. In Koroleff et al., Stability as factor in efficient forest management. Pulp Paper Res. Inst. Can., pp. 211–218.

Roe, E. I. 1935. Forest soils — the basis of forest management. Lake St. For. Exp. Sta. 9 pp.

———. 1950. Effect of time of foliage spraying with 2,4-D and 2,4,5-T on initial kill of hazel. (Abstr.) Res. Rep. 7th Ann. Nth. Cent. Weed Control Conf., p. 247.

———. 1951. Effect of time of spraying with 2,4-D and 2,4,5-T on resprouting of hazel. (Abstr.) Res. Rep. 8th Ann. Nth. Cent. Weed Control Conf., p. 162.

———. 1952a. Understory balsam fir responds well to release. Tech. Note Lake St. For. Exp. Sta. 377. 1 p.

———. 1952b. Effect of spraying with 2,4-D on hazel. (Abstr.) Res. Rep. 9th Ann. Nth. Cent. Weed Control Conf., p. 72.

———. 1953. Resprouting of mountain maple after basal spraying with 2,4,5-T. (Abstr.) Res. Rep. 11th Ann. Nth. Cent. Weed Control Conf., pp. 73–74.

Rowe, J. S. 1959. Forest regions of Canada. Bull. For. Br. Can. 123. 71 pp.

———. 1961. Critique of some vegetational concepts as applied to forests of northwest Alberta. Can. J. Bot. 39:1007–1017.

Rudolf, P. O., and R. F. Watt. 1956. Chemical control of brush and trees in the Lake States. Sta. Pap. Lake St. For. Exp. Sta. 41. 58 pp.

Sisam, J. W. B. 1939. Forest development on the Goulais River watershed, 1910–33. Silv. Res. Note For. Serv. Can. 55. 55 pp.

Stoeckeler, J. H. 1948a. Second year results of alder brush eradication tests, Argonne Experimental Forest, Wisconsin. (Abstr.) Res. Rep. 5th Ann. Nth. Cent. Weed Control Conf. Sect. 6, 33.

―――. 1948b. Results of 1948 tests of sprays along telephone lines to control sprout and seedling growth of native trees and other vegetation in northern Wisconsin. (Abstr.) Res. Rep. 5th Ann. Nth. Cent. Weed Control Conf. Sect. 6, 34.

―――. 1949. Control of weeds in conifer nurseries by mineral spirits. Sta. Pap. Lake St. For. Exp. Sta. 17. 23 pp.

―――, and M. L. Heinselman. 1950. The use of herbicides for the control of alder brush and other swamp shrubs in the Lake States. J. For. 48:870–874.

Sutton, R. F. 1958. Chemical herbicides and their uses in the silviculture of forests of eastern Canada. Tech. Note For. Br. Can. 68. 54 pp.

Thomson, C. C. 1949a. Reduced yield in a dense balsam fir stand. Silv. Leafl. For. Serv. Can. 32. 2 pp.

―――. 1949b. Releasing conifers in a young hardwood stand. Silv. Leafl. For. Serv. Can. 34. 2 pp.

Trimble, G. R. 1942. Logging damage in partial cutting of spruce-fir stands. Tech. Note Ntheast. For. Exp. Sta. 51. 2 pp.

Turner, K. B. 1952. The relation of mortality of balsam fir (*Abies balsamea* (L.) Mill.) caused by the spruce budworm (*Choristoneura fumiferana* (Clem.)), to forest composition in the Algoma district of Ontario. Publ. Dep. Agric. Can. 875. 107 pp.

Vincent, A. B. 1953. Mountain maple. Silv. Leafl. For. Br. Can. 80. 3 pp.

―――. 1954. Release of balsam fir and white spruce reproduction from shrub competition. Silv. Leafl. For. Br. Can. 100. 4 pp.

―――. 1955. Development of balsam fir and white spruce forest in northwestern New Brunswick. Tech. Note For. Br. Can. 6. 27 pp.

―――. 1962. Development of balsam fir thickets in the Green River watershed following the spruce budworm outbreak of 1913–1919. Tech. Note Dep. For. Can. 119. 20 pp.

Westveld, M. 1930a. Girdling hardwoods to release spruce and balsam fir. J. For. 28:101.

―――. 1930b. Suggestions for the management of spruce stands in the Northeast. Circ. U.S. Dep. Agric. 134. 24 pp.

―――. 1931. Reproduction on pulpwood lands in the Northeast. Tech. Bull. U.S. Dep. Agric. 223. 52 pp.

―――. 1937. Increasing growth and yield of young spruce pulpwood stands by girdling hardwoods. Circ. U.S. Dep. Agric. 431. 19 pp.

―――. 1946. Forest management as a means of controlling the spruce budworm. J. For. 44: 949–953.

―――. 1953. Ecology and silviculture of the spruce-fir forests of eastern North America. J. For. 51:422–430.

Westveld, R. H. 1933. Making spruce-fir lands profitable: care in logging is an important step in growing pulpwood. Quart. Bull. Mich. Agric. Exp. Sta. 15:172–174.

Wilde, S. A., F. G. Wilson, and D. P. White. 1949. Soils of Wisconsin in relation to silviculture. Publ. Wis. Conserv. Dep. 525-49. 171 pp.

Woodford, E. K., and S. A. Evans, eds. 1963. Weed control handbook issued by the British Weed Control Council. 3rd ed. Blackwell Scientific Publications. Oxford. 356 pp.

Zehngraff, P. H. 1948. Killing woody and herbaceous vegetation in preparation for forest planting on brushy areas. (Abstr.) Res. Rep. 5th Ann. Nth. Cent. Weed Control Conf. Sect. 6, 10.

―――, and J. von Bargen. 1949. Chemical brush control in forest management. J. For. 47: 110–112.

Zon, R. 1914. Balsam fir. Bull. U.S. For. Serv. 55. 68 pp.

Chapter 8. Growth and Yield

Allison, J. H., and R. M. Brown. 1946. Management of the Cloquet Forest. Tech. Bull. Minn. Agric. Exp. Sta. 171. 95 pp.

Allison, J. H., and R. N. Cunningham. 1939. Timber farming in the Cloquet district. Bull. Minn. Agric. Exp. Sta. 343. 35 pp.

Bakuzis, E. V., and H. L. Hansen. 1962. Distribution of balsam fir reproduction and basal area in the edaphic field of forest communities in the Central Pine section of Minnesota. Minn. For. Notes 120. 2 pp.

Baskerville, G. L. 1960. Conversion to periodic selection management in a fir, spruce, and birch forest. Tech. Note For. Br. Can. 86. 19 pp.

———. 1961. Development of immature balsam fir following crown release. Tech. Note Dep. For. Can. 101. 15 pp.

Bedell, G. H. D., and A. B. Berry. 1955. A method of determining approximate merchantable volumes. Tech. Note For. Br. Can. 14. 4 pp.

Bedell, G. H. D., and D. W. MacLean. 1952. Nipigon growth and yield survey. Silv. Res. Note For. Br. Can. 101. 51 pp.

Behre, C. E. 1927. Form-class taper curves and volume tables and their application. J. Agric. Res. 35:673–744.

Betts, H. S. 1945. Balsam fir (*Abies balsamea*). U.S. For. Serv. Amer. Woods. 8 pp.

Bickford, C. A., F. R. Longwood, and R. Bain. 1961. Average growth rates in the spruce-fir region of New England. Sta. Pap. Ntheast. For. Exp. Sta. 140. 23 pp.

Bonner, E. 1941. Balsam fir in the Clay Belt of northern Ontario. Master's thesis. Univ. Toronto. 102 pp.

Bowman, A. B. 1944. Growth and occurrence of spruce and fir on pulpwood lands in northern Michigan. Tech. Bull. Mich. Agric. Exp. Sta. 188. 82 pp.

Brown, R. M., and S. R. Gevorkiantz. 1934. Volume, yield and stand tables for tree species in the Lake States. Tech. Bull. Minn. Agric. Exp. Sta. 39. 208 pp.

Brown, R. M., and F. H. Kaufert. 1936. Volume tables. Balsam fir (*Abies balsamea*). Minnesota and Wisconsin. Mimeo. Minn. Agric. Exp. Sta. 6 pp.

Burgar, R. J. 1961. The relative growth rates of white spruce and balsam fir trees in the Port Arthur district of Ontario. For. Chron. 37:217–223.

Canada Bureau of Statistics. 1960. Canadian Forestry Statistics, 1959. Dom. Bureau Stat., Ottawa. 24 pp.

Canada Forest Service. 1930. Form-class volume tables. Can. For. Serv. 200 pp.

———. 1948. Form class volume tables. 2nd ed. Can. For. Serv. 261 pp.

Candy, R. H. 1938. Growth and regeneration surveys in Canada. Bull. For. Serv. Can. 90. 50 pp.

Cary, A. 1918. A manual for northern woodsmen. 2nd ed. Harvard Univ. Press. Cambridge, Mass. 302 pp.

Clark, J. F. 1903. On the form of the bole of the balsam fir. For. Quart. 1:56–61.

Cunningham, R. N., A. G. Horn, and D. N. Quinney. 1958. Minnesota's forest resources. For. Res. Rep., U.S. Dep. Agric. 13. 52 pp.

Dana, S. T., and W. B. Greeley. 1930. Timber growing and logging practice in the Northeast. Tech. Bull. U.S. Dep. Agric. 166. 112 pp.

Donahue, R. L. 1940. Forest-site quality studies in the Adirondacks. I. Tree growth as related to soil morphology. Mem. Cornell Agric. Exp. Sta. 229. 44 pp.

Doyle, J. A. 1952. Logging waste survey — 1951 (Gaspé Peninsula, New Brunswick, and Nova Scotia). Mimeo. For. Serv. Can. 0-165. 14 pp.

Duerr, A., chmn. 1953. Tree-volume tables based on forest site. Adirondack and Catskill regions of New York State. Bull. N.Y. St. Coll. For. 29. 8 pp.

Ferguson, R. H., and F. R. Longwood. 1960. The timber resources of Maine. For. Surv. Rep. Upper Darby, Pa. 74 pp.

Ferree, M. J. 1953. Estimating timber volume from aerial photographs. Tech. Publ. N.Y. St. Coll. For. 75. 50 pp.

———, and R. K. Hagar. 1956. Timber growth rates for natural forest stands in New York State. Tech. Publ. N.Y. St. Coll. For. 78. 56 pp.

Findell, V. E., R. E. Pfeifer, A. G. Horn, and C. H. Tubbs. 1960. Michigan's forest resources. Sta. Pap. Lake St. For. Exp. Sta. 82. 46 pp.

Funking, D. L., and H. E. Young. 1955. Balsam fir volume table. Central Maine. Tech. Note Univ. Maine 42. 7 pp.

Gagnon, J. D. 1963. Weekly radial increment of balsam fir in Quebec as related to McLintock's tree classification. For. Chron. 39:318–321.

Gevorkiantz, S. R. 1945. A composite cordwood volume table for pulpwood species in the Lake States. Tech. Note Lake St. For. Exp. Sta. 241. 1 p.

———. 1956. Site index curves for balsam fir in the Lake States. Tech. Note Lake St. For. Exp. Sta. 465. 2 pp.

———. 1957a. Site index curves for black spruce in the Lake States. Tech. Note Lake St. For. Exp. Sta. 473. 2 pp.

———. 1957b. Site index curves for white spruce in the Lake States. Tech. Note Lake St. For. Exp. Sta. 474. 2 pp.

———, and L. P. Olsen. 1950. Growth and yield of upland balsam fir in the Lake States. Sta. Pap. Lake St. For. Exp. Sta. 22. 24 pp.

———. 1951. Bark percent in Lake State trees. Tech. Note Lake St. For. Exp. Sta. 362. 1 p.

———. 1955. Composite volume tables for timber and their application in the Lake States. Tech. Bull. U.S. Dep. Agric. 1104. 51 pp.

Halliday, W. E. D. 1937. A forest classification for Canada. Bull. For. Serv. Can. 89. 50 pp.

———, and A. W. A. Brown. 1943. The distribution of some important forest trees in Canada. Ecology 24:353–373.

Heimburger, C. C. 1934. Forest-type studies in the Adirondack region. Mem. Cornell Agric. Exp. Sta. 165. 122 pp.

Hubbard, J. W. 1956. Growth and mortality in a northern Minnesota forest. Minn. For. Notes 50. 2 pp.

Jarvis, J. M. 1960. Forty-five years' growth on the Goulais River watershed. Tech. Note For. Br. Can. 84. 31 pp.

Kittredge, J. 1931. The interrelation of habitat, growth rate, and associated vegetation in the aspen community of Minnesota and Wisconsin. Ph.D. thesis. Univ. Minn. 175 pp.

Lake States Forest Experiment Station. 1935. Forest research digest. Rep. Lake St. For. Exp. Sta. 7 pp.

———. 1943. A composite cordwood volume table for the Lake States. Tech. Note Lake St. For. Exp. Sta. 202. 1 p.

Linteau, A. 1955. Forest site classification of the northeastern section, Boreal forest region, Quebec. Bull. For. Br. Can. 118. 85 pp.

Loucks, O. L. 1956. Site classification, Acadia Forest Experiment Station. Progr. Rep. Dep. North. Aff. & Nat. Res. For. Br. Maritime Distr. M-226. 48 pp.

McGuire, J. R., and R. B. Wray. 1952. Forest statistics for Vermont. Ntheast. For. Exp. Sta. 47 pp.

MacLean, D. W. 1960. Some aspects of aspen-birch-spruce-fir type in Ontario. Tech. Note For. Br. Can. 94. 24 pp.

———, and G. H. D. Bedell. 1955. Northern Clay Belt growth and yield survey. Tech. Note For. Br. Can. 20. 31 pp.

McLintock, T. F. 1948. Evaluation of the tree risk in the spruce-fir region of the Northeast. Sta. Pap. Ntheast. For. Exp. Sta. 16. 7 pp.

Meyer, W. H. 1929. Yields of second-growth spruce and fir in the Northeast. Tech. Bull. U.S. Dep. Agric. 142. 52 pp.

Morawski, Z. J. R., J. T. Basham, and K. B. Turner. 1958. A survey of pathological conditions in the forests of Ontario. Ontario Dep. Lds. & For. 96 pp.

Mulloy, G. A. 1944. Empirical stand density yield tables. Silv. Res. Note For. Serv. Can. 73. 22 pp.

———. 1947. Empirical stand density yield. Silv. Res. Note For. Serv. Can. 82. 54 pp.

———, and T. J. Kearney. 1941. Supplementary form-class volume tables including hardwood volume tables. Misc. Ser. For. Serv. Can. 2. 26 pp.

Northeastern Forest Experiment Station. 1947. Forest survey field manual. Mimeo. Ntheast. For. Exp. Sta. 92 pp.

———. 1950. Forest statistics for New Hampshire. For. Surv. Release. Ntheast. For. Exp. Sta. 9. 56 pp.

———. 1955. Forest statistics for New York. Ntheast. For. Exp. Sta. 63 pp.

Prielipp, D. O. 1956. Balsam fir tree volume in the Lake States. For. Sci. 2:92–99.

Ray, R. G. 1956. Site-types, growth and yield at the Lake Edward Forest Experiment Area, Quebec. Tech. Note For. Br. Can. 27. 53 pp.

Reineke, L. H. 1933. Perfecting a stand density index for even-aged forest. J. Agric. Res. 46:627–638.

Sayn-Wittgenstein, L. 1960. The recognition of tree species on air photographs by crown characteristics. Tech. Note Dep. For. Can. 95. 56 pp.

———. 1961. Phenological aids to species identification on air photographs. Tech. Note Dep. For. Can. 104. 26 pp.

Schenck, C. A. 1939. Fremdländische Wald- und Parkbäume. Parey. Berlin. Vol. 1–2.

Spurr, S. H. 1948. Aerial photographs in forestry. Ronald Press. New York, N.Y. 340 pp.

———. 1952. Forest inventory. Ronald Press. New York, N.Y. 476 pp.

Stone, R. N., and H. W. Thorne. 1961. Wisconsin's forest resources. Sta. Pap. Lake St. For. Exp. Sta. 90. 52 pp.

United States Forest Service. 1958. Timber resources for America's future. For. Resource Rep., U.S. Dep. Agric. 14. 713 pp.

Vezina, P. E. 1962. Crown width-d.b.h. relationships for open-grown balsam fir and white spruce in Quebec. For. Chron. 38:463–473.

Vincent, A. B. 1961. Is height/age a reliable index of site? For. Chron. 37:144–150.

Westveld, M. 1941. Yield tables for cut-over spruce-fir stands in the Northeast. Occ. Pap. Ntheast. For. Exp. Sta. 12. 18 pp.

———. 1953. Empirical yield tables for spruce-fir cut-over lands in the Northeast. Sta. Pap. Ntheast. For. Exp. Sta. 55. 64 pp.

Wilson, F. G. 1946. Numerical expression of stocking in terms of height. J. For. 44:758–761.

Wright, W. G. 1923. Investigations of taper as a factor in measurement of standing timber. J. For. 21:569–581.

———. 1927. Taper as a factor in the measurement of standing timber. Bull. For. Serv. Can. 79. 132 pp.

Zon, R. 1914. Balsam fir. Bull. U.S. For. Serv. 55. 68 pp.

Chapter 9. Utilization

Allison, J. H., and R. M. Brown. 1946. Management of the Cloquet Forest. Tech. Bull. Minn. Agric. Exp. Sta. 171. 95 pp.

Aries, R. S. 1947. Wood flour production. Bull. Ntheast. Wood Util. Counc. 17. 4 pp.

Babcock, H. M., and J. E. Nicolaiff. 1958. The Christmas tree industry in Canada. Misc. Publ. For. Br. Can. 10. 16 pp.

Bailey, L. H. 1933. The cultivated conifers in North America. Macmillan. New York, N.Y. 404 pp.

Balch, R. E. 1952. Studies of the balsam woolly aphid, *Adelges piceae* (Ratz.), and its effects on balsam fir (*Abies balsamea* (L.) Mill.). Publ. Dep. Agric. Can. 867. 76 pp.

———. 1956. Effects of balsam fir woolly aphid on quality of pulp. Bi-m. Progr. Rep. Div. For. Biol. Dep. Agric. Can. 12(4):1.

Belotelkin, K. T., L. H. Reineke, and M. Westveld. 1942. Spruce-fir selective logging costs. J. For. 40:326–336.

Bender, F. 1959. Spruce and balsam bark as a source of fibre products. Pulp Paper Mag. Can. 60(9):T275–T278.

Betts, H. S. 1945. Balsam fir (*Abies balsamea*). U.S. For. Serv. Amer. Woods. 8 pp.

Bickford, C. A., S. C. Clarke, and E. C. Jahn. 1934. A study of the nature of certain fir (*Abies*) oleoresins. Proc. Pacific Sci. Congr. Victoria and Vancouver, B.C. 5:3941–3948.

Bracken, J. R. 1932. The artistic aspect of trees. Proc. Nat. Shade Tree Conf. 8:35–42.

Bray, M. W., E. R. Schafer, and J. N. McGovern. 1944. Utilization of less commonly used species and waste and the improvement of yield in pulp manufacture. Rep. U.S. For. Prod. Lab. R1451. 16 pp.

Brown, H. P., A. J. Panshin, and C. C. Forsaith. 1949. Textbook of wood technology. Vol. 1. McGraw-Hill. New York, N.Y. 652 pp.

———. 1952. Textbook of wood technology. Vol. II. McGraw-Hill. New York, N.Y. 783 pp.

Canada Bureau of Statistics. 1906–55. The Canada yearbook. Dom. Bur. Stat., Ottawa.

———. 1962. Canadian forestry statistics, 1961. Ann. Cat. Dom. Bur. Stat. 25-202. 24 pp.

Canada Forestry Branch. 1956. Amendments 1956 to forest and forest products statistics of Canada. Bull. Econ. Sect. For. Br. Can. 106. 15 pp.

Canada Forest Products Laboratories. 1956. Strength and related properties of woods grown in Canada. Tech. Note For. Prod. Lab. Can. 3. 7 pp.

Canada Forest Products Research Branch. 1961. Span tables for wood joists and rafters for housing. Tech. Note For. Res. Br. Can. 30. 45 pp.

Canadian Lumbermen's Association. 1955. Standard grading rules. Eastern spruce, balsam fir, jack pine. Ottawa. 23 pp.

Casey, J. P. 1952. Pulp and paper chemistry and chemical technology, Vol. 1. Interscience. New York, N. Y. 795 pp.

Chandler, R. F. 1944. Amount and mineral nutrient content of freshly fallen needle litter of some northeastern conifers. Proc. Soil Sci. Soc. Amer. 8:409–411.

Chidester, G. H., J. F. Laundrie, and E. L. Keller. 1960. Chemimechanical pulps from various softwoods and hardwoods. Tappi 43:876–880.

Clark, J., and R. D. Gibbs. 1957. Studies in tree physiology. IV. Further investigations of seasonal changes in moisture content of certain Canadian trees. Can. J. Bot. 35:219–253.

Clermont, L. P., and H. Schwartz. 1948. Studies on composition of bark. Paper Tr. J. 126(19):57–60.

———. 1951. The chemical composition of Canadian woods. Pulp Paper Mag. Can. 52(13):103–105.

Coffin, M. C. 1940. Trees and shrubs for landscape effects. Scribner's Sons. New York, N.Y. 169 pp.

Cope, J. A. 1949. Christmas tree farming. Bull. Cornell Univ. 704. 32 pp.

Demmon, E. L. 1951. Reducing wood waste in the Lake States. Tech. Note Lake St. For. Exp. Sta. 354. 2 pp.

Donahue, R. L. 1940. Forest-site quality studies in the Adirondacks. I. Tree growth as related to soil morphology. Mem. Cornell Agric. Exp. Sta. 229. 44 pp.

Doyle, J. A. 1957. Effect of tree size of spruce and balsam fir on harvesting and conversion to lumber in Nova Scotia. Tech. Note For. Prod. Lab. Can. 5. 30 pp.

Duncan, D. P., E. T. Sullivan, C. J. Shiue, and R. I. Beazley. 1960. A study of consumer preference in Christmas trees. J. For. 58:537–542.

Fernow, B. E. 1910. The care of trees. Holt & Co. New York, N.Y. 392 pp.

Gevorkiantz, S. R., and L. P. Olsen. 1951. Bark percent in Lake States trees. Tech. Note Lake St. For. Exp. Sta. 362. 1 p.

Gibbs, R. D. 1935. Studies of wood. Canad. J. Res. 12:715–787.

Glennis, D. W., and H. Schwartz. 1952. Pulpwood decay: effect on yield and quality. Paper Industry 34:738–740.

Godwin, G. 1960. The characteristics of tree species used as pulpwood in Ontario and the effect on wood cost. Pulp Paper Mag. Can. 61(6):158–176.

Graham, W. E., and A. Rose. 1938. Tannins and non-tannins of the barks of some eastern Canadian conifers. Can. J. Res. 16:369–379.

Hajny, G. J., and G. J. Ritter. 1941. Holocellulose research in cooperation with the Technical Association of the Pulp and Paper Industry. Paper Tr. J. 113(13):83–87.

Hale, J. D. 1950. Factors that affect the buoyancy of pulpwood logs. I. Sapwood. Mimeo. For. Br. Can. 0-157. 8 pp.

———, and J. B. Prince. 1936. A study of variation in density of pulpwood. Pulp Paper Mag. Can. 37(6):458–459.

Hale, J. D., and J. B. Prince. 1940. Density and rate of growth in the spruces and balsam fir of eastern Canada. Bull. Dom. For. Serv. 94. 43 pp.

Hansen, H. L. 1963. Balsam fir — past, present, future. *In* Abstract, Christmas Tree Management Short Course. Dep. Agric. Short Courses, Univ. Minnesota. Pp. 9–13.

Hawley, L. F., and L. E. Wise. 1926. The chemistry of wood. Chemical Catalog. New York, N.Y. 334 pp.

Hay, K. D., and H. F. Lewis. 1940. The chemical composition of balsam bark. Paper Tr. J. 111(25):39–43.

Holt, L. 1950. Cutting methods in pulpwood operations of eastern Canada. Woodl. Res. Index Pulp Pap. Res. Inst. Can. 84. 58 pp.

Horn, A. G. 1950. Commodity drain from forests of the Lake States, 1948. Sta. Pap. Lake St. For. Exp. Sta. 20. 36 pp.

———. 1952a. Commodity drain from forests of the Lake States, 1950. Misc. Rep. Lake St. For. Exp. Sta. 18. 24 pp.

———. 1952b. Ten years' pulpwood production in the Lake States (1942–1951). Tech. Note Lake St. For. Exp. Sta. 384. 2 pp.

———. 1954. Commodity drain from forests of the Lake States. Misc. Rep. Lake St. For. Exp. Sta. 26. 25 pp.

———. 1956. Some highlights of pulpwood production in the Lake States, 1946–1955. Tech. Note Lake St. For. Exp. Sta. 457. 2 pp.

———. 1957. A record of the timber cut from forests of the Lake States, 1954. Sta. Pap. Lake St. For. Exp. Sta. 53. 47 pp.

———. 1959. Wisconsin Christmas tree harvest estimated at nearly 1⅛ million trees, 1957. Tech. Note Lake St. For. Exp. Sta. 553. 2 pp.

———. 1962. Pulpwood production in Lake States counties, 1960. Sta. Pap. Lake St. For. Exp. Sta. 94. 28 pp.

Hunt, G. M., and G. A. Garratt. 1953. Wood preservation. McGraw-Hill. New York, N.Y. 417 pp.

Hyttinen, A., and E. R. Schafer. 1958. The groundwood pulping of balsam fir and jack pine. Rep. U.S. For. Prod. Lab. 2139. 6 pp.

Irwin, D., and B. E. Lauer. 1961. Dirt contribution values for natural dirt in coniferous woods. Balsam, alpine fir, and black spruce. Tappi 44:33–35.

James, L. M. 1958. Resurvey of a Christmas tree marketing in Michigan. Spec. Bull. Mich. Agric. Exp. Sta. 419. 42 pp.

———. 1959. Production and marketing of plantation grown Christmas trees in Michigan. Spec. Bull. Mich. Agric. Exp. Sta. 423. 31 pp.

————, and L. E. Bell. 1954. Marketing Christmas trees in Michigan. Spec. Bull. Mich. Agric. Exp. Sta. 393. 38 pp.

Jenness, L. C., and J. G. L. Caulfield. 1941. Recovery of Maine balsam-needle oil. Bull. Me. Technol. Exp. Sta. 38:3–16.

Jensen, V. S. 1940. Cost of production of pulpwood of farm woodlands of the upper Connecticut River valley. Occ. Pap. Ntheast. For. Exp. Sta. 9. 17 pp.

Kaufert, F. H. 1935. Heart rot of balsam fir in the Lake States, with special reference to forest management. Tech. Bull. Minn. Agric. Exp. Sta. 110. 27 pp.

Keen, R. F. 1963. Weights and centers of gravity involved in handling pulpwood trees. Woodl. Res. Ind. Pulp Paper Res. Inst. Can. 147. 93 pp.

Kellogg, R. S. 1923. Pulpwood and wood pulp in North America. McGraw-Hill. New York, N.Y. 273 pp.

Kincaid, D. H. 1945. Groundwood from spruce and balsam tree tops. Pulp Paper Mag. Can. 46(3):155–156.

Koehler, A. 1924. The properties and uses of wood. McGraw-Hill. New York, N.Y. 354 pp.

Kress, O., C. J. Humphrey, C. A. Richards, M. W. Bray, and J. A. Staidl. 1925. Control of decay in pulp and pulpwood. Bull. U.S. Dep. Agric. 1298. 80 pp.

Kumlien, L. L. 1946. The friendly conifers. Hill Nursery. Dundee, Ill. 237 pp.

Kuphaldt, G. 1927. Die Praxis der angewandten Dendrologie in Park und Garten. Parey. Berlin. 389 pp.

Kutscha, N. P., and I. B. Sachs. 1962. Color tests for differentiating heartwood and sapwood in certain softwood tree species. Rep. U.S. For. Prod. Lab. 2246. 15 pp.

Lauer, B. E., and M. A. Youtz. 1933. The use of 40–60 mesh sawdust in the chemical evaluation of pulpwood. Paper Trade J. 96(3):36–37.

Laundrie, J. F. 1961. Sulfate pulping of balsam fir. Rep. U.S. For. Prod. Lab. 2225. 9 pp.

Lawrence, J. E. 1957. How to grow and sell Christmas trees. Deposit Courier. Deposit, N.Y. 175 pp.

Marie-Victorin, Fr. des E. C. 1919. Notes recueillies dans la region du Temiscamingue. Naturaliste Canadien 45:163–169.

Markwardt, L. S., and T. R. C. Wilson. 1935. Strength and related properties of woods grown in the United States. Tech. Bull. U.S. Dep. Agric. 479. 55 pp.

Marriott, F. G., and C. Greaves. 1947. Canada balsam, its preparation and uses. Mimeo. For. Prod. Lab. Can. 123. 5 pp.

Marshall, H. B. 1955. Utilization Abstracts, conifer needle oils. Econ. Bot. 9:299.

McGovern, J. N., and E. L. Keller. 1948. Some experiments in sodium sulphite pulping. Pulp Paper Mag. Can. 49(9):93–100.

McGovern, J. N., E. R. Schafer, and J. S. Martin. 1947. Pulping characteristics of available Lake States and northeastern woods. Tappi Monogr. 4:130–152.

McGuire, J. J., H. L. Flint, and E. P. Christopher. 1962. Cold storage of Christmas trees. Bull. Rhode Island Agric. Exp. Sta. 362. 12 pp.

McIntosh, D. C. 1948. Chemical treatment of trees. Pulp Paper Mag. Can. 49(7):117–118, 120.

————. 1949a. Treatment of trees with chemicals. Transpiration experiment. Mimeo. For. Br. Can. 0-141. 16 pp.

————. 1949b. Treatment of trees with chemicals. Effect of chemical treatment of trees on changes in weight of wood. Mimeo. For. Br. Can. 0-144. 44 pp.

————. 1951. Effects of chemical treatment of pulpwood trees. Bull. For. Br. Can. 100. 30 pp.

————, and J. D. Hale. 1949. Effect of chemical treatment of trees on ease of peeling. Mimeo. For. Br. Can. 0-140. 18 pp.

Millikin, D. E. 1955. Determination of bark volumes and fuel properties. Pulp Paper Mag. Can. 56(13):106–108.

Mitchell, G. H., and K. L. Quigley. 1960. Retailing of Christmas trees in three selected Ohio markets. Res. Circ. Ohio Agric. Exp. Sta. 81. 17 pp.

Mott, D. G. 1954. Secondary insects in chemically debarked trees. Bi-m. Progr. Rep. Div. For. Biol. Dep. Agric. Can. 10(2):1.

Mulloy, G. A. 1937. The balsam fir, its place in the forest economy of Canada. Pulp Paper Mag. Can. 38(2):210, 220, 221.

Northern Hemlock and Hardwood Manufacturers' Association. 1947. Official grading rules. Oshkosh, Wis. 55 pp.

Northern Pine Manufacturers' Association. 1939. Standard grading rules. Minneapolis, Minn. 36 pp.

Oliver, R. W. 1957. Trees for ornamental planting. Publ. Can. Dep. Agric. 995. 28 pp.

Ostrowski, H. J. 1943. Wood influence on pulp quality. Pulp Paper Mag. Can. 44(11):807–809.

Panshin, A. J., E. S. Harrar, W. J. Baker, and P. B. Proctor. 1950. Forest products. McGraw-Hill. New York, N.Y. 549 pp.

Parry, E. J. 1918. Gums and resins. Pitman & Sons. London. 106 pp.

Penhallow, D. P. 1907. A manual of North American gymnosperms. Ginn & Co. Boston, Mass. 374 pp.

Perry, H. J. 1960. History of mechanical pulping. Tappi Monogr. 21:1–8.

Peterson, T. A. 1962. Wisconsin forest products price review, November, 1962. Univ. Wisconsin, Agric. Ext. Serv. 2 pp.

Picard, J. 1924. Report of the Royal Commission on pulpwood. Sess. Paper Parliam. Can. 310. 292 pp.

Risi, J., and D. F. Arseneau. 1957. Dimensional stabilization of wood. Part I. Acetylation. For. Prod. J. 7:210–213.

———. 1958. Dimensional stabilization of wood. Part V. Phthaloylation. For. Prod. J. 8:252–255.

Risi, J., and M. Brûlé. 1945a. Étude des huiles essentielles tirées des feuilles de quelques coni-fères du Québec. Bull. Minist. Terres For. Québec 9. 52 pp.

———. 1945b. Étude des huiles essentielles tirées des feuilles de quelques conifères du Québec. Canad. J. Res. 23:199–207.

———. 1946. Utilization of conifer branches. Perfum. Essent. Oil Rec. 37(3):78–79.

Rochester, G. H. 1933. The mechanical properties of Canadian woods together with their physical properties. Bull. For. Serv. Can. 82. 88 pp.

Rogers, J. S. 1952. Potential tannin supplies from domestic barks. Bull. Ntheast. Wood Util. Counc. 39:17–28.

Rue, J. D., and E. P. Gleason. 1924. Utilization of pulpwood bark for fuel. Paper Trade J. 78(16):40–50.

Sanyer, N., E. L. Keller, and G. H. Chidester. 1962. Multistage sulfite pulping of jack pine, balsam fir, spruce, oak, and sweetgum. Tappi 45:90–104.

Sargent, C. S. 1884. Forests of North America. U.S. Dep. Int., Census Off. Washington, D.C. 612 pp.

Schafer, E. R. 1960. Effect of condition and kind of wood on groundwood pulp quality. Tappi Monogr. 21:13–32.

———, J. S. Martin, and E. L. Keller. 1955. Pulping characteristics of Lake States and north-eastern woods. Rep. U.S. For. Prod. Lab. 1675. 20 pp.

Shaw, A. C. 1953. The essential oils of *Abies balsamea* (L.) Mill. Can. Jour. Chem. 31:193–199.

Shaw, C. L. 1962. Canada. *In* Pulp and paper world review. Pulp Paper 26(15):131–135.

Shema, B. F. 1955. The microbiology of pulpwood. Tappi Monogr. 15:28–54.

Simonds, W. W. 1953. Growing Christmas trees in Pennsylvania. Circ. Pa. Agric. Exp. Sta. 415. 22 pp.

Skolko, A. J. 1947. Deterioration of fire-killed pulpwood stands in eastern Canada. For. Chron. 23:128–145.

Snyder, L. C., R. J. Wood, C. M. Christensen, and A. C. Hodson. 1948. Evergreens. Ext. Bull. Univ. Minn. Agric. Exp. Sta. 258. 25 pp.

Sowder, A. M. 1949. Christmas trees — the industry. *In* Trees. Yearb. Agric. U.S. Dep. Agric. Washington, D.C. Pp. 248–251.

———. 1956. 1955 Christmas tree data. J. For. 54:843–844.

———. 1962. 1960 Real Christmas tree data. Amer. Christmas Tree Growers' J. 6(1):5–6.

Spaulding, P. 1956. Diseases of North American forest trees planted abroad. Agric. Handb. U.S. Dep. Agric. 100. 144 pp.

Steer, H. B. 1948. Lumber production in the United States 1799–1946. Misc. Publ. U.S. Dep. Agric. 669. 233 pp.

Stewart, C. G. 1943. Pulpwood procurement problems. J. For. 41:138–140.

Stillwell, M. A. 1958. Deterioration of fire-killed balsam fir and spruce stands in Newfoundland and in New Brunswick. Bi-m. Progr. Rep. For. Biol. Div. Can. 14(4):1.

Sudworth, G. B. 1916. The spruce and balsam fir trees of the Rocky Mountain region. Bull. U.S. Dep. Agric. 327. 43 pp.

Sullivan, E. T. 1959. Trends in growing Christmas trees in Minnesota. Minn. For. Notes 82. 2 pp.

Taylor, N. 1952. The permanent garden. Van Nostrand. New York, N.Y. 128 pp.

Thickens, J. H. 1916. Groundwood pulp. Bull. U.S. Dep. Agric. 343. 150 pp.

Timell, T. E. 1957. Carbohydrate determination of ten North American species of wood. Tappi 40:568–572.

Trenk, F. B. 1960. Winter shearing of balsam fir and the spruces. Amer. Christmas Tree Growers' J. 4(1):12.

Truman, A. B. 1959. Variation of moisture content of pulpwood piled in the bush. Pulp Paper Mag. Can. 60(11):151–159.

Tryon, E. H., A. W. Goodspeed, R. P. True, and C. J. Johnson. 1951. Christmas trees — their profitable production in West Virginia. Circ. W. Va. Agric. Exp. Sta. 82. 28 pp.

United States Agricultural Marketing Service. 1957. United States standards for Christmas trees. U.S. Dep. Agric. 12 pp.

United States Bureau of the Census. 1918–32. Pulpwood consumption and pulpwood production. Ann. Rep. U.S. Dep. Comm., Washington, D.C.

———. 1906–62. Statistical abstracts of the United States. Ann. Rep. U.S. Dep. Comm., Washington, D.C.

United States Forest Products Laboratory. 1941. Treating spruce and balsam fir Christmas trees to reduce fire hazard. Tech. Note 250. 3 pp.

———. 1953. Density, fiber length, and yields of pulp for various species of wood. Tech. Note 191. 2 pp.

———. 1955. Wood handbook. Handb. U.S. Dep. Agric. 72. 528 pp.

———. 1961. Bark structure. Rep. 1666-5. 11 pp.

Wagner, C. E. van. 1961. Moisture content and inflammability in spruce, fir and Scots pine Christmas trees. Tech. Note For. Res. Br. Can. 109. 16 pp.

Wallin, W. W. 1951. Differences between balsam fir heartwood and sapwood. Unpubl. Rep. Univ. Minnesota, School of Forestry.

Waterman, A. M., R. U. Swingle, and C. S. Moses. 1949. Shade trees for the Northeast. *In* Yearb. Agric. U.S. Dep. Agric., Washington, D.C. Pp. 48–60.

Watson, R. 1955. The case for cellulose forestry. Proc. Soc. Amer. For. 1954:78–79.

Webb, W. L., E. M. Rosaco, and S. V. R. Simkins. 1956. The effect of chemical debarking on forest wildlife. Tech. Publ. N.Y. St. Coll. For. 70:35–41.

Wells, S. D., and J. D. Rue. 1927. The suitability of American woods for paper pulp. Bull. U.S. Dep. Agric. 1485. 102 pp.

Whitman, F. A. 1955. The effect of pulpwood aging on groundwood brightness. Pulp Paper Mag. Can. 56(11):152–155.

———. 1957. The effect of pulpwood aging on groundwood brightness. Tappi 40:20–40.

Wilcox, H., F. J. Czabator, G. Girolami, D. E. Moreland, and R. F. Smith. 1956. Chemical debarking of some pulpwood species. Tech. Publ. N.Y. St. Coll. For. 77. 34 pp.

Wilde, S. A. 1953. Trees of Wisconsin, their ecological and silvicultural silhouettes. Misc. Publ. Soil Dep. Univ. Wis. and Wis. Conserv. Dep. 44 pp.

Witham, G. S. 1942. Modern pulp and paper making. Reinhold. New York, N.Y. 705 pp.

Wyman, D. 1951. Trees for American gardens. Macmillan. New York, N.Y. 376 pp.

Wynne-Roberts, R. I. 1937. Grinding characteristics of various woods. Paper Tr. J. 104(6): 46–48.

Zon, R. 1914. Balsam fir. Bull. U.S. For. Serv. 55. 68 pp.

INDEX

THE Index is divided into four sections: Part I, General, comprises entries dealing directly with balsam fir as a tree and as a component of the vegetation, the biome, the landscape, and the human economy. Part II, Species, lists the Latin and common English names of plant and animal taxa associated directly or indirectly with balsam fir. (Collective terms such as "birds," "fungi," "insects," "cankers," and "mycorrhiza" appear in Part I.) Part III, Communities, lists all plant community names of different ranks discussed or mentioned in the text; it includes in addition a few combined plant-animal communities. Part IV, Localities, lists more or less distinctly defined geographical units. The Index does not include material which appears in the introductory part of the book, tables, figures, appendixes, or the list of literature cited.

PART I. GENERAL

Accretion, *see* Growth

Adaptation: to man's needs, 1, 246, 261, 293, 328; to environment, 6, 21, 36, 39, 46, 54, 59, 69, 80, 82, 88, 89, 234; to competition, 84, 88, 122, 123; to total environment, 92, 122, 211, 213, 225, 244. *See also* Succession; Tolerance

Advance growth, *see* Reproduction

Age, stand: at maturity, 13, 59, 140, 189, 214, 325; uneven, 60, 148, 193, 211, 214, 226, 244, 268, 271, 274, 277; measurements, 138, 262, 264, 265, 274, 278, 291; even, 226, 244, 246, 271, 277; and site index, 261, 262, 265, 291. *See also* Growth; Rotation; Time

Aerial photographs: in insect survey, 166, 173; species on, 257; measurements from, 255, 257, 258. *See also* Crown; Survey; Volume tables

Alpine forest: location of, 39, 69, 74, 79; elevation of, 39, 73, 74, 76, 79; vegetation of, 69, 75, 79, 80. *See also* Range

Arboretum: Moscow, USSR, 28; Hoyt, Portland, Ore., 39; Secrest, Ohio, 39; Wind River, Carson, Wash., 39, 40, 172

Ash, mineral: in litter, 108, 123; in wood, 299, 300; in bark, 319

Auxin: in bark and needles, 23; and compression wood, 171

Bacteria: relationships with, 124, 125; and spruce budworm, 125, 161, 169; and wood preservation, 294

Bark: anatomy, 9, 19, 20; morphology, 13; blisters, 13, 19, 20, 27, 321; development of, 20, 125; auxin in, 23; and fire, 101; and wildlife, 112, 119, 120; and various organisms, 125; and fungi, 126, 128, 132, 309; and insects, 156, 170, 172, 173, 176, 177, 309, 310; and logging, 235, 310; and stem form, 247; and tree volume, 253, 254, 259, 282, 290, 318; utilization, 258, 293, 318; thickness, 259; and pulp, 302, 305, 318, 319, 320; mechanical removal of, 302, 309, 310, 320; composition, 302, 318, 319; chemical removal of, 309, 310; weight, 311, 318; extractives, 319, 320. *See also* Canada balsam; Oils; Resin

Basal area, stand: and pollen deposits, 35; and reproduction, 71, 185, 207, 208, 213; and edaphic coordinates, 88, 90, 213; measurement, 210, 268, 291; and competition index, 227, 228, 262; growth of, 227, 228, 268, 270, 271; normal, 262, 268, 270, 274; as reference measure, 268, 276. *See also* Density, stand; Diameter, breast high; Growth; Volume growth; Yield tables

Baskerville, formula by, 271

Biota, postglacial, centers of, 32, 33. *See also* Fossils; Glaciation; History

Biotic factors: including cultural, 84, 178; wildlife as, 84, 110, 123, 218, 293; microorganisms as, 124; insects as, 149; and reproduction, 178, 181; and stand development, 219, 225, 234, 235; and growth measurements, 246, 247, 264. *See also* Bacteria; Fungi; Insects; Silvicultural management; Wildlife

Birds: populations of, 118, 120, 121, 122; and spruce budworm, 122, 159. *See also* Biotic factors; Wildlife

Bitterlich, angle-count method by, 210, 268, 291

Blisters, bark, *see* Bark: blisters

Bog, peat: pollen deposits in, 34, 35; communities on, 38, 50, 51, 59, 63, 65, 102, 105; planting on, 40; succession, 46, 60, 71, 81; growth relationships on, 123; and aeration, 125. *See also* Drainage; Flooding; Moisture; Soil: moisture; Swamp; Water

Branches: development of, 8, 13, 15, 22; and fungi, 8, 125, 126, 127, 129, 131, 140, 142, 143, 144; shedding of, 116, 127, 147, 225, 226, 296; effect on water in trees, 125, 309; volume of, 250, 258; and stump-culture, 323; shearing of, 323, 325. *See also* Bark; Foliage; Knots; Stump

Breeding, tree, *see* Genetic variability

Buds: development of, 9, 14, 22, 23, 26, 27; mortality of, 97; and wildlife, 115, 119, 120; and spruce budworm, 156, 157, 158; and balsam woolly aphid, 171. *See also* Strobili

Burning, *see* Fire

Calcium: in wood, 18, 19; in bark, 20; in humus, 59; in litter, 108, 123. *See also* Nutrients; Soil

Cambium: bark, 19; activity of, 27

Canada balsam: and bark, 13, 19; as product, 318, 321, 322. *See also* Bark; Resin

Canadian Lumbermen's Association, grading rules, 312

Canker: and hail, 97; and lichens, 125; on saplings, 126; description of, 131; and rots, 133; and insects, 143. *See also* Diseases; Fungi

Canopy, *see* Crown

Carbohydrates: in wood, 300, 301; in bark, 320

Cartier, Jacques, explorer, 2

Cellulose: hydrolized by fungi, 132; in wood, 299, 300, 301; in bark, 320. *See also* Lignin; Wood

Christmas trees: and double balsam, 8; breeding for, 28; and gall midge, 175; planted, 196, 197, 323, 324, 325, 328; in natural stands, 228, 323, 325, 328; properties of, 293, 322, 323, 328; species of, 322, 324, 325, 328; handling and marketing, 323, 324, 325; outdoor, 326. *See also* Recreational use

Chromosomes: number of, 8, 9, 10, 11; effect of colchicine, 8, 9; and fertilization, 11, 12. *See also* Fertilization; Meiosis; Mitosis

Classification of forest vegetation: and generalizations, 31, 41, 123, 148, 199, 202, 209, 210, 232, 244, 245, 246, 292; by Clements, 41, 42, 45, 89, 199; by Society of American Foresters, 42, 46, 47, 75, 76, 106, 199; by Braun, 42, 50, 82, 199; hierarchic, 44; genetic, 44, 73, 76, 85; multidimensional, 44, 45, 46, 47, 66, 80, 81, 82, 85, 86, 87, 88, 89, 90, 91; by Linteau, 44, 50, 54, 59, 199, 202, 203, 262; by Roe, 46; by Hills, 46, 50, 199; by Halliday, 50, 73, 82, 199; by Rowe, 50, 51, 61, 73, 86, 199; by Hustich, 52; by Raup, 53; by Hosie, 62; by Wilde, 62, 65, 66; by Heimburger, 62, 63, 65, 76, 94, 199, 203, 204, 219, 240; by Curtis, 62, 66, 85, 86, 107, 199; by Cajander, 73; by Morozov, 73; by Graves, 73; by Westveld, 75, 76, 82, 199, 207; by Sukachev, 76; by Long, 80; by Loucks, 80, 87, 88; by Pogrebniak, 88; by Merriam, 111; by Shelford, 111; by cover type groups in Canada, 199; by cover-structure types, 202, 216, 290. *See also* Community; Continuum, vegetation; Cover types; Ecosystem; Forest types; Models; Regions

Cleaning, *see* Competition; Release

Clearcutting: and reproduction, 183, 184, 185, 190, 194, 206, 209, 227, 233; strips, 184, 185, 191, 209, 232, 233, 244; commercial, 185, 190, 191, 192, 194, 206, 209, 232, 233, 236; and shrubs, 190, 209; and seed trees, 191, 209; and selection system, 191; and diameter limit cut, 191, 192, 236; and shelterwood system, 191, 209; and forest types, 206; and partial cutting, 232; in management, 235, 244, 245, 308, 327. *See also* Cutover; Cutting; Logging

Climatic conditions: outside natural range, 40; in natural range, 50, 51, 54, 61, 62, 73, 79, 91, 93, 100; and reproduction, 93, 97, 100, 178, 181, 202, 203, 208; and fungi, 131, 148; and spruce budworm, 160, 162, 163, 167. *See also* Drought; Fire; Precipitation; Range; Regions; Snow; Temperature; Weather; Wind

Climax: climatic, 41, 42, 43, 44, 45, 60, 69, 70, 71, 72, 111; polyclimax, 44, 69, 70, 75, 178, 189. *See also* Classification of forest vegetation; Regions; Succession

Climographs, of species range, 50, 91, 94

Community: genetic relationships of, 31, 44, 76, 85; mesic, 53, 59, 62, 69, 72, 218; drymesic, 59, 62, 66, 68, 80; lowland, 62, 70, 89, 137, 218; dwarf, 75; layer, 178. *See also* Cover types; Ecosystem; Forest types; Succession

Competition: in different regions, 60, 61, 73, 82, 218, 219; with tolerant hardwoods, 69, 72, 90, 138, 188, 191, 204, 206, 213, 218, 219, 226, 228, 230, 236, 237; and response to physical environment, 84, 88, 90, 92, 94, 122; root, 86, 94; and site index, 123, 213, 265; and fungi, 129, 138, 140, 148, 234; and insects, 141, 163, 171, 177, 218; and reproduction, 186, 188, 195, 208, 218, 219, 220, 221, 222, 223, 224, 227, 232, 244, 262; with shrubs, 188, 190, 197, 208, 218, 221; with spruce, 189, 204, 206, 292; with pine, 189, 206, 265; control of, 189, 220, 221, 222, 223, 224, 225, 227, 232, 244, 245; with intolerant hardwoods, 196, 213, 218, 225, 230, 234, 265; and age, 218, 220, 224, 225, 231, 234, 245, 262, 271, 278, 279. *See also* Basal area, stand; Factor: competition; Growth; Height growth; Index: competition; Release; Shrubs; Succession; Thinning; Volume growth

Composition of species: and age, 61, 72, 214, 225, 274, 278; and forest types, 63, 65, 74, 75, 178; and soil, 86, 107, 108, 110; of reproduction, 183, 186, 206, 207, 208. *See also* Cover types; Growth; Hardwoods; Insects; Litter; Overstory; Softwood; Soil; Succession; Wildlife; Wind

Cone: of bracted balsam fir, 6; morphological characteristics of, 14; disintegration of, 24, 26; and glaze, 98; collection of, 179; production, 234; on color photographs, 257. *See also* Strobili

Conifers: xylotomy of, 18; flora of Mexico, 36; slash of, 194; tolerance to 2,4-D of, 222; cutting of, 233; tracheids of, 294; wood chemical composition of, 299; sulfite pulp of, 305; pulp of bark of, 319; needle oil of, 322; fragrance of, 326. *See also* Softwood; Trees; Wood

Continuum, vegetation: idea of, 44, 66; in Wisconsin, 62, 66, 85, 86; in New Brunswick, 87. *See also* Classification of forest vegetation; Gradient; Importance values; Models; Ordination

Correlation: of flowers and shoots, 27; of seeds and blisters, 27; of roots and stem, 27; of synecological coordinates, 89; climate and growth, 91; of heat requirements, 92; of light requirements, 94; of moisture requirements, 107; fire and decay, 143; of growth elements, 257, 258, 271. *See also* Growth: correlations

Cost: pulpwood storage, 144; of spruce budworm control, 168; cleaning, 221; pulping, 305, 306, 307; pulpwood, 308, 310; logging, 308; of chemical debarking, 309. *See also* Economic problems; Management, general forest

Cover: edge, 111, 123; moose, 112; caribou,

113; deer, 115, 116; for varying hare, 117; for ruffed grouse, 121. *See also* Crown; Ground cover; Overstory; Stand structure; Understory

Cover types: other than forest, 35, 61, 210; geographic distribution of, 41, 42, 91; and moisture gradient, 106, 107; and reproduction, 199, 202; stand development by, 210, 216, 236, 240. *See also* Classification of forest vegetation; Community; Ecosystem; Forest type; Ground cover; Growth; Hardwoods; Softwood

Crown: characteristics of, 13; fire, 101; and soil temperature, 104; and spruce budworm, 167, 217; and balsam woolly aphid, 171; and reproduction, 184, 185, 196, 209; and stand development, 210, 221, 227, 228, 230, 232; diameter, 257, 258, 271, 311; shape, 257, 271, 322, 326; and tree volume, 258; and stem radial growth, 271; volume, 271, 311; closure and tree growth, 290; and loss of water of stem, 309. *See also* Aerial photographs; Cover; Density, stand; Foliage; Reproduction; Stand structure

Cull: after fire, 101, 143; determination of, 138, 139, 140, 218, 261; after windfall, 142; after clearcutting, 191; and tree size, 259, 310, 314. *See also* Damage; Volume, stand; Volume, stem; Wood deterioration

Cultivars: 'Macrocarpa,' 5, 6; list of, 7, 8; in landscaping, 326

Cutover: frost in, 93; and reproduction, 190, 199, 200, 201, 203, 206; fire in, 201, 203; swampland, 206; growth on, 216; yield on, 274, 277. *See also* Clearcutting; Cutting

Cutting: reproduction, 60, 61, 184, 185, 186, 189, 190, 191, 192, 193, 194, 195, 199, 200, 201, 202, 203, 204, 206, 207, 208, 209, 227, 233, 234, 235, 236, 237, 243, 244; and wind, 98, 99, 100, 193, 204, 206, 207, 227, 234, 236, 244; and wildlife, 112, 115, 116, 118; and stand age, 193, 224, 235, 240, 243; and site conditions, 193, 207; systems, 193, 194, 209, 235, 236, 244, 308; and species composition, 228, 234, 235, 236, 239, 240, 242, 243; and stand development, 233, 234, 236, 242, 243, 244; integrated, 233, 243, 244; minimum amount of, 235, 243; annual amount of, 288, 314. *See also* Clearcutting; Cutover; Diameter limit cut; Logging; Partial cutting; Selection system; Shelterwood system; Succession

Cutting cycle: and fungi, 138; and balsam woolly aphid, 173; and wind, 207; and stand development, 210; and partial cutting, 233, 235, 236; determination of, 235, 236; and mortality, 236, 244. *See also* Partial cutting; Rotation; Selection system

Damage: by fungi, 23, 126, 128, 130, 131, 132, 140, 144, 145, 146, 147, 258, 304, 309,

310; by logging, 30, 140, 141, 194, 234, 235; by atmospheric conditions, 41, 91, 92, 93, 94, 97, 98, 122, 123, 140, 141, 148, 183, 184, 209, 258, 262; by wildlife, 60, 112, 113, 114, 115, 116, 117, 118, 119, 120, 123, 140, 148, 183, 198, 202; by fire, 101, 123, 200, 201; by insects, 143, 152, 153, 154, 155, 165, 166, 170, 171, 173, 174, 175, 176, 258, 262, 304, 309, 310; by mites, 150; by phytocides, 222. *See also* Cull; Decay; Defoliation; Mortality; Salvage; Wind; Wood deterioration

Decay: and root grafting, 23; and site, 75, 135, 137, 138; and atmospheric conditions, 93, 98, 99, 126, 132, 135, 139, 140, 142, 143, 148; and fire, 101, 140, 143, 148, 310, 327; and bear, 120; and age, 124, 126, 132, 135, 137, 138, 139, 140, 142, 209, 233, 259, 262; and other fungi, 125, 126, 127, 131; entrance courts of, 125, 131, 132, 133, 140, 141, 142, 143, 144, 145, 146, 147, 148, 165, 171, 173, 175, 176, 296; and pruning, 127, 147; and tree vigor, 133, 138, 234; and insects, 141, 143, 144, 149, 165, 171, 173, 175, 176; and pulp, 145, 301, 304, 305; and seedbed, 201; in virgin forest, 213; and physical properties of wood, 296; and chemical composition of wood, 301; and chemical debarking, 310; and pulpwood grading, 314. *See also* Damage: by fungi; Diseases; Rot; Wood deterioration

Decomposition, *see* Decay; Litter; Wood deterioration

Defoliation: by winter drying, 97; by snow-mold, 128; by spruce budworm, 150, 153, 154, 155, 163, 165, 166, 181, 190, 216, 217, 218, 222, 234; by phytocides, 222. *See also* Foliage; Needles

Density, stand: and insect control, 164, 166, 173, 216; and stem form, 250; effect on growth and yield, 225, 226, 227, 228, 245, 262, 264, 272, 274, 275, 276, 277, 278, 279, 280, 281, 290, 292; and basal area, 262; and diameter, 267, 268, 274, 276; measures of, 267, 268, 274, 275, 276, 291; changes of, 267, 268, 275, 276, 278, 279, 280; and height, 268; of spruce-fir in U.S.A., 285. *See also* Basal area, stand; Cover; Crown; Index: stand density; Mortality; Reproduction: density, stocking; Trees: number of; Volume, stand

Density, wood, *see* Wood: density

Diameter: at half of tree, 247; stump, 250; top, 250, 252, 253, 282, 314; crown, 311. *See also* Diameter, breast high; Stem form; Stump

Diameter, breast high (dbh): at maturity, 13, 212; and tree height, 213, 247, 256, 281, 291, 311; and age, 245, 256, 281, 291; and stem form, 247, 249, 250; and bark, 247, 249, 259; and stump diameter, 250; and

top diameter, 250, 252, 253, 282; and volume, 250, 252, 253, 256, 258, 281, 282, 285, 291, 311; and crown diameter, 257; of sawtimber and poletimber, 282; and logging cost, 308; and weight of tree, 311. *See also* Density, stand; Diameter; Diameter growth; Diameter limit cut; Stem form

Diameter growth: rings of, 17, 296; and stem expansion, 25, 27; and defoliation, 152, 165, 217, 218; at different heights, 165, 234, 245; response to release, 227, 228, 230, 231, 234, 242; and cutting cycle, 235; pattern of, 265, 268, 272; and increment cores, 270; and tree classes, 271. *See also* Diameter; Growth; Yield tables

Diameter limit cut: and other systems, 192, 209, 235, 236; and reproduction, 192, 193, 194, 204, 208, 227; and stand development, 214, 234, 235, 236, 239, 240, 244. *See also* Cutting cycle; Partial cutting

Diseases: resistance to, 28, 30, 126; classification of fungous, 126, 128, 130, 132; survey of, 135, 148; and forest types, 137, 148; and stand development, 140, 148, 150, 211, 216, 218, 244; and reproduction, 189, 202. *See also* Damage; Decay; Fungi; Insects; Rot; Salvage

Distribution of species, *see* Range

Douglas, Robert, discovery of *Abies balsamea* var. *macrocarpa*, 5

Drainage: and succession, 60; and growth, 105; and spruce budworm, 163; and reproduction, 195. *See also* Bog, peat; Flooding; Moisture; Soil: moisture; Swamp; Water

Drought: and photosynthesis, 21; summer, 40, 97, 184; a climatic factor, 91; and shade, 94; and needle blight, 97; winter, 97, 122, 262; and partial cutting, 234. *See also* Climatic conditions; Moisture; Weather

Dynamics, *see* Succession; Time

Eberhard, shelter-wedge system by, 193

Ecoclimate: in Ontario, 46, 53; in New Brunswick, 80, 87

Ecographs, of species, 66, 89, 94, 110. *See also* Models

Economic problems: of small mammals, 118; of pathogenic fungi, 124, 130, 148; of salvage and storage, 144; of insect pests, 150, 152, 164; of silviculture, 178, 193, 198, 218, 221, 223, 224, 225, 227, 232, 244; of cutting, 235, 236, 245; of utilization, 322, 327, 328; of Christmas trees, 324, 325. *See also* Cost; Management, general forest; Range: economic

Ecosystem: concept, 31, 44, 84, 85; models, 31, 85, 89, 90, 122; matter-energy budget, 84, 85, 106, 210, 291; and soils, 102, 106; and wildlife, 123; and microorganisms, 147, 151; and insects, 151, 152, 173; and synusiae and niches, 151, 178. *See also* Classification of forest vegetation; Environmental factors; Models

Ecotone, *see* Regions: transitional

Ecotype, *see* Genetic variability

Edaphic conditions: by regions, 50, 51, 62, 63, 73, 74, 102, 225; and diseases and insects, 148, 233; and reproduction, 178, 181, 208, 209, 213; and stand development, 213, 214, 216, 219, 234, 235; and growth and yield, 261, 264, 265. *See also* Environmental factors; Moisture; Nutrients; Physiographic conditions; Soil; Water

Edaphic coordinates, *see* Models: and synecological coordinates

Edaphic field, *see* Models: moisture-nutrient

Embryo: development of, 11, 12; differentiation of, 13; dormancy in, 179, 180. *See also* Fertilization; Germination; Seed

Engelmann, classification of firs by, 4

Engler, classification by, 4

Environmental factors: morphological features of, 44; evaluation of, 83, 84, 85, 86, 87, 88, 123; changes of, 83, 209; classification of, 84, 106, 178; modification by man of, 84, 100, 121, 162, 177, 178, 181, 191; uniformity of, 225, 246. *See also* Adaptation; Biotic factors; Climatic conditions; Edaphic conditions; Heat; Light; Models; Moisture; Site

Esthetic values, *see* Recreational use

Evaporation, from soil, 97

Evapotranspiration, in different regions, 51, 62, 73

Exposure: and fire, 100; and reproduction, 194, 233, 245; and residual stand, 233, 234, 237, 239, 240, 243, 244, 245. *See also* Damage; Drought; Frost; Heat; Release; Sunscald; Wind

Factor (numerical): competition, 227, 228; form, 247; conversion of cords, 253, 278; correction, for composite volume, 254; correction, for volume index, 262; for normal yield, 264; burst and tear, of pulp, 304. *See also* Index

Fertilization: of ovum, 11, 12; double, 12

Fiber, wood, *see* Wood: fibers

Fire: and reproduction, 52, 61, 72, 183, 189, 191, 199, 200, 201, 203, 206, 208, 219, 227; and succession, 60, 61, 72, 81, 100, 211, 213, 214, 244; a climatic factor, 91, 100, 101; edaphic effect of, 100, 101; and lightning, 100, 101; use of, 100, 173, 194; control of, 101; resistance to, 101, 323; and decay, 101, 140, 142, 143, 148, 310, 327; and wildlife, 112, 113, 117; and slash, 177, 194, 201; and virgin forest, 189, 211, 244; and stand development, 213, 214, 219, 233, 244. *See also* Damage; Reproduction; Salvage; Slash: as fuel; Succession

Flooding: effect of, 91; lake shore, 105; by

beaver, 119. *See also* Bog, peat; Swamp; Water

Flora: Cretaceous, 32; Tertiary, 32; Cenozoic, 32; Mexican coniferous, 36; Pacific, 53; Atlantic, 53; at Apple River Canyon, Wis., 68. *See also* Alpine forest; Classification of forest vegetation; Prairie; Range; Regions

Flowering: conditions of, 24, 25, 26, 27, 179; periodicity of, 25, 179, 181; and spruce budworm, 156, 162, 163, 167. *See also* Fertilization; Megagametophyte; Microgametophyte; Pollen; Strobili

Foliage: and wind, 99; and flooding, 105; and diseases, 128, 129, 130; and mites and spiders, 150; and spruce budworm, 156, 158, 163; and spraying, 222, 223; amount of, 250, 258, 259; on aerial photographs, 257; pattern of Christmas tree, 323; esthetic contrasts of, 326. *See also* Crown; Defoliation; Litter; Needles; Slash

Forest site types, *see* Biotic factors; Classification of forest vegetation; Climatic conditions; Edaphic conditions; Forest types; Site

Forest types: and polyclimax, 44; and soil, 59, 63, 65, 68, 74, 75, 76, 94, 103, 105, 107, 262; permanent, 178, 206; biological equivalence of, 178; and subtypes, 178, 199; and reproduction, 199, 201, 202, 203, 206, 207, 233; and site index, 262. *See also* Classification of forest vegetation; Cover types; Ecosystem; Site

Formula: basal area increment, 228; competition factor, 228; tree volume, 254; stand volume, 256; competition index, 262; stand-density index, 267

Fossils: and history of *Abies*, 31; macrofossils, 32, 33; pollen, 33, 34, 35. *See also* History

Franco, classification of firs by, 4

Frost: spring, 40, 160, 262; hardiness, 40, 41; frost-free season, 51, 62, 73, 91, 281; damage by, 92, 93, 94, 197, 262; and insect relationships, 160. *See also* Climatic conditions; Heat; Temperature; Weather

Fungi: parasitic, 124, 125, 126, 127, 128; saprophytic, 124, 126, 127, 128, 147; symbiotic, 126; identification of, 126, 133, 134, 135, 144, 145, 146, 147; and diseases, 128–144 *passim*, 176, 177; control of, 128, 130, 132, 148, 194; and insects, 143, 144, 161, 172; succession of, 144, 147, 148, 184. *See also* Canker; Decay; Diseases; Rot; Wood deterioration

Gametophyte, *see* Megagametophyte; Microgametophyte

Genetic variability: and taxonomy, 1, 4, 5, 6, 7, 8, 29; and breeding, 27, 28, 30, 178, 299; and range, 37; and stagnation, 225. *See also* Identification of species; Taxonomy

Germination: conditions, 13, 121, 179, 181,

183, 186, 190, 209, 232; and stratification, 179, 180; tests, 180, 196. *See also* Seed; Seedbed; Seeding; Seedling

Girdling: by mice, 118; by bear, 120; and control of balsam woolly aphid, 173; release by, 221, 228, 230, 231, 232; and growth, 231, 232, 236, 237, 244; for moisture loss, 309, 310. *See also* Competition; Hardwoods; Release

Glaciation: reduction of gene pool by, 6; in history of *Abies*, 31, 32, 33; periods, 33, 34; deposits, 51, 62, 63, 74, 102. *See also* Biota; Fossils; History; Physiographic conditions

Glaze, *see* Damage: by atmospheric conditions

Gradient: clinal, 6; and chronosequences, 72, 81, 84; environmental, 85, 88; of light, 86, 94; reproduction, 199, 200, 203. *See also* Continuum, vegetation; Models; Ordination; Succession

Grading: sulfate pulp, 306; pulpwood, 314, 327, 328; Christmas trees, 325; trees, logs, and lumber, 327

Grafting: root, 23, 24, 127, 145; stock, 28; greenhouse and field, 28; heteroplastic, 29. *See also* Propagation

Ground cover: of different communities, 54, 60, 63, 65, 66, 68, 70, 74, 75, 79, 80, 81, 262, 264; and other cover, 86, 226, 245; and reproduction, 183, 186, 188, 196, 208, 209. *See also* Cover; Crown; Overstory; Shrubs; Stand structures; Understory

Groundwood pulp, *see* Pulping: groundwood

Growing stock: in different regions, 52, 73, 281, 283, 285, 287, 288, 291. *See also* Hardwoods; Softwood; Volume, stand; Yield

Growth: optimum, 13, 46, 50, 59, 87, 91, 102, 107, 108, 123, 178, 189, 213, 225, 234, 244, 264; of needles, 21; substances, 23, 30; of roots, 23, 27, 29, 74, 183, 196; and grafting, 24, 29; correlations, 27, 163, 165, 217, 234, 245, 246, 247, 264, 270, 271, 281, 291, 296; in different regions, 40, 52, 59, 73, 74, 75, 91, 103, 213, 214, 216, 227, 228, 230, 234, 255, 261, 271, 272, 281, 283, 287, 288, 289, 290, 291, 292; and forest types, 59, 75, 102, 103, 137, 139, 140, 214, 216, 234, 240, 242, 262, 264, 265, 274, 278, 281, 289, 290, 291, 296; and light, 94, 198, 221; and soil properties, 103, 105, 106, 107, 108, 123, 183, 213, 227, 274; and fungi, 124, 130, 131, 137, 138, 139, 140; and competition, 138, 186, 194, 218, 219, 220, 221, 225, 226, 227, 228, 230, 231, 232, 233; and partial cutting, 204, 234, 235, 236, 237, 239, 240, 242, 243, 244, 245; of different species, 210, 225, 246, 261, 262, 264, 265, 272, 274, 278, 281, 283, 290, 291, 292; in undisturbed conditions, 211, 212, 213; and natural disturbances, 213, 214, 216, 217, 218, 227, 244, 278; pattern of growth, 246, 264, 272;

intercept method, 262; culmination age, 265, 272; accretion, 282, 290; rate, and wood properties, 296. *See also* Basal area, stand; Density, stand; Diameter growth; Growing stock; Height growth; Ingrowth; Overstory; Regions; Stagnation; Volume growth; Yield; Yield tables

Hail, *see* Damage: by atmospheric conditions

Hardwoods: reproduction of, 60, 61, 202, 204, 234; and fire, 72; and light, 94; and wind, 99, 207; and deer, 113; and conifer reproduction, 199, 201, 203, 204, 206; and growth, 225, 264, 278, 290; utilization of, 233, 236, 243; in Canada, 283; and chemical debarking, 310; and esthetics, 326. *See also* Competition; Composition of species; Cover types; Girdling; Overstory; Release; Softwood

Hartig, net of mycorrhiza by, 126

Heat: factor in models, 85, 86, 87, 88, 92; destructive effect of, 91, 101; as climatic factor, 91, 92, 93; resistance of bark, 101; mode of action of, 106; and germination, 209; reaction of wood to, 294; value of wood, 318. *See also* Frost; Temperature

Height: maximum, 13, 212, 213, 225, 264, 265; and seed production, 25; classes, 59, 290, 311; and stem form, 247, 249; and volume tables, 250, 254, 255; and tariff methods, 256, 291; on aerial photographs, 257, 258; and age, 261, 262, 264, 265, 278, 281; measurements of, 262; and density, 268; and radial growth, 271. *See also* Diameter, breast high; Growth; Height growth; Site index; Tariff

Height growth: pattern of, 25, 26; of cuttings, 29; and frost, 92, 262; and hail, 98; and squirrel damage, 119; and spruce budworm, 166, 262; and competition, 190, 191, 219, 220, 221, 225, 228, 233, 262, 265, 274; in afforestation, 198, 323; and release, 220, 224, 230; of reproduction and site index, 264; and volume production, 265; culmination age, 265, 272. *See also* Height; Growth; Yield tables

Herbaceous cover, *see* Ground cover

Herbicides, *see* Phytocides

History: records of, 1, 2, 3, 4, 29; of *Abies*, 31; of distribution, 32, 33, 34, 35; of spruce budworm outbreaks, 152, 153, 154, 155; of balsam woolly aphid distribution, 168; stand, 210; of aspen-birch-spruce-fir cover type, 213, 214; of stand, and growth, 265, 281; of lumber production, 311; of pulpwood production, 316. *See also* Biota; Fossils; Glaciation; Succession

Humus: and root development, 23; in Dryopteris-Oxalis type, 59; and nutrient scalar, 87; decomposition, 125; of needles, 183;

as seedbed, 184, 190; and bark, 320. *See also* Seedbed; Slash; Soil: organic matter

Hybrids, *see* Genetic variability

Identification of species: historical, 2, 3, 4; botanical, 4, 5, 6, 7, 8, 18; on aerial photographs, 257. *See also* Genetic variability; Taxonomy

Importance values: by Wisconsin school, 85, 87, 88. *See also* Continuum, vegetation

Index: site, 59, 123, 213, 225, 261, 262, 264, 265, 270, 271, 272, 276, 291; similarity, 85, 87, 88; vegetation moisture (VMI), 86; Paterson's, 91; spruce budworm vulnerability, 164; competition, 262, 278, 279, 280; volume, 262; stand density, 267, 276, 278, 291; ingrowth, 274, 281, 282; significance of, 291. *See also* Correlation; Factor; Site index; Stem form: class, quotient; Tariff

Ingrowth: and reproduction, 202; and disturbance, 216; and cutting cycle, 235; and total growth, 282, 290. *See also* Growth; Index: ingrowth

Insecticides: DDT, 122, 168, 177; BHC, 169, 177; aldrin, dieldrin, endrin, 169; oils, nicotine, lime sulfur, 172; benzene hexachloride, 177. *See also* Insects

Insects: resistance to, 28, 30, 40, 164, 169, 177; general considerations, 84, 149, 151, 173, 177; and wind, 98, 100; and birds, 121; control of, 125, 151, 159, 164, 168, 169, 172, 173, 174, 177, 178, 194; and bacteria, 125, 169; and fungi, 126, 131, 132, 140, 142, 143, 144, 149, 173; survey of, 148, 166, 167; composition of species and, 150, 151, 164, 169, 216; classification of, 150, 173, 176; parasites of, 160, 168; on dead trees, 175, 176, 177; and reproduction, 189, 206, 227; in virgin forests, 211; of competing species, 218. *See also* Biotic factors; Damage; Insecticides; Silvicultural management: and protection; Salvage

Josselyn, John, early observations by, 2

Kansas, University of, tree-breeding program by, 28

Kerner, A. von Marilaun, early ideas by, 44

Knots: and strength of wood, 296, 298; and pulp quality, 304, 305; and use of wood, 308, 327; and pulpwood grading, 314. *See also* Branches

Köppen, classification of climates by, 50, 62, 73

Layering: ground, 24, 69, 80, 206; air, 29. *See also* Propagation

Leaf, *see* Needles

Light: and photosynthesis, 20, 21; and flowering, 24; a climatic factor, 51, 62, 91, 93, 94, 106; factor in models, 85, 86, 88, 89, 90,

93, 94; under canopy, 86, 94; and growth, 94, 123, 219, 220, 234; and insects, 94, 157, 163, 164; and moisture, 94, 309; and reproduction, 180, 183, 198, 209. *See also* Adaptation; Environmental factors; Models: multidimensional; Sunshine; Tolerance

Lightning, *see* Fire: and lightning

Lignin: and fungi, 132; in wood, 299, 300, 304; in bark, 320. *See also* Cellulose; Wood

Linné (Linnaeus), first scientific name by, 2

Litter: and forest type, 59, 75, 105, 123; and species composition, 86, 183; accumulation of, 108, 123, 196; and insects, 151; and seed storage, 180, 183; and seedbed, 183, 185, 186, 209; and reproduction, 183, 184, 186, 196; and scarification, 196; and bark, 320. *See also* Humus; Needles; Slash

Log rule: Bangor, Doyle, Dimick, Maine, New Brunswick, Quebec, Scribner, 253; Scribner Decimal C, 253, 254; International ¼-inch, 253, 254, 282. *See also* Stem form; Volume tables

Logging: mechanization of, 190, 194, 195, 209, 227, 244, 245, 308, 309, 310, 311, 327; waste, 261. *See also* Cost; Cutting; Damage; Mortality; Seedbed; Shrubs; Slash; Succession

Lumber: and wood properties, 293, 296, 299, 307, 309, 312, 314; production of, 308, 309; history of production, 311, 312, 314, 328; use of, 312; grading, 312, 327. *See also* Logging; Pulpwood; Sawtimber; Wood

McComb, A. L., information on range by, 39

McLintock, tree classification by, 271

Management, general forest: and species characteristics, 1; pathological effect on, 140; spruce budworm effect on, 152; and reproduction, 189, 190, 193, 203; and silvicultural techniques, 193, 195, 208, 209, 210, 218, 225, 227, 232, 235, 242, 244, 245, 327. *See also* Cost; Economic problems; Rotation; Silvicultural management

Maps: of species range, 37, 41, 42, 110; significance of, 81, 246; of sunshine hours, 93; of animal ranges, 110; of spruce budworm outbreaks, 153, 164. *See also* Aerial photographs; Range; Regions; Survey

Mayo Institute, plantations at Rochester, Minn., 197

Megagametophyte, development of, 11. *See also* Cone; Fertilization; Flowering; Microgametophyte; Strobili

Megasporangium: development of, 11; structure of, 12; embryo development in, 12, 13

Meiosis: in microspore mother cells, 10; in megaspore mother cells, 11

Mesification, and primary succession, 42, 45, 46, 68, 71. *See also* Climax; Paludification; Succession

Microgametophyte, development of, 9, 10, 11.

See also Flowering; Megagametophyte; Pollen; Strobili

Microsporangium: number in microsporophylls, 9; development of, 9, 10, 11; structure of, 9

Minnesota Department of Conservation, planting by, 116

Mitosis, in zygote, 12

Models: development of, 31, 41, 42, 66, 82, 84, 85, 122, 210, 291; moisture-nutrient, 42, 44, 45, 46, 47, 66, 71, 80, 81, 85, 87, 88, 89, 90, 103, 108, 110, 213; multidimensional, 44, 45, 46, 82, 85, 86, 87, 88, 89, 90, 91; macroclimatic, 46, 53, 54; landform, 46, 53, 54; and continuum, 66, 85, 86; and scalars, 87; and synecological coordinates, 88, 89, 90; triangular, 90, 91, 108; of insect populations, 151, 152, 167; of selection forest, 235. *See also* Classification of forest vegetation; Ecographs; Ecosystem; Environmental factors; Succession

Moisture: and photosynthesis, 20, 21; and planting, 40, 41, 323, 326; coordinates, 42, 44, 46, 47, 52, 66, 68, 80, 85, 86, 87, 88, 89, 90, 94, 103, 106, 107, 213; and reproduction, 105, 106, 181, 183, 184, 208, 209, 213, 218; and cone ripening, 179. *See also* Bog, peat; Drainage; Drought; Edaphic conditions; Environmental factors; Flooding; Models; Precipitation; Soil: moisture; Succession; Swamp; Water; Wood moisture

Mortality: by hail, 97; by flooding, 105, 119; by browsing, 113, 115; of birds by DDT, 122; by needle cast and blight, 130; and logging, 141; and defoliation by spruce budworm, 153, 163, 164, 165, 176, 216, 217; of spruce budworm, 157, 167; by balsam woolly aphid, 170, 171; of reproduction, 181, 183, 184, 186, 201, 213; and site, 213, 258; and stagnation, 225; and partial cutting, 233, 236, 243, 244; in virgin and managed forest, 258; and stocking, 279, 280; in different regions, 281, 282, 288, 290. *See also* Damage; Decay; Salvage; Wood deterioration; Yield

Munger, T. T., information on plantations by, 40

Mycorrhiza, on roots, 126, 127, 165. *See also* Fungi; Roots

Names: and botanical nomenclature, 1, 2, 7; common names of fir, 3, 4; of animals, 111, 112

Needles: length of, 6, 14, 21, 22, 23, 97; genetic variability of, 6; anatomy of, 9, 15; morphological characteristics of, 14; persistence of, 14, 21; shade and light, 15, 20; photosynthesis and respiration of, 20, 21; volume of, 21; weight of, 21, 22, 322; growth of, 22, 97; abscission of, 23; of aberrant growth, 23; phenology of, 27; fos-

sil, 32; and frost, 92, 97; and drought, 97; chemical composition of, 108, 114; as deer food, 114; and varying hare, 117; as grouse food, 120; rusts on, 126; fungi on, 126, 127, 129, 130; mites on dead, 150; and spruce budworm, 150, 156, 158; and balsam woolly aphid, 170; of Christmas trees, 293; and chemical debarking, 310; oil of, 318, 322. *See also* Foliage; Litter; Oils; Slash

New York State College of Forestry at Syracuse University: grafting experiments at, 28, 29; development of tariff methods, 255

Nitrogen: in topsoil, 59, 91, 107; in litter, 108, 123; and N-P-K coordinates, 108. *See also* Nutrients; Soil

Nomenclature, *see* Names

Northeastern Forest Experiment Station, tree-breeding program by, 28

Northern Hemlock and Hardwood Association, lumber grading rules by, 312

Northern Pine Manufacturers' Association, lumber grading rules by, 312

Nursery: diseases, 126, 128; practice, 196; weed control, 222; ornamental forms in, 325. *See also* Christmas trees; Planting; Seeding

Nutrients: in edaphic coordinates, 44, 46, 85, 86, 87, 88, 106, 110, 209, 213; and forest types, 59, 68, 80; in topsoil, 107; in litter, 108. *See also* Edaphic conditions; Environmental factors; Models; Soil; Succession

Oils: essential, 318, 322; in bark, 319; in oleoresin, 321. *See also* Bark; Canada balsam; Foliage; Needles; Resin

Oleoresin, *see* Resin

Ordination, of vegetation, 81, 85, 86, 87, 88. *See also* Continuum, vegetation; Gradient; Importance values; Models

Ostwald, ideas in forest measurements by, 268

Outlying communities, *see* Regions: outliers of

Overstory: and wind, 72; and undergrowth, 86; and spruce budworm, 163, 165; and reproduction, 183, 186, 208, 209; seedbed for, 188, 208, 209, 226, 245; and growth, 225, 226, 227, 228, 230, 232, 236, 237, 276, 277, 278; species composition, 225, 226, 228, 232; cutting in, 228, 230; girdling of, 230, 231, 232. *See also* Stand structure; Understory

Oxygen: deficiency and reproduction, 106; and algae, 125. *See also* Bog, peat; Swamp

Pacific Northwest Forest Experiment Station, tree-breeding program by, 28

Paludification, in humid areas, 46, 66, 81. *See also* Climax; Mesification; Succession

Paper: and wood properties, 293, 296, 308, 314, 320, 327; and pulping process, 302, 305, 307; and species, 317; and waste, 318,

320. *See also* Bark; Cellulose; Lignin; Pulp; Pulping; Pulpwood; Wood

Parenchyma cells: in needles, 15, 23; in wood, 17, 18, 19; in bark, 20. *See also* Bark; Needles; Wood

Partial cutting: and reproduction, 60, 185, 190, 192, 206, 227, 233, 236, 237, 243, 244; and stand development, 60, 228, 233, 234, 235, 236, 237; and wind, 99, 234, 244; and growth, 204, 234, 235, 236, 237, 239, 240, 242, 243, 244, 245; and exposure, 233, 234, 235, 236, 243, 244, 245; minimum amount of, 235; and forest types, 240, 242, 243; integrated, 244. *See also* Cutting; Cutting cycle; Mortality; Release

Pattern: in coordinates, 46, 85, 88, 89, 90, 108, 213; of reproduction, 66, 71, 72, 178, 181, 184, 189, 208, 209, 225, 271; of stand development, 210, 223, 225, 228, 232, 233, 245, 246, 265, 271; in landscaping, 325. *See also* Models; Stand structure

Phenology: observations in, 2, 25, 26, 27, 30; and heat factor, 86; and spruce budworm, 156, 157, 166. *See also* Climatic conditions; Precipitation; Temperature; Weather

Phosphorus, in topsoil, 59, 91, 107. *See also* Nutrients; Soil

Photosynthesis: of light and shade foliage, 20; and light intensity, 20; and age of needles, 21; and moisture, 21. *See also* Respiration

Physiographic conditions: development of, 32, 33, 34; in regions, 51, 62, 74, 102. *See also* Bog, peat; Edaphic conditions; Environmental factors; Glaciation; Range; Regions; Rocks; Site; Slope; Soil; Swamp

Physiology: nature of study, 1, 20, 30, 147, 177, 246, 291; and factor classification, 84; of microorganisms, 124, 126, 147; of insects, 149, 151, 177. *See also* Flowering; Germination; Growth; Phenology; Photosynthesis; Respiration

Phytocides: tolerance to 2,4-D and 2,4,5-T, 116, 186, 222, 224, 232; and resprouting, 116, 223, 224; and reproduction, 186, 195, 222, 223, 224, 228, 232; and vegetation changes, 224; and thinning, 225, 227, 232, 245. *See also* Release; Shrubs; Thinning

Planting: outside natural range, 39, 40, 41; in shelterbelts, 40, 97, 197, 326; for wildlife, 116; sites, 191, 193, 197, 224, 323; results of, 197, 198; for landscaping, 325, 326. *See also* Christmas trees; Nursery; Recreational use; Reproduction; Seeding

Podzols: in humid areas, 46; and species range, 50; in different regions, 51, 63, 65, 74, 102, 108. *See also* Paludification; Soil

Poletimber: and ingrowth index, 274, 282; dbh of, 282. *See also* Sapling; Sawtimber; Trees

Pollen: buds, 9; mother cells, 10; and micro-

spores, 10; structure of, 10; release of, 11, 24, 27; fossil, 33, 34, 35; analysis, 34, 35; and spruce budworm, 156, 158. *See also* Fertilization; Flowering; Microgametophyte; Strobili

Potassium, in topsoil, 59, 91, 101. *See also* Nutrients; Soil

Prairie: shelterbelts in, 40, 97, 197; forest transition, 53, 65, 188, 203, 219; openings and deer, 113. *See also* Planting; Range; Regions

Precipitation: and stem expansion, 27; in regions, 50, 51, 62, 73, 79, 211, 212; and species distribution, 91, 93, 123; distribution of, 94, 96, 97; and wind, 98, 100. *See also* Climatic conditions; Moisture; Snow; Water; Weather

Propagation: by grafting, 28, 29; by cuttings, 29; by air layering, 29; by stump-culture, 323. *See also* Grafting; Layering; Nursery; Planting; Seeding

Pulp: and fungi, 145; quality, 293, 294, 296, 302, 303, 304, 305, 306, 307, 310, 327; yield, 293, 302, 304, 305, 306, 307, 308; and pitch, 302, 314, 320; of compression wood, 304; of insect damaged wood, 304; and dirt, 305; and chemical debarking, 310; of charred wood, 310; of bark, 319. *See also* Paper; Pulping; Pulpwood

Pulping: and decay, 142; and pulp quality, 294, 327; processes, 302; groundwood, 302, 303, 304, 310; sulfite, 305, 310, 320; sulfate, 305, 306, 310; semichemical, 306, 307; neutral sulfite, 307; soda, 307; chemigroundwood, 307. *See also* Paper; Pulp; Pulpwood

Pulpwood: rotation age, 139; water transport, 142, 309; storage, 144, 304; decay, 145, 301, 304, 305; production, 224, 226, 227, 228, 237, 240; as primary product, 293, 310, 312, 314, 316, 317, 318; moisture, 295, 303, 309; chemically debarked, 309; grading, 314, 327, 328; bark amount of, 318. *See also* Cost; Logging; Paper; Pulping; Wood

Quetico-Superior Wilderness Research Center, investigations by, 27, 224

Range: of *Abies*, 2, 36; natural geographical, 5, 6, 31, 35, 36, 37, 38, 39, 52, 80, 82, 102; historical, 31, 32, 33, 34, 35; outside plantations, 31, 39, 40, 41, 97, 197; of timberline, 35, 50, 52, 53, 75, 79, 80; ecological, 46, 51, 52, 53, 54, 81, 82, 86, 87, 89, 90, 91, 93, 94, 96, 98, 103, 106, 108, 110, 206, 213, 219; of animals, 110–122 *passim*; of spruce budworm, 152, 153, 154, 155; of balsam woolly aphid, 170; of other insects, 173, 174, 175; and reproduction, 189, 200, 203; economic, 283–289 *passim*. *See also* Climatic conditions; Flora; Maps; Physiographic conditions; Prairie; Regions

Ray, John, early descriptions by, 2

Recreational use: early, 3; of cultivars, 7, 8, 326; planting for, 196, 293, 325, 326; of natural environment, 293, 325, 326, 327, 328; of managed forest, 327, 328. *See also* Christmas trees; Planting

Regions: description of, 31, 50–81 *passim*; transitional, 35, 50, 53, 69, 71, 86, 111, 121; classification of, 41, 42, 80, 82; synecological coordinates and, 44, 47, 88, 89; outliers of, 50, 68, 69, 70; and reproduction, 59, 61, 186, 189, 199–208 *passim*, 214, 218, 233; species performance in, 82, 135, 138, 218, 244, 246; forest classification and, 199, 244; and growth and yield, 261, 268, 270, 271, 272, 281, 283–291 *passim*. *See also* Classification of forest vegetation; Climatic conditions; Climax; Edaphic conditions; Flora; Maps; Physiographic conditions; Prairie; Range

Reineke, L. H., stand-density index by, 267, 276

Release: and growth, 94, 227, 228, 230, 234; and reproduction, 191, 218, 220, 221, 222, 223, 224, 225, 227, 231, 233; response and reaction to, 233, 234, 244, 245. *See also* Competition; Cutting; Exposure; Girdling; Partial cutting; Phytocides; Shrubs; Thinning

Reproduction: analysis in models, 90, 91; and soil, 91, 106, 184, 185, 190, 195, 196, 198, 206, 208, 209, 225; process defined, 178; density, 181, 183, 186, 192, 195, 196, 198, 199, 200, 201, 203, 204, 206, 208, 209, 227; promotion of, 184, 185, 186, 189, 195, 196, 209; stocking, 194, 195, 199, 200, 203, 206; and stand development, 202, 210, 225. *See also* Germination; Layering; Nursery; Planting; Propagation; Seed; Seedbed; Seeding; Seedling

Reproduction, subsequent: and scarification, 186, 195; and cutting system, 189, 190, 191; after spruce budworm, 190; in different regions, 199, 200, 201, 203, 206, 233. *See also* Planting; Propagation; Reproduction; Seeding

Resin: in bark, 13, 19, 20, 320; in needles, 14, 15; in bud scales, 14; in seed coat, 15; in traumatic canals of wood, 17, 19, 29; and pulp, 305; and peeling, 309; and lumber, 312; composition of, 321; synthetic, 322. *See also* Bark; Canada balsam; Oils

Respiration: rates of foliage, 20, 21; problems of, 30. *See also* Photosynthesis; Physiology

Rocks: and layering, 24; exposed, and range of species, 50, 51, 63, 66, 69, 103; Precambrian, 51, 62, 66, 74, 102; limestone, 51, 62, 68, 74, 102, 108; Cretaceous, 53; Paleozoic, 62, 68, 74, 102; cooling effect of, 68, 69; wounded roots by, 140; exposed as seedbed, 190; gardens with, 326; and natural landscape, 327. *See also* Glaciation; Physiographic conditions; Soil

Roe, E. I., information on range by, 38, 39

Roots: morphology of, 13, 29; and water, 19, 88, 106; development of, 23, 60, 74, 147, 183, 190, 196; grafting of, 24, 28, 29; and layering, 24; and stem growth, 27; and cuttings, 29; competition of, 86, 94; and frost, 93; and wind, 98, 99, 100, 141; amount of, 103, 259; and mice, 118; and mycorrhiza, 126, 127; and fungi, 128, 132, 141; and insects, 141, 143, 165, 175. *See also* Layering; Rot: root

Rot: heart, 23, 123, 126, 127, 132, 133, 134, 135, 140, 141, 142, 144, 213, 218, 296; butt, 98, 99, 132, 133, 135, 137, 139, 141, 143, 145, 146, 147, 218, 262, 274; root, 99, 125, 126, 127, 133, 135, 137, 140, 141, 146, 175; branch, 127; kinds of, 132, 133, 135; brown, 132, 133, 135, 146; white, 132, 133, 135, 145, 146; sapwood, 132, 133, 142; trunk, 135, 137, 139, 141; red, 144. *See also* Decay; Diseases; Fungi

Rotation: and rots, 137, 138, 139, 140, 148, 209, 233; and spruce budworm, 169; and balsam woolly aphid, 173; and reproduction, 190, 201, 208; of different species, 204, 233, 242; and change of species, 242, 274; of Christmas tree culture, 323. *See also* Cutting; Management, general forest

Royal Society of London, early communications, 2

Saeman, method of wood analysis by, 301

Sampling: of soil moisture, 104; continuous, 148, 291; of insect populations, 151, 166, 167, 168; for seed testing, 180; of reproduction, 181, 199; of yield, 276, 278, 289; of wood properties, 297, 299, 311, 320. *See also* Aerial photographs; Models; Survey

Salvage: after fire, 143, 144, 310, 327; after windfall, 144, 310; and insect attack, 144, 173, 310, 327; and diseases, 310. *See also* Cull; Damage; Diseases; Fire; Insects; Fungi; Mortality; Wind; Wood deterioration

Sapling: diseases of, 126, 128; growth, 138; weevil on, 175; competition, 218, 219; and partial cutting, 237; esthetic value of, 326. *See also* Poletimber; Seedling; Trees

Sargent, classification of firs by, 4

Sawtimber: and thinning, 224; and understory, 226, 239, 240, 242; volume tables, 253; ingrowth index, 274; dbh, 282; floatability, 309; prices, 310; resources in Canada, 312, 314; utilization, 314. *See also* Logging; Lumber; Poletimber; Trees

Schiffel, form-quotient by, 247

Seed: endosperm, 11, 12, 13; coat, 13, 15, 179;

wings, 14, 15, 179; production age, 24, 25; crop periodicity, 25, 179, 181; dispersal, 26, 61, 100, 119, 179, 181, 191, 192; crop and growth correlation, 27; source, 28, 40, 196, 209, 227, 232; and rodents, 119; and seed chalcid, 175; handling of, 179, 180, 183, 196; in litter, 179, 180, 183; properties, 179, 180, 196; and budworm, 181; crop and reproduction, 181, 184; trees, 191, 192, 200, 206, 207. *See also* Embryo; Fertilization; Flowering; Genetic variability; Germination

Seedbed: information, 179; mor, 183; litter, 183, 185, 186, 209; mineral, 184, 185; slash, 185, 194; windfall and, 190; and logging, 184, 185, 186, 191, 194, 195; changes, 191, 194; preparation, 196, 198; in nursery, 196; and logging and fire, 201; and understory, 208, 226, 245; and weed control, 222. *See also* Germination; Humus; Litter; Slash

Seeding: ability, 81, 208; and rodents, 118, 198; time, 179, 196, 198; experiments, 185, 197; and cover, 185, 198; from margin, 191; and planting, 198. *See also* Germination; Nursery; Planting; Reproduction; Seedling

Seedling: development of, 13, 23, 181, 183; growth pattern, 25; survival, 90, 179, 181, 184, 194, 213; shade tolerance, 94, 138, 183; and wildlife, 114, 117; and fungi, 124, 126, 128; and insects, 165, 175; abundance, 183, 184, 194, 203, 213, 233, 237; and natural landscape, 326. *See also* Competition; Damage; Diseases; Release; Reproduction; Sapling; Seeding; Trees

Selection system: and forest type, 60, 204, 207; and moose range, 112; and fungi, 148; and control of balsam woolly aphid, 173; and mechanization, 190, 209; and clearcutting, 191; and diameter limit cut, 192, 193; conversion to, 193; and girdling, 232; and cutting cycle, 235; and yield and profit, 235; amount cut under, 236. *See also* Cutting; Cutting cycle; Logging; Management, general forest; Partial cutting

Shade, *see* Light

Shelterwood system: and mechanization, 190, 209, 245; and clearcutting, 191, 209; and diameter limit cut, 192, 209; cutting sequence, 192, 193; and reproduction, 194; and spruce budworm, 217; and management, 244, 245. *See also* Cutting; Logging; Management, general forest; Partial cutting

Shrubs: in different regions, 52, 188, 213, 218, 219; and forest types, 60, 204, 214, 219, 220, 234, 262; in succession, 60, 61, 81; and spruce budworm, 151, 166; and cutting, 185, 219, 220, 234; and reproduction, 186, 190, 208; and site conditions, 188, 213,

218; and selection system, 193; and forest tent caterpillar, 218; control of, 218, 219, 220, 221, 222, 223, 224, 226, 244; and understory, 226, 245; and herbaceous plants, 245; decorative, 326. *See also* Competition; Phytocides; Release

Silvicultural management: species characteristics and, 1, 29, 30, 41, 219; and forest classification, 59, 178, 203, 210, 232, 246; and protection, 100, 112, 123, 132, 147, 148, 164, 168, 169, 172, 173, 177, 178; and reproduction, 193, 195, 202, 203, 209, 222; in virgin and second-growth forest, 209, 244; and stand development, 213; and forest inventory, 225; of two-storied stands, 226, 227, 228; and yield, 261, 264, 289. *See also* Management, general forest

Site: upland, 24, 53, 59, 62, 68, 69, 70, 71, 72, 89, 137, 140, 218; regions, 46, 53, 54; lowland, 53, 59, 62, 70, 71, 89, 137, 218; productivity, 59, 220, 261, 270, 272, 277, 278, 280, 281; site class, 59, 262; and wind effect, 98, 99; and fungi, 135, 137, 138, 140, 218, 259; and insects, 163, 171, 177; and seed production, 181; and reproduction, 183, 184, 196, 198, 202, 203, 204, 207, 208, 224; and mortality, 213, 258; and stand development, 218, 225; indicators, 264; broad groups, 264, 274, 275, 278; and stand-density index, 267, 276, 278, 291; and competition index, 278, 280; and ornamental planting, 325, 326. *See also* Environmental factors; Forest types; Planting: sites; Site index; Soil; Yield

Site index: range of, 59, 213, 261, 265, 270; of spruce-fir, 59, 261, 262, 276; and soil, 59, 123, 262, 264; in mixed stands, 123, 213, 225, 265, 271; of individual species, 213, 261, 265; indicators, 225, 261, 265; reference age, 262, 265; classes, 262; nature of curves of, 264, 265, 272. *See also* Site; Tariff; Yield tables

Slash: as fuel, 101, 177, 194, 201; and fungi, 145, 146, 177, 194; and insects, 151, 177; and reproduction, 177, 183, 185, 186, 191, 192, 194, 195; amount of, 186, 194, 311; decomposition, 194; disposal, 194, 195, 209, 310; and girdling, 230. *See also* Humus; Litter; Logging; Seedbed; Wood deterioration

Slope: cooling effect of, 40, 68, 69; and environmental scalars, 87; and wind, 98; and compression wood, 296. *See also* Physiographic conditions; Soil

Snow: and tree line, 79; damage by, 91, 98, 123; interception, 96; and fungi, 128; and balsam woolly aphid, 172; and seed dispersal, 192. *See also* Alpine forest; Climatic conditions; Precipitation

Softwood: cover type, 199, 290; reproduction, 199, 201, 203, 206; cutting, 236, 239, 240,

242, 243; and hardwood cutting, 242, 243; site group, 264, 274, 275, 278; stand density, 274, 275; productivity, 278; area, 283; volume, 287; mortality, 288, 290; growth, 290, 291. *See also* Composition of species; Conifers; Cover type; Growing stock; Hardwoods

Soil: moisture, 20, 21, 23, 40, 41, 46, 53, 59, 87, 88, 94, 104, 105, 106, 185, 206, 208, 227, 232, 244, 252, 323, 325, 326; high lime, 23, 51, 68; texture, 23, 46, 100, 103, 104; sand, 24, 59, 90, 102, 184; depth, 24, 68, 87, 98, 207, 274; lowland, 24, 218; fertility, 42, 43, 44, 71, 81, 87, 188, 193, 225; and succession, 44, 71; arctic, 50; humic-gley, 50, 63, 65, 102; gray-brown podzolic, 50, 63, 74; gray-wooded, 50, 51, 63, 104; peat, 50, 51, 102, 107; brown-wooded, 51; alluvial, 51; nutrients, 59, 87, 107, 110; silt-clay, 59, 87, 90, 103, 104, 323; brown podzolic, 63, 65, 74, 104; organic matter, 63, 65, 87, 102, 104, 151; brown forest, 65, 102; on talus slopes, 68, 81, 102; triangular coordinates, 88, 90, 91, 108; hardpan, 98, 102, 227; classification, 101, 102; clay, 102, 103; rendzina, 102; loam, 102, 105, 108; profile, 103, 104; acidity, 103, 107, 108, 123; physical properties, 104, 105, 106, 125, 128; chemical properties, 107, 108, 110, 123; sterilant, 232. *See also* Bog, peat; Composition of species; Edaphic conditions; Forest types; Growth; Humus; Litter; Moisture; Nutrients; Physiographic conditions; Podzols; Rocks; Site; Site index; Swamp; Water

Spacing, *see* Pattern

Stagnation, stand: and needle casts and blights, 130; and treatment, 192, 208, 245; and uniformity, 225; and site quality, 225, 227, 232. *See also* Growth; Mortality

Stand and stock tables, description of, 256. *See also* Volume tables; Yield tables

Stand structure: canopy levels in, 61, 86, 210, 216, 226, 245; controlling factors of, 86, 245; and insects, 164, 216, 217, 218; changes of, 216, 217, 218, 225, 230, 278; treatment effect on, 226, 230, 232, 233, 234, 237, 240; analysis of, 237, 240, 279; and stand development, 245, 246; and stem development, 250, 256. *See also* Age, stand; Composition of species; Density, stand; Overstory; Pattern; Tariff; Understory; Virgin forest

Stem form: significance of, 9; as morphological feature, 13, 98; adjustments to conditions, 234, 246, 250; quotient, 247, 251; class, 247, 249, 250, 251, 253; and butt swell, 247, 250; and taper, 247; and forest type, 249; and utilization, 293; ornamental, 325, 326. *See also* Log rule; Stump; Tariff; Trees; Volume, stem; Volume tables

Stocking, *see* Density, stand; Reproduction: stocking

Stoddard, solvent by, 222

Strobili: buds, 9; structure of, 9, 11, 14, 24; development of, 11; arrangement of, 14, 24. *See also* Cone; Flowering; Megasporangium; Microsporangium; Pollen

Stump: and fungi, 127, 145; and insects, 175; and reproduction, 184; and phytocides, 223, 224; diameter, 234, 235; height, 250, 282; volume, 250, 251, 252, 259; and dbh taper, 250; unused, 258, 259, 261; age at height of, 262, 264; annual rings and site index, 264; culture of Christmas trees, 323. *See also* Stem form; Trees; Volume, stem

Succession: significance of, 31, 82, 83; and ecosystem models, 31, 85; primary, 31, 42, 44, 45, 46, 69, 70, 71, 72, 73, 81, 85, 89, 108, 122; and nutrients, 42, 60, 108; stages of, 43, 44, 46, 60, 61, 72, 81, 189; and edaphic field, 44, 45, 46, 71, 85, 89; and moisture, 46, 60, 71, 81; secondary, 60, 61, 72, 100, 111, 122, 189, 190, 204, 211, 214, 244; and fire, 60, 61, 72, 81, 100, 113, 214, 244; and wildlife, 61, 110, 111, 113, 122; and spruce budworm, 61, 190, 216; and cutting, 61, 72, 81, 196, 204; on spoil banks, 72; and landslides, 81; and land abandonment, 81; and stand development, 210, 216, 225. *See also* Adaptation; Classification of forest vegetation; Climax; Community; Competition; Composition of species; Fungi: succession of; Gradient; Mesification; Models; Paludification

Sunscald, damage by, 41, 93. *See also* Heat; Temperature; Weather

Sunshine, hours of, 51, 62, 93. *See also* Light; Weather

Survey: forest, 36, 38; of moose browse, 112; early land, 113; of fungi, 135, 148; insect, 149, 150, 152, 166; reproduction, 199, 209; timber resource, 285. *See also* Aerial photographs; Maps; Sampling

Swamp: planting in, 41; and wildlife, 113, 117; growth in, 123; aeration in, 125; seeding at margin of, 198; reproduction in, 206. *See also* Bog, peat; Drainage; Flooding; Moisture; Soil: moisture; Water

Taper, *see* Stem form

Tariff: and standard dbh-height curves, 255, 256, 291; and region and site, 255, 281; volume table, 258; yield table, 274, 281, 282. *See also* Diameter, breast high; Height; Site index; Stand structure; Stem form; Volume tables; Yield tables

Taxonomy: and classification, 1, 4, 5, 7; and life cycle, 9; descriptions in, 29; of wildlife, 111, 112, 120; of fungi, 124, 126, 132; of insects, 149, 150, 155, 169, 170. *See also* Genetic variability; Identification of species

Temperature: and photosynthesis, 20; and layering, 24; and phenology, 27; by regions, 50, 51, 62, 73, 79, 91, 191; and species distribution, 86, 92, 93, 122; in models, 86, 87; and winter drying, 97; soil, 104, 105, 128; and spruce budworm, 156, 157, 161, 162; and balsam woolly aphid, 170, 171; and germination, 180, 181; and cutting system, 191; and storage of plants, 197; inversion and spraying, 223. *See also* Climatic conditions; Frost; Heat; Weather

Theophrastus, recording of information by, 2

Thinning: and fungi, 147, 148; and spruce budworm, 164, 217; defined, 224; chemical, 225, 227, 232, 245; mechanical, 225, 227, 245; and stand characteristics, 225, 226, 227, 228, 230, 245; studies, 225, 228, 245; techniques, 226, 227, 228, 232; and partial cutting, 228; and mortality, 258. *See also* Competition; Growth; Management, general forest; Mortality; Phytocides; Release; Silvicultural management; Stagnation, stand

Thornthwaite, classification of climates by, 50, 62, 73

Time: in models, 42, 45; and stand development, 82, 83, 178, 209, 210, 221, 225, 244, 245; and microbiological complex, 135, 138, 139, 140, 147, 148; and insect populations, 149, 151, 152, 153, 162, 166, 167, 168; and reproduction, 195, 196, 199, 206, 209, 221. *See also* Age, stand; Gradient: and chronosequences; History; Rotation; Succession

Tolerance: moisture, 86, 87, 88, 89, 90, 94, 96, 97, 104, 105, 106, 107; nutrient, 86, 87, 88, 89, 90, 91, 107, 108, 110; heat, 86, 87, 88, 89, 90, 91, 92, 93; light, 86, 87, 88, 89, 90, 93, 94. *See also* Adaptation; Heat; Light; Moisture; Nutrients; Phytocides

Tree breeding, *see* Genetic variability

Trees: checklist of, 2, 3, 4, 6; number of species of, 4, 5, 35, 36, 107; dominance classes of, 24, 271, 291; witness, 113; vigor of, 133, 138, 234, 271; number of, 199, 216, 221, 228, 232, 239, 267, 268, 274, 276, 282; juvenile stage of, 234; open-grown, 246, 271; centers of gravity of, 310, 311; weight and volume relation of, 310. *See also* Poletimber; Sapling; Sawtimber; Seed: trees; Seedling; Stem form; Volume, stem; Volume tables; Yield tables

Understory: development of, 68, 218; decadent, 72, 188; vegetation, 84, 86, 105; in deer yards, 116; and spruce budworm, 163; and pine reproduction, 208, 218, 226, 245; and forest tent caterpillar, 218; relations to overstory, 226, 228, 239, 240, 242, 245; yield table of, 226, 268, 274, 276; on aerial photographs, 257, 258. *See also* Ground cover; Overstory; Stand structure

Virgin forest: structure of, 60, 61, 211, 244; reproduction in, 61, 190, 202, 209, 211, 213, 233, 236; wind in, 98, 99, 189, 211; fire in, 189, 211, 244; decay in, 189, 213; growth and yield in, 209, 211, 212, 213, 233, 244; cutting in, 234, 236; mortality in, 258. *See also* Silvicultural management; Succession

Volume, stand: after cutting, 185, 190, 191, 192, 194, 204, 243, 244, 274, 278; and stem form, 250; computations, 255; from aerial photographs, 255, 257; formula, 256; correlations, 258; unused, and defects, 258, 259, 261; index, 262; per unit basal area, 268. *See also* Cull; Density, stand; Growing stock; Volume growth; Volume tables; Yield; Yield tables

Volume, stem: and conditions, 79, 246, 250, 251, 252, 254; and form, 247; total and merchantable, 252; formula, 254; unused, and defects, 259. *See also* Stem form; Trees; Volume tables

Volume growth: and volume, 73, 281; and volume index, 262; and competition index, 262, 278, 279, 280; in relation to height, 265; and basal area, 268, 274, 276, 277; culmination age, 272, 278, 280, 281, 291; and effective age, 274, 278. *See also* Density, stand; Growth; Volume tables; Yield; Yield tables

Volume tables: and growth, 210; construction of, 246, 255; composite-species, 246, 254; form-class, 247, 250, 251; board-foot, 247, 250, 253; and taper, 247; cubic-foot, 250; cord, 250, 252; regional, 250; local, 250, 251, 252; total-stem, 250; merchantable volume, 250, 254; stump and, 250, 251; top and, 250, 251; foliage and, 250; pulpwood, 253, 254; sawlog, 253, 254; of bolts, 254; and height curves, 255, 291; by tariffs, 255, 256; aerial photographs, 257, 258. *See also* Log rule; Stem form; Tariff; Volume, stand; Volume, stem; Yield tables

Water: and roots, 19, 88, 106; table, 46, 87, 88, 104, 105, 106; cooling by, 68, 69; holding capacity, 87; stagnant, 105, 106; and wetwood, 125; changes in level of, 244; and esthetics, 327. *See also* Bog, peat; Drainage; Flooding; Moisture; Precipitation; Snow; Wood moisture

Weather: and spruce budworm, 98, 156, 157, 158, 167; and reproduction, 181, 183, 208. *See also* Climatic conditions; Drought; Frost; Precipitation; Temperature; Wind

Weeding, *see* Release

Wildlife: and reproduction, 52, 112, 113, 114, 117, 121, 122, 183, 202; and composition of species, 61, 110, 111, 113, 118, 119, 122; cutting and, 112, 113, 114, 115, 116, 118; fire and, 112, 113, 117; food preference,

114, 120; resistance to browsing, 114. *See also* Biotic factors; Birds; Damage

Wind: swaying, 24, 100, 141; shelterbelts, 40, 97, 197, 326; windfall, 66, 98, 99, 123, 126, 132, 139, 142, 148, 189, 190, 193, 204, 207; and reproduction, 66, 71, 72, 189, 190, 191, 201, 227, 233; uprooting, 72, 99, 188; velocity, 79, 100; and scrub forest, 79, 80; a climatic factor, 91; breakage, 98, 99, 123; and rots, 98, 99, 126, 132, 135, 139, 140, 142, 143, 148; composition and, 98, 99; age and, 98, 100; and virgin forest, 98, 99, 189, 211; seed dispersal by, 100, 191; insect dispersal by, 100, 156, 162; and fire, 100, 142, 201; study of, 123, 244; and seed trees, 192, 206, 207; and compression wood, 296; salvage of damage, 310. *See also* Climatic conditions; Cutting: and wind; Damage; Exposure; Mortality; Salvage; Weather

Winter-drying, *see* Drought: winter

Winthrop, Governor John, communications by, 2

Wisconsin, University of, tree-breeding program by, 28

Wisconsin school, of ecological thought, 66, 85, 88

Witches' brooms, and fungi, 129, 130. *See also* Diseases; Fungi

Wood: properties in general, 13, 15, 29, 30, 41, 124, 132, 293, 294, 297, 299, 307, 308, 312, 318, 327; parenchyma, 17, 18, 19; rays, 17, 18, 19; tracheids, 17, 18, 19, 172, 294; xylem, 19, 171; compression, 19, 296, 298, 304; heartwood and sapwood in, 125, 126, 132, 140, 294, 296, 309; fibers, 293, 294, 302, 304, 305, 308; color, 293, 294, 302; products and size, 293, 294, 304, 307, 308, 312; density, 293, 294, 296, 302, 307, 308, 312, 314; strength, 293, 294, 297, 298, 299, 312, 327; taste, 294, 312; odor, 294, 312; springwood and summerwood in, 294, 296; weight, 295, 296, 309, 310, 311; specific gravity of, 295, 296, 297, 298, 310; paint holding, 297; products, 307, 308, 312, 314, 318. *See also* Lumber; Paper; Pulpwood

Wood deterioration: on ground, 108; and microorganisms, 124; and fungi, 126, 127, 131, 132, 142, 143, 144, 148, 214, 234, 309; resistance to, 126, 234, 296; and utilization, 126, 132, 145, 327; fire-killed, 142, 143, 148; insect-killed, 142, 143, 144, 148; windthrown, 142, 143, 144, 148; by insects, 149, 152, 171, 172, 176, 177, 310; and reproduction, 184, 214. *See also* Cull; Damage; Decay; Mortality; Rot; Salvage; Slash

Wood moisture: and wetwood, 125, 294, 295; in heartwood and sapwood, 125; and floatability, 142, 309; throughout season, 295; and swelling, 297; and seasoning, 297, 309; and pulp quality, 303; and green weight, 311. *See also* Moisture; Water

Wood preservation: and insects, 177; and permeability, 294, 308; and decay, 296; methods and results, 297, 308; of heartwood and sapwood, 297. *See also* Decay; Salvage

Yale University, tree-breeding program at New Haven, Conn., 28

Yield: pulpwood, 225, 226; and cutting system, 235, 236, 277; in general, 246, 261, 264, 291; per unit basal area, 268; by volume and weight, 296. *See also* Growing stock; Growth; Mortality; Site; Virgin forest; Volume, stand; Yield tables

Yield tables: and productivity, 210, 246; of understory, 226, 268, 274, 276, 278; and diameter growth, 235; and mortality, 258; normal, 261, 264, 277; composite-species, 261; special-use, 261, 262, 267, 272, 274, 275, 276, 277, 278, 280, 281, 282; construction of, 261, 268, 272; and site index, 261, 264, 291; stand-density, 264, 276, 278; and species, 264, 268, 274; and tariff classes, 281, 282. *See also* Age, stand; Basal area, stand; Diameter, breast high; Height; Site index; Tariff; Trees: number of; Volume growth

Zürich-Montpellier school, of ecological thought, 44

PART II. SPECIES

Abies (Fir): general information on, 1; historical knowledge of, 2; nomenclature of, 2, 3; classification of, 4; in cultivation, 4; polyploidy in, 8, 9; life cycle of, 9, 10, 11, 12, 13; embryogeny of, 12; arrangement of needles of, 14; anatomy of needles of, 15; classification of woods of, 18, 19; structure of xylem of, 19; hybridization of, 28; grafting of, 28, 29; history of, 31, 32, 33, 34, 35; fossils of, 32, 33, 34, 35; refugium of, 33; in spruce-fir types, 39; in Wind River Arboretum, Wash., 40; and mycorrhiza, 126; and damping-off, 128; and rusts, 130; and Adelginae, 170; in British Columbia and Alberta, 287; wood properties of, 294, 297, 298; as Christmas tree, 322; in ornamental planting, 326

Abies alba (Fir, European silver): hybrid with, 28; fossils of, 32; in Secrest Arboretum, Ohio, 39; and evaporation, 97; and fungi, 126; and spruce budworm, 155; and aphids, 169, 171, 172; in cultivation, 197

Abies amabilis (Fir, Pacific silver): refugium of, 33; wood properties of, 296, 299

Abies americana, name of balsam fir, 3

Abies arizonica, see *A. lasiocarpa* var. *arizonica*

Abies aromatica, name of balsam fir, 3

Abies balsamea Bigel, name of Fraser fir, 3

Abies balsamea (Linnaeus) Miller, accepted scientific name of balsam fir, 3

Abies balsamea Miller, name of balsam fir, 2

Abies balsamea ssp. *balsamea* var. *balsamea* (incl. f. *hudsonica*), balsam fir in Boivin's classification, 7

Abies balsamea ssp. *balsamea* var. *phanerolepis* (incl. f. *aurayana*), bracted balsam fir in Boivin's classification, 7

Abies balsamea ssp. *lasiocarpa* var. *arizonica*, corkbark fir in Boivin's classification, 7

Abies balsamea ssp. *lasiocarpa* var. *lasiocarpa* (incl. f. *compacta*), subalpine fir in Boivin's classification, 7

Abies balsamea var. *balsamea*, typical variety of balsam fir, 37

Abies balsamea var. *brachylepis*, name of balsam fir, 3

Abies balsamea var. *fraseri*, name of Fraser fir, 3

Abies balsamea var. *hudsonia*, see *A. b. f. hudsonica*

Abies balsamea var. *hudsonica*, see *A. b. f. hudsonica*

Abies balsamea var. *macrocarpa*, see *A. b. 'Macrocarpa'*

Abies balsamea var. *phanerolepis* (Bracted balsam fir): description of, 6; as double balsam, 8; and *A. intermedia*, 28; range of, 37

Abies balsamea f. *hudsonia*, see *A. b. f. hudsonica*

Abies balsamea f. *hudsonica*: dwarf balsam fir, 6, 7; grafting of, 28; in decorative planting, 326

Abies balsamea f. *phanerolepis*, see *A. b.* var. *phanerolepis*

Abies balsamea x *A. lasiocarpa*, hybrid, 28

Abies balsamea 'Angustata,' cultivar, 7

Abies balsamea 'Argentea,' cultivar, 7

Abies balsamea 'Brachylepis,' cultivar, 7

Abies balsamea 'Coerulea,' cultivar, 7

Abies balsamea 'Columnaris,' cultivar, 7, 28

Abies balsamea 'Compacta,' cultivar, 326

Abies balsamea 'Glauca,' cultivar, 8

Abies balsamea 'Globosa,' cultivar, 8

Abies balsamea 'Longifolia,' cultivar, 8

Abies balsamea 'Lutescens,' cultivar, 8, 28

Abies balsamea 'Macrocarpa,' cultivar, 5, 8

Abies balsamea 'Marginata,' cultivar, 8

Abies balsamea 'Nana,' cultivar, 8, 326

Abies balsamea 'Nudicaulis,' cultivar, 8

Abies balsamea 'Paucifolia,' cultivar, 8

Abies balsamea 'Prostrata,' cultivar, 8, 326

Abies balsamea 'Pyramidalis,' cultivar, 8

Abies balsamea 'Variegata,' cultivar, 8

Abies balsamea 'Versicolor,' cultivar, 8

Abies balsamifera, name of balsam fir and Fraser fir, 2, 38

Abies concolipites, fossil, 32

Abies concolor (Fir, white): and fossil, 32; and spruce budworm, 152; at Cloquet, Minn., 197; wood properties of, 296, 297, 298; ornamental use of, 325

Abies conica, hybrid, 28

Abies duplex, *see* Double balsam

Abies excelsa fraseri, name of balsam fir, 3

Abies fraseri (Fir, Fraser): as she-balsam, 2; distinction of, 3, 82; taxonomy of, 4, 5; and bracted balsam fir, 6; and double balsam, 8; wood anatomy of, 18; fossil, 33; range of, 35, 36, 37, 38; in arboretum, 39, 172; and balsam gall midge, 174

Abies fraseri hudsoni, name of balsam fir, 3

Abies grandis (Fir, grand): refugium, 33; in Wind River Arboretum, 40; in Germany, 41; seed moisture of, 179; at Cloquet, Minn., 197; and phytocides, 222; properties of wood of, 296, 298, 299; oleoresin of, 321

Abies guatemalensis, range of, 36, 39

Abies hibrida, hybrid with *A. sibirica*, 28

Abies holophylla, and spruce budworm, 28

Abies hudsonia, name of balsam fir, 3

Abies intermedia, taxonomic status of, 6, 28

Abies kamakamii, range of, 36

Abies lasiocarpa (Fir, subalpine): taxonomic status, 4, 5, 6, 7; wood of, 18; refugium, 33; range of, 36, 37, 39, 53; at Cloquet, Minn., 197; in British Columbia and Alberta, 287; specific gravity of wood of, 296; oleoresin of, 321

Abies lasiocarpa var. *arizonica* (*A. arizonica*), specific gravity of wood of, 296

Abies magnifica: range of, 39; specific gravity of wood of, 296

Abies minor, name of balsam fir, 3

Abies nana, hybrid, 28

Abies nephrolepis, and spruce budworm, 28

Abies nobilis, see *Abies procera*

Abies parvula, hybrid, 28

Abies pendula, hybrid, 28

Abies pinsapo, embryogeny of, 12

Abies procera (*A. nobilis*, Fir, noble): and phytocides, 222; wood properties of, 296, 298

Abies pyramidalis, hybrid, 28

Abies ramesi, fossil, 32

Abies religiosa, range of, 39

Abies sibirica: and spruce budworm, 28; hybrids with, 28; range of, 28, 36; in Finland, 40

Abies sibirica x *A. balsamea columnaris*, hybrid, 28

Abies venusta, embryogeny of, 12

Abietaceae, *see* Pinaceae

Abietineae, xylem of roots of, 19

Abietipites, fossil, 32

Abietipites antiquus, fossil, 32

Abietites cretacea, fossil, 32
Abietites linkii, fossil, 32
Abietoidea, subfamily of species, 4
Abietoxylon, fossil, 32
Acanthostigma parasiticum, on needles, 128
Acarina (Mites): damage by, 150; and spruce budworm, 160
Acer (Maple): grafting of, 29; refugium of, 33; pollen, 35; in spruce-fir types, 39; and porcupine, 120; in Viburnum-Oxalis type, 204; in virgin forest, 211
Acer negundo (Boxelder), in Apple River Canyon, Wis., 68
Acer pensylvanicum (Maple, striped): in Lake Edward Forest, Que., 65, 219; in Adirondacks, N.Y., 219
Acer rubrum (Maple, red): fossil, 33; in different regions, 53, 63, 65, 68, 69, 70, 74, 75, 80, 219; in succession, 71, 72; and moisture, 105, 106; and moose, 112; and deer, 114; and spruce budworm, 190; and phytocides, 223; overstory of, 230, 231
Acer saccharum (Maple, sugar): in different regions, 53, 54, 65, 66, 68, 69, 70, 75, 80, 87, 88, 219; in succession, 72; and temperature, 92; and light, 93, 94; and deer, 114; and porcupine, 120; in edaphic coordinates, 213; and phytocides, 223; overstory of, 226, 230, 231, 236, 264
Acer spicatum (Maple, mountain): in Viburnum type, 65; in Itasca Park, Minn., 70; and reproduction, 189, 218, 219, 221, 223; and spruce budworm, 189, 190; and birch dieback, 202; in edaphic coordinates, 213; and aspen, 214; and phytocides, 223, 224
Acleris variana (*Peronea variana*; Black-headed budworm): resistance to, 27; common insect, 150; description of, 173, 174
Actea rubra, in Dryopteris-Oxalis type, 60
Actinomycetes, and root rot, 125
Adelges (Aphids): taxonomy of, 169, 170; life history of, 170, 171, 172; control of, 172, 173
Adelges merkeri, and *A. piceae*, 169
Adelges nüsslini: and *A. piceae*, 169; at Pacific Coast, 170
Adelges piceae (Balsam woolly aphid): and "compression wood," 19, 171, 296; resistance to, 28; in Wind River Arboretum, 40, 172; and light, 94; and fungi, 143, 171; significance of, 149, 150, 169, 173; taxonomy of, 169; life cycle of, 169, 170, 171, 172; distribution of, 170; feeding of, 170, 171, 172; "gout" disease of, 170, 171; and temperature, 170, 171, 172; and host, 170, 171, 172, 176; predators of, 172; control of, 172, 173; and secondary insects, 176; and stand development, 217, 218
Adelginae, hosts and forms of, 170
Adelopus balsamicola, see *A. nudus*
Adelopus nudus, on needles, 128

Alces alces (Moose): habitat, 111, 112, 116, 123; eastern and northwestern, 112; population changes of, 113
Alder, see *Alnus*
Aleurodiscus amorphus, canker, 131
Algae: relationships with, 124; and oxygen, 125
Alnus (Alder): pollen, 35; and oxygen supply, 106; in lowland shrub complex, 213; and site index, 264
Alnus crispa: at Lake Superior, 68, 69; and reproduction, 218
Alnus rugosa: in different regions, 60, 65, 70; and reproduction, 186, 218, 223
Ambrosia beetle, see *Trypodendron bivittatum*
Amelanchier bartramiana, in different regions, 60, 68, 70, 219
Amelanchier canadensis, in Adirondacks, N.Y., 219
American silver fir, name of balsam fir, 4
Amphisphaeria thujina, see *Kirschteiniella thujina*
Apanteles fumiferana, parasite on spruce budworm, 160
Aphids, see *Adelges*
Aplomyia caesar, parasite on spruce budworm, 160
Arachnida (Spiders), predators on spruce budworm, 150, 160
Aralia nudicaulis, in different regions, 60, 66, 68, 70
Arctostaphylos uva-ursi: on Beaver Isl., Mich., 68; and ruffed grouse, 121
Argyresthia (Bud-miners), damage by, 173
Armillaria mellea: white stringy butt rot, 134, 135, 137, 143, 147
Arthropods: other than insects, 149, 150; as predators on spruce budworm, 160
Ascomycetes: number of species, 126; and decay, 132
Ash, see *Fraxinus*
Ash, black, see *Fraxinus nigra*
Ash, green, see *Fraxinus pennsylvanica*
Aspen, see *Populus*
Aspen, bigtooth, see *Populus grandidentata*
Aspen, quaking, see *Populus tremuloides*
Aspidium, and reproduction, 188
Aster acuminatus: in Dryopteris-Oxalis type, 60; in Sphagnum-Oxalis type, 65
Aster lowrianus, in Dryopteris-Oxalis type, 60
Aster macrophyllus: in Dryopteris-Oxalis type, 60; in Sphagnum-Oxalis type, 65
Athyrium filix-femina, in Itasca Park, Minn., 70

Bacillus cereus, on spruce budworm, 161
Bacillus thuringiensis, and spruce budworm, 161, 169
Balm of fir, name of balsam fir, 4
Balsam woolly aphid, see *Adelges piceae*
Balsameae, section of *Abies*, 4

Basidiomycetes: number of species, 126; and decay, 132

Basswood, *see Tilia americana*

Bear, *see Euarctos americanus*

Beauveria, fungus on spruce budworm, 161

Beaver, *see Castor canadensis*

Beech, American, *see Fagus grandifolia*

Betula (Birch): refugium of, 33; pollen, 35; in spruce-fir type, 39; on spoil banks, 72; and hail, 97; and porcupine, 120; dieback, 150, 163, 189, 202, 216; and reproduction, 196; and development of aspen-birch-spruce-fir stands, 214; overstory of, 225

Betula alleghaniensis (Birch, yellow): in different regions, 53, 54, 63, 65, 66, 68, 69, 70, 72, 74, 75, 76, 80, 204, 219, 228; and light, 93; as deer food, 114; and porcupine, 120; in virgin forest, 211; overstory of, 228, 230, 231, 236; in stand and stock tables, 256; and site index, 264; heating value of bark of, 318

Betula papyrifera (Birch, paper; white): refugium of, 33; in different regions, 52, 53, 59, 63, 66, 68, 69, 70, 75, 76, 79, 80, 218, 219; in succession, 61, 71, 72; in synecological coordinates, 89; and heat, 92; and light, 93; damage to, 98, 99; and drainage, 106; and wildlife, 113, 120; reproduction of, 195, 204; in virgin forest, 211; and stand dynamics, 227, 230, 231, 233, 242, 243; and girdling, 230, 231; in stand and stock tables, 256; pulpwood weight of, 295; heating value of bark of, 318

Betula papyrifera var. *minor*, in White Mts., N.H., 80

Betula populifolia (Birch, gray), in Acadian Region, 80

Bifusella faullii, needle cast, 127, 128, 129

Birch, *see Betula*

Birch, gray, *see Betula populifolia*

Birch, paper, *see Betula papyrifera*

Birch, white, *see Betula papyrifera*

Birch, yellow, *see Betula alleghaniensis*

Black-headed budworm, *see Acleris variana*

Blarina brevicauda, short-tailed shrew, 118

Blister pine, name of balsam fir, 4

Blueberry, *see Vaccinium*

Bonasa umbellus (Grouse, ruffed): habitat of, 116, 120, 121, 123; diet of, 121

Bothrodiscus, see *Godronia abietis*

Braconidae, parasites on spruce budworm, 160

Bracted balsam fir, *see Abies balsamea* var. *phanerolepis*

Bud-miner, *see Argyresthia*

Caeoma arcticum, rust, 130

Calamagrostis canadensis, in Itasca Park, Minn., 70

Calliergon schreberi, see *Pleurozium schreberi*

Calliergonella schreberi, see *Pleurozium schreberi*

Caltha palustris, in Itasca Park, Minn., 70

Calyptospora, rusts, 130

Camponotus herculeanus (Carpenter ant): common insect, 150; secondary insect, 176

Canachites canachites canace, subspecies of spruce grouse, 120

Canachites canachites canachites, subspecies of spruce grouse, 120

Canachites canachites osgoodi, subspecies of spruce grouse, 120

Canachites canadensis (Grouse, spruce): habitat, 120, 121; diet of, 120, 121

Canada balsam, name of balsam fir, 4

Canadian fir, name of balsam fir, 4

Canis lupus (Wolf), relationships with, 111

Carex: in Itasca Park, Minn., 70; and reproduction, 186

Carex trisperma, in Lake Edward Forest, Que., 65

Caribou, woodland, *see Rangifer caribou*

Carpenter ant, *see Camponotus herculeanus*

Carya (Hickory), fossil, 33

Carya cordiformis (*C. amara*; Hickory, bitternut), and heat, 92

Castanea dentata (Chestnut, American), and heat, 92

Castor canadensis (Beaver): flooding by, 105, 119; food of, 119

Catoplaca elegans, at Lake Superior, 69

Cenangium abietis, twig blight, 127

Cenangium balsameum, see *Dermea balsamea*

Cephalosporium album, canker, 131

Cerambicidae, secondary insects, 176

Cerastium, and rusts, 130

Ceratocystis bicolor, yeast, 144

Cervus (Elk), and reproduction, 61, 115

Chalcidae, parasites on spruce budworm, 160

Chamaedaphne calyculata, on Beaver Isl., Mich., 68

Chermes, see *Adelges*

Cherry, black, *see Prunus nigra*

Cherry, choke, *see Prunus virginiana*

Cherry, pin, *see Prunus pensylvanica*

Chickadee, *see Parus hudsonicus*

Chipmunk, *see Tamias*

Choristoneura fumiferana (Spruce budworm): resistance to, 27, 28; and light, 94, 156, 157; and birds, 122, 159; and viruses, 124, 125, 161; and bacteria, 125, 161, 169; and rots, 143, 146, 148, 149, 327; significance of, 149, 151, 152, 153, 166, 173, 177, 314; taxonomy of, 150, 155; and predators, 150, 151, 159, 160, 167, 168, 169; and parasites, 150, 151, 160, 161, 167, 168, 169; and reproduction, 150, 160, 165, 166, 181, 189, 190, 206, 216, 227; population dynamics of, 151, 152, 153, 154, 159, 160, 161, 162, 163, 165, 166, 167, 168; and tree and stand conditions, 151, 156, 158, 160, 162, 163, 164, 165, 166, 167, 169, 177, 216, 217, 234, 239, 244, 262; control of,

152, 153, 166, 168, 169, 177; cycle of, 152, 162, 163, 167; adults of, 155, 156, 159; dispersal of, 155, 156, 157, 158, 160, 162, 164, 167; eggs of, 155, 156, 159, 160, 166, 167; pupa of, 155, 158, 159, 167; larva of, 155, 156, 157, 158, 159, 160, 161, 166, 167, 168; feeding of, 155, 156, 158, 159, 162, 163, 166, 168; oviposition of, 156; hibernation of, 156, 162; diapause of, 156, 157; and temperature, 156, 157, 162; emergence of, 157, 159; survival of, 157, 167; life tables of, 166, 167; and management, 169, 209, 233, 244; and secondary insects, 176

Choristoneura muriana, in relation to *C. fumiferana*, 155

Choristoneura pinus (Budworm, jack pine), and spruce budworm, 155

Cladosporium herbarum, in succession of fungi, 127

Clethrionomys gapperi (Vole, red-backed): population of, 117; diet of, 118; and spruce budworm, 159, 160

Clintonia borealis: in Dryopteris-Oxalis type, 60; in Sphagnum-Oxalis type, 65; in White Mts., N.H., 80

Coniferales, proembryo of, 12

Coniophora cerebella, see *Coniophora puteana*

Coniophora puteana, brown cubical rot, 127, 134, 144, 146

Coptis groenlandica: in Kalmia-Ledum type, 65; on Isle Royale, Mich., 68

Coptis trifolia, in White Mts., N.H., 80

Cornus alternifolia, in balsam fir–basswood type, 70

Cornus baileyi, on Beaver Isl., Mich., 68

Cornus canadensis (Dogwood, dwarf; Bunchberry): in different regions, 60, 65, 66, 68, 69, 70, 80; and reproduction, 186; and site index, 264

Cornus stolonifera, in different regions, 68, 70, 218, 219

Corticium confluens, decay of bark, 127

Corticium fuscostratum, root and butt rot, 135

Corticium galactinum, white stringy butt rot, 134, 135, 145, 147

Corticium laeve, trunk rot, 135

Cortinarius, mycorrhiza, 126

Corylus (Hazel): and deer, 114; and reproduction, 188, 197; and moisture, 213, 219; and phytocides, 223, 224

Corylus americana (Hazel, American), and phytocides, 224

Corylus cornuta (Hazel, beaked): in different regions, 70, 188, 218, 219; and reproduction, 219, 223, 224

Coryne sarcoides, inhibiting fungus, 127

Cottonwood, *see Populus deltoides*

Crataegus rotundifolia, in balsam fir–basswood type, 70

Creonectria, see *Nectria*

Crossbill, red, *see Loxia curvirostra*

Crossbill, white-winged, *see Loxia leucoptera*

Cylindrocarpon cylindroides, see *Nectria cucurbitula* var. *macrospora*

Cytospora abietis, see *Valsa abietis*

Dalibarda repens, in Lake Edward Forest, Que., 65

Dasyneura balsamicola (Gall midge, balsam): common insect, 150; primary insect, 173; description, 175

Deer, white-tailed, *see Odocoileus virginianus borealis*

Dendroica castanea (Warbler, bay-breasted), and spruce budworm, 122, 159

Dermea balsamea, dieback, 131, 143

Dicranum undulatum: in Dryopteris-Oxalis type, 60; in Kalmia-Ledum type, 65

Diervilla lonicera: on Beaver Isl., Mich., 68; in Adirondacks, N.Y., 219

Dioryctria reniculella (Spruce needle worm): common insect, 150; and spruce budworm, 160

Diprion hercynia (Sawfly, European spruce): and succession, 61, 150; primary insects, 173

Dirca palustris, in balsam fir–basswood type, 70

Dogwood, dwarf, *see Cornus canadensis*

Double balsam ("*Abies* duplex"), meaning of, 8

Douglas fir, *see Pseudotsuga menziessii*

Dreyfusia, see *Adelges*

Dryopteris, and rusts, 130

Dryopteris disjuncta, in Dryopteris-Oxalis type, 60

Dryopteris noveboracensis, in Dryopteris-Oxalis type, 60

Dryopteris phegopteris, in Dryopteris-Oxalis type, 60

Dryopteris spinulosa: in Dryopteris-Oxalis type, 60; on Beaver Isl., Mich., 68

Eastern fir, name of balsam fir, 4

Elderberry, *see Sambucus*

Elk, *see Cervus*

Elm, *see Ulmus*

Elm, American, *see Ulmus americana*

Empusa, fungus on spruce budworm, 161

Epilobium, and rusts, 130

Erethizon dorsatum (Porcupine), damage by, 119, 120

Euarctos americanus (Bear), damage by, 120

Fagus americana, see *F. grandifolia*

Fagus grandifolia (Beech, American): pollen, 35; in succession, 44; in different regions, 65, 68, 75, 80, 204; and heat, 92; and light, 93; bark resistance to fire, 101; and porcupine, 120; and pulpwood production, 226, 264; girdling of, 230, 231, 236

Fern, bracken, *see Pteridium aquilinum*
Fern moss, *see Hylocomium proliferum*
Ferns, *see Polypodiaceae*
Fir, *see Abies*
Fir, European silver, *see Abies alba*
Fir, Fraser, *see Abies fraseri*
Fir, grand, *see Abies grandis*
Fir, Pacific silver, *see Abies amabilis*
Fir, pine, name of balsam fir, 4
Fir, subalpine, *see Abies lasiocarpa*
Fir tree, name of balsam fir, 4
Fir, white, *see Abies concolor*
Firre, name of balsam fir, 2
Fisher, *see Martes pennanti*
Fomes annosus, root rot, 125, 127
Fomes pini, heart rot, 135
Fomes pinicola, and spruce budworm, 143, 144
Forest tent caterpillar, *see Malacosoma disstria*
Fraxinus (Ash): and grafting, 29; and pollen, 35; in Itasca Park, Minn., 69; and phytocides, 224
Fraxinus nigra (Ash, black): in different regions, 52, 53, 66, 70, 72, 74, 80, 218; and flooding, 105; and oxygen, 106; and mortality, 213
Fraxinus pennsylvanica (Ash, green): in Itasca Park, Minn., 70; and ice storms, 98
Fungi Imperfecti: number of species, 126; and rusts, 130; and decay, 132
Fura, and fir, 3
Fusarium, damping-off, 128
Fusicoccum abietinum, canker, 131
Fusoma parasiticum, damping-off, 128

Galium triflorum, in Itasca Park, Minn., 70
Gall midge, balsam, *see Dasyneura balsamicola*
Gaultheria hispidula, in Kalmia-Ledum type, 65
Geum peckii, on White Mts., N.H., 80
Gloeosporium balsameae, rust, 130
Glypta fumiferanae, parasite on spruce budworm, 160
Gnathotrichus materiarius, secondary insects, 176
Godronia abietina, canker, 132
Godronia abietis, canker, 132
Grouse, ruffed, *see Bonasa umbellus*
Grouse, spruce, *see Canachites canadensis*
Gulo luscus (Wolverine), relationships with, 111
Gypsy moth, *see Porthetria dispar*

Hare, American varying, *see Lepus americanus*
Hare, European, *see Lepus europaeus*
Hare, Virginia varying, *see Lepus americanus virginianus*
Hazel, *see Corylus*
Hazel, American, *see Corylus americana*
Hazel, beaked, *see Corylus cornuta*
He-balsam, name of red spruce, 2

Hemerocampa leucostigma, common insect, 150
Hemlock, eastern, *see Tsuga canadensis*
Hemlock, western, *see Tsuga heterophylla*
Hemlock looper, *see Lambdina fiscellaria fiscellaria*
Hirsutella, fungus on spruce budworm, 161
Hophornbeam, *see Ostrya virginiana*
Hyalopsora, rusts, 130
Hylobini, on dead trees, 175
Hylobius (Weevils): primary insects, 173; description of, 175
Hylobius pales, root weevil, 175, 176
Hylobius pinicola: and decay, 141, 143; root weevil, 175
Hylocomium, and seedbed, 186
Hylocomium proliferum, on Isle Royale, Mich., 68
Hylocomium splendens: in Dryopteris-Oxalis type, 60; in Kalmia-Ledum type, 65
Hylocomium triquetrum, on Isle Royale, Mich., 68
Hypnum, and seedbed, 186
Hypnum crista-castrensis, in different regions, 60, 68, 70
Hypodermella, needle cast, 129
Hypodermella mirabilis, in succession of fungi, 127
Hypodermella nervata, on needles, 128
Hysteriaceae, needle casts, 127, 129

Ichneumonoidae, parasites on spruce budworm, 160
Impatiens capensis, in Itasca Park, Minn., 70
Ironwood, *see Ostrya virginiana*
Itoplectis conquisitor, parasite on spruce budworm, 160

Jack pine budworm, *see Choristoneura pinus*
Juglans, fossil, 33
Juglans cinerea (Butternut): in Apple River Canyon, Wis., 68; and heat, 92
Juniperus, fossil, 33

Kalmia, *Corticium galactinum* on, 145
Kalmia angustifolia, in Kalmia-Ledum type, 65
Kinglet, *see Regulus satrapa*
Kirschteiniella thujina: bluestain, 127; and *Stereum sanguinolentum*, 128

Labrador tea, *see Ledum groenlandicum*
Lachnella agassizii, canker, 132
Lachnella arida, canker, 132
Lachnella hahniana, canker, 132
Lachnella resinaria, canker, 132
Lambdina fiscellaria fiscellaria (Hemlock looper): and siricids and fungi, 144; common insect, 150; control of, 168, 174; primary insect, 173; description of, 174
Larch, *see Larix*
Laricobius erichsonii, predator of aphids, 172

Larix (Larch): fossil, 33; and grouse, 120; and porcupine, 120; and spruce budworm, 152; host of Adelginae, 170; and balsam fir sawfly, 174; cones of European, 179

Larix laricina (Tamarack): refugium of, 33; pollen, 35; in different regions, 38, 39, 50, 52, 53, 65, 70; in succession, 60; and ice storms, 98; and moisture, 105, 106; in nursery, 196; on aerial photographs, 257

Larix occidentalis (Larch, western), refugium, 33

Lasiocarpae, series of *Abies*, 4

Ledum groenlandicum (Labrador tea): in different regions, 65, 68, 80, 186; and site index, 264

Lenzites saepiaria, decay, 143

Lepidoptera: Ontario species of, 150; pathogens of, 161; bud-mining, 173

Leptosphaeria, in succession of fungi, 127

Lepus americanus (Hare, American varying): damage by, 115, 117; habitat of, 116, 117

Lepus americanus virginianus (Hare, Virginia varying), habitat of, 116, 117

Lepus europaeus (Hare, European), introduced, 116

Lichens: on White Mts., N.H., 80; relationships with, 124, 125; and spruce budworm, 156; and balsam woolly aphid, 172

Linnea borealis: on Isle Royale, Mich., 66; on White Mts., N.H., 80

Lonicera canadensis, in Viburnum type, 219

Lophodermium, needle cast, 128, 129

Lophodermium autumnale, in succession of fungi, 127

Lophodermium lacerum, needle cast, 129

Lophodermium nervisequum, needle cast, 129

Lophodermium piceae, needle cast, 127, 129

Loxia curvirostra (Crossbill, red), and seed, 122

Loxia leucoptera (Crossbill, white-winged), and seed, 122

Lychnis alba, host of *Corticium galactinum*, 145

Lycopodium annotinum, in different regions, 60, 68, 70

Lycopodium annotinum var. *pungens*, on White Mts., N.H., 80

Lycopodium clavatum, in Dryopteris-Oxalis type, 60

Lynx canadensis (Lynx), relationships with, 111

Maianthemum canadense, in different regions, 60, 65, 80

Malacosoma disstria (Forest tent caterpillar), and spruce budworm, 163, 218

Maple, *see Acer*

Maple, mountain, *see Acer spicatum*

Maple, red, *see Acer rubrum*

Maple, striped, *see Acer pensylvanicum*

Maple, sugar, *see Acer saccharum*

Martes americana (Marten), relationships with, 111

Martes pennanti (Fisher), relationships with, 111

Megastigmus specularis (Seed chalcid): primary insect, 173; description of, 175

Melampsora, rusts, 130

Melampsoraceae, rusts, 130

Melampsorella, rusts, 130

Melampsorella caryophyllacearum, witches' broom, 130

Melanconiaceae, rusts, 130

Mertensia paniculata, at Wykoff, Minn., 69

Merulius, brown cubical rot, 134

Merulius americanus, brown butt rot, 134, 135

Merulius himantioides, see *M. americanus*

Meteorus trachyotus, parasites on spruce budworm, 160

Microtus pennsylvanicus (Vole, meadow): population of, 117; diet of, 118

Milesia, rusts, 130

Milesia kriegeriana, rust, 128

Milesia polypodophila, witches' broom, 130

Mindarus abietinus (Twig aphid, balsam): primary insect, 173; description of, 174

Mitella nuda: on Isle Royale, Mich., 66; in Itasca Park, Minn., 70

Mites, *see* Acarina

Mnium, in Itasca Park, Minn., 70

Moneses uniflora, in Dryopteris-Oxalis type, 60

Monochamus (Woodborers): secondary insects, 173, 176; and chemical debarking, 310

Monochamus marmorator (Sawyer, balsam fir), description of, 176

Monochamus scutellatus, damage of wood by, 176

Moose, *see Alces alces*

Moss, in spruce-fir types, 74, 75

Mountain ash, *see Sorbus americana*

Mouse, deer, *see Peromyscus maniculatus*

Mouse, white-footed, *see Peromyscus leucopus*

Myxomycetes (Slime molds): relationships with, 124, 125, 126

Nectria, cankers, 131

Nectria balsameae, canker, 131, 143

Nectria cucurbitula, canker, 131, 143

Nectria cucurbitula var. *macrospora*, canker, 131

Needle worm, spruce, *see Dioryctria reniculella*

Nemophanthus mucronata, in Cornus type, 219

Neodiprion abietis (Sawfly, balsam fir): common insect, 150; primary insect, 173; description of, 174

Neoleucopis obscura, predator on aphids, 172

Northern white-cedar, *see Thuja occidentalis*

Nothophacidium abietinellum, needle blight, 128, 129

Oak, *see Quercus*

Oak, bur, *see Quercus macrocarpa*

Oak, red, *see Quercus rubra*

Odocoileus virginianus borealis (Deer, white-tailed): and reproduction, 52, 61, 114, 115, 198; habitat of, 113, 114, 115, 116, 123; food of, 114, 115, 223

Odontia bicolor, decay, 134, 135, 145, 146, 147

Omphalia campanella, see *Xeromphalina campanella*

Ophiostoma bicolor, see *Ceratocystis bicolor*

Orgyia antiqua, common insect, 150

Osmunda claytoniana, in balsam fir–basswood type, 70

Ostrya virginiana (Hophornbeam; Ironwood): pollen, 35; in different regions, 68, 69, 70

Oxalis montana, in different regions, 60, 65, 80

Parus hudsonicus (Chickadee), and spruce budworm, 122

Penicillium glaucum, and wood-rotting fungi, 127

Peniophora gigantea, and *Fomes annosus*, 127

Perezia fumiferana, protozoa on spruce budworm, 161

Peridermium balsameum, see *Milesia kriegeriana*

Peromyscus leucopus (Mouse, white-footed), and seed, 118, 119

Peromyscus maniculatus (Mouse, deer), populations of, 117, 118, 159, 160

Peronea variana, see *Acleris variana*

Pestalotia hartigii, on seedlings, 128

Peuce balsamea, name of balsam fir, 3

Phacidiaceae, blights, 127, 129

Phacidiella, see *Potebniamyces balsamicola*

Phacidium abietinellum, snow mold, 128, 129

Phacidium balsameae, snow mold, 127, 128, 129

Phacidium infestans, snow mold, 128, 129

Phegopteris, and rusts, 130

Phomopsis, see *Fusicoccum abietinum*

Phryxe pecosensis, parasites on spruce budworm, 160

Phycomycetes, in nursery, 126

Phymatodes dimidiatus, secondary insect, 176

Phytophthora, damping-off, 128

Picea (Spruce): name of, 3; embryogeny of, 12; roots, 13, 23; grafting of, 29; fossil, 33, 34, 35; pollen, 35; reproduction of, 35, 183, 184, 186, 188, 190, 195, 197, 201, 202, 203, 204, 206, 207, 208, 214, 226; in spruce-fir types, 39; and hail, 97; and ice storms, 98; and wind, 98, 99; and lightning, 101; and deer, 115; and hare, 117; and red squirrel, 119; and beaver, 119; and grouse, 120; and spruce budworm, 152, 166; and balsam woolly aphid, 173; and sawfly, 174; in virgin forest, 211; dynamics of, 204, 214, 216, 220, 221, 225, 228, 230, 233, 237, 239, 240, 242, 243, 244; growth and yield of, 211, 214,

220, 233, 240, 244, 261, 270, 274, 278; composite volume tables for, 254; stand and stock tables of, 256; on aerial photographs, 257; length of tracheids in wood of, 294; moisture in trees of, 295; pulpwood weight, 295; paint and wood of, 297; strength of wood of, 299; chemical composition of wood of, 300; pulp yield of, 302; pulping of, 303, 305, 320; logging of, 308; lumber of, 309, 312; price of material, 310; pulpwood production of, 316; Christmas trees, 322, 323

Picea abies (Spruce, Norway): early names of, 3; root rot of, 125; cones of, 179; shelterwood cut of, 192, 193; root growth of, 196; volume of, 254

Picea balsamea, name of balsam fir, 2

Picea balsamea var. *longifolia*, name of balsam fir, 3

Picea balsamifera, name of balsam fir, 3

Picea engelmannii, refugium of, 33

Picea fraseri, name of balsam fir, 3

Picea glauca (Spruce, white): and pruches, 2; name of balsam fir, 4; photosynthesis, 20; and moisture, 21; refugium of, 33; pollen of, 34; in different regions, 35, 50, 52, 53, 59, 65, 66, 68, 69, 70, 75, 76, 80, 88; in succession, 61, 72; and heat, 92; tolerance of, 94; and ice storms, 98; and wind, 99; and flooding, 105; and moose, 113; and deer, 116; and small mammals, 118; and logging, 141; and spruce sawfly, 150; and spruce budworm, 156, 158, 159; seed production of, 179, 181; and reproduction, 185, 192, 193, 198, 206, 227; in virgin forest, 212, 213; growth and yield of, 213, 234, 243, 244, 261, 264, 265, 268, 272; dynamics of, 214, 218, 227, 233, 234, 243, 244, 265; and phytocides, 222; and composite volume tables, 254; on aerial photographs, 257; specific gravity of wood of, 296; and wood density, 302; pulpwood of, 303; stem center of gravity of, 311; bark, 318, 319

Picea mariana (Spruce, black): and pruches, 2; layering of, 24; refugium of, 33; pollen, 34; in different regions, 38, 50, 52, 53, 54, 65, 68, 70, 74, 75, 79, 80; in succession, 60, 61; and heat, 92; tolerance of, 94; and drought, 97; and hail, 98; and ice storms, 98; and moisture, 105, 106; and moose, 113; and deer, 116; logging damage to, 141; and spruce budworm, 156, 158, 159; and balsam woolly aphid, 173; seed production of, 181; reproduction of, 185, 193, 198, 201, 203; stand dynamics, 193, 213, 214, 218, 227, 233, 234; and phytocides, 222; composite volume tables of, 254; on aerial photographs, 257; and site index, 261, 264; productivity of, 268, 296; wood properties of, 294, 296; pulping of, 302, 303, 304; chemical peeling of, 310; stem gravity center of, 311; bark of, 318, 319; needle oil, 322

Picea pungens (Spruce, blue), and phytocides, 222

Picea rubens (Spruce, red): as he-balsam, 2; roots of, 23; refugium of, 33; in different regions, 35, 63, 65, 74, 75, 76, 80; and light, 93, 94; winter drying, 97; and deer, 114; and hare, 117; and small mammals, 118; and beaver, 120; and spruce budworm, 156; reproduction of, 183, 193, 204, 206, 208, 230; stand development of, 204, 206, 208, 219, 221, 228, 230, 235, 236, 242, 244; growth and yield of, 208, 235, 236, 242, 244, 261, 264, 265, 268, 272; composite volume tables for, 254; and tariff classes, 256; on aerial photographs, 257; cull of, 259; wood density of, 294, 302; bark of, 318, 320

Picea sitchensis, refugium of, 33

Pinaceae (Abietaceae): taxon, 4; life cycle of, 9

Pine, *see Pinus*

Pine, jack, *see Pinus banksiana*

Pine, lodgepole, *see Pinus contorta*

Pine, ponderosa, *see Pinus ponderosa*

Pine, red, *see Pinus resinosa*

Pine, Scotch, *see Pinus sylvestris*

Pine, southern: pulp of, 303, 306; pulpwood production of, 317

Pine, white, *see Pinus strobus*

Pinus (Pine): and true firs, 3, 4; pollen, 10, 35; embryogeny of, 12; fossil, 32, 33; reproduction of, 35, 188, 196, 206, 208, 226; range, 52; in Adirondacks, N.Y., 76; and hare, 117; and porcupine, 120; and *Corticium galactinum*, 145; as host of Adelginae, 170; height of, 225; replacement of, 244; on aerial photographs, 257; and Christmas trees, 323, 324, 328

Pinus abies balsamea, name of balsam fir, 2

Pinus balsamea, name of balsam fir, 2

Pinus balsamea var. *longifolia*, name of balsam fir, 2

Pinus banksiana (Pine, jack): and layering, 24; refugium of, 33; range of, 52, 80; in succession, 72; and heat, 92; and light, 94; and drought, 97; and hail, 97; and ice storms, 98; and deer, 114, 116; overstory of, 218; and phytocides, 222; on aerial photographs, 257; length of tracheids in wood of, 294; pulpwood weight of, 295; chemical debarking of, 310; gravity center of stem of, 311; lumber grading of, 312

Pinus contorta (Pine, lodgepole): refugium of, 33; strength of wood, 299

Pinus echinata, and *Corticium galactinum*, 145

Pinus monticola, refugium of, 33

Pinus ponderosa (Pine, ponderosa): refugium of, 33; and *Cenangium abietis*, 127; strength of wood of, 299

Pinus resinosa (Pine, red): refugium of, 33; in different regions, 53, 66, 69, 80; in succession, 72; and light, 94; and drought, 97; and flooding, 105; and deer, 116; and small mammals, 118; storage of plants of, 197; reproduction of, 198; overstory of, 218; and phytocides, 222; on aerial photographs, 257; height growth of, 265; as Christmas tree, 323, 325

Pinus strobus (Pine, white): roots of, 23; and layering, 24; refugium of, 33; in different regions, 38, 53, 63, 65, 66, 68, 69, 74, 75, 80; in succession, 71, 72; and heat, 92; and light, 93, 94; drought effect on, 97; and hail, 97; and wind, 99; and flooding, 105; and small mammals, 118; and *Corticium galactinum*, 145; storage of plants of, 197; reproduction of, 198; overstory of, 204, 218, 230, 239, 243, 244; in virgin forest, 211; and phytocides, 222; on aerial photographs, 257; height growth of, 265; needle oil of, 322

Pinus sylvestris (Pine, Scotch): as fir, 3; and mice, 118; and needle cast, 129; cones of, 179; as Christmas trees, 323, 324, 325

Pinus taxifolia, name of balsam fir, 2

Pissodes dubius, secondary insect, 175, 176

Pityokteines sparsus, description of, 176

Plagiothecium denticulatum, in Dryopteris-Oxalis type, 60

Pleroneura borealis, common insect, 150

Pleurozium (*Calliergon, Calliergonella, Hypnum*) *schreberi*: in different regions, 65, 68, 70; as seedbed, 183, 186

Podocarpaceae, fossils, 32

Polygraphus rufipennis, secondary insect, 176

Polypodiaceae (Ferns): in spruce-fir types, 74, 75; and rusts, 130

Polyporus abietinus, decay, 143, 144, 146

Polyporus balsameus, heart rot, 133, 134, 135, 137, 143, 146

Polyporus schweinitzii, heart rot, 127, 133

Polyporus tomentosus, and *Coryne sarcoides*, 127

Polytrichum, as seedbed, 186, 190

Poplar, balsam, *see Populus balsamifera*

Populus (Aspen; Poplars): in different regions, 37, 52, 53, 69; sites, 53, 69; in succession, 60, 61, 71, 72, 213, 214; damage to, 97, 98, 105; and wildlife, 112, 113, 114, 116, 117; overstory, 116, 163, 189, 198, 213, 214, 218, 225, 230, 244, 265; and rusts, 130; and reproduction of, 184, 196, 224, 234, 244; productivity of, 213, 218, 225, 261, 264; and phytocides, 224; and integrated cutting, 243, 244; pulpwood weight of, 295; wood chemical composition of, 300; pulp brightness of, 302; moisture in wood of, 309; pulpwood production of, 317

Populus balsamifera (Poplar, balsam): pollen, 35; in different regions, 52, 53, 70, 80, 218;

in succession, 61, 218; and heat, 92; and hail, 97; and ice storms, 98; and flooding, 105; and site index, 213, 264; height growth of, 265; brightness of pulp, 302

Populus deltoides (Cottonwood): on Beaver Isl., Mich., 68; durability of wood of, 296; brightness of pulp of, 302

Populus grandidentata (Aspen, bigtooth): refugium of, 33; in Apple River Canyon, Wis., 68; litter, 108; girdling of, 230

Populus tremuloides (Aspen, quaking; trembling): in different regions, 53, 70, 218; and light, 93; and ice storms, 98; and drainage, 106; composition of wood of, 299

Porcupine, *see Erethizon dorsatum*

Poria asiatica, root and butt rot, 135

Poria cocos, brown cubical rot, 134, 135

Poria subacida, heart rot, 133, 134, 135, 137, 143, 145, 146, 147

Porthetria dispar (Gypsy moth): common insect, 150; primary insect, 173

Potebniamyces balsamicola, blight, 129

Potentilla fruticosa, on Beaver Isl., Mich., 68

Prunus nigra (Cherry, black), in balsam fir–basswood type, 70

Prunus pensylvanica (Cherry, pin): and light, 204; in Boreal Region, 218; and moisture, 219; competition by, 226

Prunus virginiana (Cherry, choke): in Itasca Park, Minn., 70; in Boreal Region, 218

Pseudotorreya, taxon, 4

Pseudotsuga, host of Adelginae, 170

Pseudotsuga menziessii (Douglas fir): and true firs, 4; refugium of, 33; in Wind River Arboretum, Wash., 40; in Germany, 41; and mice, 118; and spruce budworm, 152; cones of, 179; strength of wood of, 299; as Christmas tree, 324

Pteridium aquilinum (Fern, bracken): in balsam fir–basswood type, 70; and reproduction, 186, 188; and phytocides, 224; and site index, 264

Pteridium latiusculum, in Dryopteris-Oxalis type, 60

Pteromalidae, parasites on spruce budworm, 160

Pucciniaceae, rusts, 130

Pucciniastrum, rust, 130

Pyrola secunda: in Sphagnum-Oxalis type, 65; at Wykoff, Minn., 69

Pyrola virens, at Wykoff, Minn., 69

Pyrus (*Sorbus*) *decora*, at Lake Superior, 69

Pythium, damping-off, 128

Quercus (Oak): refugium of, 33; pollen, 35; and *Corticium galactinum*, 145

Quercus borealis var. *maxima*, in balsam fir–basswood type, 69. *See also Q. rubra*

Quercus macrocarpa (Oak, bur): in Itasca Park, Minn., 70; and flooding, 105

Quercus rubra (Oak, red): in different re-

gions, 66, 68, 80; and heat, 92; and light, 93. *See also Q. borealis* var. *maxima*

Rabbit, snowshoe, *see* Hare, American varying

Rangifer caribou (Caribou, woodland), relationships with, 111, 112, 113

Regulus satrapa (Kinglet), and spruce budworm, 122

Rehmiellopsis, blights, 129

Rhamnus alnifolia, in different regions, 68, 69, 70, 219

Rhizoctonia, damping-off, 128

Ribes, in different regions, 68, 218, 219

Ribes americanum, in Itasca Park, Minn., 70

Ribes glandulosum, in spruce-fir type, 219

Ribes triste, in different regions, 70, 219

Rosa acicularis, on Beaver Isl., Mich., 68

Rubus: Corticium galactinum on, 145; and reproduction, 188

Rubus canadensis, in Dryopteris-Oxalis type, 60

Rubus pubescens: at Wykoff, Minn., 68; in balsam fir–basswood type, 70

Rubus strigosus: on Beaver Isl., Mich., 68; and phytocides, 224

Salix (Willow): on White Mts., N.H., 80; and rusts, 130; and phytocides, 224

Salix bebbiana: on Beaver Isl., Mich., 68; in balsam fir–basswood type, 70

Sambucus (Elderberry), invasion of, 190

Sambucus pubens: in Dryopteris-Oxalis type, 60; on Beaver Isl., Mich., 68

Sapindus, taxon, 4

Sapsucker, yellow-bellied, *see Sphyrapicus varius*

Sawfly, balsam fir, *see Neodiprion abietis*

Sawfly, European spruce, *see Diprion hercynia*

Sawyer, balsam fir, *see Monochamus marmorator*

Scleroderris, see *Godronia abietina*

Seed chalcid, *see Megastigmus specularis*

Semiothisa granitata, common insect, 150

Serropalpus barbatus, secondary insect, 176

She-balsam, name of Fraser fir, 2

Shrew, cinereous, *see Sorex cinereus*

Shrew, short-tailed, *see Blarina brevicauda*

Silver-fir, female fir, 2

Silver pine, name of balsam fir, 4

Single spruce, name of balsam fir, 4

Sirex, secondary insects, 176

Sirex cyaneus: and fungi, 144, 173; secondary insect, 176

Siricidae, and fungi, 173

Slime molds, *see* Myxomycetes

Smilacina trifolia, in Dryopteris-Oxalis type, 60

Solidago macrophylla: in Dryopteris-Oxalis type, 60; in White Mts., N.H., 80

Sorbus americana (Mountain ash): in different regions, 60, 65, 76, 79, 93, 219; and moose, 112

Sorbus decora, see *Pyrus decora*

Sorex cinereus (Shrew, cinereous), population of, 118

Sphaeriaceae, blights, 129

Sphagnum: in different regions, 70, 75, 80; and reproduction, 186; and site index, 264

Sphagnum acutifolium, in Lake Edward Forest, Que., 65

Sphagnum capillaceum, in Dryopteris-Oxalis type, 60

Sphagnum palustre, in Dryopteris-Oxalis type, 60

Sphyrapicus varius (Sapsucker, yellow-bellied), diet of, 122

Spiders, *see* Arachnida

Spirea alba, on Beaver Isl., Mich., 68

Spirea latifolia var. *septentrionalis,* on White Mts., N.H., 80

Spruce, *see Picea*

Spruce, black, *see Picea mariana*

Spruce, blue, *see Picea pungens*

Spruce fir, name of spruce, 3, 4

Spruce, Norway, *see Picea abies*

Spruce, red, *see Picea rubens*

Spruce, white, *see Picea glauca*

Squirrel, red, *see Tamiasciurus hudsonicus*

Stegopezizella balsameae, see *Phacidium balsameae*

Stellaria, and rusts, 130

Stereum chailletii, trunk rot, 120, 135, 142, 143, 144

Stereum purpureum, root and butt rot, 127, 135

Stereum sanguinolentum, red top rot, 120, 127, 133, 134, 135, 140, 141, 142, 143, 144, 145, 173, 177

Streptopus roseus, in Dryopteris-Oxalis type, 60

Tachinidae, parasites on spruce budworm, 160

Tamarack, *see Larix laricina*

Tamias (Chipmunk), feeding on seed, 119

Tamiasciurus hudsonicus (Squirrel, red): habitat of, 111, 116, 119; food of, 119

Taxus canadensis: on Isle Royale, Mich., 66; in Iowa, 68

Tetropium cinnamopterum, secondary insects, 176

Thalictrum dioicum, in Itasca Park, Minn., 70

Thelypteris spinulosus var. *intermedia,* in Sphagnum-Oxalis type, 65

Thuidium delicatulum, in different regions, 60, 70

Thuja occidentalis (White-cedar, northern): pollen, 35; in spruce-fir types, 39; in different regions, 52, 53, 63, 65, 66, 68, 69, 74, 75, 80; in succession, 60, 72; in New Brunswick models, 87, 88; and heat, 92; and light, 93; and ice storms, 98; and wind,

99; and moisture, 105, 106; and deer, 114; and hare, 117; and porcupine, 120; and bear, 120; in virgin forest, 211; and stand development, 213, 218; on aerial photographs, 257; use of lumber of, 312; needle oil of, 322

Thuja plicata, refugium of, 33

Thyronectria, see *Nectria*

Tilia americana (Basswood, American): pollen, 35; in different regions, 66, 68, 69, 70, 104; in succession, 72; and light, 93; and flooding, 105; and deer, 114; in edaphic coordinates, 213; durability of wood, 296

Trichoderma viridis, and *Fomes annosus,* 127

Trichomonascus mycophagus, parasitic on fungi, 127

Trichosphaeria parasitica, see *Acanthostigma parasiticum*

Trientalis borealis, in different regions, 60, 66, 80

Tryblidiaceae, cankers, 132

Trypodendron bivittatum (Ambrosia beetle): secondary insect, 173; description of, 176, 177

Tsaria, fungus on spruce budworm, 161

Tsuga: and true firs, 4; fossil, 32; in spruce-fir types, 39

Tsuga canadensis (Hemlock, eastern): and true firs, 4; refugium of, 33; pollen, 35; in succession, 44, 71, 72; in different regions, 53, 65, 74, 75, 76, 80; and heat, 92; and light, 93; and drought, 97; and wind, 99; bark resistance to fire, 101; and deer, 114; and small mammals, 118; and porcupine, 120; and birds, 122; and Adelginae, 170; and shrubs, 219; and thinning, 230; strength of wood, 299; needle oil of, 322

Tsuga heterophylla (Hemlock, western): refugium of, 33; strength of wood, 299

Twig aphid, balsam, *see Mindarus abietinus*

Tympanis abietina, canker, 132

Tympanis truncatula, canker, 132

Ulmus (Elm): pollen, 35; in different regions, 53, 54, 66

Ulmus americana (Elm, American): in different regions, 52, 68, 69, 80; and light, 93; and flooding, 105

Uredinopsis, rusts, 130

Urocerus albicornis, secondary insect, 173, 176

Vaccinium (Blueberry): on Beaver Isl., Mich., 68; on White Mts., N.H., 80; and rusts, 130; and site index, 264

Vaccinium myrtilloides, in Kalmia-Ledum type, 65

Valsa abietis, dieback, 131, 143

Veratrum viride, on White Mts., N.H., 80

Viburnum alnifolium: in Lake Edward Forest, Que., 65; in Adirondacks, 219

Viburnum cassinoides: in Dryopteris-Oxalis type, 60; in Boreal Region, 218
Viburnum edule, in Boreal Region, 218
Viburnum lantanoides, see *V. alnifolium*
Viburnum trilobum, in balsam fir–basswood type, 70
Viola incognita: in Dryopteris-Oxalis type, 60; on Beaver Isl., Mich., 68
Viola pallens, in Itasca Park, Minn., 70
Vireo (Vireo), and spruce budworm, 122
Vole, meadow, see *Microtus pennsylvanicus*
Vole, red-backed, see *Clethrionomys gapperi*

Warbler, see *Dendroica castanea*
Weevils, see *Hylobius*

White-cedar, see *Thuja occidentalis*
White fir, name of balsam fir, 4
Willow, see *Salix*
Wolf, see *Canis lupus*
Wolverine, see *Gulo luscus*
Woodborers, see *Monochamus*

Xeris spectrum: and fungi, 144; secondary insect, 173
Xeromphalina campanella, white stringy butt rot, 134
Xylotrechus undulatus, secondary insect, 176

Yponomeutidae, primary insects, 173

PART III. COMMUNITIES

Abies fraseri–Picea rubens–Betula alleghaniensis, in southern Appalachians, 35
Abietum balsameae, in Gaspé Peninsula, Que., 105
Aspen-birch, in Minnesota, 230
Aspen-birch-spruce-fir: in Boreal Region, Ont., 60, 61, 214, 233, 234, 274; in Canada and USA, 213
Aspidium-Oxalis, in Adirondacks, N.Y., 219

Balsam fir–basswood, in Itasca C., Minn., 69, 70, 71, 102
Balsam fir flat, in northeastern USA, 264. See *also* Spruce flat
Balsam fir–paper birch–white spruce, on Isle Royale, Mich., 71
Balsam fir–white spruce: at Athabaska–Great Slave Lake, Alta., 53; in Newfoundland, 236
Balsam fir–yellow birch, in New Brunswick, 87
Basswood–sugar maple, in Itasca C., Minn., 71. See *also* Sugar maple–basswood
Beech-maple: in Region 7 by Braun, 42; in Lake Forest Region, 71
Beech–sugar maple–yellow birch, on Cape Breton Isl., N.S., 81
Birch-poplar, in northeastern USA, 74
Black ash–American elm–basswood, in Itasca Park, Minn., 70
Black spruce, in Boreal and Northern Region by Soc. Amer. For., 46
Black spruce–Chamaedaphne, in Acadia Forest, N.B., 80
Black spruce–Sphagnum, in Acadia Forest, N.B., 80
Boreal coniferous forest biome, sensu Shelford, 111
Boreal forest: formation sensu Clements, 41, 42, 326; in Wisconsin, 86
Boreal-hardwood transition, in Great Lakes area, 71, 86

Cedar-tamarack-spruce, in USA, 289

Cladonia-Ledum, in Boreal Region, Que., 59
Cladonia-Vaccinium, in Boreal Region, Que., 59
Clintonia-Lycopodium, in Wisconsin, 65, 219
Cornus: in Lake Edward Forest, Que., 63, 204, 219, 240, 242; in Newfoundland, 201
Cornus-Maianthemum, in Petawawa Forest, Ont., 228

Deciduous forest: formation sensu Clements, 41, 42; in Minnesota, 110; in Algonquin Park, Ont., 122
Dryopteris-Oxalis, in Boreal Region, Que., 59, 137, 233, 242

Fir-pine-birch, in forest zone of Maritime Provinces, 80
Flat, see Spruce flat

Galium-Equisetum, in Wisconsin, 66
Gaultheria-Maianthemum, in Wisconsin, 66

Hardwood: in Adirondacks, N.Y., 103; in Canada, 199, 201, 203, 283; in Minnesota, 206; in New Hampshire, 236, 290; in Maine, 290
Hardwood land, see Hardwood slope
Hardwood slope, in northeastern USA, 74, 75, 94, 183, 249, 265, 289. See *also* Spruce-hardwood
Hardwood-spruce-fir, in New England, 75, 191, 206, 207, 220
Hemlock-hardwood, in Argonne Forest, Wis., 99
Hylocomium-Cornus, in Adirondacks, N.Y., 219
Hylocomium-Oxalis, in Boreal Region, Que., 135, 137, 233, 242
Hypnum-Cornus, in Laurentides Park, Que., 233

Jack pine, in Great Lakes–St. Lawrence Region, 218

Kalmia-Ledum, in Lake Edward Forest, Que., 65

Lake forest, formation sensu Clements, 41

Maple-basswood, *see* Sugar maple–basswood
Mixed mesophytic, in Region *1* Sect. B by Braun, 42
Mixed-wood: in Canada, 53, 137, 185, 193, 199, 201, 202, 203, 206, 214, 239, 283, 291; in Minnesota, 206; in New Hampshire, 236, 290; in Maine, 290
Mountain top, *see* Spruce slope

Oak-beech, in Deciduous Region sensu Clements, 41
Oak-chestnut: in Deciduous Region sensu Clements, 41; in Region *4* by Braun, 42
Oak-hickory: in Deciduous Region sensu Clements, 41, 71; in Region *3* Sect. C by Braun, 42
Old-field spruce, in northeastern USA, 74, 184, 191, 208, 225, 264
Oxalis-Coptis, in Wisconsin, 65
Oxalis-Cornus: in Lake Edward Forest, Que., 63, 102, 107, 204, 208, 219, 228, 240, 242; in Adirondacks, N.Y., 63, 208, 232
Oxalis-Dicranum, in Adirondacks, N.Y., and White Mts., N.H., 76
Oxalis-Hylocomium, in Adirondacks, N.Y., 219

Pine-spruce-Cladonia, in Acadia Forest, N.B., 80
Pinetum myrtillosum, at Leningrad, USSR, 122
Poplar-birch, in Canada, 117

Red pine–moose, in Boreal-Deciduous ecotone, 111
Red spruce, in Northern Region by Soc. Amer. For., 47, 102
Red spruce–balsam fir–paper birch, in northeastern USA, 264
Red spruce–yellow birch, in northeastern USA, 264
Rubus parviflorus, in Wisconsin, 66

Smilacina-Polygonatum, in Wisconsin, 66
Softwood: in Canada, 137, 199, 200, 201, 202, 203, 206, 239, 283, 291; in Minnesota, 206; in New Hampshire, 236, 290; in Maine, 290
Sphagnum, in Newfoundland, 201
Sphagnum-Oxalis, in Lake Edward Forest, Que., 65
Spruce, in Pennsylvania and New Hampshire, 96
Spruce–balsam fir, *see* Spruce-fir
Spruce-fir: in southern and northern Appalachians, 35, 117; in USA, 39, 42, 63, 91, 285;

in Great Lakes–St. Lawrence Region, 42, 69, 94, 120, 154, 186, 188, 213, 219, 226, 268, 276, 278, 288, 289, 290; in Boreal Region, 60, 111, 120, 121, 122, 150, 152, 195, 202, 216, 217, 236, 262; in Acadian-Appalachian Region, 73, 74, 75, 76, 96, 98, 114, 116, 118, 184, 191, 192, 193, 220, 225, 226, 230, 235, 261, 264, 274, 278, 281, 285, 287, 289, 290; in Canada, 203, 285
Spruce-fir-birch: in Minnesota, 69, 206; in Green River area, N.B., 221
Spruce-fir-hardwood: in Acadian-Appalachian Region, 75, 191; in Boreal Region, 214; in Great Lakes–St. Lawrence Region, 258, 289, 290; in USA, 285
Spruce flat, in northeastern USA, 74, 103, 114, 206, 207, 208, 232, 249, 265, 289, 296
Spruce-hardwood, in northeastern USA, 103, 207, 208, 289. *See also* Hardwood slope
Spruce-larch-Carex, in Acadia Forest, N.B., 80
Spruce-poplar-fir, in northwestern Alberta, 61
Spruce slope, in northeastern USA, 75, 116, 206, 207, 208, 264
Spruce swamp, in northeastern USA, 74, 94, 103, 203, 206, 247, 249, 255, 264, 265, 289, 296
Sugar maple–basswood, in Great Lakes–St. Lawrence Region, 35, 42, 70, 183
Sugar maple–beech–yellow birch, in Acadian-Appalachian Region, 35
Sugar maple–spruce, in Acadian-Appalachian Region, 75
Swamp, *see* Spruce swamp

Urtica-Thalictrum, in Wisconsin, 66

Vaccinium-Cornus, in Wisconsin, 219
Vaccinium-Cornus-Rubus, in Wisconsin, 65
Viburnum: in Lake Edward Forest, Que., 65, 219, 221, 240, 242; in Adirondacks, N.Y., 65, 219
Viburnum-Oxalis: in Lake Edward Forest, Que., 63, 204, 208, 219, 240, 242; in Adirondacks, N.Y., 65, 208, 219, 232

White pine–red pine, in northeastern Minnesota, 71
White spruce, in Alberta, 42
White spruce–balsam fir: in Minnesota and Wisconsin, 42; in Boreal Region, 60, 102; in Duck Mts., Man., 202
White spruce–balsam fir–paper birch, in Northern Region by Soc. Amer. For., 47, 102
Willow-alder swamp, in Canada, 117

Yellow birch–spruce, in northeastern USA, 75, 232

PART IV. LOCALITIES

Abitibi, Lake, Site Region, Ont., 53
Acadia Experimental Forest, N.B., 28, 29, 80, 87, 186, 227, 228
Acadia National Park, Maine, 326
Acadian-Appalachian Forest Region, 35, 47, 73, 102, 206, 218, 219
Acadian Forest Region, 73, 80, 91, 94
Adirondack Mts., N.Y., 13, 44, 63, 74, 76, 94, 97, 98, 103, 108, 116, 119, 129, 175, 183, 188, 194, 219, 230, 247, 253, 259, 261, 281, 289, 326
Agassiz, Glacial Lake, Man., Minn., N.D., 53
Aleutian Isl., Alaska, 41
Algoma District, Ont., 153
Algoma Section (L-10), 62, 63, 72, 211, 214
Algonquin-Pontiac Section (L-4b), 63, 71
Algonquin Provincial Park, Ont., 71, 122, 327
Allamakee C., Iowa, 39, 68
Alvin, Wis., 198
Anticosti Isl., Que., 51, 52, 61, 115
Anticosti Section (B-28c), 52
Antigonish, N.S., 28
Appalachian Mts., 2, 33, 35, 36, 51, 74, 102, 112, 121
Apple River Canyon, Wis., 68
Argonne Experimental Forest, Wis., 99, 115, 198
Asia, eastern, 35
Athabaska, Lake, Alta. and Sask., 36
Athabaska–Great Slave Lake, Alta., Sask., NW Terr., 53
Athabaska River, Alta., 6, 36, 53
Athabaska South Section (B-22b), 53
Australia, 41, 172, 325
Austria, 115, 172
Avalon Section (B-30), 52

Badoura, Hubbard C., Minn., 197
Baltic States, 40
Banks Island (Banksland), NW Terr., 32
Basswood Lake, Minn., 223
Baxter State Park, Maine, 326
Beaver Island, Lake Superior, Mich., 68
Beech-Maple Forest Region (Region 7), 42
Bering Sea, 32, 33
Berkshire C., Mass., 37
Black Forest of Gaspé, Que., 189
Black River Experimental Forest, Ont., 181, 184, 185, 190, 191
Black Sturgeon Lake, Ont., 181
Blue Ridge Mts., Va., 6, 38
Boreal Forest Region, 42, 47, 50, 91, 94, 102, 110, 155, 191, 199, 218, 220, 225, 261
Brainerd, Crow Wing C., Minn., 104
British Isles, 325
Brownington Pond, Vt., 35
Bruce Peninsula, Ont., 68

Canadian Life Zone, 111, 117
Canadian Shield, 51, 62, 102

Cape Breton Island, N.S., 39, 74, 80, 98, 131, 153, 170, 174, 327
Cape Breton–Antigonish Section (A-7), 80
Cape Breton Plateau Section (A-6), 80
Cape Harrison, Lab., 36
Cascade Range, 40
Catskill Mts., N.Y., 36, 38, 74, 253, 326
Cedar Creek Preserve, Minn., 106
Cedar Lake, Ont., 25
Central Forest Region, 42
Central Hardwoods–Hemlock Zone, 75
Central Hardwoods–Hemlock–White Pine Zone, 75
Central Lowlands Physiographic Province, 62
Central Plateau Section (B-8), 60, 181, 184, 196, 214, 234
Charlton Island, James Bay, NW Terr., 37
Chernigovsk, USSR, 41
Chibougamau-Natashquan Section (B-1b), 54, 242. *See also* Northeastern Coniferous Section (B-1)
China, southwest, 36, 39
Churchill, Man., 36
Clay Belt, Ont., 214, 220, 259. *See also* Northern Clay Section (B-4)
Clearwater River, Alta., 53
Cloquet Experimental Forest, Minn., 97, 99, 121, 188, 197, 289
Colesville, Md., 38
Cook C., Minn., 69
Corbin Park, N.H., 230
Cordillera Range, 33
Cumberland House, Sask., 36

Decorah, Winneshiek C., Iowa, 38
Denmark, 125, 145
Dodge C., Minn., 38
Door C., Wis., 68
Douglas Lake, Cheboygan C., Mich., 43, 71, 105
Duck Mts., Man., 202, 212
Dummer, N.H., 184
Dunbar Experimental Forest, Mich., 228
Durham, N.C., 25

East Atlantic Shore Section (A-5b), 80
East James Bay Section (B-6), 52
East Prussia, 41
Eastern Townships Section (L-5), 63, 73
Edmonton, Alta., 37
Elkader, Clayton C., Iowa, 39
England, 222
Epaule River, Que., 188, 189, 242
Erie-Ontario-Lakes Site Region, Ont., 54
Europe, 126, 145, 146, 160, 170, 172, 191, 192, 193, 325

Fairfield, St. John C., N.B., 231
Faribault, Rice C., Minn., 33
Fillmore C., Minn., 33, 38, 68

Finland, 40, 325
Fir-Pine-Birch Forest Zone, 80
Forest Tundra Section (B-32), 52
Fort Assiniboine, Alta., 36
Fort George Section (B-13b), 52
Frankfort-on-Main, Germany, 32
Franklin C., N.Y., 219
Fredericton, N.B., 160

Gaspé Peninsula, Que., 32, 61, 104, 150, 261
Gaspé Section (B-2), 51, 52, 59, 60, 194, 202, 216, 221
George Islands, James Bay, NW Terr., 37, 39
Georgian Bay Section (L-4d), 63, 72
Georgian Bay Site Region, Ont., 53
Germany, 4, 41, 128
Gouin Section (B-3), 52
Goulais River, Ont., 72, 211, 214, 239, 240, 256
Grand Lake, Lab., 52
Grand' Mère, Que., 231
Grand Rapids, Minn., 104
Grant C., W. Va., 38
Grayson C., Vt., 38
Great Bear Lake, NW Terr., 36
Great Lakes–St. Lawrence Region, 47, 51, 61, 91, 94, 102, 203, 218
Great Slave Lake, NW Terr., 36
Great Whale River, Que., 36
Green Mts., Vt., 74
Green River, Colo., 32
Green River, N.B., 13, 21, 24, 25, 26, 60, 152, 159, 160, 166, 167, 176, 181, 190, 193, 202, 216, 221, 223, 227, 234, 271
Griswold, Conn., 38
Guatemala, 36, 39
Gulf of St. Lawrence, 32

Haileybury Clay Section (L-8), 63
Hamilton and Eagle Valleys Section (B-12), 52
Harburg, Germany, 41
Harpoon Experimental Forest, Newf., 192
Hay River Section (B-18b), 51, 199
Hemlock–White Pine–Northern Hardwoods Region, 42, 61, 73
Himalaya Mts., 39
Hoyt Arboretum, Portland, Ore., 39
Hudson Bay, 35, 36, 37, 39, 51
Hudson Bay Lowlands Section (B-5), 51, 52
Hudson Bay Site Region, Ont., 53
Hudsonian Life Zone, 111, 117
Hungary, 41

India, 172
Interior Plains Physiographic Region, 51
Iron C., Mich., 93
Isle Royale National Park, Mich., 24, 66, 71, 112, 113, 326
Itasca C., Minn., 69, 102

Itasca State Park, Minn., 35, 69, 94, 99, 104, 183, 224, 326

James Bay, NW Terr., 37, 51
James Bay Site Region, Ont., 53
Japan, 172
Joseph Lake, Ont., 11

Kedgwick River, N.B., 144
Keewatin, NW Terr., 102
Kenscoff, Haiti, 41
Keweenaw Peninsula, Mich., 72
Kickapoo River, Vernon C., Wis., 38
Knob Lake, Que., 37
Koksoak River, Ungava Bay, Que., 37
Koochiching C., Minn., 119, 258

Labrador Peninsula, 36, 52, 61, 102
LaCrosse C., Wis., 38
Lac Seul, Ont., 153, 163, 165
Lac Tremblant, Que., 237
Lake Clear, N.Y., 159
Lake Edward Experimental Forest, Que., 62, 72, 102, 107, 120, 203, 204, 219, 228, 231, 240, 256
Latvia, 40
Laurentian Section (L-4a), 63, 73, 204, 219, 221, 228, 231
Laurentian Upland, 32, 39, 62
Laurentide-Onatchiway Section (B-1a), 54, 188. *See also* Northeastern Coniferous Section (B-1)
Laurentides Park, Que., 60, 93, 97, 128, 129, 135, 137, 141, 175, 233, 327
Leningrad, USSR, 41
Lesser Slave Lake, Alta., 6, 36, 37
Liard River, Alta., 37
Little Pabos River, Que., 194
Lower English River Section (B-14), 53
Lower Foothills Section (B-19a), 51, 53, 200
Lower Peninsula, Mich., 38

Mackinac Island, Lake Michigan, Mich., 68, 108
Madison C., Va., 38
Manitoba Lowlands Section (B-15), 51, 53
Mantorville, Dodge C., Minn., 38
Maple-Basswood Forest Region, 42
Matane, Que., 191
Mediterranean area, 4, 36
Melville, Lake, Lab., 52
Mexico, 36, 39
Middle Ottawa Section (L-4c), 63
Middle St. Lawrence Section (L-3), 63
Minnesota Central Pine Section, 88, 90, 103, 105, 107
Mixed-Wood Section (B-18a), 51, 53, 199, 202, 212
Moose River, Ont., 52
Moscow, USSR, 28, 41, 325
Mount Desert Island, Maine, 108

Mount Katahdin, Maine, 32, 74
Mount Rogers, Va., 38, 39
Mount Tremblant, Que., 327
Mount Washington, N.H., 6, 32, 39, 79
Mower C., Minn., 33

Nelson River Section (B-21), 53
New Albin, Allamakee C., Iowa, 39
New England Section (9G), 75
Newcomb, Adirondacks, N.Y., 235, 236
Newfoundland–Labrador Barrens Section (B-31), 51
Nipigon, Lake, 97, 153, 165, 168
Nipigon Section (B-10), 214
North River, N.S., 144
Northeastern Coniferous Section (B-1), 44, 50, 54, 59, 199, 200, 202, 220. *See also* Laurentide-Onatchiway Section (B-1a); Chibougamau-Natashquan Section (B-1b)
Northeastern Transition Section (B-13a), 52
Northern Appalachian Highland Division, 73
Northern Clay Section (B-4), 51, 52, 60, 93, 196, 214, 234
Northern Coniferous Section (B-22a), 53
Northern Forest Region, 42, 47, 61
Northern Hardwood Region, 46
Northern Hardwoods–Hemlock–White Pine Zone, 75
Northern Interior, Alta., 42
Northern Peninsula Section (B-29), 51
Norway, 40, 131

Old Woman Bay, Lake Superior, Ont., 69
Ottawa, Ont., 197
Ozark Mts., 145

Page C., Va., 38
Pakistan, 172
Pasquia Hills, Sask., 115
Patricia District, Ont., 99
Peace River, Alta., 37
Penobscot Experimental Forest, Maine, 100, 117, 118, 119, 153, 232, 244
Petawawa Experimental Forest, Ont., 28, 228
Piedmont, S.C., 33
Pike Bay Experimental Forest, Minn., 194, 230
Pitch Pine–Oak Zone, 75
Pocahontas C., W. Va., 38
Pocomoke City, Md., 38
Porcupine Mts., 72, 326
Port Arthur District, Ont., 97, 265
Portland, Maine, 36
Portland, Ore., 39, 40
Portugal, 4, 325
Postville, Allamakee C., Iowa, 39
Presidential Range, White Mts., N.H., 79

Quetico Provincial Park, Ont., 97
Quetico Section (L-11), 63
Quetico-Superior Wilderness, 27, 68, 224, 327

Rainy River District, Ont., 318
Randolph C., W. Va., 38
Red Lake, Minn., 104
Red Spruce–Hemlock–Pine Zone, 80
Richmond Gulf, Hudson Bay, 35, 37
Riding Mountain Park, Man., 61, 115, 327
Roan Mt., N.C., 35
Rocky Mts., 36
Rosemount Experimental Farm, Minn., 197

Saguenay Section (L-7), 63
Saddle Hills, Alta., 37
St. Lawrence River, 2
St. Maurice Valley, Que., 318
Sandiland Forest Reserve, Man., 98
Sangamon, Ill., 33
Sault Ste. Marie, Ont., 144, 152
Scarborough, Ont., 33
Schoales Dam, Kings C., N.B., 231
Secrest Arboretum, Ohio, 39
Simcoe-Rideau, Lakes Site Region, Ont., 53
Slate Islands, Lake Superior, 113
Smoky River, Alta., 37
Soviet Union (USSR), 40, 41, 46
Spring Valley, Fillmore C., Minn., 38
Springfield, Brown C., Minn., 33
Spruce-Fir Coast Zone, 80
Spruce–Fir–Intolerant Hardwoods Subzone, 75
Spruce–Fir–Northern Hardwoods Zone, 75
Spruce Taiga Zone, 80
Spruce Woods Forest Reserve, Man., 154
Star Island, Cass Lake, Minn., 42, 43
Sturgeon Lake, Alta., 37
Sudbury–North Bay Section (L-4e), 63
Sugar Maple–Yellow Birch–Fir Zone, 80
Superior, Lake, North Shore, 27, 69, 116, 318
Superior National Forest, Minn., 26, 192
Superior Physiographic Province, 62
Superior Section (B-9), 60, 196, 214, 234
Sweden, 40
Switzerland, 193
Syracuse, N.Y., 25

Tama C., Iowa, 33
Tamarac National Wildlife Refuge, Minn., 115
Tamarack Creek, Trempeleau C., Wis., 70
Tambov, USSR, 41
Temiscouata-Restigouche Section (L-6), 63, 73, 203
Timagami, Lake, Ont., 35, 278
Timagami, Lake, Site Region, Ont., 53
Timagami Section (L-9), 63, 203
Tolerant Hardwood–Spruce–Fir Subzone, 75
Toronto, Ont., 33, 35
Transition Hardwood–White Pine–Hemlock Zone, 75
Trempeleau C., Wis., 38, 70
Trostjanc, USSR, 41
Tucker C., W. Va., 38
Tyrol Alps, Austria, 172, 173

Ukraine, 41
Ungava Bay, Que., 37, 52
Upper Churchill Section (B-20), 53
Upper Embarrass River, Alta., 37
Upper Iowa River, Winneshiek C., Iowa, 39
Upper Lièvre Valley, Que., 216
Upper Mackenzie Section (B-23a), 53
Upper Peninsula, Mich., 106, 114, 153, 206, 223
Upper St. Lawrence Section (L-2), 63
Uppsala, Sweden, 40

Valcour Island, Lake Champlain, N.Y., 23
Vernon C., Wis., 38
Voronesh, USSR, 41

Waukegan, Ill., 5
Wealden, Germany, 32

West Indies, 145
Western Interior Lowlands Physiographic Region, 102
White Mountains, N.H., 24, 39, 74, 76, 81, 184, 257
White Spruce–Balsam Fir Region, 46
Whitney Park, N.Y., 235
Wilderness State Park, Mich., 114
Wind River Arboretum, Carson, Wash., 39, 40, 172
Winneshiek C., Iowa, 39, 68
Winnipeg, Lake, Man., 51
Wisconsin Driftless Section, 32, 62
Wolf River, Wis., 5
Worcester C., Mass., 37
Wykoff, Fillmore C., Minn., 38, 68

Yellow River, Allamakee C., Iowa, 39
Yukon Territory, 32, 33, 37